UNIVERSITY OF ROCHESTER LIBRARIES
3 9087 02374117 8

D1714214

Handbook of Experimental Pharmacology

Volume 158

Springer

Berlin
Heidelberg
New York
Hong Kong
London
Milan
Paris
Tokyo

The Macrophage as Therapeutic Target

Contributors
J. M. Aerts, R. Boot, E. J. Brown, G. D. Brown, P. C. Calder, T. J. Chambers, W. J. S. de Villiers, F. Di Virgilio, M. G. Espey, D. Ferrari, A. Frauenschuh, T. Ganz, C. K. Glass, S. Gordon, D. R. Greaves, A. Groener, D. N. J. Hart, P. M. Henson, S. R. Himes, C. Hollak, D. A. Hume, Z. Johnson, G. Kraal, R. I. Lehrer, J. D. MacMicking, J. Mahoney, L. Martinez-Pomares, J. D. McKinney, J. L. Miller, S. M. Moghimi, V. H. Perry, A. Proudfoot, H. Rosen, D. R. van der Westhuyzen, N. van Rooijen, S. Vuckovic, J. S. Welch, J. A. Willment, S. Wong, P. Yaqoob

Editor
Siamon Gordon

Springer

Professor
Siamon Gordon
Sir William Dunn School of Pathology
University of Oxford
South Parks Road
Oxford OX1 3RE
United Kingdom
e-mail: christine.holt@pathology.oxford.ac.uk

With 27 Figures and 9 Tables

ISBN 3-540-44250-2 Springer-Verlag Berlin Heidelberg New York

Cataloging-in-Publication Data applied for

Bibliographic information published by Die Deutsche Bibliothek
Die Deutsche Bibliothek lists this publication in the Nationalbibliografie;
detailed bibliographic data is available in the Internet at
<http://dnb.ddb.de>.

Springer-Verlag Berlin Heidelberg New York
a member of BertelsmannSpringer Science+Business Media GmbH

Printed in Germany

Cover design: design & production GmbH, Heidelberg
Typesetting: Stürtz AG, 97080 Würzburg

Printed on acid-free paper 27/3150 hs – 5 4 3 2 1 0

Preface

During the past decade, the rapid growth of molecular and cellular knowledge of macrophages (MØ), as a specialized host defence and homeostatic system, has begun to offer attractive targets for therapeutic intervention. MØ play a central role in a wide range of disease processes, from genetically determined lysosomal storage diseases, to acute sepsis, chronic inflammation and repair, tissue injury and cell death. Under- or overactivity of MØ clearance, immune effector functions and responses to metabolic abnormalities contribute to common disorders such as autoimmunity, atherosclerosis, Alzheimer's disease and major infections including AIDS and tuberculosis.

The discovery of tumour necrosis factor (TNF)-α and the development of specific inhibitors that are highly effective clinically in diseases such as rheumatoid arthritis illustrate the way scientific advances have already been translated into practice. Development of powerful molecular genetic methods made it possible to clone and express specific cytokines produced by MØ and potent growth factors acting on MØ and other myeloid cells, such as granulocyte–macrophage colony-stimulating factor (GM-CSF), to modulate and boost MØ activities. Specific plasma membrane receptors control cell recruitment, adhesion, endocytosis and activation of innate and acquired immune functions. Discovery of the Toll-like receptors has already served as a major spur to uncover signalling pathways that might yield future therapeutic targets. Phagocytic uptake of apoptotic cells and micro-organisms by MØ contribute to host defences, as well as providing a potential niche for intracellular pathogens.

Whilst the goals of therapeutic intervention based on improved understanding of MØ functions and their contribution to pathogenesis may seem self evident, there are considerable difficulties in producing useful new agents. The MØ of the body constitute a distributed cellular system also known as the mononuclear phagocyte, or reticulo-endothelial system, with great heterogeneity in cell differentiation and activation in different organs and disease states. Common progenitors give rise to tissue MØ which differ considerably in their properties, e.g. in the nervous system and liver, to dendritic cells, specialized for antigen presentation to naive T lymphocytes and to osteoclasts, multinucleated cells able to resorb living bone. Their receptors and versatile biosynthetic and secretory responses result in adaptation to very different microenvironments. Their systemic actions help to integrate many physiologic functions and pathologic processes. It should therefore not come as a surprise if MØ contribute beneficial as well as deleterious roles to the diseased host. They represent the classic two-edged sword. It is a formidable challenge to aim selective intervention at subpopulations of cells or subsets of their gene products without affecting their vital normal functions or other cells, directly or indirectly. A further difficulty is our present-day ignorance of basic mechanisms of MØ functions, e.g. in vaccine development (their role as natural adjuvants or immunosuppressants) and their complex life history in vivo. Gene ablation and transgenesis have confirmed some hypotheses and left others open. Cell culture models are useful, but imperfect, in not mimicking complexity within the host. Classic pharmacologic approaches need to be aligned with newer knowledge of MØ gene expression and regulation.

The present volume covers a range of subjects and provides opportunities for a more focused MØ-targeted approach. The individual chapters review selected topics briefly, to place cellular processes and molecular targets in perspective. These are grouped broadly. Section I deals with general issues of cell differentiation, routes of delivery and of MØ-specific gene targeting, with particular emphasis on the living host. Section II deals with selected plasma membrane receptors, uptake processes, regulation of cellular responses and different categories of secretory products. Finally, Sect. III considers specialized cell types, environments and examples of cell-pathogen interactions. Overall, the volume should provide a broad sample of the state of the art. Useful reviews and references in the literature are cited within individual chapters.

I would like to acknowledge the excellent assistance of Christine Holt, and of the editorial staff at Springer-Verlag.

Oxford, Spring 2003 Siamon Gordon

List of Contributors

(their addresses can be found at the beginning of their respective chapters)

List of Contents

Part 1. General

Part 2. Macrophage Targets in Inflammation

Part 3. Modulation of Specialised Macrophage Activities

Part 1
General

The Macrophage as a Validated Pharmaceutical Target

H. Rosen

Department of Immunology and The Committee for Advanced Human Therapeutics,
The Scripps Research Institute, ICND-118,
10550 North Torrey Pines Road, La Jolla, CA 92037, USA
e-mail: hrosen@scripps.edu

Abstract "Therapeutic validation" is a utilitarian classification of the application of the basic science base of the macrophage. Reduction to therapeutic practice represents the cutting edge of therapy, but rests upon a decades-old basic science foundation. Macrophage-targeted therapeutics have now added significant value to the lives and quality of life of patients, without undue adverse effects in multiple disease settings. These are exemplified by the impact of macrophage enzyme replacement for a lysosomal storage disease (Gaucher's), the modulation of osteoclast-dependent bone destruction by bisphosphonates, and revolutionary impact of TNF sequestrants on both rheumatoid arthritis, as well as the delineation of new mechanisms in the understanding of Crohn's diseases. The macrophage, as a cell, is now beginning to reach a full measure of therapeutic maturity in the application of the understanding of the particular rate-limiting roles that it plays in the maintenance of health or the induction of diseases.

Keywords Bisphosphonates, Crohn's disease, Enzyme replacement, Gaucher's disease, Macrophage, Osteoporosis, Rheumatoid arthritis, TNF sequestration

The concept of therapeutic target validation bears disproportionate weight in the conduct of applied science in the pharmaceutical and biotechnology sectors, where research decisions are significantly influenced by perceptions of market size. These organizations measure the success of their scientific output by products launched and the expansion of revenues earned and not by the successful testing of meaningful hypotheses and the resulting high-impact publications. To do this, however, the products must add significant value to the lives and quality of life of patients, without undue adverse effects. The ability to regard the macrophage as a "validated" therapeutic target suggests the macrophage, as a cell, is now beginning to reach a full measure of maturity in the application of the understanding of the particular rate-limiting roles that it plays in the maintenance of health or the induction of diseases.

The notion of "therapeutic validation" is simply a utilitarian classification of the application of the basic science base of the macrophage. In many respects, the reduction to therapeutic practice may represent the cutting edge of therapy, but rests upon a decades-old basic science foundation. It provides an objective, although retrospective, validation of the importance of the study of macrophages to advancing human therapeutics. The view is, however, skewed towards the major medical needs that drive economic activity, and does not necessarily reflect the full contribution of the study of macrophages, where the advancing science base has had major impacts upon the understanding of immunity, vaccination, tissue remodeling, and embryology, although in areas that have not yet been brought to therapeutic fruition. In this essay, I have used a broad definition of macrophage study that includes cells of the monocytic lineage such as monocytes, macrophages, osteoclasts, and have also included discussion of inherited metabolic defects in which macrophages are central to the expression of tissue pathology and dysfunction. I will also show how the quality of the basic science base was prospectively predictive of both efficacy and the adverse experience profiles associated with therapeutic approaches to molecules affecting macrophage function.

1 Modification of Macrophage Function Can Now Be Said to Modify Disease Outcome, Rather than Provide Symptomatic Relief

The validation of the macrophage as a proven therapeutic target has crossed multiple treatment modalities. In the macrophage-specialized function of lysosomal degradation, the lysosomal storage defects of Gaucher's disease have led to the relatively successful replacement of a heritable defective enzyme, glucocerebrosidase, using recombinant enzyme, as an orphan disease. In contrast, the treatment of rheumatoid arthritis and Crohn's disease has been revolutionized by the advent of tumor necrosis factor (TNF) sequestrants. These have not only improved patients' lives but have provided unique insights into human disease mechanisms not readily modeled in animal models (Podolsky 2002). The successful treatment of osteoporosis and Paget's disease of bone by the inhibition of osteoclast-dependent bone resorption represents the successful use of small molecules (bisphosphonates) in the alteration of outcome of an event entirely dependent upon cells of the monocytic lineage. These all represent highly specific macrophage-related events in pathology usefully modulated therapeutically, and are much more compelling evidence for the specific role of macrophages in pathology than is, for example, the expression and upregulation of cyclooxygenase 2 in macrophages, where therapeutic efficacy is only partly attributable to macrophage effects, and where such efficacy is largely symptomatic, rather than disease modifying in terms of long-term impact upon patient outcomes.

2
Enzyme Replacement in Gaucher's Disease

Gaucher's disease is an autosomal recessive disease with its highest prevalence (type I) in Ashkenazi Jews (Balicki and Beutler 2002). It is a typical lysosomal storage disease caused by a deficiency in glucocerebrosidase, an essential step in the cleavage of glucose from ceramide (Inoue and Lupski 2002). Glucocerebroside substrate, therefore, accumulates within cells of the mononuclear phagocyte system and within the central nervous system. (Dwek et al. 2002) Three clinical subtypes of Gaucher's disease are described. The adult or type I Gaucher's is a non-cerebral form of storage disease, with some residual enzymatic activity, that accounts for its emphasis on splenic, hepatic, and skeletal involvement. Infantile Gaucher's (type II) is an acute disease dominated by cerebral accumulation of substrate, as well as hepatosplenic involvement. No enzyme activity is usually detected. Type III Gaucher's is a juvenile disease intermediate in signs and symptoms from types I (adult) and II (infantile). The involvement of the CNS and pre-existing skeletal muscle lesions has significant impact on the efficacy of enzyme replacement therapy.

The disease pathology results directly from either CNS dysfunction, or the physical effects of the accumulation of distended macrophage "Gaucher's cells" in spleen, liver, marrow, lymph nodes, thymus, and Peyer's patches. These long-term effects include anemia and thrombocytopenia. Longitudinal studies with 2–5 years of clinical follow-up have been published in type I Gaucher's disease, using recombinant glucocerebrosidase targeted to macrophage lysosomes through the endocytic route (Weinreb et al. 2002). Enzyme replacement is highly effective in ameliorating extra-CNS disease manifestations. Patients show long-term improvements and maintenance of improvement in anemia and thrombocytopenia. There is also a measurable benefit in bone erosions and bone pain, with more than 50% of patients showing measurable improvement even after the presence of radiologically documented bone lesions and bone pain (Bembi et al. 2002; Poll et al. 2002). This is a pleasing and rational approach to the correction of a primary heritable peripheral defect of macrophages, and illustrates the therapeutic accessibility of macrophages that have access to the circulation. Large molecules equilibrate inefficiently across the blood–brain barrier in the absence of specific transcytotic mechanisms or leakage, and therefore the primary CNS defect is not amenable to approach via the circulation, and will likely require the development of effective gene therapies or alternate approaches to the modification of glycosylation pathways (Dwek et al. 2002). The disease, because of its low prevalence, will always be considered an orphan. This ultimately limits the resources that could be brought to therapeutic approaches, and reflects that market forces will always favor the highly prevalent, and patients and physicians will need to rely upon small companies seeking market niches, or academic institutions with governmental or charitable funding to approach these diseases.

3 TNF Sequestrant Therapy Improves the Outcome of RA and Sheds Critical Mechanistic Insights into Crohn's Disease

The central role of the macrophage in the mechanism of human disease has emerged from the clinical evaluation of TNF sequestrants in rheumatoid arthritis and Crohn's disease. The impact of the elucidation of the pathophysiology of TNF on therapeutics, sheds light into the long lag phase between the leading edge of macrophage biology and its therapeutic application. TNF-α was discovered in the 1970s by Old and colleagues (Feldmann and Maini 2001) and cloned in the early 1980s, when it was shown that cells of the monocytic lineage were the major source of TNF production (Tracey and Cerami 1994). TNF is a rapidly produced proinflammatory cytokine, with serum levels detectable within 30 min of lipopolysaccharide (LPS) stimulus, and likely reflects the cleavage of preformed membrane-bound TNF-α by TNF-α-converting enzyme (TACE). Blockade of TNF release blunts the release of other proinflammatory cytokines such as interleukin (IL)-1 and IL-6, suggesting that TNF plays a role as a molecular trip wire for the activation of stress responses to noxious stimuli, that engage a cascade of events culminating in spatially and temporally regulated recruitment of inflammatory and immune leukocytes at sites of injury. In keeping with these data was the demonstration of the pivotal role of TNF-α in the early control of intracellular bacterial pathogens. The work of Havell in listeriosis (Havell 1989) and Vassalli (Kindler et al. 1989) in early granuloma formation and the control of Bacillus Calmette-Guerin (BCG) infection in mice, both accurately showed the role of TNF-α as an essential organizer of the granulomatous response, and foreshadowed the efficacy of TNF sequestration in Crohn's disease, (Sandborn and Targan 2002) as well as the deleterious potentiation of bacterial diseases including tuberculosis (Tb) by TNF sequestration. This remains one of the few areas of basic biology where early studies have so accurately prospectively predicted the adverse effects expected upon clinical usage. The lag phase between characterization of the role of TNF and the central role of the macrophage as organizer and participant in the granuloma, and the clinical exploitation of these data was almost 15 years. The features that contributed to the gap between the basic discovery and the clinical exploitation included developing adequate methods of production and manufacture, as well as the lengthy requirements for clinical trials in chronic diseases with complex clinical endpoints such as rheumatoid arthritis (Feldmann and Maini 2001; Bresnihan 2002; Jenkins and Hardy 2002; Kalden 2002; Scott 2002; Weisman 2002) and Crohn's disease. In addition, the essential role of TNF in host defense coupled with the long half-lives of TNF sequestrants required significant empirical approaches to dose ranging. The choice of dose in attempting to validate a pharmacological mechanism is always difficult, in this case especially so because of the need to walk a tightrope between suppression of the host response sufficient for efficacy, and suppression of macrophage function to the point where host defense impairment results in catastrophic potentiation of rapid bacterial infections. The advantage of the rheumatoid arthritis field is that standard therapies including methotrexate and corticosteroids also have well-documented risks of potentiation of infection. With appropriate patient exclusion criteria, the clinical safety of TNF sequestration has been acceptable, with a somewhat higher prevalence of upper respiratory infection in First World usage. Recent reports have included the reactivation of miliary Tb in patients by TNF sequestrants, and more than 70 reports of Tb reactivation can now be found within the Adverse Experience database maintained by the Food and Drug Administration (FDA) (Keane et al. 2001; Mayordomo et al. 2002).

An unexpected clinical finding has been the development of demyelinating disease in a small number of patients receiving TNF sequestrant therapy for adult or juvenile rheumatoid arthritis (Sicotte and Voskuhl 2001). Trials of lenercept in multiple sclerosis (Lenercept 1999) showed that there were no significant differences between groups on any magnetic resonance imaging (MRI) study measure, but the number of lenercept-treated patients experiencing exacerbations was significantly increased compared with patients receiving placebo (p=0.007) and their exacerbations occurred earlier (p=0.006). These findings suggest that one of the roles of TNF may be in the control of demyelination, and that the macrophage, by extension, can play pleiotropic roles in human pathology, exacerbating rheumatoid arthritis and Crohn's disease, while playing a suppressive role in multiple sclerosis. Another hint towards a suppressive role for TNF in disease is the small incidence of drug-induced lupus on TNF sequestrant therapy. This again is in keeping with the predictive value of basic studies, where TNF-α deletant mice also developed elevated levels of anti-double stranded (DS) DNA antibodies (Ettinger and Daniel 2000).

The findings that TNF sequestration can have beneficial effects upon the expression of Crohn's disease but not ulcerative colitis, has again sharpened the focus on the role of the macrophage in Crohn's disease, and strongly differentiated the pathogenesis of Crohn's disease from ulcerative colitis (Podolsky 2002; Sandborn and Targan 2002). One of the strengths of the pharmacological approach is this ability, through controlled clinical trials, to provide insights into human as opposed to model diseases. It is, however, salutary that in this field at least, disease models in lower species including rodents have in large part been prospectively predictive of both efficacy and adverse experience.

4
Bisphosphonates Enhance Bone Mass Retention and Protect Against Fractures by Inhibition of Osteoclast Formation

Whereas the two previous examples deal with macrophage enzyme replacement using recombinant enzyme, or sequestration of a macrophage-secretory product TNF with antibodies or receptor fusion proteins, the efficacy of small synthetic chemical moieties, specifically nitrogen-containing bisphosphonates, impact upon the resorption of bone mass, by inducing osteoclast and macrophage apoptosis.

Bisphosphonates are the most effective inhibitors of bone resorption and are extensively used for the treatment of systemic or local bone loss including postmenopausal osteoporosis and tumor bone disease. Bisphosphonates are pyrophosphate analogs, which bind to bone. Bisphosphonates are then removed from the mineral matrix in the acidic compartment formed between the osteoclast and the bone surface, analogous to the sealed compartments described in macrophages upon immune complexes or complement ligands (Wright and Silverstein 1984). Labeled bisphosphonates accumulate in osteoclasts and inhibit further bone resorption (Reszka et al. 1999). Evidence has accumulated that all bisphosphonates that inhibit the resorption of bone induce the caspase-dependent formation of pyknotic nuclei and the cleavage of Mst1 kinase. This cleavage of Mst1 kinase and caspase activation is dependent upon the bisphosphonate inhibition of the mevalonate pathway, and is specifically blocked by the addition of geranylgeraniol, a key precursor for geranylgeranyl diphosphate (Benford et al. 1999; Reszka et al. 1999). It emerges from these studies that the flux through the mevalonate pathway to geranylgeraniol is essential for the formation of osteoclasts from macrophages, and for the long-term maintenance of the osteoclast population (Fisher et al. 1999; Fisher et al. 2000). In the absence of protein geranylgeranylation, osteoclasts fail to differentiate from macrophages in murine, rabbit, and chicken systems, and that this failure can be attributed to the effects of bisphosphonates on geranylgeranylation and not farnesylation of proteins (Coxon et al. 2000). The effects of bisphosphonates are therefore complex. Effective pharmacodynamic inhibition of bone resorption that can be measured both as the retention of bone mass, as well as by the inhibition of fractures in long-term clinical use (Karpf et al. 1997) can be achieved by the modulation of osteoclast differentiation from macrophages, as well as by the inhibition of osteoclast function through the activation of caspase-mediated apoptotic events.

5
Future Directions

These areas have shown the therapeutic feasibility of macrophage enzyme replacement, secretory product sequestration, and inhibition of osteoclast differentiation and function. Efficacious therapies with direct effects upon macrophages now target rheumatoid arthritis, Crohn's disease, osteoporosis, tumor disease of bone, and inherited metabolic defects of the mononuclear phagocyte system. The important role of the macrophage as a regulator and remodeler of tissue, and regulator of the migratory and differentiative events of tissue cells has not yet been exploited. That the macrophage plays a central role in atherosclerosis is clear. The therapies effectively directed towards the role of the macrophage in that disease still requires reduction to practice. It may be that many of the therapeutically advantageous effects of peroxisome proliferator-activated receptors (PPAR) agonists in vascular disease will prove to be useful probes of macrophage function in human pathology (Berger and Moller 2002).

6
References

Balicki D, Beutler E (2002) Gene therapy of human disease. Medicine 81:69

Bembi B, Ciana G, Mengel E, Terk M, Martini C, Wenstrup R (2002) Bone complications in children with Gaucher disease. Br J Radiol 75:A37

Benford HL, Frith JC, Auriola S, Monkkonen J, Rogers MJ (1999) Farnesol and Geranylgeraniol Prevent Activation of Caspases by Aminobisphosphonates: Biochemical Evidence for Two Distinct Pharmicological Classes of Bisphosphonate Drugs. Molecular Pharmacology 56:131

Berger J, Moller DE (2002) The Mechanisms of Action of PPARS. Annu Rev Medicine 53, no. 1:409

Bresnihan B (2002) Preventing joint damage as the best measure of biologic drug therapy. J Rheumatol 65, no. 29 Sept. Suppl.:39

Coxon FP, Helfrich MH, Van'T Hof R, Sebti S, Ralston HSH A, Rogers MA (2000) Protein Geranylgeranylation is Required for Osteoclast Formation, Function, and Survival: Inhibition by Bisphosphonates and GGTI-298. Journal of Bone and Mineral Research 15, no. 8:1467

Dwek R, Butters T, Platt F, Zitzmann N (2002) Targeting glycosylation as a therapeutic approach. Nature Reviews Drug Discovery 1:65

Ettinger R, Daniel N (2000) TNF-deficient mice develop anti-nuclear autoantiboies. Scand J Immunol 51:S88

Feldmann M, Maini RN (2001) anti-TNF-α therapy of RHEUMATOID arthritis: what have we learned? Annu Rev Immunol 19, no. 1:163

Fisher JE, Rodan GA, Reszka AA (2000) In vivo Effects of Bisphosphonates on the Osteoclast Mevalonate Pathway. Endocrinology 141, no. 12:4793

Fisher JE, Rogers MJ, Halasy JM, Luckman SP, Hughes DE, Masarachia PJ, Wesolowski G, Russell RGG, Rodan GA, Reszka AA (1999) Alendronate mechanism of action: geranylgerniol, and intermediate in the mevalonate pathway, prevents inhibition of osteoclast formation, bone resorption, and kinase activation in vitro. Proc Natl Acad Sci USA 96:133

Havell E (1989) Evidence that tumor necrosis factor has an important role in antibacterial resistance. J Immunol 143:2894

Inoue K, Lupski JR (2002) Molecular mechanisms for genomic disorders. Annu Rev Genom Hum Genet 3, no. 1:199

Jenkins JK, Hardy KJ (2002) Biological modifier therapy for the treatment of rheumatoid arthritis. Am J Med Sci 323, no. 4:197

Kalden JR (2002) Emerging role of anti-tumor necrosis factor therapy in rheumatic diseases. Arthritis Res 2, no. 4 Suppl:S34

Karpf D, Shapiro D, Seeman E, Ernsrud K, Johnston CAS, Harris S, Santora A, Hirsch L, Oppenheimer L, Thompson D (1997) Prevention of nonvertebral fractures by alendronate: a meta-analysis. J Am Med Assoc 277:1159

Keane J, Gershon S, Wise R, Mirabilie-Levens E, Kasnica J, Schwieterman W, Siegel J, Braun M (2001) Tuberculosis associated with infliximab, a tumor necrosis factor alpha-neutralizing ag4ent. N Engl J Med 345:1098

Kindler V, Sappino A, Grau G, Piguet P, Vassalli P (1989) The Inducing Role of Tumor Necrosis Factor In the Development of Bactericidal Granulomas during BCG Infection. Cell 56:731

Lenercept Msg (1999) TNF neutralization in MS: Results of a randomized, placebo-controlled multicenter study. Neurology 53, no. 3:457

Mayordomo L, Marenco J, Gomez-Mateos J, Rejon E (2002) Pulmonary miliary tuberculosis in a pateint with anti-TNF-alpha treatment. Scand J Rheumatol 31:44

Podolsky D (2002) Inflammatory bowel disease. N Engl J Med 347:417

Poll L, Maas M, Terk M, Roca-Espiau M, Bembi B, Ciana G, Weinreb R (2002) Response of Gaucher bone disease to enzyne replacent therapy. Br J Radiol 75:A25

Reszka AA, Halasy-Nagy J, Masarachia PJ, Rodan GA (1999) Bisphosphonates Act Directly on the Osteoclast to Induce Caspase Cleavage of Mst1 Kinase Apoptosis. The Journal of Biological Chemistry 274, no. 49:34967

Sandborn WJ, Targan SR (2002) Biologic therapy of inflammatory bowel disease. Gastroenerology 122, no. 6:1592

Scott DL (2002) Advances in the medical management of rheumatoid arthritis. Hosp Medicine 63, no. 5:1227

Sicotte NL, Voskuhl RR (2001) Onset of multiple sclerosis associated with anti-TNF therapy. Neurology 57, no. 10:1885

Tracey KJ, Cerami A (1994) Tumor necrosis factor: a pleiotropic cytokine and therapuetic target. Annu Rev Medicine 45, no. 1:491

Weinreb N, Charrow J, Andersson H, Kaplan P, Kolodny E, Mistry P, Pastores G, Rosenbloom B, Scott C, Wappner R, Zimran A. (2002) Effectiveness of enzyme replacement therapy in 1028 pateinest with type I Gaucher disease after 2 to 5 years of treatment. Am J Med 113:112

Weisman MH (2002) What are the risks of biologic therapy in rheumatoid arthritis? An update on safety. J Rheumatol 65, no. 29 Sept. Suppl:33

Wright S, Silverstein S (1984) Phagocytosing macrophages exclude proteins from the zones of contact with opsonized targets. Nature 309:359

Transcription Factors That Regulate Macrophage Development and Function

D. A. Hume · S. R. Himes

Department of Microbiology, Molecular Bioscience Building,
University of Queensland, Q4072, Australia
e-mail: D.Hume@imb.uq.edu.au

Abstract The transcription factors that play a major role in the development and function of the monocyte/macrophage lineage are outlined in this review. Hematopoiesis proceeds through the binding of specific combinations of transcription factors leading to a temporal and lineage-restricted pattern of gene expression. We summarize current knowledge of how transcription factors are able to drive monocyte differentiation from early pluripotent progenitors. These transcription factors include numerous homeobox family proteins, AML-1, Pu.1, MZF-1, Egr-1, ICSBP, and STAT family proteins. The transcription factors that control inducible gene expression in mature macrophages are also covered. Macrophages are able to respond to a range of bacterial products and play a critical role in both the innate and adaptive immune response to infection. The inducible transcription factors that regulate inflammatory and anti-microbial gene products covered here include NF-κB/Rel, IRF, STAT, and C/EBP family proteins. This activation of macrophages must be precisely regulated to prevent damage to host tissue. Many transcription factors can act as repressors of gene expression in macrophages and the signals that down-regulate inflammatory responses such as activation of TGFβ, PPARγ and glucocorticoid receptors are discussed. Macrophages play a fundamental role in adaptive immunity and are likely to influence the formation of either a Th1 or Th2 pattern of immune reactivity. We briefly review the current understanding of how activation of different transcription factors can influence the profile of cytokines expressed in macrophages and contribute to the formation of distinct immune responses.

Keywords Gene regulation, Transcription factors, Macrophage development, Inflammation, Immune response

1
Introduction

The mononuclear phagocyte family is present throughout development and in every organ system, where they can comprise 10%–15% of the total cell mass. Progenitor cells in the bone marrow give rise to blood monocytes that can enter tissue and further develop into distinct tissue macrophages. Macrophages are vital to both the innate and acquired immune response by direct endocytosis and cytotoxicity, antigen presentation, and production of biologically active molecules such as cytokines and chemokines. Their destructive potential underlies many aspects of the pathology of acute and chronic inflammation.

Macrophages share many constitutively expressed genes, such as the endocytic receptors responsible for recognition of microorganisms. The expression of other genes is acutely regulated by external stimuli such as microbial products, lymphokines, and growth factors. Ultimately, the pattern of gene expression is determined by specific transcription factors. This chapter deals with the control of transcription in cells of the macrophage lineage.

2
Transcription Factors in Macrophage Development

Hematopoiesis branches into distinct pathways through a temporal and lineage-specific pattern of transcription factor expression. Lineage commitment appears to take place through rapid changes in the expression or function of specific transcription factors. This pattern of activation may arise extrinsically, from activation by growth factors or intrinsically, from either genetically coded or stochastic events. Lineage commitment can still occur in the absence of many cytokines known to influence myelopoiesis. Although the hematopoietic growth factors, granulocyte colony-stimulating factor (G-CSF), interleukin (IL)-3 and granulocyte macrophage (GM)-CSF can regulate growth and differentiation from bone marrow precursors, they do not appear to be required for steady-state hematopoiesis (Nishinakamura et al. 1995). In contrast, macrophage (M)-CSF (CSF-1) is required for macrophage production as determined by the phenotype of mice with a targeted deletion of the CSF-1 receptor or a natural mutation in the CSF-1 gene (Cecchini et al. 1997; Dai et al. 2002).

Molecular events involved in early myelopoiesis have been extensively studied due to their importance in the formation of myeloid leukemias. Most studies have focused on either transcription factors that are targets for gene translocation in leukemia or interacting proteins that act in concert to effect proliferation or differentiation of progenitor cells. Leukemia-associated proteins include several members of the homeobox superfamily of genes, acute myeloid leukemia protein-1 (AML-1), promyelocytic leukemia zinc finger protein (PLZF) and myeloid zinc finger protein (MZF-1). Many transcription factors that effect myeloid development are also required for gene expression in differentiated macrophages. These proteins include the Ets protein family transcription factor PU.1, the bZIP proteins of the CCAAT enhancer binding protein family (C/EBP), c-Maf, AML-1 and the zinc finger proteins, specificity protein 1 (Sp1) and the early growth response factor (Egr)-1. A number of transcription factors can respond to the extrinsic stimuli that regulate myelopoiesis. Retinoic acid receptors (RAR) are important in induction of terminal granulocyte differentiation and signal transducers and activators of transcription (STAT) transcription factors control cytokine-induced differentiation. This review will briefly describe early events in myeloid development with emphasis on transcription factors associated with monocyte/macrophage development.

2.1 Early Myeloid Development

One of the most common targets for translocation in acute myeloid leukemia is within the gene that encodes the AML-1 (Runx1) transcription factor. Many translocations disrupt AML-1 function by formation of a fusion protein that replaces the *trans*-activation domain with a histone deacetylase repressor protein (ETO, MTG16) (Hiebert et al. 2001). Interaction between AML-1 and the cofactor, CBFβ allows high-affinity binding to DNA and forms the core binding factor (CBF) protein complex (Bushweller 2000). AML-1 and CBFβ are normally expressed in all hematopoietic tissues during myeloid differentiation and in mature macrophages (Tracey and Speck 2000). The CBF protein complex appears to act as a master regulator of hematopoiesis since targeted deletion of either AML-1 or CBFβ in transgenic mice resulted in a complete loss of fetal liver hematopoiesis (Lutterbach and Hiebert 2000). CBF is thought to act as an organizing factor to facilitate the actions of other transcription factors. AML-1 is able to bind to C/EBP proteins and PU.1, and these interactions are important for expression of CSF-1 receptor on myeloid progenitors and mature macrophages (Petrovick et al. 1998). C/EBP proteins and PU.1 can both synergistically activate the human CSF-1 receptor promoter with AML-1 (Zhang et al. 1996a). Like many of the transcription factors mentioned here, AML-1 may lie downstream of hematopoietic growth factor signaling pathways. CSF-1 receptor signaling activates the ras-raf-MAPK pathway that leads to serine phosphorylation of AML-1 by the extra-cellular signal regulated kinase-1 (ERK1). ERK-dependent phosphorylation potentiates the *trans*-activation ability of AML-1 and may be important for differentiation of progenitor cells and activation of mature macrophages (Tanaka et al. 1996).

The function of the *HOX* family of homeobox genes in hematopoiesis was also identified through chromosomal abnormalities associated with certain leukemias. These *HOX* genes are expressed in somewhat lineage-restricted patterns and are an important component of the coded pattern of gene expression during hematopoiesis. There is evidence that HOXA10, HOXA9, HOXA5, HOXB3, HOXB8, and HOXB7 can influence myeloid development (Lawrence et al. 1997). Targeted deletion of HOXA9 resulted in approximately 30% reduction in total leukocytes and lymphocytes, accompanied by atrophy of the spleen and thymus. Myeloid/erythroid and pre-B progenitors in the marrow were significantly reduced with little or no disruption of earlier progenitors (Lawrence et al. 1997). Overexpression of HOXB3 in bone marrow cells resulted in loss of nearly all B-cell progenitors, with elevated numbers of granulocyte-macrophage colony forming cells in the spleen and bone marrow (Sauvageau et al. 1997). HOXA5 may also regulate lineage restriction. HOXA5 expression in CD34($^+$) multipotent progenitors shifted differentiation toward myelopoiesis and away from erythropoiesis (Crooks et al. 1999). Down-regulation of HOXA10 may be required to allow cells to progress to later stages of myeloid development (Thorsteinsdottir et al. 1997). Targeted deletion of HOXA10 resulted in a twofold increase in peripheral blood neutrophils and monocytes and a fivefold increase in myeloid progenitors (Tenen et al. 1997). The expression of HOXA10 is highest in CD34$^+$ progenitors and is down-regulated during myeloid development (Lawrence et al. 1995).

Some HOX proteins appear to specifically promote monocyte development, but this activity may be a result of ectopic expression. Further analysis will be required to determine if these effects are consistent with the normal pattern of expression during hematopoiesis. HOXB8 expression in the 32Dcl3 progenitor cell line inhibited granulocyte differentiation in response to G-CSF but was required for GM-CSF induced monocyte differentiation (Krishnaraju et al. 1997). Similarly, HOXB7 expression in the HL60 cell line inhibited retinoic acid-induced granulocyte differentiation but not vitamin D3-induced monocyte differentiation (Lill et al. 1995). It should be noted that HOXB7 does not appear to be expressed in monocytes and was not induced by treatment with CSF-1 (Lill et al. 1995).

Several transcription factors from the zinc finger family are potentially important in myeloid development. These proteins contain cysteine and histidine residues that bind zinc ions to form a protein loop capable of binding a range of specific DNA sequences. Zinc finger proteins involved in myeloid differentiation belong to the *Drosophila Kruppel*-related proteins and include Sp1, MZF-1, and Egr-1.

MZF-1 transcription factors appear to be important for maintenance of myeloid progenitor cells. MZF-1 expression was detected in the myeloid lineage from myeloblasts to metamyelocytes and not in other bone marrow cell types (Bavisotto et al. 1991). Consistent with this expression pattern, the MZF-1 promoter contains binding sites for MZF-1, PU.1, and retinoic acid receptors (Hui et al. 1995). Expression of MZF-1 in the IL-3-dependent myeloid cell line,

FDCP1, decreased apoptosis after withdrawal of IL-3 and led to tumor formation in 70% of transduced cells. Similarly, MZF-1 decreased retinoic acid-induced apoptosis in HL-60 cells (Hromas et al. 1996; Robertson et al. 1998). The neoplastic and anti-apoptotic characteristics of MZF-1 suggest it may function to expand numbers of myeloid precursors before they proceed to terminally differentiate.

The Sp1 and Egr-1 transcription factors are important regulators of both basal and inducible gene expression in mature macrophages and are discussed in more detail later. Although they do not show lineage-restricted expression, there is significant evidence that both can play a limited role in monocyte development. A number of studies have shown that Sp1 can mediate specificity and inducibility during cell differentiation. There is evidence of enhanced Sp1 binding to sites in myeloid cells possibly due to cell-specific phosphorylation of Sp1 (Chen et al. 1993; Zhang et al. 1994a,b). Egr-1 is an early response gene that can be induced by a variety of stimuli and is up-regulated during monocytic but not granulocytic differentiation. Expression and anti-sense inhibition of Egr-1 provided evidence that Egr-1 acts as a specific inducer of monocyte differentiation (Nguyen et al. 1993; Krishnaraju et al. 1995; Lee et al. 1996). Disruption of the Egr-1 gene in transgenic mice failed to show any abnormality in monocyte development, but this result may be due to compensation by the other Egr-1 family members, Egr-2 and Egr-3 (Lee et al. 1996).

The STAT transcription factors are a family of seven proteins important for cytokine-regulated gene expression (Ward et al. 1999). STAT transcription factors are latent until tyrosine phosphorylated by receptor-associated Janus kinases (JAK) (Parganas et al. 1998). STAT proteins are important for induction of genes during immune responses and this activity will be discussed later in the review. STAT3 plays a crucial role in proliferation and survival of many cell types and targeted deletion of STAT3 is embryonic lethal (Takeda et al. 1997). IL-6 induced macrophage differentiation of the M1 cell line required STAT3 activation (Hirano et al. 2000). STAT5a and STAT5b can both contribute to myeloid cell development by mediating activation through GM-CSF signaling. Targeted deletion of both STAT5a and STAT5b, among other defects, showed a decrease in colony forming ability in response to GM-CSF (Teglund et al. 1998). Both STAT3 and STAT5 are able to activate important proliferation and anti-apoptosis genes and are a common target for viral oncogenes such as *v-abl* (Kieslinger et al. 2000; Nosaka et al. 1999). STAT5 proteins are likely to act on myeloid progenitors by enhancing proliferation and cell survival. Signaling through the CSF-1 receptor is able to activate STAT1, STAT3 and STAT5 and activation of STAT1 and STAT3 is augmented by co-stimulation with γ-interferon (γ-IFN) (Novak et al. 1996). Since CSF-1 receptor expression is low in progenitor cells and increases as cells differentiate to monocyte/macrophages, induction of STAT proteins could be critical for proliferation, survival and differentiation of macrophages in response to CSF-1 (Eilers and Decker 1995; Eilers et al. 1995).

2.2
Monocyte Differentiation

The mechanism that allows myelopoiesis to branch into the monocytic and granulocytic pathways is not well defined. One theory is that the level of specific transcriptional activators determines the pathway of cell development. Many of these transcription factors regulate their own expression, and small changes in expression may push cells towards lineage commitment. The transcription factors most closely tied to monocyte development are PU.1, interferon consensus sequence binding protein (ICSBP or IRF-8), and c-Maf.

PU.1 is a member of the Ets family of transcription factors and is the product of the *Spi*-1 oncogene identified in Friend virus-induced erythroleukemias. PU.1 is expressed at low levels during early hematopoietic development and specifically up-regulated during myeloid development (Cheng et al. 1996). PU.1 is expressed at high levels in myeloid and B cells but is not found in T cells (Chen et al. 1995a). This expression pattern suggests that PU.1 can play a role in commitment to early myeloid lineages as well as the later stages of differentiation. Consistent with this role is the down-regulation of PU.1 during erythropoiesis and the inhibition of erythropoiesis observed upon retroviral overexpression in bone marrow cultures (Schuetze et al. 1993). Gene knockout of PU.1 in mice gave conflicting results. One knockout resulted in embryonic lethality at day E16 (Scott et al. 1994) and another resulted in viable animals that could be kept alive for days under sterile conditions (Mckercher et al. 1996). Although the first knockout suggested PU.1 disruption could affect early multipotent progenitors, the second demonstrated a more restricted phenotype, consistent with the pattern of PU.1 expression. Recent research has demonstrated that this difference in transgenic animals is due to the genetic background of the mice, suggesting that other transcription factors can compensate to a varying extent for PU.1 deficiency (Luchin et al. 2001). Viable transgenic animals showed an absence of monocytes and mature B cells but were still capable of producing B-cell progenitors. T cells and neutrophilic cells were present although neutrophils were reduced in number and altered in function (Mckercher et al. 1996; Anderson et al. 1998).

The PU.1 protein contains an 80 amino acid DNA-binding domain, characteristic of Ets factors, located at the carboxyl terminal end (Klemsz et al. 1990). Like other Ets factors, PU.1 binds to purine-rich sequences characterized by a loose consensus core motif GGAA (Klemsz et al. 1990), but binding is distinct from the other Ets family members, Ets-1, Ets-2, Elf-1, and Fli-1 (Voso et al. 1994). The PU.1 gene promoter itself contains a PU.1-binding site important for function in myeloid cells. A significant up-regulation of PU.1 coincides with the first detection of early myeloid progenitors, suggesting amplification of PU.1 occurs through an autocrine loop and this enhanced expression may play a major role in commitment to the myeloid lineage (Voso et al. 1994; Chen et al. 1995a). A model of how PU.1 can act as an essential component of monocyte development yet play a more restricted role in B-lymphocyte and granulocyte development is emerging.

At one level, PU.1 expression is much higher in macrophages than B cells and over-expression of PU.1 in B cells can actually suppress activity of B cell-specific enhancers (Ross et al. 1994; Dekoter and Singh 2000). Recent research in myeloid development has focused on proteins that interact with PU.1 or compete for DNA binding, and these two mechanisms also contribute to the restricted activity of PU.1 in the formation of certain cell types. The Ets domain of PU.1 can interact with the C-terminal finger region of the transcription factor GATA-1, leading to repression of PU.1-mediated gene expression (Nerlov et al. 2000). GATA-1 is down-regulated in myeloid development but plays an important role in development of non-myeloid lineages (Voso et al. 1994). In a similar manner, B cell-specific activator protein (BSAP) PAX5 can interact with PU.1 through the *trans*-activation domain and inhibit PU.1 activity (Maitra and Atchison 2000). The expression of BSAP is restricted to B lymphopoiesis. Another lymphocyte-specific inhibitor of PU.1 is a second Ets-family protein, PU.1-related factor (Prf), that can compete for DNA binding. Prf is expressed during B-lymphocyte development only and since Prf does not appear to function as a transcriptional activator, it may function as an antagonist of PU.1 activity (Hashimoto et al. 1999).

These mechanisms suggest that PU.1 activity is of limited importance in early B-lymphocyte development. Although PU.1 is required for final maturation and expression of genes in differentiated B lymphocytes, the role of PU.1 during lymphopoiesis is less defined. The reduction of PU.1 activity during B lymphopoiesis would also be expected to have a minimal impact due to partial redundancy with the closely related Spi-B transcription factor (Garrett-Sinha et al. 2001). Spi-B is expressed in B lymphocytes and not in myeloid cells and Spi-B gene knockout in PU.1 +/− mice resulted in a reduction of immature and mature B lymphocytes (Chen et al. 1995b; Su et al. 1996).

The ability of PU.1 to interact with members of the interferon regulatory factor (IRF) family may also be significant in directing monocyte differentiation. The IRF family member, ICSBP, has been shown to regulate myeloid differentiation (Holtschke et al. 1996; Scheller et al. 1999; Tamura et al. 2000). The IRF family members are able to bind the interferon-stimulated response elements (ISRE) found in many interferon-inducible genes (Contursi et al. 2000). ICSBP is expressed in monocyte/macrophage lineages, B lymphocytes, and activated T lymphocytes and is induced by γ-interferon (IFN) (Darnell et al. 1994). Although γ-IFN alone is myelosuppressive, in combination with CSF-1 or GM-CSF it is an activator of monocytopoiesis (Breen et al. 1991). ICSBP forms a complex with PU.1 and activates composite elements in the myeloid-specific gp91(phox) p67(phox) and Toll-like receptor 4 genes (Eklund and Kakar 1999; Rehli et al. 2000). Targeted deletion of the ICSBP gene resulted in a significant increase in granulocytes and a reduction in mature macrophages with a significant impairment of the ability of GM-CSF or CSF-1 to form macrophage colonies (Scheller et al. 1999). Expression of ICSBP in a bipotential cell line, derived from ICSBP$^{-/-}$ mice, showed that

ICSBP drives myeloid progenitors to differentiate into macrophages but inhibits granulocyte differentiation (Tamura et al. 2000).

The PU.1 gene knockout described above also proposed a limited role for PU.1 in early granulocyte development. Similar to B cells, PU.1 appeared to be significant mainly in the final stages of granulocyte differentiation. This stage-specific role is consistent with considerable evidence that commitment to the granulocyte lineage is closely associated with expression of C/EBP proteins and not PU.1 (Gombart and Koeffler 2002). The C/EBP family is a subfamily of bZIP (leucine zipper) proteins. Relevant C/EBP family members include C/EBPα, C/EBPβ, C/EBPδ, and C/EBPε, all of which bind to similar DNA sequences (Antonson et al. 1996). Targeted deletion of either the C/EBPα or C/EBPε genes resulted in severe impairment of granulocyte development (Yamanaka et al. 1997a; Zhang et al. 1997). Expression of both C/EBPα and C/EBPε increases during granulocyte differentiation but decreases during monocyte differentiation (Morosetti et al. 1997; Scott et al. 1992; Yamanaka et al. 1997b). Interestingly, the ectopic expression of C/EBPε alone in 32Dcl3 progenitor cells was sufficient for differentiation to mature granulocytes, indicating a certain level of functional redundancy (Khanna-Gupta et al. 2001).

This expression pattern is not true for a third member of the family, C/EBPβ (NF-IL6) which increases during monocyte but not granulocyte differentiation (Natsuka et al. 1992). In the mouse, C/EBPβ is a strong transactivator of the CSF-1 receptor (*c-fms*) gene promoter (Xie et al. 2002). C/EBPβ shows low activity in myeloid cells until cells are activated by inflammatory mediators and is generally associated with inducible gene expression that is discussed later in the review. Gene knockout of C/EBPβ resulted in no abnormalities in myeloid differentiation, although there were deficiencies in macrophage activation at later stages (Screpanti et al. 1995). The possibility that C/EBP family members can substitute in pairs has not been eliminated.

Another bZIP family member that affects monocyte development is c-Maf. Overexpression of c-Maf induces HL60 and U937 cells to terminally differentiate to macrophages (Hegde et al. 1999). c-Maf is thought to promote macrophage differentiation by forming complexes with c-Myb and inhibiting specific c-Myb regulated targets (Hedge et al. 1998). Gene knockout of c-Myb resulted in embryonic lethality and a failure of fetal liver hematopoiesis (Mucenski et al. 1991). c-Myb is thought to act early in hematopoiesis, and its expression is down-regulated during differentiation (Gewirtz and Calabretta 1988). This down-regulation may be important for monocyte differentiation since c-Myb has been shown to inhibit the CSF-1 receptor gene (Reddy et al. 1994). Inhibition of c-Myb activity appears to be of lesser importance for granulocyte development. c-Myb can synergize with C/EBP proteins and a number of granulocyte-specific genes are *trans*-activated by c-Myb (Oelgeschlager et al. 1996; Verbeek et al. 1999).

The microphthalmia transcription factor (MiTF) family is a set of four related members of the bHLH-ZIP class of proteins. The MiTF family member is required for development of osteoclast cells, a closely related cell lineage to macrophages (Moore 1995). Evidence of their role in monocytopoiesis arises from the dominant phenotype of a subset of MiTF mutations. MiTF transcription factors bind DNA as homodimers or heterodimers with other family members and there is evidence that MiTF proteins interact with PU.1 (Steingrimsson et al. 1994; Luchin et al. 2001). Targeted deletion of individual MiTF family members was not associated with any deficiency in macrophage development; however, this probably reflects the redundant expression in macrophages of all four members, MiTF, TFE-3, TFE-B, and the myeloid-specific member, TFE-C (Rehli et al. 1999).

3
Common Regulatory Elements in Macrophage-Specific Promoters

Regulation of macrophage-specific genes requires the coordinate assembly of multiple transcription factors onto distinct regulatory elements. Promoters that direct lineage-specific expression in macrophages are different from many other tissue-specific promoters. Macrophage-specific promoters generally lack a defined initiator sequence, such as a TATAA box. Transcription initiation can occur at multiple sites within the promoter or at a single strong start site. Transfection of tissue culture cells showed that macrophage restricted expression in some genes required only a small region upstream of the major transcription start site. Deletional analysis of the PU.1 and human CSF-1R promoters showed that minimal, tissue-specific promoter activity was encoded in an approximately 90-bp region (Zhang et al. 1994a; Chen et al. 1995a; Kistler et al. 1995; Ross et al. 1998). The core promoter elements that allow transcription initiation in the absence of a TATAA box are a major component of macrophage-restricted gene expression as outlined below.

Several transcription factors have been shown to interact with TATAA-binding protein (TBP), and this interaction may allow certain transcription factor binding sites to substitute for a TATAA sequence. The presence of functional Sp1 sites in the initiation sequence of many promoters may overcome the absence of a TATAA sequence. Sp1 has been shown to interact with a number of proteins that influence TFIID binding and hence transcription initiation. Sp1 can interact with TBP as well as TAF110 and CRSP, a cofactor complex that can directly bind TFIID (Emili et al. 1994; Ryu et al. 1999). Although Sp1 expression is ubiquitous, the DNA-binding activity can be cell-type specific as outlined above. An in vivo study showed that Sp1 binds the CD11b promoter specifically in myeloid cells and was required for myeloid-specific promoter activity (Chen et al. 1993).

Many macrophage-specific promoters, however, do not contain Sp1 sites within the core upstream promoter sequence. These promoters show a consistent pattern of elements near the start of transcription as outlined in Fig. 1. The purine-rich elements within these regions have been repeatedly identified as PU.1 binding sites. Indeed, nearly all of the macrophage-specific promoters analyzed so far contain a PU.1 site near the start of transcription. An interesting ex-

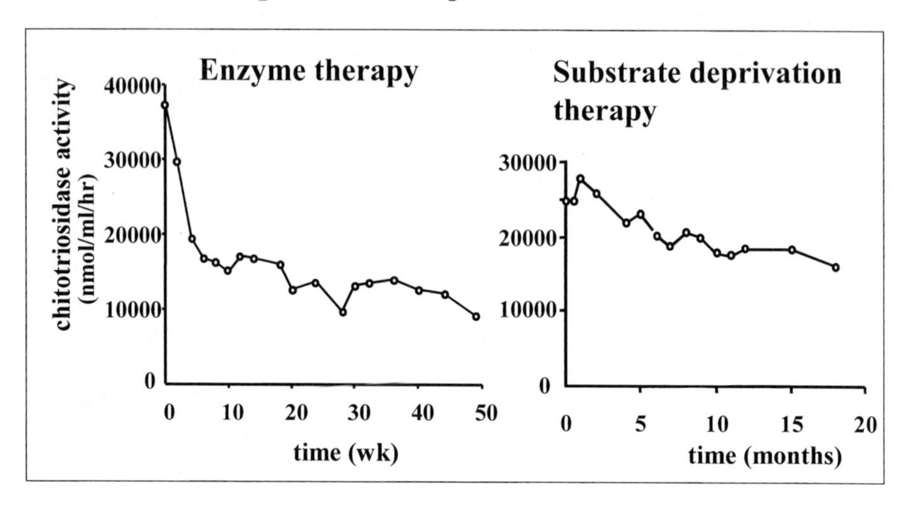

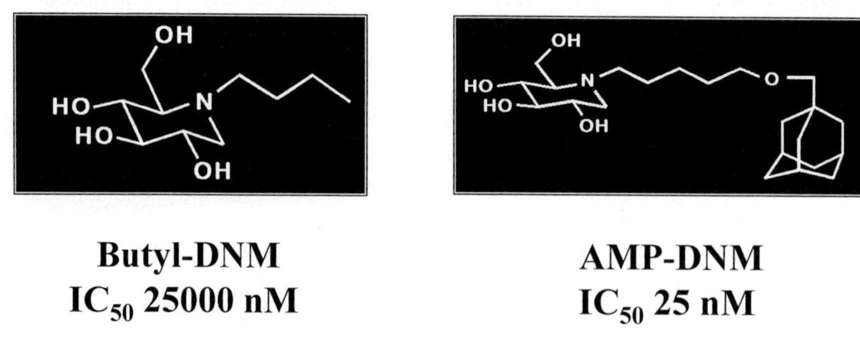

Fig. 1 Sequence comparison of the proximal promoter regions of macrophage-restricted genes. Sequences are aligned by an E box-like motif present in many macrophage-restricted promoters lacking a TATAA box element. *Arrows* mark published transcription initiation sites. PU.1/Ets-like GGAA motifs are shaded in *gray* and *dark shading* indicates motifs where mutation resulted in a significant loss of myeloid-specific promoter activity

ception is the CD14 promoter, which contains a TATAA box initiation site but no PU.1 binding site (Zhang et al. 1994b). The CD14 promoter is capable of directing cell-specific expression only in transient assays and not in transgenic mice (Ferrero et al. 1993). Similarly, the chicken lysozyme gene also lacks a PU.1 site in its promoter but contains one in an upstream enhancer important for cell-restricted expression (Ahne and Stratling 1994). PU.1 probably acts to recruit the basal transcription complex. A purine-rich repeat alone can function as a macrophage-specific promoter (Ross et al. 1998) and PU.1 has been shown to bind directly to TBP in vitro (Hagemeier et al. 1993). Further evidence that PU.1 can play a role in macrophage-specific transcription initiation was provided by study of FcγreceptorI gene expression. Basal and γ-INF-induced expression was abolished by mutation of the PU.1 site, and replacement of the mutated site with a TATAA box restored activity (Eichbaum et al. 1994).

Other Ets family transcription factors are likely to be important for macrophage promoter activity. Macrophage-specific genes typically contain several purine-rich sequence elements, not all of which bind PU.1. The other Ets factors expressed in myeloid cells are Ets-2, Fli-1, Elf-1, and myeloid Elf-1-like factor (MEF) (Klemsz et al. 1993; Kola et al. 1993; Voso et al. 1994). The Ets-2 transcription factor mediates activation of genes in response to CSF-1 receptor signaling such as urokinase plasminogen activator (υPA) and scavenger receptor (Fowles et al. 1998). Activation of the ras-raf-MEK1-ERK kinase pathway by CSF-1 results in phosphorylation of Ets-2 and increased transcriptional activity (Fowles et al. 1998). Ets-2 has also been shown to *trans*-activate the CSF-1 receptor promoter (Ross et al. 1998). Targeted deletion of Ets-2 was embryonic lethal at E8.5, but other studies showed Ets-2 can play an important role in basal and inducible gene expression in macrophages (Reddy et al. 1994; Henkel et al. 1996; Yang et al. 1996).

Binding sites for C/EBP transcription factors are also common in myeloid-specific promoters. Although C/EBP proteins are critical for granulocyte development, there is evidence for a role in macrophage gene expression. The C/EBPβ family member is active in macrophages and is important for inducible gene expression. Targeted deletion of C/EBPβ resulted in impaired macrophage function and mice were immunodeficient (Screpanti et al. 1995). As mentioned above, the level of C/EBPβ specifically increased during monocyte differentiation and C/EBPβ is activated by phosphorylation with ERK kinases and hence a target for CSF-1 receptor signaling. Although clearly not required for lineage development, C/EBPβ can *trans*-activate the CSF-1 receptor promoter with AML-1 and PU.1. It is not clear how this interaction, perhaps in response to infection, can potentiate cell-specific gene expression or differentiation of macrophages (Zhang et al. 1996b; Xie et al. 2002).

Many of the protein complexes that bind regulatory elements in macrophage-specific genes are still not identified. A comprehensive analysis of at least one macrophage-specific gene may be required to catalogue required transcription factors. Advances in array technologies such as gene chips may allow identification of transcription factors and the pathways that regulate them. Changes in chromatin structure are an important determinant of the pattern of gene expression and many transcription factors recruit co-factors that can acetylate or deacetylate histones. Studies on how transcription factors restructure chromatin will also be needed to understand the complex architecture of protein binding that enables cell-specific expression.

4 Regulation of Inducible Gene Expression

Regulation of inducible gene expression in macrophages is of crucial importance to the pathogenesis of several diseases including inflammation, septic shock, atherosclerosis, rheumatoid arthritis, pulmonary fibrosis, and inflammatory bowel disease. During the course of infection, the activation of macrophages must be strictly controlled to prevent damage to host cells. The precise regulation of macrophages is also required to direct host defenses towards specific pathogens and sites of infection. Adaptive immunity has been loosely categorized into two distinct T-cell responses, based on the profile of cytokines expressed. The response of T helper (Th)-1 cells is mainly directed towards systemic infection by viruses and bacteria, and Th2 cell responses are directed towards mucosal infections, primarily in response to infection by parasites. Macrophage or T-cell activation by two distinct subsets of cytokines directs these responses. Autocrine or paracrine stimulation with IL-4 or IL-10 promotes Th2 cell responses and stimulation with IL-12 or γ-IFN promotes Th1 cell responses (Schulze-Koops and Kalden 2001). The role of macrophages in the formation of these two T cell subsets is generally understated. Macrophages are the most sensitive cell type to the microbial products that activate distinct immune responses. For example, the ability of bacterial DNA to specifically potentiate the Th1 immune response is well established yet T cells are not capable of responding to bacterial DNA (Weiner 2000). Additionally, macrophages develop into distinct lineages in tissue and are likely to mediate organ-associated immune responses such as activation of distinct T cell subsets or immune tolerance in the intestinal mucosa. This review will concentrate on recent studies of transcription factors that control gene expression in response to pro-inflammatory and anti-inflammatory signals.

4.1 Pro-inflammatory Signals

NF-κB/Rel Proteins. Nuclear factor-κB/Rel proteins are a small family of transcription factors that are latent in the cytoplasm by virtue of their association with IκB inhibitor proteins. A large number of extracellular signals can trigger a variety of distinct signal transduction pathways that lead to the degradation of IκB. The nuclear localization signal on NF-κB/Rel proteins is then unmasked, allowing translocation to the nucleus and DNA binding as hetero or homodimer protein complexes. NF-κB/Rel proteins are important for the induction of nearly all of the proteins associated with inflammation (Silverman and Maniatis 2001). The ubiquitination and proteolytic degradation of IκBs is activated by the IκB kinases, IKK$\alpha/\beta/\gamma$ (Cheng et al. 1996; Lee and Rikihisa 1998; Silverman and Maniatis 2001). Although activated by similar signals, the NF-κB/Rel family proteins possess distinct functions. The p50 and p52 members do not contain the C-terminal *trans*-activation domain, and *trans*-activation requires dimerization with the p65 (RelA), RelB, or c-Rel proteins (Silverman and Maniatis 2001). The prototypic protein complex that mediates inducible gene expression is the NF-κB, p50/p65 heterodimer. The primary activation signals in macrophages are microbial products such as lipopolysaccharide (LPS) and bacterial DNA, IL-1, IL-6, and tumor necrosis factor-α (TNF-α) (Zhang and Ghosh 2000). The mechanisms of signal transduction and activation of NF-κB/Rel is not covered here, since they have been reviewed elsewhere (Silverman and Maniatis 2001).

Several studies have shown that transcriptional activation by Rel proteins required both the degradation of IκB and Rel protein phosphorylation and that these two activation signals can be uncoupled (Schmitz et al. 2001). NF-κB, p65 has a protein kinase A phosphorylation site on serine 276 (Zhang et al. 1997). Phosphorylation at this site was dependent on IκB degradation and was required for binding to the transcriptional co-activator, CREB-binding protein (CBP) (Zhong et al. 1998). A second site on p65, serine 529, was phosphorylated in response to TNF-α stimulation and was also dependent on IκB degradation (Wang and Baldwin 1998; Wang et al. 2000a). Phosphorylation of s529 appeared to be mediated by casein kinase II (Wang et al. 2000a) and resulted in enhanced transcriptional activation but not nuclear translocation or DNA binding. IL-1 has also been shown to induce phosphorylation of p65, and this phosphorylation required phosphotidylinositol-3 kinase and Akt (Sizemore et al. 1999). This signaling appears to involve IKKβ and required serines 529 and 536 of p65 (Madrid et al. 2001). PI3 K and Akt stimulation was again found to enhance transcriptional activity and not nuclear translocation or DNA binding (Madrid et al. 2001). Considerably less work has been performed on activation of c-Rel proteins; however, many of the phosphorylation sites in p65 occur in the *rel* homology domain. There is also evidence that uncoupling of IκB degradation and transcriptional activation occurs with c-Rel. Targeted deletion of c-Rel resulted in enhanced expression of GM-CSF, IL-6, and TNF-α in peritoneal macrophages treated with LPS but reduced expression of GM-CSF and IL-2 in activated T cells (Gerondakis et al. 1996).

The evidence that s276 phosphorylation allowed binding to the co-activator CBP may in part explain the anti-inflammatory activity of PPARγ signaling and cAMP activation. Both PPARγ, discussed below, and cAMP-response element binding protein (CREB) may compete for limiting amounts of CBP and interfere with p65 transcriptional activation (Li et al. 2000). This activity highlights the potential therapeutic value of compounds that target the phosphorylation of p65 or the associated binding proteins during inflammation, particularly since targeted deletion of p65 or IκB produced a lethal phenotype (Silverman and Maniatis 2001).

STAT Proteins. Many cytokines that are important for cellular activation in the macrophage lineage utilize the JAK-STAT signaling pathway. The ligand-activated γ-IFN receptor recruits JAK1 and JAK2 kinases, leading to phosphorylation, nuclear translocation, and binding of STAT1 homodimers to GAS (γ-IFN activated site) elements. The activation of STAT proteins is transient and several constitutive and inducible pathways down-modulate STAT activity (Darnell 1997). Constitutive pathways that down-modulate STAT activity include dephosphorylation, proteolytic degradation, and binding to inhibitor molecules (Haspel et al. 1996; Kim and Maniatis 1996; Chung et al. 1997a,b). Induction of the suppressors of cytokine signaling molecules (SOCS) by a large number of cytokines is important for feedback inhibition of JAK kinase activity and allows cross-inhibition between cytokine receptors (Endo et al. 1997; Starr et al. 1997). Signaling through the mitogen-activated protein kinase (MAPK) pathway has also been shown to inhibit STAT activation (Sengupta et al. 1998). Phorbol ester, ionomycin or GM-CSF activation of ERK kinase has been shown to inhibit IL-6-mediated JAK1 and JAK2 phosphorylation of STAT3 (Petricoin et al. 1996; Sengupta et al. 1996). Other pathways that inhibit STAT activation in macrophages include, crosslinking of Fc or complement CR3 receptors (Feldman et al. 1995; Marth and Kelsall 1997) and activation of the signal transducers, protein kinase A and protein kinase C (Bhat et al. 1995; Lee and Rikihisa 1998). STAT proteins are also regulated through serine phosphorylation by kinases yet to be identified. The functional significance of serine phosphorylation varies and can either activate or inhibit tyrosine phosphorylation or DNA binding (Eilers et al. 1995; Wen et al. 1995; Beadling et al. 1996).

The precise regulation of STAT proteins in macrophages may be instrumental in directing either Th1 or Th2 cell responses. For example, the cytokine IL-10 is a potent growth factor for activated B cells, but down-modulates T-cell responses by suppressing expression of major histocompatibility complex class II and B7 on macrophages (Ding et al. 1993). The induction of SOCS3 by IL-10 can block γ-IFN mediated STAT1 activation by binding to phosphorylated residues in the tyrosine kinase domain of JAK kinases (Ito et al. 1999). STAT1 along with IRFs are required for expression of IL-12, an important cytokine for directing Th1 responses (Durbin et al. 2000). Strong inflammatory signals from bacteria can also alter cell signaling. Treatment of macrophages with LPS, IL-1, and TNF-α inhibited IL-6 and IL-10 but not γ-IFN activation of STAT phosphorylation and DNA binding (Ahmed and Ivashkiv 2000).

IL-4 can induce expression of a distinct subset of genes in macrophages, including Fcεreceptor IIb (CD23), 15-lipoxygenase, and IL-1 receptor antagonist (Vercelli et al. 1988; Conrad et al. 1992; Fenton et al. 1992). The induction of STAT6 by IL-4 signaling is important in both activation and inhibition of gene expression (Kaplan et al. 1996; Shimoda et al. 1996; Takeda et al. 1996). Pretreatment of macrophages with γ-IFN or β-IFN can suppress activation of STAT6 by IL-4. This repression is mediated by interferon-induced expression of the JAK–STAT inhibitor SOCS1 (Dickensheets et al. 1999). IL-4 activation of STAT6 binding can, in turn, suppress γ-IFN-induced transcription. STAT6 appears to inhibit interferon-induced gene expression by competition for occupancy of promoter binding sites with STAT1 (Ohmori and Hamilton 1998). STAT6$^{-/-}$ mice showed decreased susceptibility to septic peritonitis due to enhanced bacterial clearance. This enhanced anti-microbial activity was associated with increased levels of IL-12, TNF-α, and macrophage-derived chemokine (Matsukawa et al. 2001).

IRF Proteins IRF Protein. IRFs are a family of at least nine transcription factors with a broad range of activities (Taniguchi et al. 2001). All members are expressed in most cell types except IRF-4 and ICSBP, which are expressed in lymphoid and myeloid cells only (Driggers et al. 1990; Rosenbauer et al. 1999). The members that effect macrophage activation the most are IRF-1, interferon-stimulated gene factor-3γ (ISGF3γ, IRF-9), and ICSBP (IRF-8). IRF-1 expression is strongly enhanced in macrophages by treatment with γ-IFN (Flodstrom and Eizirik 1997). Targeted disruption of IRF-1 resulted in a wide range of defects in macrophage activation and consequent resistance to several models of chronic inflammatory disease. Macrophages from IRF$^{-/-}$ mice failed to induce iNOS (inducible nitric oxide synthetase) in response to LPS or β-IFN (Kamijo et al. 1994; Martin et al. 1994). The iNOS enzyme catalyzes production of nitric oxide, important for intracellular killing of pathogens and may explain the severe pathology of *Mycobacterium bovis* infection in IRF-1$^{-/-}$ mice (Kamijo et al. 1994). IRF-1 was also required for expression of the p40 subunit if IL-12, essential for activation of the Th1-cell subset in the immune system (Lohoff et al. 1997).

ICSBP (IRF-8) shows selective expression in myeloid and lymphoid cells and is important for macrophage development as discussed above. ICSBP is induced by γ-IFN but not by α-IFN or β-IFN (Kanno et al. 1993). ICSBP can bind to composite elements by formation of activating transcriptional complexes with IRF-1, IRF-2, and PU.1 (Eklund et al. 1998). Transcriptional activation by ICSBP is critical for the Th1 immune response since ICSBP$^{-/-}$ mice failed to produce IL-12 and γ-IFN and showed increased susceptibility to virus infection (Wang et al. 2000b).

ISGF3γ/IRF-9 also forms activating complexes upon stimulation with α and β-IFNs and exerts its transcriptional effects exclusively by association with STAT1 and STAT2 (Veals et al. 1992; Bluyssen et al. 1996; Darnell 1997). The resulting trimolecular IRF-9/STAT1,2 complex, referred to as ISGF3, is able to activate transcription of many genes by binding to ISREs (Stark et al. 1998). γ-IFN treatment of macrophages can strongly enhance transcriptional activation through ISREs by up-regulating IRF-9 and STAT1 expression. There is also evidence that γ-IFN together with spontaneously produced α- or β-IFN allows physical association between the IFN receptors and STAT1/2 docking and activation (Bandyopadhyay et al. 1990; Levy et al. 1990).

C/EBP ProteinsC/EBP Protein. The principle C/EBP family members that activate gene expression in response to inflammatory stimuli are C/EBPβ and C/EBPδ. The level of C/EBPε is extremely low in mature macrophages and C/EBPα levels generally decrease in response to inflammatory signals (Tengku-Muhammad et al. 2000). C/EBPβ is predominantly expressed in monocyte/macrophages, hepatocytes, keratinocytes, and adipocytes and is the principle mediator of C/EBP-induced inflammatory gene expression in macrophages (Akira et al. 1990; Cao et al. 1991; Lekstrom-Himes and Xanthopoulos 1998; Maytin and Habener 1998). The expression and activity of C/EBPβ is induced by many inflammatory mediators such as LPS, IL-1, IL-6, γ-IFN, and TNF-α (Akira et al. 1990; Pope et al. 1994; Darville and Eizirik 2001; Hu et al. 2001). γ-IFN can stimulate the transcriptional activity of C/EBPβ through activation of ERK1 and ERK2 kinases. This phosphorylation and activation of C/EBPβ was required for induction of ISGF3γ/IRF-9 expression, allowing formation of the ISGF3 complex on ISREs (Hu et al. 2001). Induction of C/EBPβ plays a central role in acute-phase responses by activating expression of a number of inflammatory mediators, including TNF-α, IL-1β, IL-6, IL-8, MIP-1α, MCP-1, and MMP-1 (Grove and Plumb 1993; Shirakawa et al. 1993; Stein and Baldwin 1993; Pope et al. 1994). A role for C/EBP proteins has been established in a number of inflammatory diseases (Poli 1998). Macrophages that express high levels of C/EBPβ were found in the synovial lining of patients with rheumatoid arthritis and the level of expression strongly correlated to lining thickness (Pope et al. 1999).

Egr-1 ProteinsEgr-1 Protein. A number of studies have shown that Egr-1 transcription factors are important for macrophage activation during chronic or acute inflammation (Mcmahon and Monroe 1996). Many recent studies have specifically analyzed the role of Egr-1 in the pathogenesis of atherosclerotic lesions (Khachigian 2001). Egr-1 is induced not only by inflammatory cytokines and LPS, but also by enzymatically degraded low-density lipoprotein (LDL), hypoxia, physical force, and injury (Silverman and Collins 1999). Egr-1 activation influences expression of many stress-response genes including platelet-derived growth factor (PDGF), transforming growth factor (TGF)β, TNF-α, intracellular adhesion molecule (ICAM)-1, urokinase-type plasminogen activator, and metalloproteinases (Mccaffrey et al. 2000). Elevated levels of Egr-1 can be found at early stages of atherosclerosis with progressive increase in expression during formation of lesions, particularly in areas of macrophage infiltration (Mccaffrey et al. 2000). Activation of Egr-1 may be important for infiltration of macrophage into lesions. Egr-1 expression and binding was required for activation of the CD44 promoter in response to IL-1 (Maltzman et al. 1996). Expression of the adhesion molecule, CD44 is important for recruitment of leukocytes to inflammatory sites, and CD44 signaling results in expression of pro-inflammatory chemokines in macrophages (Ariel et al. 2000; Stoop et al. 2001).

4.2 Anti-inflammatory Signals

SMAD ProteinsSMAD Protein. The cytokine TGFβ can act as a potent anti-inflammatory agent in macrophages, and up-regulation of TGFβ in gastrointestinal tissue is essential for induction of oral tolerance (Letterio and Roberts 1998). Targeted deletion of the TGFβ1 gene resulted in systemic inflammation and early death (Shull et al. 1992). SMADs are a family of proteins that transduce signals from type I and type II TGFβ receptors (Massague and Wotton

2000). TGFβ activation of the type I receptor results in serine phosphorylation of SMAD2 and SMAD3 which associates with SMAD4 during nuclear transport forming complexes that bind DNA and recruit transcription factors (Letterio and Roberts 1998; Massague and Wotton 2000). Inhibition of LPS activation in macrophage by TGFβ treatment required SMAD3 and appeared to involve the ability of SMAD3 to compete for the co-activator p300 (Werner et al. 2000).

The pro-inflammatory cytokines, TNF-α and γ-IFN can both inhibit the TGFβ/SMAD signaling pathway. These cytokines inhibit SMAD activity by inducing the expression of the repressor, SMAD7 that can occupy ligand activated TGFβ1 R1 and block SMAD2 and SMAD3 phosphorylation (Hayashi et al. 1997; Nakao et al. 1997). Disruption of the SMAD signaling pathway is implicated in the pathogenesis of inflammatory bowel disease. SMAD7 expression was up-regulated in mucosal tissue from patients with Crohn's disease and ulcerative colitis, and analysis of lamina propria mononuclear cells from diseased tissue showed extremely low levels of phosphorylated SMAD3 and these cells did not respond to TGFβ treatment (Monteleone et al. 2001).

PPARγ. Recent research has provided evidence that the peroxisome proliferator-activated receptor (PPAR) family of nuclear receptors can act as a down-modulator of inflammation. PPAR proteins are capable of both positive and negative regulation of gene expression in response to ligand binding. PPAR proteins contain a highly conserved DNA-binding domain and ligand-dependent and -independent *trans*-activation domains (Nolte et al. 1998; Xu et al. 1999). PPAR positively regulates gene expression by binding as a complex with retinoid X receptors to composite elements within target genes (Direnzo et al. 1997). The PPARγ family member is expressed in macrophages and appears to be involved in lipid accumulation and inflammatory responses. Both the natural ligand, 15-deoxy-$\Delta^{12,14}$prostaglandin J_2 and the thiazolidinedione (TZD) class of synthetic ligands have anti-inflammatory effects on macrophages and TZDs show early signs of efficacy against inflammatory bowel disease (Delerive et al. 2001; Lewis et al. 2001). Treatment of macrophages with PPARγ ligands reduced expression of the inflammatory cytokines TNF-α, IL-1β and IL-6 (Jiang et al. 1998). Treatment of elicited macrophages induced a resting phenotype and suppressed iNOS, gelatinase B and scavenger receptor A (Ricote et al. 1998). PPARγ inhibition of the iNOS promoter did not involve DNA binding. PPARγ appears to inhibit gene expression by suppressing the activity of NF-κB, AP-1, and STAT1 transcription factors through competition for recruitment of the shared co-activators p300 and CBP (Ricote et al. 1998; Li et al. 2000). IL-4-mediated inhibition of iNOS synthesis partially involves PPARγ by inducing the coordinate expression of PPARγ and its ligands 13 HODE and 15 HETE (Ricote et al. 2000).

Evidence that PPARγ may influence lipid accumulation in macrophages has led to concerns over the use of TZDs in the treatment of diabetes. PPARγ-activated THP-1 cells showed increased expression of the scavenger receptor CD36 and increased uptake of oxidized LDL (Chawla et al. 2001). Targeted deletion and retroviral expression of PPARγ confirmed that CD36 expression is enhanced by PPARγ (Chawla et al. 2001). Other reports, however, have shown that PPARγ does not promote formation of macrophage foamy cells and can activate cholesterol efflux by induction of apo A1-mediated cholesterol transporter (ABCA1) gene expression (Chinetti et al. 2001).

Glucocorticoid ReceptorGlucocorticoid Receptor Activation. Glucocorticoids (GC) are used as immunosuppressive agents in organ transplantation, immune diseases, and inflammatory disorders. GCs suppress many functions in macrophages, the most significant of which is inhibition of cytokine production, including suppression of TNF-α production. Administration of GCs results in adverse effects in many tissues and causes systemic monocytopenia in humans (Joyce et al. 1997). GCs have been shown to interfere with the function of members of the NF-κB and AP-1 family of transcription factors (Auphan et al. 1995). The decreased binding of NF-κB seen with glucocorticoid treatment may result from either interaction between ligand-activated glucocorticoid receptor and p65/RelA or induction of IκB synthesis (Ramdas and Harmon 1998; Costas et al. 2000). Recent evidence suggests that glucocorticoids can alter the function of histone acetylases by inhibiting p65-associated acetylase transferase or CBP recruitment by NF-κB (Ito et al. 2001; Kagoshima et al. 2001). There is, however, evidence that post-transcriptional mechanisms contribute significantly to glucocorticoid inhibition of cytokine production. In humans, glucocorticoids reduce the stability of IL-1β, IL-1α, and IL-6 mRNA (Keffer et al. 1991; Amano et al. 1993).

GCs also have a profound effect on the regulation of Th1/Th2 cytokine responses (Almawi et al. 1999). Treatment with the GC, dexamethasone resulted in specific inhibition of IL-12 activation of STAT4 with no effect on IL-4 activation of STAT6 (Franchimont et al. 2000). The loss of IL-12 activity resulted in reduced γ-IFN expression from T cells and natural killer cells and a subsequent decrease in the ability of macrophages to direct Th1 cell responses. The resulting Th2 cytokine profile with GC treatment is believed to be a result of this inhibition of Th1 cell responses. GCs were not shown to directly enhance Th2 cytokine responses and addition of γ-IFN abrogated the dexamethasone enhanced Th2 cytokine profile (Agarwal and Marshall 2001; Miyaura and Iwata 2002).

5 Future Studies

Macrophage-specific gene expression involves the temporal and lineage-specific assembly of transcription factor complexes on DNA. A critical step in transcription factor access to DNA is the remodeling of chromatin, and both cell differentiation and macrophage activation require extensive changes in chromatin structure. Many developmental transcription factors such as NF-κB, AML-1, homeobox family proteins, and HMG-box proteins can modify chromatin either directly or by binding specific co-factors. Identifying the DNA-binding proteins that allow rapid changes in chromatin structure is an important area for future study. Finally, it is clear that a large number of important transcription factors are still not identified. A detailed analysis of protein binding has been performed on a few select promoters. Even within these genes, the transcription factors that bind many of the conserved elements have not been identified. These studies may require technology that allows rapid identification of proteins assembled on promoter or enhancer sequences. Recent advances in microchip technology, used for analysis of protein-protein interaction, may help to identify proteins that associate with transcription factors or structural proteins in chromatin. Expression profiling by microarray can determine the spectrum of transcription factors present during macrophage development and activation. These analyses may reveal new targets for novel therapeutics that act at the level of gene transcription.

6 References

Agarwal SK, Marshall GD Jr (2001) Dexamethasone promotes type 2 cytokine production primarily through inhibition of type 1 cytokines. J Interferon Cytokine Res 21:147–155

Ahmed ST, Ivashkiv LB (2000) Inhibition of IL-6 and IL-10 signaling and Stat activation by inflammatory and stress pathways. J Immunol 165:5227–5237

Ahne B, Stratling WH (1994) Characterization of a myeloid-specific enhancer of the chicken lysozyme gene. Major role for an Ets transcription factor-binding site. J Biol Chem 269:17794–17801

Akira S, Isshiki H, Sugita T, Tanabe O, Kinoshita S, Nishio Y, Nakajima T, Hirano T, Kishimoto T (1990) A nuclear factor for IL-6 expression (NF-IL6) is a member of a C/EBP family. Embo J 9:1897–1906

Almawi WY, Melemedjian OK, Rieder MJ (1999) An alternate mechanism of glucocorticoid anti-proliferative effect: promotion of a Th2 cytokine-secreting profile. Clin Transplant 13:365–374

Amano Y, Lee SW, Allison AC (1993) Inhibition by glucocorticoids of the formation of interleukin-1 alpha, interleukin-1 beta, and interleukin-6: mediation by decreased mRNA stability. Mol Pharmacol 43:176–182

Anderson KL, Smith KA, Conners K, McKercher SR, Maki RA, Torbett BE (1998) Myeloid development is selectively disrupted in PU.1 null mice. Blood 91:3702–3710

Antonson P, Stellan B, Yamanaka R, Xanthopoulos KG (1996) A novel human CCAAT/enhancer binding protein gene, C/EBPepsilon, is expressed in cells of lymphoid and myeloid lineages and is localized on chromosome 14q11.2 close to the T-cell receptor alpha/delta locus. Genomics 35:30–38

Ariel A, Lider O, Brill A, Cahalon L, Savion N, Varon D, Hershkoviz R (2000) Induction of interactions between CD44 and hyaluronic acid by a short exposure of human T cells to diverse pro-inflammatory mediators. Immunology 100:345–351

Auphan N, DiDonato JA, Rosette C, Helmberg A, Karin M (1995) Immunosuppression by glucocorticoids: inhibition of NF-kappa B activity through induction of I kappa B synthesis. Science 270:286–290

Bandyopadhyay SK, Kalvakolanu DV, Sen GC (1990) Gene induction by interferons: functional complementation between trans-acting factors induced by alpha interferon and gamma interferon. Mol Cell Biol 10:5055–5063

Bavisotto L, Kaushansky K, Lin N, Hromas R (1991) Antisense oligonucleotides from the stage-specific myeloid zinc finger gene MZF-1 inhibit granulopoiesis in vitro. J Exp Med 174:1097–1101

Beadling C, Ng J, Babbage JW, Cantrell DA (1996) Interleukin-2 activation of STAT5 requires the convergent action of tyrosine kinases and a serine/threonine kinase pathway distinct from the Raf1/ERK2 MAP kinase pathway. Embo J 15:1902–1913
Bhat GJ, Thekkumkara TJ, Thomas WG, Conrad KM, Baker KM (1995) Activation of the STAT pathway by angiotensin II in T3CHO/AT1A cells. Cross-talk between angiotensin II and interleukin-6 nuclear signaling. J Biol Chem 270:19059–19065
Bluyssen AR, Durbin JE, Levy DE (1996) ISGF3 gamma p48, a specificity switch for interferon activated transcription factors. Cytokine Growth Factor Rev 7:11–17
Breen FN, Hume DA, Weidemann MJ (1991) Interactions among granulocyte-macrophage colony-stimulating factor, macrophage colony-stimulating factor, and IFN-gamma lead to enhanced proliferation of murine macrophage progenitor cells. J Immunol 147:1542–1547
Bushweller JH (2000) CBF—a biophysical perspective. Semin Cell Dev Biol 11:377–382
Cao Z, Umek RM, McKnight SL (1991) Regulated expression of three C/EBP isoforms during adipose conversion of 3T3-L1 cells. Genes Dev 5:1538–1552
Cecchini MG, Hofstetter W, Halasy J, Wetterwald A, Felix R (1997) Role of CSF-1 in bone and bone marrow development. Mol Reprod Dev 46:75–83; discussion 83–74
Chawla A, Barak Y, Nagy L, Liao D, Tontonoz P, Evans RM (2001) PPAR-gamma dependent and independent effects on macrophage-gene expression in lipid metabolism and inflammation. Nat Med 7:48–52
Chen H, Ray-Gallet D, Zhang P, Hetherington CJ, Gonzalez DA, Zhang DE, Moreau-Gachelin F, Tenen DG (1995a) PU.1 (Spi-1) autoregulates its expression in myeloid cells. Oncogene 11:1549–1560
Chen HM, Pahl HL, Scheibe RJ, Zhang DE, Tenen DG (1993) The Sp1 transcription factor binds the CD11b promoter specifically in myeloid cells in vivo and is essential for myeloid-specific promoter activity. J Biol Chem 268:8230–8239
Chen HM, Zhang P, Voso MT, Hohaus S, Gonzalez DA, Glass CK, Zhang DE, Tenen DG (1995b) Neutrophils and monocytes express high levels of PU.1 (Spi-1) but not Spi-B. Blood 85:2918–2928
Cheng T, Shen H, Giokas D, Gere J, Tenen DG, Scadden DT (1996) Temporal mapping of gene expression levels during the differentiation of individual primary hematopoietic cells. Proc Natl Acad Sci USA 93:13158–13163
Chinetti G, Lestavel S, Bocher V, Remaley AT, Neve B, Torra IP, Teissier E, Minnich A, Jaye M, Duverger N, Brewer HB, Fruchart JC, Clavey V, Staels B (2001) PPAR-alpha and PPAR-gamma activators induce cholesterol removal from human macrophage foam cells through stimulation of the ABCA1 pathway. Nat Med 7:53–58
Chung CD, Liao J, Liu B, Rao X, Jay P, Berta P, Shuai K (1997a) Specific inhibition of Stat3 signal transduction by PIAS3. Science 278:1803–1805
Chung J, Uchida E, Grammer TC, Blenis J (1997b) STAT3 serine phosphorylation by ERK-dependent and -independent pathways negatively modulates its tyrosine phosphorylation. Mol Cell Biol 17:6508–6516
Conrad DJ, Kuhn H, Mulkins M, Highland E, Sigal E (1992) Specific inflammatory cytokines regulate the expression of human monocyte 15-lipoxygenase. Proc Natl Acad Sci USA 89:217–221
Contursi C, Wang IM, Gabriele L, Gadina M, O'Shea J, Morse HC 3rd, Ozato K (2000) IFN consensus sequence binding protein potentiates STAT1-dependent activation of IFNgamma-responsive promoters in macrophages. Proc Natl Acad Sci USA 97:91–96
Costas MA, Muller Igaz L, Holsboer F, Arzt E (2000) Transrepression of NF-kappaB is not required for glucocorticoid-mediated protection of TNF-alpha-induced apoptosis on fibroblasts. Biochim Biophys Acta 1499:122–129
Crooks GM, Fuller J, Petersen D, Izadi P, Malik P, Pattengale PK, Kohn DB, Gasson JC (1999) Constitutive HOXA5 expression inhibits erythropoiesis and increases myelopoiesis from human hematopoietic progenitors. Blood 94:519–528

Dai XM, Ryan GR, Hapel AJ, Dominguez MG, Russell RG, Kapp S, Sylvestre V, Stanley E R (2002) Targeted disruption of the mouse colony-stimulating factor 1 receptor gene results in osteopetrosis, mononuclear phagocyte deficiency, increased primitive progenitor cell frequencies, and reproductive defects. Blood 99:111–120
Darnell JE Jr (1997) STATs and gene regulation. Science 277:1630–1635
Darnell JE Jr, Kerr IM, Stark GR (1994) Jak-STAT pathways and transcriptional activation in response to IFNs and other extracellular signaling proteins. Science 264:1415–1421
Darville MI, Eizirik DL (2001) Cytokine induction of Fas gene expression in insulin-producing cells requires the transcription factors NF-kappaB and C/EBP. Diabetes 50:1741–1748
DeKoter RP, Singh H (2000) Regulation of B lymphocyte and macrophage development by graded expression of PU.1. Science 288:1439–1441
Delerive P, Fruchart JC, Staels B (2001) Peroxisome proliferator-activated receptors in inflammation control. J Endocrinol 169:453–459
Dickensheets HL, Venkataraman C, Schindler U, Donnelly RP (1999) Interferons inhibit activation of STAT6 by interleukin 4 in human monocytes by inducing SOCS-1 gene expression. Proc Natl Acad Sci USA 96:10800–10805
Ding L, Linsley PS, Huang LY, Germain RN, Shevach EM (1993) IL-10 inhibits macrophage costimulatory activity by selectively inhibiting the up-regulation of B7 expression. J Immunol 151:1224–1234
DiRenzo J, Soderstrom M, Kurokawa R, Ogliastro M H, Ricote M, Ingrey S, Horlein A, Rosenfeld MG, Glass CK (1997) Peroxisome proliferator-activated receptors and retinoic acid receptors differentially control the interactions of retinoid X receptor heterodimers with ligands, coactivators, and corepressors. Mol Cell Biol 17:2166–2176
Driggers PH, Ennist DL, Gleason SL, Mak WH, Marks MS, Levi BZ, Flanagan JR, Appella E, Ozato K (1990) An interferon gamma-regulated protein that binds the interferon-inducible enhancer element of major histocompatibility complex class I genes. Proc Natl Acad Sci USA 87:3743–3747
Durbin JE, Fernandez-Sesma A, Lee CK, Rao TD, Frey AB, Moran TM, Vukmanovic S, Garcia-Sastre A, Levy DE (2000) Type I IFN modulates innate and specific antiviral immunity. J Immunol 164:4220–4228
Eichbaum QG, Iyer R, Raveh DP, Mathieu C, Ezekowitz RA (1994) Restriction of interferon gamma responsiveness and basal expression of the myeloid human Fc gamma R1b gene is mediated by a functional PU.1 site and a transcription initiator consensus. J Exp Med 179:1985–1996
Eilers A, Decker T (1995) Activity of Stat family transcription factors is developmentally controlled in cells of the macrophage lineage. Immunobiology 193:328–333
Eilers A, Georgellis D, Klose B, Schindler C, Ziemiecki A, Harpur AG, Wilks AF, Decker T (1995) Differentiation-regulated serine phosphorylation of STAT1 promotes GAF activation in macrophages. Mol Cell Biol 15:3579–3586
Eklund EA, Jalava A, Kakar R (1998) PU.1, interferon regulatory factor 1, and interferon consensus sequence-binding protein cooperate to increase gp91(phox) expression. J Biol Chem 273:13957–13965
Eklund EA, Kakar R (1999) Recruitment of CREB-binding protein by PU.1, IFN-regulatory factor-1, and the IFN consensus sequence-binding protein is necessary for IFN-gamma-induced p67phox and gp91phox expression. J Immunol 163:6095–6105
Emili A, Greenblatt J, Ingles CJ (1994) Species-specific interaction of the glutamine-rich activation domains of Sp1 with the TATA box-binding protein. Mol Cell Biol 14:1582–1593
Endo TA, Masuhara M, Yokouchi M, Suzuki R, Sakamoto H, Mitsui K, Matsumoto A, Tanimura S, Ohtsubo M, Misawa H, Miyazaki T, Leonor N, Taniguchi T, Fujita T, Kanakura Y, Komiya S, Yoshimura A (1997) A new protein containing an SH2 domain that inhibits JAK kinases. Nature 387:921–924

Feldman GM, Chuang EJ, Finbloom DS (1995) IgG immune complexes inhibit IFN-gamma-induced transcription of the Fc gamma RI gene in human monocytes by preventing the tyrosine phosphorylation of the p91 (Stat1) transcription factor. J Immunol 154:318–325

Fenton MJ, Buras JA, Donnelly RP (1992) IL-4 reciprocally regulates IL-1 and IL-1 receptor antagonist expression in human monocytes. J Immunol 149:1283–1288

Ferrero E, Jiao D, Tsuberi BZ, Tesio L, Rong GW, Haziot A, Goyert SM (1993) Transgenic mice expressing human CD14 are hypersensitive to lipopolysaccharide. Proc Natl Acad Sci USA 90:2380–2384

Flodstrom M, Eizirik DL (1997) Interferon-gamma-induced interferon regulatory factor-1 (IRF-1) expression in rodent and human islet cells precedes nitric oxide production. Endocrinology 138:2747–2753

Fowles LF, Martin ML, Nelsen L, Stacey KJ, Redd D, Clark YM, Nagamine Y, McMahon M, Hume DA, Ostrowski MC (1998) Persistent activation of mitogen-activated protein kinases p42 and p44 and ets-2 phosphorylation in response to colony-stimulating factor 1/c-fms signaling. Mol Cell Biol 18:5148–5156

Franchimont D, Galon J, Gadina M, Visconti R, Zhou Y, Aringer M, Frucht DM, Chrousos GP, O'Shea JJ (2000) Inhibition of Th1 immune response by glucocorticoids: dexamethasone selectively inhibits IL-12-induced Stat4 phosphorylation in T lymphocytes. J Immunol 164:1768–1774

Garrett-Sinha LA, Dahl R, Rao S, Barton KP, Simon MC (2001) PU.1 exhibits partial functional redundancy with Spi-B, but not with Ets-1 or Elf-1. Blood 97:2908–2912

Gerondakis S, Strasser A, Metcalf D, Grigoriadis G, Scheerlinck JY, Grumont RJ (1996) Rel-deficient T cells exhibit defects in production of interleukin 3 and granulocyte-macrophage colony-stimulating factor. Proc Natl Acad Sci USA 93:3405–3409

Gewirtz AM, Calabretta B (1988) A c-myb antisense oligodeoxynucleotide inhibits normal human hematopoiesis in vitro. Science 242:1303–1306

Gombart AF, Koeffler HP (2002) Neutrophil specific granule deficiency and mutations in the gene encoding transcription factor C/EBP(epsilon). Curr Opin Hematol 9:36–42

Grove M, Plumb M (1993) C/EBP, NF-kappa B, and c-Ets family members and transcriptional regulation of the cell-specific and inducible macrophage inflammatory protein 1 alpha immediate-early gene. Mol Cell Biol 13:5276–5289

Hagemeier C, Bannister AJ, Cook A, Kouzarides T (1993) The activation domain of transcription factor PU.1 binds the retinoblastoma (RB) protein and the transcription factor TFIID in vitro: RB shows sequence similarity to TFIID and TFIIB. Proc Natl Acad Sci USA 90:1580–1584

Hashimoto S, Nishizumi H, Hayashi R, Tsuboi A, Nagawa F, Takemori T, Sakano H (1999) Prf, a novel Ets family protein that binds to the PU.1 binding motif, is specifically expressed in restricted stages of B cell development. Int Immunol 11:1423–1429

Haspel RL, Salditt-Georgieff M, Darnell JE Jr (1996) The rapid inactivation of nuclear tyrosine phosphorylated Stat1 depends upon a protein tyrosine phosphatase. Embo J 15:6262–6268

Hayashi H, Abdollah S, Qiu Y, Cai J, Xu YY, Grinnell BW, Richardson MA, Topper JN, Gimbrone MA Jr, Wrana JL, Falb D (1997) The MAD-related protein Smad7 associates with the TGFbeta receptor and functions as an antagonist of TGFbeta signaling. Cell 89:1165–1173

Hedge SP, Kumar A, Kurschner C, Shapiro LH (1998) c-Maf interacts with c-Myb to regulate transcription of an early myeloid gene during differentiation. Mol Cell Biol 18:2729–2737

Hegde SP, Zhao J, Ashmun RA, Shapiro LH (1999) c-Maf induces monocytic differentiation and apoptosis in bipotent myeloid progenitors. Blood 94:1578–1589

Henkel GW, McKercher SR, Yamamoto H, Anderson KL, Oshima RG, Maki RA (1996) PU.1 but not ets-2 is essential for macrophage development from embryonic stem cells. Blood 88:2917–2926

Hiebert SW, Lutterbach B, Amann J (2001) Role of co-repressors in transcriptional repression mediated by the t(8;21), t(16;21), t(12;21), and inv(16) fusion proteins. Curr Opin Hematol 8:197–200

Hirano T, Ishihara K, Hibi M (2000) Roles of STAT3 in mediating the cell growth, differentiation and survival signals relayed through the IL-6 family of cytokine receptors. Oncogene 19:2548–2556

Holtschke T, Lohler J, Kanno Y, Fehr T, Giese N, Rosenbauer F, Lou J, Knobeloch KP, Gabriele L, Waring JF, Bachmann MF, Zinkernagel RM, Morse HC 3rd, Ozato K, Horak I (1996) Immunodeficiency and chronic myelogenous leukemia-like syndrome in mice with a targeted mutation of the ICSBP gene. Cell 87:307–317

Hromas R, Boswell S, Shen RN, Burgess G, Davidson A, Cornetta K, Sutton J, Robertson K (1996) Forced over-expression of the myeloid zinc finger gene MZF-1 inhibits apoptosis and promotes oncogenesis in interleukin-3-dependent FDCP.1 cells. Leukemia 10:1049–1050

Hu J, Roy SK, Shapiro PS, Rodig SR, Reddy SP, Platanias LC, Schreiber R D, Kalvakolanu DV (2001) ERK1 and ERK2 activate CCAAAT/enhancer-binding protein-beta-dependent gene transcription in response to interferon-gamma. J Biol Chem 276:287–297

Hui P, Guo X, Bradford PG (1995) Isolation and functional characterization of the human gene encoding the myeloid zinc finger protein MZF-1. Biochemistry 34:16493–16502

Ito K, Jazrawi E, Cosio B, Barnes PJ, Adcock IM (2001) p65-activated histone acetyltransferase activity is repressed by glucocorticoids: mifepristone fails to recruit HDAC2 to the p65-HAT complex. J Biol Chem 276:30208–30215

Ito S, Ansari P, Sakatsume M, Dickensheets H, Vazquez N, Donnelly RP, Larner AC, Finbloom DS (1999) Interleukin-10 inhibits expression of both interferon alpha- and interferon gamma- induced genes by suppressing tyrosine phosphorylation of STAT1. Blood 93:1456–1463

Jiang C, Ting AT, Seed B (1998) PPAR-gamma agonists inhibit production of monocyte inflammatory cytokines. Nature 391:82–86

Joyce DA, Steer JH, Abraham LJ (1997) Glucocorticoid modulation of human monocyte/macrophage function: control of TNF-alpha secretion. Inflamm Res 46:447–451

Kagoshima M, Wilcke T, Ito K, Tsaprouni L, Barnes PJ, Punchard N, Adcock IM (2001) Glucocorticoid-mediated transrepression is regulated by histone acetylation and DNA methylation. Eur J Pharmacol 429:327–334

Kamijo R, Harada H, Matsuyama T, Bosland M, Gerecitano J, Shapiro D, Le J, Koh SI, Kimura T, Green SJ, et al. (1994) Requirement for transcription factor IRF-1 in NO synthase induction in macrophages. Science 263:1612–1615

Kanno Y, Kozak CA, Schindler C, Driggers PH, Ennist DL, Gleason SL, Darnell JE Jr, Ozato K (1993) The genomic structure of the murine ICSBP gene reveals the presence of the gamma interferon-responsive element, to which an ISGF3 alpha subunit (or similar) molecule binds. Mol Cell Biol 13:3951–3963

Kaplan MH, Schindler U, Smiley ST, Grusby MJ (1996) Stat6 is required for mediating responses to IL-4 and for development of Th2 cells. Immunity 4:313–319

Keffer J, Probert L, Cazlaris H, Georgopoulos S, Kaslaris E, Kioussis D, Kollias G (1991) Transgenic mice expressing human tumour necrosis factor: a predictive genetic model of arthritis. Embo J 10:4025–4031

Khachigian LM (2001) Catalytic oligonucleotides targeting EGR-1 as potential inhibitors of in-stent restenosis. Ann N Y Acad Sci 947:412–415

Khanna-Gupta A, Zibello T, Sun H, Lekstrom-Himes J, Berliner N (2001) C/EBP epsilon mediates myeloid differentiation and is regulated by the CCAAT displacement protein (CDP/cut). Proc Natl Acad Sci USA 98:8000–8005

Kieslinger M, Woldman I, Moriggl R, Hofmann J, Marine JC, Ihle JN, Beug H, Decker T (2000) Antiapoptotic activity of Stat5 required during terminal stages of myeloid differentiation. Genes Dev 14:232–244

Kim TK, Maniatis T (1996) Regulation of interferon-gamma-activated STAT1 by the ubiquitin-proteasome pathway. Science 273:1717–1719

Kistler B, Pfisterer P, Wirth T (1995) Lymphoid- and myeloid-specific activity of the PU.1 promoter is determined by the combinatorial action of octamer and ets transcription factors. Oncogene 11:1095–1106

Klemsz MJ, Maki RA, Papayannopoulou T, Moore J, Hromas R (1993) Characterization of the ets oncogene family member, fli-1. J Biol Chem 268:5769–5773

Klemsz MJ, McKercher SR, Celada A, Van Beveren C, Maki RA (1990) The macrophage and B cell-specific transcription factor PU.1 is related to the ets oncogene. Cell 61:113–124

Kola I, Brookes S, Green AR, Garber R, Tymms M, Papas TS, Seth A (1993) The Ets1 transcription factor is widely expressed during murine embryo development and is associated with mesodermal cells involved in morphogenetic processes such as organ formation. Proc Natl Acad Sci USA 90:7588–7592

Krishnaraju K, Hoffman B, Liebermann DA (1997) Lineage-specific regulation of hematopoiesis by HOX-B8 (HOX-2.4): inhibition of granulocytic differentiation and potentiation of monocytic differentiation. Blood 90:1840–1849

Krishnaraju K, Nguyen HQ, Liebermann DA, Hoffman B (1995) The zinc finger transcription factor Egr-1 potentiates macrophage differentiation of hematopoietic cells. Mol Cell Biol 15:5499–5507

Lawrence HJ, Helgason CD, Sauvageau G, Fong S, Izon DJ, Humphries RK, Largman C (1997) Mice bearing a targeted interruption of the homeobox gene HOXA9 have defects in myeloid, erythroid, and lymphoid hematopoiesis. Blood 89:1922–1930

Lawrence HJ, Sauvageau G, Ahmadi N, Lopez AR, LeBeau MM, Link M, Humphries K, Largman C (1995) Stage- and lineage-specific expression of the HOXA10 homeobox gene in normal and leukemic hematopoietic cells. Exp Hematol 23:1160–1166

Lee EH, Rikihisa Y (1998) Protein kinase A-mediated inhibition of gamma interferon-induced tyrosine phosphorylation of Janus kinases and latent cytoplasmic transcription factors in human monocytes by Ehrlichia chaffeensis. Infect Immun 66:2514–2520

Lee SL, Wang Y, Milbrandt J (1996) Unimpaired macrophage differentiation and activation in mice lacking the zinc finger transplantation factor NGFI-A (EGR1). Mol Cell Biol 16:4566–4572

Lekstrom-Himes J, Xanthopoulos KG (1998) Biological role of the CCAAT/enhancer-binding protein family of transcription factors. J Biol Chem 273:28545–28548

Letterio JJ, Roberts AB (1998) Regulation of immune responses by TGF-beta. Annu Rev Immunol 16:137–161

Levy DE, Lew DJ, Decker T, Kessler DS, Darnell JE Jr (1990) Synergistic interaction between interferon-alpha and interferon-gamma through induced synthesis of one subunit of the transcription factor ISGF3. Embo J 9:1105–1111

Lewis JD, Lichtenstein GR, Stein RB, Deren JJ, Judge TA, Fogt F, Furth EE, Demissie EJ, Hurd LB, Su C G, Keilbaugh SA, Lazar MA, Wu GD (2001) An open-label trial of the PPAR-gamma ligand rosiglitazone for active ulcerative colitis. Am J Gastroenterol 96:3323–3328

Li M, Pascual G, Glass CK (2000) Peroxisome proliferator-activated receptor gamma-dependent repression of the inducible nitric oxide synthase gene. Mol Cell Biol 20:4699–4707

Lill MC, Fuller JF, Herzig R, Crooks GM, Gasson JC (1995) The role of the homeobox gene, HOX B7, in human myelomonocytic differentiation. Blood 85:692–697

Lohoff M, Ferrick D, Mittrucker HW, Duncan GS, Bischof S, Rollinghoff M, Mak TW (1997) Interferon regulatory factor-1 is required for a T helper 1 immune response in vivo. Immunity 6:681–689.

Luchin A, Suchting S, Merson T, Rosol TJ, Hume DA, Cassady AI, Ostrowski MC (2001) Genetic and physical interactions between Microphthalmia transcription factor and

PU.1 are necessary for osteoclast gene expression and differentiation. J Biol Chem 276:36703–36710
Lutterbach B, Hiebert SW (2000) Role of the transcription factor AML-1 in acute leukemia and hematopoietic differentiation. Gene 245:223–235
Madrid LV, Mayo MW, Reuther JY, Baldwin AS Jr (2001) Akt stimulates the transactivation potential of the RelA/p65 Subunit of NF-kappa B through utilization of the Ikappa B kinase and activation of the mitogen-activated protein kinase p38. J Biol Chem 276:18934–18940
Maitra S, Atchison M (2000) BSAP can repress enhancer activity by targeting PU.1 function. Mol Cell Biol 20:1911–1922
Maltzman JS, Carman JA, Monroe JG (1996) Role of EGR1 in regulation of stimulus-dependent CD44 transcription in B lymphocytes. Mol Cell Biol 16:2283–2294
Marth T, Kelsall BL (1997) Regulation of interleukin-12 by complement receptor 3 signaling. J Exp Med 185:1987–1995
Martin E, Nathan C, Xie QW (1994) Role of interferon regulatory factor 1 in induction of nitric oxide synthase. J Exp Med 180:977–984
Massague J, Wotton D (2000) Transcriptional control by the TGF-beta/Smad signaling system. Embo J 19:1745–1754
Matsukawa A, Kaplan MH, Hogaboam CM, Lukacs NW, Kunkel SL (2001) Pivotal role of signal transducer and activator of transcription (Stat)4 and Stat6 in the innate immune response during sepsis. J Exp Med 193:679–688
Maytin EV, Habener JF (1998) Transcription factors C/EBP alpha, C/EBP beta, and CHOP (Gadd153) expressed during the differentiation program of keratinocytes in vitro and in vivo. J Invest Dermatol 110:238–246
McCaffrey TA, Fu C, Du B, Eksinar S, Kent KC, Bush H Jr, Kreiger K, Rosengart T, Cybulsky MI, Silverman ES, Collins T (2000) High-level expression of Egr-1 and Egr-1-inducible genes in mouse and human atherosclerosis. J Clin Invest 105:653–662
McKercher SR, Torbett BE, Anderson KL, Henkel GW, Vestal DJ, Baribault H, Klemsz M, Feeney AJ, Wu GE, Paige CJ, Maki RA (1996) Targeted disruption of the PU.1 gene results in multiple hematopoietic abnormalities. Embo J 15:5647–5658.
McMahon SB, Monroe JG (1996) The role of early growth response gene 1 (egr-1) in regulation of the immune response. J Leukoc Biol 60:159–166
Miyaura H, Iwata M (2002) Direct and indirect inhibition of Th1 development by progesterone and glucocorticoids. J Immunol 168:1087–1094
Monteleone G, Kumberova A, Croft NM, McKenzie C, Steer HW, MacDonald TT (2001) Blocking Smad7 restores TGF-beta1 signaling in chronic inflammatory bowel disease. J Clin Invest 108:601–609
Moore KJ (1995) Insight into the microphthalmia gene. Trends Genet 11:442–448
Morosetti R, Park DJ, Chumakov AM, Grillier I, Shiohara M, Gombart AF, Nakamaki T, Weinberg K, Koeffler HP (1997) A novel, myeloid transcription factor, C/EBP epsilon, is upregulated during granulocytic, but not monocytic, differentiation. Blood 90:2591–2600
Mucenski ML, McLain K, Kier AB, Swerdlow SH, Schreiner CM, Miller TA, Pietryga DW, Scott WJ Jr, Potter SS (1991) A functional c-myb gene is required for normal murine fetal hepatic hematopoiesis. Cell 65:677–689
Nakao A, Afrakhte M, Moren A, Nakayama T, Christian JL, Heuchel R, Itoh S, Kawabata M, Heldin NE, Heldin CH, ten Dijke P (1997) Identification of Smad7, a TGFbeta-inducible antagonist of TGF-beta signalling. Nature 389:631–635
Natsuka S, Akira S, Nishio Y, Hashimoto S, Sugita T, Isshiki H, Kishimoto T (1992) Macrophage differentiation-specific expression of NF-IL6, a transcription factor for interleukin-6. Blood 79:460–466
Nerlov C, Querfurth E, Kulessa H, Graf T (2000) GATA-1 interacts with the myeloid PU.1 transcription factor and represses PU.1-dependent transcription. Blood 95:2543–2551

Nguyen HQ, Hoffman-Liebermann B, Liebermann DA (1993) The zinc finger transcription factor Egr-1 is essential for and restricts differentiation along the macrophage lineage. Cell 72:197–209

Nishinakamura R, Nakayama N, Hirabayashi Y, Inoue T, Aud D, McNeil T, Azuma S, Yoshida S, Toyoda Y, Arai K et al. (1995) Mice deficient for the IL-3/GM-CSF/IL-5 beta c receptor exhibit lung pathology and impaired immune response, while beta IL3 receptor-deficient mice are normal. Immunity 2:211–222

Nolte RT, Wisely GB, Westin S, Cobb JE, Lambert MH, Kurokawa R, Rosenfeld MG, Willson TM, Glass CK, Milburn MV (1998) Ligand binding and co-activator assembly of the peroxisome proliferator-activated receptor-gamma. Nature 395:137–143

Nosaka T, Kawashima T, Misawa K, Ikuta K, Mui AL, Kitamura T (1999) STAT5 as a molecular regulator of proliferation, differentiation and apoptosis in hematopoietic cells. Embo J 18:4754–4765

Novak U, Nice E, Hamilton JA, Paradiso L (1996) Requirement for Y706 of the murine (or Y708 of the human) CSF-1 receptor for STAT1 activation in response to CSF-1. Oncogene 13:2607–2613

Oelgeschlager M, Nuchprayoon I, Luscher B, Friedman AD (1996) C/EBP, c-Myb, and PU.1 cooperate to regulate the neutrophil elastase promoter. Mol Cell Biol 16:4717–4725

Ohmori Y, Hamilton TA (1998) STAT6 is required for the anti-inflammatory activity of interleukin-4 in mouse peritoneal macrophages. J Biol Chem 273:29202–29209

Parganas E, Wang D, Stravopodis D, Topham DJ, Marine JC, Teglund S, Vanin EF, Bodner S, Colamonici OR, van Deursen JM, Grosveld G, Ihle JN (1998) Jak2 is essential for signaling through a variety of cytokine receptors. Cell 93:385–395

Petricoin E 3rd, David M, Igarashi K, Benjamin C, Ling L, Goelz S, Finbloom DS, Larner AC (1996) Inhibition of alpha interferon but not gamma interferon signal transduction by phorbol esters is mediated by a tyrosine phosphatase. Mol Cell Biol 16:1419–1424

Petrovick MS, Hiebert SW, Friedman AD, Hetherington CJ, Tenen DG, Zhang DE (1998) Multiple functional domains of AML1: PU.1 and C/EBPalpha synergize with different regions of AML1. Mol Cell Biol 18:3915–3925

Poli V (1998) The role of C/EBP isoforms in the control of inflammatory and native immunity functions. J Biol Chem 273:29279–29282

Pope RM, Leutz A, Ness SA (1994) C/EBP beta regulation of the tumor necrosis factor alpha gene. J Clin Invest 94:1449–1455

Pope RM, Lovis R, Mungre S, Perlman H, Koch AE, Haines GK 3rd (1999) C/EBP beta in rheumatoid arthritis: correlation with inflammation, not disease specificity. Clin Immunol 91:271–282

Ramdas J, Harmon JM (1998) Glucocorticoid-induced apoptosis and regulation of NF-kappaB activity in human leukemic T cells. Endocrinology 139:3813–3821

Reddy MA, Yang BS, Yue X, Barnett CJ, Ross IL, Sweet MJ, Hume DA, Ostrowski MC (1994) Opposing actions of c-ets/PU.1 and c-myb protooncogene products in regulating the macrophage-specific promoters of the human and mouse colony-stimulating factor-1 receptor (c-fms) genes. J Exp Med 180:2309–2319

Rehli M, Lichanska A, Cassady AI, Ostrowski MC, Hume DA (1999) TFEC is a macrophage-restricted member of the microphthalmia-TFE subfamily of basic helix-loop-helix leucine zipper transcription factors. J Immunol 162:1559–1565

Rehli M, Poltorak A, Schwarzfischer L, Krause SW, Andreesen R, Beutler B (2000) PU.1 and interferon consensus sequence-binding protein regulate the myeloid expression of the human Toll-like receptor 4 gene. J Biol Chem 275:9773–9781

Ricote M, Li AC, Willson TM, Kelly CJ, Glass CK (1998) The peroxisome proliferator-activated receptor-gamma is a negative regulator of macrophage activation. Nature 391:79–82

Ricote M, Welch JS, Glass CK (2000) Regulation of macrophage gene expression by the peroxisome proliferator-activated receptor-gamma. Horm Res 54:275–280

Robertson KA, Hill DP, Kelley MR, Tritt R, Crum B, Van Epps S, Srour E, Rice S, Hromas R (1998) The myeloid zinc finger gene (MZF-1) delays retinoic acid-induced apoptosis and differentiation in myeloid leukemia cells. Leukemia 12:690–698

Rosenbauer F, Waring JF, Foerster J, Wietstruk M, Philipp D, Horak I (1999) Interferon consensus sequence binding protein and interferon regulatory factor-4/Pip form a complex that represses the expression of the interferon-stimulated gene-15 in macrophages. Blood 94:4274–4281

Ross IL, Dunn TL, Yue X, Roy S, Barnett CJ, Hume DA (1994) Comparison of the expression and function of the transcription factor PU.1 (Spi-1 proto-oncogene) between murine macrophages and B lymphocytes. Oncogene 9:121–132

Ross IL, Yue X, Ostrowski MC, Hume DA (1998) Interaction between PU.1 and another Ets family transcription factor promotes macrophage-specific Basal transcription initiation. J Biol Chem 273:6662–6669

Ryu S, Zhou S, Ladurner AG, Tjian R (1999) The transcriptional cofactor complex CRSP is required for activity of the enhancer-binding protein Sp1. Nature 397:446–450

Sauvageau G, Thorsteinsdottir U, Hough MR, Hugo P, Lawrence HJ, Largman C, Humphries RK (1997) Overexpression of HOXB3 in hematopoietic cells causes defective lymphoid development and progressive myeloproliferation. Immunity 6:13–22

Scheller M, Foerster J, Heyworth CM, Waring JF, Lohler J, Gilmore GL, Shadduck RK, Dexter TM, Horak I (1999) Altered development and cytokine responses of myeloid progenitors in the absence of transcription factor, interferon consensus sequence binding protein. Blood 94:3764–3771.

Schmitz ML, Bacher S, Kracht M (2001) I kappa B-independent control of NF-kappa B activity by modulatory phosphorylations. Trends Biochem Sci 26:186–190

Schuetze S, Stenberg PE, Kabat D (1993) The Ets-related transcription factor PU.1 immortalizes erythroblasts. Mol Cell Biol 13:5670–5678

Schulze-Koops H, Kalden JR (2001) The balance of Th1/Th2 cytokines in rheumatoid arthritis. Best Pract Res Clin Rheumatol 15:677–691

Scott EW, Simon MC, Anastasi J, Singh H (1994) Requirement of transcription factor PU.1 in the development of multiple hematopoietic lineages. Science 265:1573–1577

Scott LM, Civin CI, Rorth P, Friedman AD (1992) A novel temporal expression pattern of three C/EBP family members in differentiating myelomonocytic cells. Blood 80:1725–1735

Screpanti I, Romani L, Musiani P, Modesti A, Fattori E, Lazzaro D, Sellitto C, Scarpa S, Bellavia D, Lattanzio G, et al. (1995) Lymphoproliferative disorder and imbalanced T-helper response in C/EBP beta-deficient mice. Embo J 14:1932–1941

Sengupta TK, Schmitt EM, Ivashkiv LB (1996) Inhibition of cytokines and JAK-STAT activation by distinct signaling pathways. Proc Natl Acad Sci USA 93:9499–9504

Sengupta TK, Talbot ES, Scherle PA, Ivashkiv LB (1998) Rapid inhibition of interleukin-6 signaling and Stat3 activation mediated by mitogen-activated protein kinases. Proc Natl Acad Sci USA 95:11107–11112

Shimoda K, van Deursen J, Sangster MY, Sarawar SR, Carson RT, Tripp RA, Chu C, Quelle FW, Nosaka T, Vignali DA, Doherty PC, Grosveld G, Paul WE, Ihle JN (1996) Lack of IL-4-induced Th2 response and IgE class switching in mice with disrupted Stat6 gene. Nature 380:630–633

Shirakawa F, Saito K, Bonagura CA, Galson DL, Fenton MJ, Webb AC, Auron PE (1993) The human prointerleukin 1 beta gene requires DNA sequences both proximal and distal to the transcription start site for tissue-specific induction. Mol Cell Biol 13:1332–1344

Shull MM, Ormsby I, Kier AB, Pawlowski S, Diebold RJ, Yin M, Allen R, Sidman C, Proetzel G, Calvin D, et al. (1992) Targeted disruption of the mouse transforming growth factor-beta 1 gene results in multifocal inflammatory disease. Nature 359:693–699

Silverman ES, Collins T (1999) Pathways of Egr-1-mediated gene transcription in vascular biology. Am J Pathol 154:665–670
Silverman N, Maniatis T (2001) NF-kappaB signaling pathways in mammalian and insect innate immunity. Genes Dev 15:2321–2342
Sizemore N, Leung S, Stark GR (1999) Activation of phosphatidylinositol 3-kinase in response to interleukin-1 leads to phosphorylation and activation of the NF-kappaB p65/RelA subunit. Mol Cell Biol 19:4798–4805.
Stark GR, Kerr IM, Williams BR, Silverman RH, Schreiber RD (1998) How cells respond to interferons. Annu Rev Biochem 67:227–264
Starr R, Willson TA, Viney EM, Murray LJ, Rayner JR, Jenkins BJ, Gonda TJ, Alexander WS, Metcalf D, Nicola NA, Hilton DJ (1997) A family of cytokine-inducible inhibitors of signalling. Nature 387:917–921
Stein B, Baldwin AS Jr (1993) Distinct mechanisms for regulation of the interleukin-8 gene involve synergism and cooperativity between C/EBP and NF-kappa B. Mol Cell Biol 13:7191–7198
Steingrimsson E, Moore KJ, Lamoreux ML, Ferre-D'Amare AR, Burley SK, Zimring DC, Skow LC, Hodgkinson CA, Arnheiter H, Copeland NG, et al. (1994) Molecular basis of mouse microphthalmia (mi) mutations helps explain their developmental and phenotypic consequences. Nat Genet 8:256–263
Stoop R, Kotani H, McNeish JD, Otterness IG, Mikecz K (2001) Increased resistance to collagen-induced arthritis in CD44-deficient DBA/1 mice. Arthritis Rheum 44:2922–2931
Su GH, Ip HS, Cobb BS, Lu MM, Chen HM, Simon MC (1996) The Ets protein Spi-B is expressed exclusively in B cells and T cells during development. J Exp Med 184:203–214
Takeda K, Noguchi K, Shi W, Tanaka T, Matsumoto M, Yoshida N, Kishimoto T, Akira S (1997) Targeted disruption of the mouse Stat3 gene leads to early embryonic lethality. Proc Natl Acad Sci USA 94:3801–3804
Takeda K, Tanaka T, Shi W, Matsumoto M, Minami M, Kashiwamura S, Nakanishi K, Yoshida N, Kishimoto T, Akira S (1996) Essential role of Stat6 in IL-4 signalling. Nature 380:627–630
Tamura T, Nagamura-Inoue T, Shmeltzer Z, Kuwata T, Ozato K (2000) ICSBP directs bipotential myeloid progenitor cells to differentiate into mature macrophages. Immunity 13:155–165
Tanaka T, Kurokawa M, Ueki K, Tanaka K, Imai Y, Mitani K, Okazaki K, Sagata N, Yazaki Y, Shibata Y, Kadowaki T, Hirai H (1996) The extracellular signal-regulated kinase pathway phosphorylates AML1, an acute myeloid leukemia gene product, and potentially regulates its transactivation ability. Mol Cell Biol 16:3967–3979
Taniguchi T, Ogasawara K, Takaoka A, Tanaka N (2001) IRF family of transcription factors as regulators of host defense. Annu Rev Immunol 19:623–655
Teglund S, McKay C, Schuetz E, van Deursen JM, Stravopodis D, Wang D, Brown M, Bodner S, Grosveld G, Ihle JN (1998) Stat5a and Stat5b proteins have essential and nonessential, or redundant, roles in cytokine responses. Cell 93:841–850
Tenen DG, Hromas R, Licht JD, Zhang DE (1997) Transcription factors, normal myeloid development, and leukemia. Blood 90:489–519
Tengku-Muhammad TS, Hughes TR, Ranki H, Cryer A, Ramji DP (2000) Differential regulation of macrophage CCAAT-enhancer binding protein isoforms by lipopolysaccharide and cytokines. Cytokine 12:1430–1436
Thorsteinsdottir U, Sauvageau G, Hough MR, Dragowska W, Lansdorp PM, Lawrence HJ, Largman C, Humphries RK (1997) Overexpression of HOXA10 in murine hematopoietic cells perturbs both myeloid and lymphoid differentiation and leads to acute myeloid leukemia. Mol Cell Biol 17:495–505
Tracey WD, Speck NA (2000) Potential roles for RUNX1 and its orthologs in determining hematopoietic cell fate. Semin Cell Dev Biol 11:337–342

Veals SA, Schindler C, Leonard D, Fu XY, Aebersold R, Darnell JE Jr, Levy DE (1992) Subunit of an alpha-interferon-responsive transcription factor is related to interferon regulatory factor and Myb families of DNA-binding proteins. Mol Cell Biol 12:3315–3324

Verbeek W, Gombart AF, Chumakov AM, Muller C, Friedman AD, Koeffler HP (1999) C/EBPepsilon directly interacts with the DNA binding domain of c-myb and cooperatively activates transcription of myeloid promoters. Blood 93:3327–3337

Vercelli D, Jabara HH, Lee BW, Woodland N, Geha RS, Leung DY (1988) Human recombinant interleukin 4 induces Fc epsilon R2/CD23 on normal human monocytes. J Exp Med 167:1406–1416

Voso MT, Burn TC, Wulf G, Lim B, Leone G, Tenen DG (1994) Inhibition of hematopoiesis by competitive binding of transcription factor PU.1. Proc Natl Acad Sci USA 91:7932–7936

Wang D, Baldwin AS Jr (1998) Activation of nuclear factor-kappaB-dependent transcription by tumor necrosis factor-alpha is mediated through phosphorylation of RelA/p65 on serine 529. J Biol Chem 273:29411–29416

Wang D, Westerheide SD, Hanson JL, Baldwin AS Jr (2000a) Tumor necrosis factor alpha-induced phosphorylation of RelA/p65 on Ser529 is controlled by casein kinase II. J Biol Chem 275:32592–32597

Wang IM, Contursi C, Masumi A, Ma X, Trinchieri G, Ozato K (2000b) An IFN-gamma-inducible transcription factor, IFN consensus sequence binding protein (ICSBP), stimulates IL-12 p40 expression in macrophages. J Immunol 165:271–279

Ward AC, Hermans MH, Smith L, van Aesch YM, Schelen AM, Antonissen C, Touw IP (1999) Tyrosine-dependent and -independent mechanisms of STAT3 activation by the human granulocyte colony-stimulating factor (G-CSF) receptor are differentially utilized depending on G-CSF concentration. Blood 93:113–124

Weiner GJ (2000) The immunobiology and clinical potential of immunostimulatory CpG oligodeoxynucleotides. J Leukoc Biol 68:455–463

Wen Z, Zhong Z, Darnell JE Jr (1995) Maximal activation of transcription by Stat1 and Stat3 requires both tyrosine and serine phosphorylation. Cell 82:241–250

Werner F, Jain MK, Feinberg MW, Sibinga N E, Pellacani A, Wiesel P, Chin MT, Topper JN, Perrella MA, Lee ME (2000) Transforming growth factor-beta 1 inhibition of macrophage activation is mediated via Smad3. J Biol Chem 275:36653–36658

Xie Y, Chen C, Stevenson MA, Hume DA, Auron PE, Calderwood SK (2002) NF-IL6 and HSF1 Have Mutually Antagonistic Effects on Transcription in Monocytic Cells. Biochem Biophys Res Commun 291:1071–1080.

Xu HE, Lambert MH, Montana VG, Parks DJ, Blanchard SG, Brown PJ, Sternbach DD, Lehmann JM, Wisely GB, Willson TM, Kliewer SA, Milburn MV (1999) Molecular recognition of fatty acids by peroxisome proliferator-activated receptors. Mol Cell 3:397–403

Yamanaka R, Barlow C, Lekstrom-Himes J, Castilla LH, Liu PP, Eckhaus M, Decker T, Wynshaw-Boris A, Xanthopoulos KG (1997a) Impaired granulopoiesis, myelodysplasia, and early lethality in CCAAT/enhancer binding protein epsilon-deficient mice. Proc Natl Acad Sci USA 94:13187–13192

Yamanaka R, Kim GD, Radomska HS, Lekstrom-Himes J, Smith LT, Antonson P, Tenen DG, Xanthopoulos KG (1997b) CCAAT/enhancer binding protein epsilon is preferentially up-regulated during granulocytic differentiation and its functional versatility is determined by alternative use of promoters and differential splicing. Proc Natl Acad Sci USA 94:6462–6467

Yang BS, Hauser CA, Henkel G, Colman MS, Van Beveren C, Stacey KJ, Hume DA, Maki RA, Ostrowski MC (1996) Ras-mediated phosphorylation of a conserved threonine residue enhances the transactivation activities of c-Ets1 and c-Ets2. Mol Cell Biol 16:538–547

Zhang DE, Fujioka K, Hetherington CJ, Shapiro LH, Chen HM, Look AT, Tenen DG (1994a) Identification of a region which directs the monocytic activity of the colony-stimulating factor 1 (macrophage colony-stimulating factor) receptor promoter and binds PEBP2/CBF (AML1). Mol Cell Biol 14:8085–8095

Zhang DE, Hetherington CJ, Meyers S, Rhoades KL, Larson CJ, Chen HM, Hiebert SW, Tenen DG (1996a) CCAAT enhancer-binding protein (C/EBP) and AML1 (CBF alpha2) synergistically activate the macrophage colony-stimulating factor receptor promoter. Mol Cell Biol 16:1231–1240

Zhang DE, Hetherington CJ, Tan S, Dziennis SE, Gonzalez DA, Chen HM, Tenen DG (1994b) Sp1 is a critical factor for the monocytic specific expression of human CD14. J Biol Chem 269:11425–11434

Zhang DE, Hohaus S, Voso MT, Chen HM, Smith LT, Hetherington CJ, Tenen DG (1996b) Function of PU.1 (Spi-1), C/EBP, and AML1 in early myelopoiesis: regulation of multiple myeloid CSF receptor promoters. Curr Top Microbiol Immunol 211:137–147

Zhang DE, Zhang P, Wang ND, Hetherington CJ, Darlington GJ, Tenen DG (1997) Absence of granulocyte colony-stimulating factor signaling and neutrophil development in CCAAT enhancer binding protein alpha-deficient mice. Proc Natl Acad Sci USA 94:569–574

Zhang G, Ghosh S (2000) Molecular mechanisms of NF-kappaB activation induced by bacterial lipopolysaccharide through Toll-like receptors. J Endotoxin Res 6:453–457

Zhong H, Voll RE, Ghosh S (1998) Phosphorylation of NF-kappa B p65 by PKA stimulates transcriptional activity by promoting a novel bivalent interaction with the coactivator CBP/p300. Mol Cell 1:661–671

Exploitation of Macrophage Clearance Functions In Vivo

S. M. Moghimi

Molecular Targeting and Polymer Toxicology Group,
School of Pharmacy and Biomolecular Sciences,
University of Brighton, Brighton, BN2 4GJ, UK
e-mail: s.m.moghimi@brighton.ac.uk

Abstract The propensity of macrophages for the phagocytic clearance of colloidal particles provides a rational approach to macrophage-specific targeting and drug delivery. Furthermore, by precision engineering, colloidal drug carriers can be targeted to selective population of macrophages in the body as well as intracellular locations. These approaches have led to the development of a number of regulatory-approved particulate formulations for delivery of therapeutic and diagnostic agents to macrophages. This article will briefly discuss selected approaches and highlight barriers in in vivo macrophage-specific targeting with colloidal carriers via intravenous, subcutaneous and oral routes of administration, and it explores avenues for selective modification of macrophage cellular activity.

Keywords Antigen presentation, Colloidal carriers, Drug delivery, Kupffer cells, Liposomes, Lymph nodes, M cells, Macrophage suicide, Nanospheres, Poloxamer, Poloxamine, Splenic macrophages, Stimulated macrophages

Abbreviations

APC	antigen-presenting cell
EO	ethylene oxide
IES	interendothelial cell slits
MHC	major histocompatibility complex
PO	propylene oxide
TAT	trans-activating transcriptional activator

1 Introduction

Macrophages are widely distributed throughout the body and perform a wide range of homeostatic, physiological, and immunological functions. Among them, phagocytosis (or endocytosis) is the macrophage's primary task, which has been well conserved throughout evolution. This is achieved by virtue of a vast array of specialised plasma membrane recognition receptors with which the macrophage can arrest and eliminate senescent and damaged cells, particulate debris and microbial invaders (Gordon 1995).

The propensity of macrophages for the phagocytic/macropinocytic clearance of foreign particles provides a rational approach to macrophage-specific targeting through suitable particulate vehicles. The concept of particulate targeting of macrophages is an attractive one, in that a wide variety of systems are available and particles with differing physicochemical properties and loading capacities can be constructed (Poznansky and Juliano 1984; Moghimi et al. 2001). The particulate entities which have been used include liposomes, niosomes, polymeric nanospheres, oil-in-water microemulsions, and even natural constructs such as lipoproteins and erythrocytes. For targeting, three criteria must be considered. The first criterion is the distribution of macrophages in tissues in terms of access from various physiological portals of entry. The second involves determinants of phagocytic recognition and ingestion, which includes macrophage phagocytic/endocytic receptors (e.g. the nature, density, and their state of activation) and the effect of environmental factors on their phagocytic functions. The last is the physicochemical characteristics of the particles to be ingested and includes particle morphology, hydrodynamic size and surface characteristics (e.g. ligand expression, bound opsonins). This article will briefly discuss selected approaches and highlight barriers in in vivo macrophage-specific targeting with colloidal carriers via intravenous, subcutaneous and oral routes of administration and explores avenues for selective modification of macrophage cellular activity.

2 The Macrophage as a Therapeutic Target for Nanocarriers

Colloidal targeting of macrophages offers a number of advantages. These include treatment of diseases and disorders of the macrophage system as well as attempts to manipulate phagocytic cell number and their functional activities. For example, although most micro-organisms are killed by macrophages, many pathogenic organisms have developed means for resisting macrophage destruction following phagocytosis. In certain cases, the macrophage lysosome and/or cytoplasm is the obligate intracellular home of the micro-organism, examples include *Toxoplasma gondii*, various species of *Leishmania*, *Mycobacterium tuberculosis*, *Mycobacterium leprae*, *Listeria monocytogenes*, and *Cryptococcus neoformans* (Alving 1988; Donowitz 1994). The targeting of antimicrobial agents encapsulated in colloidal vehicles to infected macrophages is therefore a logical strategy for effective microbial killing. The endocytic pathway will direct a colloidal carrier to lysosomes where pathogens are resident. Degradation of the carrier by lysosomal enzymes can release drug into the phagosome-lysosome vesicle itself or into the cytoplasm either by diffusion or by specific transporters, depending on the physicochemical nature of the drug molecules (Alving 1988; Lloyd 2000). On the other hand, delivery to cytosol can be further enhanced by triggering drug release in late endosomes. Examples include pH-sensitive and fusogenic vesicles (Drummond et al. 2000). A pH-sensitive vesicle maintains stable phospholipid bilayers at neutral pH or above but destabilises and becomes fusion-competent at the acidic pH of late endosomes and subsequently release their encapsulated cargo into cytoplasm (Horwitz et al. 1980). Alternatively, by mimicking the characteristics of certain toxins or viruses, cytoplasmic delivery of agents can also be enhanced. This requires co-encapsulation of bacterial pore-forming toxins (e.g. listeriolysin O) or fusion peptides found in envelope glycoproteins of certain viruses (e.g. HA2 fusion peptide of influenza virus) in particulate vehicles (Lee et al. 1996; Beauregard et al. 1997; Drummond et al. 2000). The fusion glycoprotein of such viruses displays sharp fusion profiles with pH midpoints from 5.0 to 6.6, which is in good agreement with estimates of the endosomal pH. These targeting strategies also overcome the toxicity of drugs effective against microbes. For example, the most prominent potential toxicities associated with pentavalent antimonials (e.g. sodium stibogluconate) used for leishmaniasis treatment include changes in the electrocardiogram and hepatocellular damage. Thus direct targeting allows significant reductions in the required quantity of drugs, while achieving therapeutic drug concentrations in the infected cell.

Macrophages play an important role in induction of immunity. The induction of an immune response against a protein antigen invokes the interaction of the antigen with macrophage [or an antigen-presenting cell (APC)] that partially degrades the antigen and channels peptides into the MHC molecules (class I or class II) for processing and presentation (Fling et al. 1994; Harding and Song 1994; Kovacsovics-Bankowski and Rock 1995). These highly polymorphic MHC class I and class II molecules bind and transport peptide fragments of intact proteins to the surface of APCs for interaction with either $CD4^+$ or $CD8^+$ T lymphocytes. It is generally accepted that endogenous proteins of a cell are presented via the MHC class I pathway, whereas exogenous peptides are presented via the MHC class II route. Most soluble antigens are poor at priming MHC class I-restricted cytotoxic T-lymphocyte responses because of their inability to gain access to the cytosol. Colloidal carriers and particularly liposomes, however, act as powerful adjuvants if they are physically associated with a protein antigen (Gregoriadis 1990; Rao and Alving 2000). After phagocytosis by macrophages or APCs, the entrapped antigens in liposomes are presented to either MHC class I or class II pathways. Successful cytotoxic T-cell responses may further be obtained following antigen delivery with pH-sensitive liposomes as the antigen cargo may be delivered either to group 1 CD1 molecules (which also belong to MHC class I molecules but are localised in the endosomes and are able to bind and present glycolipids and microbial lipids) or be released directly into the cytosol.

Phagocytosis is not a prerequisite for intracellular targeting. Decoration of the nanoparticulate surface with *trans*-activating transcriptional activator (TAT) protein from HIV-1, or related peptides, facilitates cytoplasmic entry, which does not involve endocytosis and occurs by an energy-independent process (Torchilin et al. 2001). So far, TAT protein movement and import do not appear to be cell-specific. Although this approach may have implications for intracellular delivery of biologicals, the intracellular fate of the carrier must also be considered. A related technology is based on the remarkable cellular trafficking properties of the 35-kDa herpesvirus-1 structural protein VP22 (Elliott and O'Hare 1997). This protein reaches the nuclear compartment, despite lacking a nuclear localisation sequence, and binds chromatin in a matter of minutes. It also can act as a soluble carrier to transport peptides and proteins to the cell nucleus and therefore it is an attractive candidate for delivery of transcriptional factors, functional genes, cell cycle control regulators and DNA vaccines to macrophages and stem cells in vitro prior to transplantation.

There are many potential dysfunctions of macrophages that may be involved directly or indirectly in pathogenesis of diseases. For example, newborn infants manifest increased susceptibility to lung infections due to deficiency in alveolar macrophage-mediated secretion of biological response modifiers (Lee et al. 2001). Other examples include autoimmune blood disorders, spinal cord injury, sciatic nerve injury, T cell-mediated autoimmune diabetes and rheumatoid arthritis (Alves-Rosa et al. 2000; Barrera et al. 2000; Liu et al. 2000; Wu et al. 2000). These conditions should be amenable to treatment or become manageable following challenge with particulate carriers containing encapsulated drugs and genes via appropriate routes of administration. Indeed, colloidal-mediated macrophage suicide (i.e. delivery of macrophage toxins) has proved to be a powerful approach in removing unwanted macrophages in various experimental situations (van Rooijen and Sanders 1997; Alves-Rosa et al. 2000; Barrera et al. 2000; Liu et al. 2000; Wu et al. 2000; Kotter et al. 2001; Polfliet et al. 2001).

Macrophages are also heterogeneous with respect to phenotype and physiological properties, even within a single organ (Gordon et al. 1992; Rutherford et al. 1993). Understanding of macrophage heterogeneity will undoubtedly provide new insights and opportunities for designing carriers that can selectively deliver agents to defined macrophage sub-populations in vivo (see also "Targeting and Therapeutics: Colloid Engineering Meets Immunobiology"). An interesting attempt is to enhance delivery to newly proliferated macrophages or newly recruited monocytes (Moghimi and Patel 2002) at selected sites rather than local resident macrophages.

Apart from therapeutic goals, colloidal carriers are also useful for assessing macrophage phagocytic and clearance functions in vivo. Similarly, particulate colloids with entrapped radiopharmaceutical or contrast agents are helpful in imaging a designated pathology through macrophage loading (Tilcock 1995; Moghimi and Bonnemain 1999).

3 Targeting and Therapeutics: Colloid Engineering Meets Immunobiology

3.1 Intravenous Route

3.1.1 Resident Macrophages in Contact with Blood

Hepatic phagocytes or Kupffer cells are the largest population of macrophages in contact with blood (approximately 90%) and are the most prominent participants in particle clearance from the circulation (Moghimi et al. 2001). In addition to Kupffer cells, splenic marginal zone and red pulp macrophages also participate considerably in particle sequestration from the blood. Particle extraction by splenic macrophages is further enhanced in phagocytic malfunctions of Kupffer cells; this is simply due to increased blood concentration of particles (Moghimi et al. 2001).

Advantage has been taken of the natural physiological fate of particulate drug carriers to deliver, albeit passively, a variety of agents, which include antimicrobials and immunomodulators, to Kupffer cells and splenic macrophages (Poznansky and Juliano 1984). This specific delivery has resulted in direct killing of microbes and in the activation of the phagocytes to a bactericidal, fungicidal and tumouricidal state. The therapeutic efficacy of particulate-encapsulated drug may be further improved by targeting of the carrier to specific macrophage plasma membrane receptors. One such example is tuftsin receptor. Tuftsin, a part of the Fc portion of the heavy chain of the leukophilic IgG, is a natural macrophage-activator peptide. The binding of tuftsin to its receptor causes macrophage activation. Intravenous injection of tuftsin-bearing liposomes to infected animals not only resulted in delivery of liposome-encapsulated drugs to the macrophage phagolysosomes, but also in the non-specific stimulation of liver and spleen macrophage functions against parasitic, fungal and bacterial infections (Agrawal and Gupta 2000). Another interesting approach for future exploration is surface decoration of liposomes and nanoparticles with mannose receptor ligands, since ligands of mannose receptor carbohydrate recognition domains may modulate macrophage function (Linehan et al. 2001).

Our understanding of splenic architecture and intrasplenic microcirculation has provided us with new opportunities for selective targeting of nanoparticles to splenic marginal zone and red pulp macrophage. To enhance splenic reten-

(A) Structure of Poloxamines:

$$
\begin{array}{ccc}
 & & CH_3 \\
 & & | \\
(EO)_b\text{–}(PO)_a & & (CH_2CHO)_a\text{–}(CH_2CH_2O)_b\text{–}H \\
\backslash & & / \quad \text{PO unit} \qquad \text{EO unit} \\
 & NCH_2CH_2N & \\
/ & & \backslash \\
(EO)_b\text{–}(PO)_a & & (PO)_a\text{–}(EO)_b
\end{array}
$$

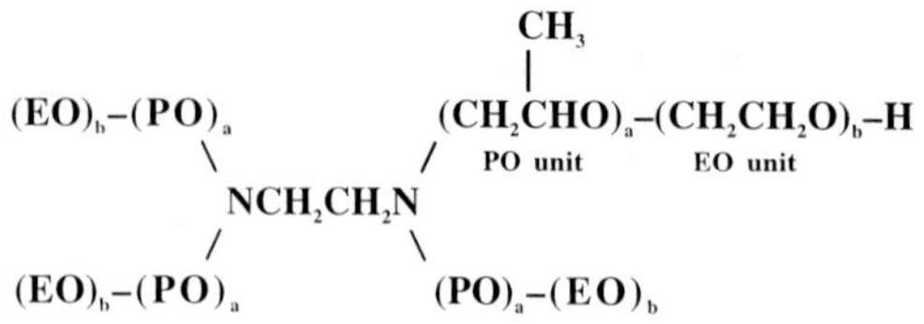

poloxamine 904	a = 17	b = 15
poloxamine 908	a = 17	b = 119

(B) Structure of Poloxamers:

$(EO)_b\text{–}(PO)_a\text{–}(EO)_b$

poloxamer 401	a = 67	b = 5
poloxamer 402	a = 67	b = 11
poloxamer 403	a = 67	b = 20
poloxamer 407	a = 67	b = 98

Fig. 1
Structure of poloxamines (**A**) and poloxamers (**B**)

tion of particles, a number of criteria must be met (Moghimi et al. 1991, 1993b,c; Moghimi and Hunter 2000). First, particles must exhibit prolonged circulation times in the blood; that is to avoid rapid clearance from the blood by Kupffer cells. Second, particles must be non-deformable and their size should be equivalent or exceed the reported width of splenic interendothelial cell slits (IES) in the sinus walls usually in the order of 200–500 nm. Indeed, IES are the sites that retain blood cells and blood-borne particles, depending on their bulk properties, size, sphericity and deformability. Finally, filtered particles must be prone to phagocytosis by the neighbouring macrophages.

Thus, to prolong particle concentration in the blood, polymeric materials such as dextrans, poloxamer 407, poloxamine 908 and poly(ethyleneglycol)-5000 have been used to camouflage particles against Kupffer cell capture (Moghimi et al. 2001; Gbadamosi et al. 2002; Moghimi 2002). For example, poloxamine 908 (Fig. 1) consists of four central hydrophobic blocks, each composed of 17 propylene oxide (PO) units, joined together by an ethylene diamine bridge. Each hydrophobic block is also attached to a relatively hydrophilic chain consisting of 119 ethylene oxide (EO) units. The hydrophilicity of EO chains arises from hydrogen bonding with water molecules. This copolymer can adsorb onto the surface of hydrophobic particles (e.g. polystyrene nanospheres) via its central hydrophobic portion, while the EO flanks extend from the particle surface and provide stability to the particle suspension by repulsion through a steric mechanism of stabilisation involving both enthalpic and entropic contribu-

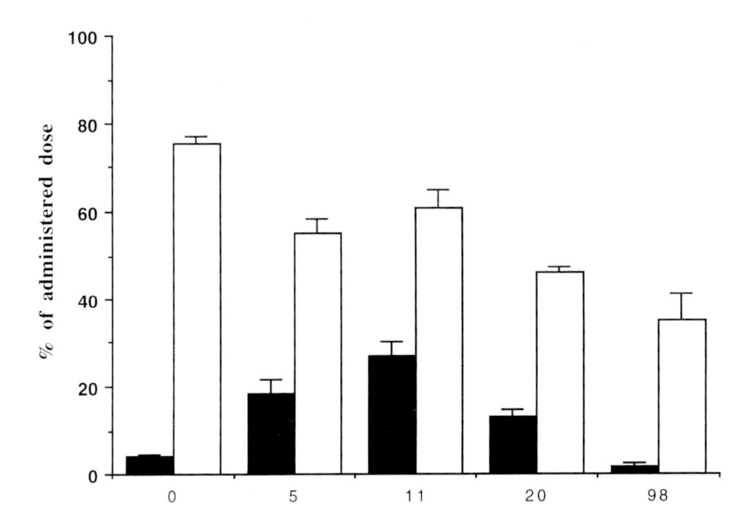

Fig. 2 Footpad and lymph node (popliteal and iliac nodes) distribution of uncoated and poloxamer-coated polystyrene nanospheres (60 nm) at 2 h after subcutaneous injection into rat footpads. Poloxamers 401, 402, 403 and 407 were used for particle coating and their structure is shown in Fig. 1. *Closed columns* represent uptake by regional lymph nodes and *open columns* represent the fraction of particles retained at footpads. For experimental details, see Moghimi et al. (1994)

tions. The steric barrier of EO chains suppresses protein adsorption on to the particle surface as well as particle interaction with plasma membrane of Kupffer cells (Moghimi et al. 1993d). If the size of the camouflaged particles is below 150 nm, they escape splenic filtration at IES and remain in the blood with reported half-lives of 24–48 h (Moghimi et al. 2001). Due to their altered pharmacokinetic properties, these particles can passively accumulate at sites of inflammation, infection and solid tumours, but the extent of particle extravasation is dependent upon the porosity of the blood vessels at such pathological sites. At such sites, activated macrophages can recognise and phagocytose extravasated long circulating particles by an unknown mechanism (Moghimi et al. 2001). On the other hand, if the size of rigid particles is in the range of 200–300 nm, splenic filtration will predominate (in excess of 50% of the injected dose within a few hours of intravenous injection with Kupffer cell uptake of <15% of the dose) (Moghimi et al. 1993c). Surprisingly, in the spleen, filtered poloxamine-coated particles are eventually phagocytosed by the red pulp macrophages; this is presumably due to gradual loss of the surface coating and subsequent opsonisation.

3.1.2
Stimulated or Activated Liver Macrophages

In contrast to resident Kupffer cells, stimulated or activated liver macrophages or newly recruited marginating monocytes can rapidly recognise and ingest surface-engineered long-circulating as well as poloxamine-based splenotropic particles from the blood by an opsonin-independent mechanism (Moghimi et al. 1993a; Moghimi and Murray 1996; Moghimi and Gray 1997). The nature of macrophage receptors that recognise such engineered particles remains to be unravelled, but a role for CD14, class A scavenger receptors and even Dectins seems possible. Nevertheless, these surface-engineered particles may find applications as diagnostic/imaging tools for stimulated or newly recruited hepatic macrophages (Fig. 2) (Laverman et al. 2001). For example, such diagnostic procedures may prove useful for patient selection or for monitoring the progress of treatment with long-circulating nanoparticles carrying anti-cancer agents, thus minimising damage to hepatic macrophages.

3.2 Subcutaneous Route

3.2.1 Interstitial and Lymph Node Macrophages

The distinct physiological function of the lymphatic capillaries opens up an opportunity for macrophage-specific targeting with colloidal carriers. In these capillaries, numerous endothelial cells overlap extensively at their margins and lack adhesion mechanisms at many points. Immediately following interstitial injection, many of the overlapped endothelial cells are separated and thus passageways are provided between the interstitium and lymphatic lumen, and particles are conveyed to the nodes via the afferent lymph. In lymph nodes, macrophages of medullary sinuses and paracortex are mainly responsible for particle capture from the lymph. However, the behaviour of particles following interstitial administration is controlled by a number of physicochemical and biological factors (Moghimi and Rajabi-Siahboomi 1996).

Physicochemical considerations include particle size and its surface characteristics. Generally, the size of the particles must be larger than 20 nm to prevent their leakage into the blood capillaries. Although larger particles (>100 nm) may carry a considerable amount of agent, they move very slowly from the site of injection into initial lymphatics; the drainage often takes a period of days. This slow transit may induce local inflammation and renders particles susceptible to phagocytosis or macropinocytosis at the injection site (Moghimi et al. 1994). However, considerable differences in drainage and lymph node uptake of particles of similar size can occur. These are explained by differences in surface characteristics of particles, which include the extent of surface hydrophobicity/hydrophilicity and the presence of macrophage-specific ligands. For example, the extent of surface hydrophobicity/hydrophilicity controls particle aggregation at the injection site as well as its interaction with the amorphous ground substance of the interstitium. Hence, the greater the tendency for particle aggregation or interaction with the interstitial structures, the slower the particle drainage and presentation to lymph node macrophages (Moghimi et al. 1994). The

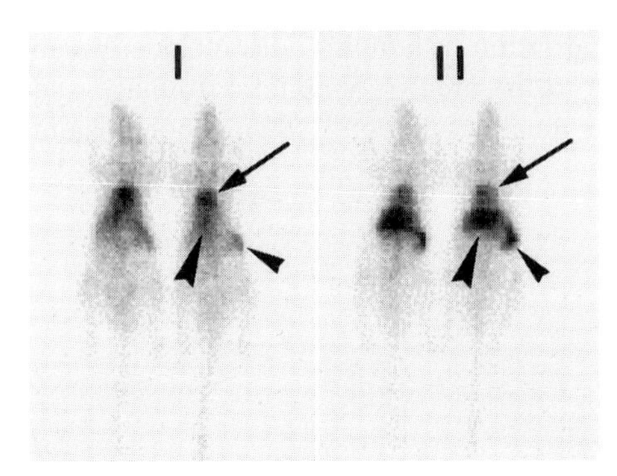

Fig. 3 Scintigraphic images of rats with normal (*I*) and enhanced (*II*) macrophage activity 4 h following intravenous injection of ($^{99m}TcO_4^-$)-labelled long-circulating liposomes. To enhance macrophage phagocytic activity, poloxamine 908 was injected intravenously (43 mg/kg) 3 days prior to liposome injection. In (*I*) the images represent the circulatory blood pool in the heart region (*arrow*) and poor localisation of liposomes in both liver (*large arrowhead*) and spleen (*small arrowhead*) regions. In (*II*) a large fraction of liposomes is captured by stimulated Kupffer cells and splenic macrophages (*arrowheads*)

concept of steric stabilisation of nanoparticles with poloxamer and poloxamine copolymers proved successful in minimising particle retention at the injection site while maximising uptake by local lymph node macrophages (Moghimi et al. 1994). The longer the EO chain (while maintaining the length of central hydrophobic segment) of the coating copolymers, the lesser the tendency for particle aggregation and interaction with interstitial elements. The resulting outcome was rapid particle drainage into initial lymphatics (Fig. 3). The choice of coating copolymer also influences macrophage recognition of particles. For example, the strong steric barrier of poloxamer 407 suppresses particle opsonisation in lymph and/or interaction with macrophage receptors. Such engineered entities drain rapidly, escape clearance by lymph node macrophages, reach the systemic circulation, and remain in the blood for prolonged periods. For medical imaging, these particles have applications for visualising the lymphatic chain, but for macrophage targeting appropriate ligands (e.g. mannose, antibodies) must be attached to the distal end of the hydrophilic chains. To enhance both rapid particle drainage and capture by regional lymph node macrophages simultaneously, particles require coating with copolymers with 5–15 EO units (Fig. 3). These nanoengineering concepts satisfy at least the requirements for lymphoscintigraphy and indirect lymphography (rapid spreading of particles from the injection site and good retention in regional lymph nodes). Other investigators have exploited the adjuvant effect of such engineered nanoparticles in vaccination protocols following interstitial injection as the coating agents (e.g. poloxamers) can manipulate a number of macrophage accessory functions (e.g. upregulation of MHC class II expression) (Hunter and Bennett 1984; Howerton et al. 1990).

Regardless of these surface engineering techniques, particle drainage from the interstitium and their subsequent capture by lymph node macrophages also depends on the potency of the lymphatic system and lymph node. Other biological factors which influence particle localisation in lymph nodes include lymph propulsion, obesity and recent surgery in the area of lymphatic drainage and are discussed elsewhere (Moghimi and Rajabi-Siahboomi 1996; Moghimi and Bonnemain 1999).

3.3 The Oral Route

3.3.1 Intestinal Macrophages

A challenging but attractive strategy is to deliver nanoparticles to macrophages in the oral cavity. The majority of the available evidence suggests the adsorption of particulates in the intestine following oral administration takes place at the Peyer's patches (Simecka 1998). The epithelial cell layer overlying the Peyer's patches contains specialised M cells. In the immunocompetent host, the M-cell apical membranes bind and take up bacteria, viruses, inert particles, etc. from the lumen and transport them to underlying macrophages and dendritic cells. The microbial sampling function of M cells is critical to the ability of the mucosal immune system to monitor the contents of the intestinal lumen. Indeed, the use of colloidal particles to deliver vaccines and adjuvants has its foundation in the M cell's ability for transcytosis. Microencapsulation of antigens in such systems also provides better protection for the antigen during intestinal transit. Numerous studies have now confirmed protective immunity induced by mucosal immunisation with polymeric particulate systems (Marx et al. 1993; Jones et al. 1997). The immune outcomes have included mucosal (secretory IgA) and serum antibody (IgG and IgM) responses as well as systemic cytotoxic T lymphocyte responses in splenocytes. Tolerance to orally administered microparticulate encapsulated antigens is another potential outcome but has received little attention. Oral tolerance could provide tremendous potential in treating autoimmune and inflammatory diseases through antigens entrapped in nanoparticles. Low-dose, transmucosal exposure to antigens may also trigger antigen-specific IgE responses (Simecka 1998). T helper-2 cell activation, which supports mucosal IgA responses, promotes isotype switching to IgE-B cells. This is believed to be through secretion of IL-4 (Finkelman et al. 1991). Such IgE responses, whether local or systemic, can mediate potentially life-threatening immune reactions. Thus, IgE responses against antigen should be carefully monitored because of the potential for hypersensitivity reactions in susceptible individuals.

Induction of an appropriate immune response following oral administration depends primarily on factors that affect uptake and particle translocation by M cells. These include particle size, dose, composition and surface chemistry as well as the region of the intestine where particles are taken up, membrane recycling from intracellular sources and the species (Ermak and Giannasca 1998). For example, studies using a number of different microparticles have established that M cells can transport these in a size range similar to that of micro-organisms (reviewed by Ermak and Giannasca 1998). A lower limit for size has yet to be determined for M-cell sampling, but the upper limit is believed to be in the region of 10 μm. Following transcytosis particles may be engulfed by phagocytic cells in the dome region or pass through the basal lamina into the sub-endothelial dome region where they gain access via efferent lymphatics to macrophages in mesenteric lymph nodes and peripheral lymphoid organs. Particles transported beyond the follicle-associated epithelium are also disseminated in a size-dependent manner, with dispersion increasing with decreasing particle size (<500 nm). The composition of particles may also affect cytokine production by macrophages and dendritic cells in the dome region. The efficiency of sampling orally delivered nanoparticles by M cells is, however, low as these cells are relatively rare in the epithelial lining (the cumulative surface area of M-cell apical membrane throughout the intestine has been estimated to be 1/10,000th of enterocytes). Therefore, strategies are needed to specifically target the M-cell apical surfaces. Indeed, the use of M cell-selective probes such as secretory IgAs, monoclonal antibodies, lectins, and M-cell attachment proteins of reovirus has enhanced the efficiency of particulate targeting and subsequent presentation to the regional macrophages (Ermak and Giannasca 1998). Such approaches could ultimately be required for optimal stimulation of immunological responses following oral delivery.

4 Current Clinical Trends and Future Prospects

It is now clear that the propensity of macrophages for the phagocytic clearance of colloidal vehicles provides a rational approach to macrophage-specific targeting and drug delivery. To date, this approach has led to the development of a number of regulatory approved particulate formulations. Examples include Ambisome (Gilead Sciences) and a recent liposomal vaccine against hepatitis A. Ambisome is a liposomal formulation of amphotericin B, which is used for treatment of visceral leishmaniasis or confirmed infections caused by specific fungal species.

Alternatively, a significant delay of macrophage involvement would extend the circulation time of intravenously injected colloidal vehicles, thus enabling them to deliver their cargo to non-macrophage sites. Evasion of binding or uptake of particles by macrophages has been achieved by numerous surface modification strategies (Moghimi et al. 2001). Several formulations of stable, long-circulating liposomes are now marketed or seeking approval as carriers of anticancer agents in the treatment of solid tumours. Examples include DaunoXome (daunorubicin, marketed by Gilead Sciences), Doxil (doxorubicin, marketed by Alza Corporation), Onco TCS (vincristine, INEX Pharmaceuticals) and NX211 (lurtotecan, OSI Pharmaceuticals). The next generation of long circulating colloidal carriers are likely to be surface decorated with suitable ligands to internalising receptors overexpressed on tumour cells (e.g. cancers overexpressing HER2/neu proto-oncogene, folate) or angiogenic endothelial cells (e.g. $\alpha_V\beta_3$ integrin complex, Tie-2) (Moghimi et al. 2001). Such approaches will improve the selectivity and anti-tumour activity of existing and newly discovered anti-cancer agents.

Intriguing progress is being made in diagnostic imaging with the development of a long-circulating ultra-small super-paramagnetic iron oxide particle (Sinerem, Guerbet, France). Following intravenous injection, a significant fraction of Sinerem reaches numerous lymph nodes in the body, particularly abdominal and mediastinal nodes, which are the main sites of metastases and are not readily accessible for microscopic evaluation. Interestingly, in the lymph nodes these particles are phagocytosed by resident macrophages. As a result, Sinerem has helped to distinguish between normal and tumour-bearing nodes or reactive and metastatic nodes with magnetic resonance in man (Moghimi and Bonnemain 1999). Sinerem can also be administered subcutaneously.

It is conceivable that future sophistication in colloidal targeting of macrophages as well as non-macrophage cells and the outcome of end-results will depend on the detailed understanding of biological barriers as well as macrophage functions and their recognition mechanisms.

5 References

Agrawal AK, Gupta CM (2000) Tuftsin-bearing liposomes in treatment of macrophage-based infections. Adv Drug Deliv Rev 41:135–146

Alves-Rosa F, Stanganelli C, Cabrera J, van Rooijen N, Palermo MS and Isturzi MA (2000) Treatment with liposome-encapsulated clodronate as a new strategic approach in the management of immune thrombocytopenic purpura in a mouse model. Blood 96: 2834–2840

Alving CR (1988) Macrophages as targets for delivery of liposome-encapsulated antimicrobial agents. Adv Drug Deliv Rev 2:107–128

Barrera P, van Lent PLEM, van Bloois L, van Rooijen N, Malefijit MCD, van de Putte LBA, Storm G, van den Berg WB (2000) Synovial macrophage depletion with clodronate-containing liposomes in rheumatoid arthritis. Arthritis Rheumat 43:1951–1959

Beauregard KE, Lee KD, Collier RJ, Swanson JA (1997) pH-dependent perforation of macrophage phagosomes by Listeriolysin O from *Listeria monocytogenes*. J Exp Med 186:1159–1163

Donowitz GR (1994) Tissue-directed antibiotics and intracellular parasites: complex interaction of phagocyte, pathogens and drugs. Clin Infect Dis 19:926–930

Drummond DC, Zignani M, Leroux JC (2000) Current status of pH-sensitive liposomes in drug delivery. Prog Lipid Res 39:409–460

Elliott G, O'Hare P (1997) Intercellular trafficking and protein delivery by a herpesvirus structural protein. Cell 88:223–233

Ermak TH, Giannasca PJ (1998) Microparticle targeting to M cells. Adv Drug Deliv Rev 34:261–283

Finkelman FD, Urbon JF, Beckman MP, Schooley KA, Homes JM, Katona IM (1991) Regulation of murine in vivo IgG and IgE responses by a monoclonal anti-IL-4 receptor antibody. Int Immunol 3:599–607

Fling SP, Arp B, Pious D (1994) HLA-DMA and -DMB genes are both required for MHC class II/peptide complex formation in antigen-presenting cells. Nature 368:554–558

Gbadamosi JK, Hunter AC, Moghimi SM (2002) PEGylation of microspheres generates a heterogeneous population of particles with differential surface characteristics and biological performance. FEBS Letters 532:338–344

Gordon S (1995) The macrophage. Bioessays 17:977–986

Gordon S, Lawson L, Rabinowitz S, Crocker PR, Morris L, Perry VH (1992) Antigen markers of macrophage differentiation in murine tissues. Curr Top Microbiol Immunol 181:137
Gregoriadis G (1990) Immunological adjuvants: a role for liposomes. Immunol Today 11:89–96
Harding CV, Song R (1994) Phagocytic processing of exogenous particulate antigens by macrophages for presentation by class I MHC molecules. J Immunol 153:4925–4933
Horwitz BA, Shintizky M, Kreutz W, Yatvin MB (1980) pH-sensitive liposomes: possible clinical implications. Science 210:1253–1255
Howerton DA, Hunter RL, Ziegler HK, Check IJ (1990) Induction of macrophage Ia expression in vivo by a synthetic block copolymer, L81. J Immunol 144:1578–1584
Hunter RL, Bennett B (1984) The adjuvant activity of nonionic block polymer surfactants. II. Antibody formation and inflammation related to the structure of triblock and octablock copolymers. J Immunol 133:3167–3175
Jones DH, Corris S, McDonald S, Clegg JCS, Farrar GH (1997) Poly(DL-lactide-co-glycolide) encapsulated plasmid DNA elicits systemic and mucosal antibody responses to encoded protein after oral administration. Vaccine 15:814–817
Kotter MR, Setzu A, Sim FJ, van Rooijen N, Franklin RJM (2001) Macrophage depletion impairs oligodendrocyte remyelination following lysolecithin-induced demyelination. GLIA 35:204–212
Kovacsovics-Bankowski M, Rock KL (1995) A phagosome-to-cytosol pathway for exogenous antigens presented on MHC class I molecules. Science 267:243–246
Laverman P, Carstens MG, Storm G, Moghimi SM (2001) Recognition and clearance of methoxypoly(ethyleneglycol)2000-grafted liposomes by macrophages with enhanced phagocytic capacity. Implications in experimental and clinical oncology. Biochim Biophys Acta 1526:227–229
Lee PT, Holt PG, McWilliam AS (2001) Ontogeny of rat pulmonary alveolar macrophage function: evidence for a selective deficiency in IL-10 and nitric oxide production by newborn alveolar macrophages. Cytokine 15:5357
Lee KD, Oh YK, Portnoy DA, Swanson JA (1996) Delivery of macromolecules into cytosol using liposome containing hemolysin from Listeria monocytogenes. J Biol Chem 271:7249–7252
Linehan SA, Martinez-Pomares L, da Silva RP, Gordon S (2001) Endogenous ligands of carbohydrate recognition domains of the mannose receptor in murine macrophages, endothelial cells and secretory cells; potential relevance to inflammation and immunity. Eur J Immunol 31:1857–1866
Liu T, van Rooijen N, Tracy DJ (2000) Depletion of macrophages reduces axonal degeneration and hyperalgesia following nerve injury. Pain 86:2532
Lloyd JB (2000) Lysosome membrane permeability: implications for drug delivery. Adv Drug Deliv Rev 41:189–200
Marx PA, Compans RW, Gettie A, Staas JK, Gilley RM, Mulligan MJ, Yamshchikov GV, Chen D, Eldridge JH (1993) Protection against vaginal SIV transmission with microencapsulated vaccine. Science 260:1323–1327
Moghimi SM (2002) Chemical camouflage of nanospheres with a poorly reactive surface: towards development of stealth and target-specific nanocarriers. Biochim Biophys Acta 1590:131–139
Moghimi SM, Bonnemain B (1999) Subcutaneous and intravenous delivery of diagnostic agents to the lymphatic system: applications in lymphoscintigraphy and indirect lymphography. Adv Drug Deliv Rev 37:295–312
Moghimi SM, Gray T (1997) A single dose of intravenously injected poloxamine-coated long-circulating particles triggers macrophage clearance of subsequent doses in rats. Clin Sci 93:371–379

Moghimi SM, Hawley AE, Christy NM, Gray T, Illum L, Davis SS (1994) Surface engineered nanospheres with enhanced drainage into lymphatics and uptake by macrophages of the regional lymph nodes. FEBS Lett 344:25–30
Moghimi SM, Hedeman H, Christy NM, Illum L, Davis SS (1993a) Enhanced hepatic clearance of intravenously administered sterically stabilized microspheres in zymosan-stimulated rats. J Leukoc Biol 54:513–517
Moghimi SM, Hedeman H, Illum L, Davis SS (1993b) Effect of splenic congestion associated with haemolytic anaemia on filtration of 'spleen-homing' microspheres. Clin Sci 84:605–609
Moghimi SM, Hedeman H, Muir IS, Illum L, Davis SS (1993c) An investigation of the filtration capacity and the fate of large filtered sterically-stabilized microspheres in rat spleen. Biochim Biophys Acta 1157:233–240
Moghimi SM, Hunter AC (2000) Poloxamers and poloxamines in nanoparticle engineering and experimental medicine. Trend Biotechnol 18:412–420
Moghimi SM, Hunter AC, Murray JC (2001) Long-circulating and target-specific nanoparticles: theory to practice. Pharmacol Rev 53:283–318
Moghimi SM, Muir IS, Illum L, Davis SS, Kolb-Bachofen V (1993d) Coating particles with a block copolymer (poloxamine-908) suppresses opsonization but permits the activity of dysopsonins in the serum. Biochim Biophys Acta 1179:157–165
Moghimi SM, Murray JC (1996) Poloxamer-188 revisited: a potentially valuable immune modulator? J Natl Cancer Inst 88:766–768
Moghimi SM, Porter CJH, Muir IS, Illum L, Davis SS (1991) Non-phagocytic uptake of intravenously injected microspheres in rat spleen: influence of particle size and hydrophilic coating. Biochem Biophys Res Commun 177:861–866
Moghimi SM, Patel HM (2002) Modulation of murine Kupffer cell clearance of liposome by diethylstilbestrol. The effect of vesicle surface charge and a role for complement receptor Mac-1 (CD11b/CD18) of the newly recruited macrophages in liposome recognition. J Control Rel 78:55–65
Moghimi SM, Rajabi-Siahboomi AR (1996) Advanced colloid-based systems for efficient delivery of drugs and diagnostic agents to the lymphatic tissues. Prog Biophys Mol Biol 65:221–249
Polfliet MMJ, Goede PH, van Kesteren-Hendrikx EML, van Rooijen N, Dijkstra CD, van der Berg TK (2001) A method for the selective depletion of perivascular and meningeal macrophages in the central nervous system. J Neuroimmunol 116:188–195
Poznansky M, Juliano RL (1984) Biological approaches to the controlled delivery of drugs: a critical review. Pharmacol Rev 36:277–336
Rao M, Alving CR (2000) Delivery of lipids and liposomal proteins to the cytoplasm and golgi of antigen-presenting cells. Adv Drug Deliv Rev 41:171–188
Rutherford MS, Witsell A, Schook LB (1993) Mechanisms generating functionally heterogeneous macrophages. Chaos revisited. J Leukoc Biol 53:602–618
Simecka JW (1998) Mucosal immunity of the gastrointestinal tract and oral tolerance. Adv Drug Deliv Rev 34:235–259
Tilcock C (1995) Imaging tools: liposomal agents for nuclear medicine, computed tomography, magnetic resonance, and ultrasound. In: Philippot JD, Schubar F (eds) Liposomes as Tools in Basic Research and Industry, CRC Press, Boca Raton, FL, pp 225–240
Torchilin VP, Rammohan R, Weissig V, Levchenko TS (2001) TAT peptide on the surface of liposomes affords their efficient intracellular delivery even at low temperature and in the presence of metabolic inhibitors. Proc Natl Acad Sci USA 98:8786–8791
van Rooijen N, Sanders A (1997) Elimination, blocking, and activation of macrophages: three of a kind? J Leukoc Biol 62:702–709
Wu GS, Korsgren O, Zhang JG, Song ZS, van Rooijen N, Tibell A (2000) Pig islet xenograft rejection is markedly delayed in macrophage-depleted mice: a study in streptozotocin diabetic animals. Xenotransplantation 7:214–220

Reaching the Macrophage: Routes of Delivery

G. Kraal · N. van Rooijen

Department of Molecular Cell Biology,
Vrije Universiteit Medical Center,
Van der Boechorststraat 7, 1081 BT Amsterdam, The Netherlands
e-mail: g.kraal@vumc.nl

Abstract The functional heterogeneity of macrophages based on anatomical localization and receptor expression as well as their versatile way through which they react to different stimuli makes them key players in many processes of the body. To stimulate or downregulate their functions in therapeutic ways using their extremely efficient phagocytosis capacity, one has to consider that they are often located at sites that are hard to reach. Local administration of drugs is therefore often the method of choice whereby liposomes seem to be an ideal vehicle through their versatility and relative ease of handling, as well as their efficient uptake by macrophages. Especially the use of the macrophage suicide technique has enabled the study of many functions of the macrophage. An overview is given of the many routes that can be used to reach the macrophage at different sites and the effects this has on the function of the cell.

Keywords Liposomes, Phagocytosis, Clodronate

1 Introduction

Macrophages are extremely versatile cells that have evolved as professional phagocytes. Their phagocytosis capacity is pivotal for the uptake and degradation of infectious agents and senescent or damaged cells of the body. This makes them central cells in tissue remodeling and repair, as well as key players in immune responses and inflammatory reactions.

Although a certain phagocytic capacity can be ascribed to a variety of cells, this capacity is more or less intermediate compared to that of macrophages, and often specialized or restricted as in the case of retinal rod internalization by the epithelial cells of the retina (Rabinovitch 1995).

The extreme efficiency of the macrophage in terms of particle uptake is due to the presence of a vast array of different receptors on the cell membrane that uniquely, or in concert, act to facilitate uptake.

As a consequence of the possible interactions of the various receptors and their signaling modes inside the cell, the process of phagocytosis can be quite complicated and leads to different subsequent actions of the macrophage (Aderem and Underhill 1999).

Many of the receptors on the macrophage that are involved in the binding to pathogenic ligands are able to recognize conserved motifs, so-called pathogen-associated molecular patterns (Janeway 1992). These include mannans found on yeast cell walls, bacterial peptides, lipopolysaccharides, lipoteichoic acid, but also self-molecules that have been modified such as oxidized low-density lipoprotein (LDL) are recognized. The pattern recognition receptors involved form distinct groups based on molecular structure and recognition profile, and they include the family of scavenger receptors, Toll-like receptors, and mannose receptor (Krieger and Herz 1994; Medzhitov and Janeway 2000; Pearson 1996; Peiser et al. 2002; Platt and Gordon 2001). In addition, members of the group of pattern recognition receptors can recognize humoral factors such as components of the complement system, antibodies and factors such as mannan-binding protein. These factors are involved in the opsonization of pathogens, thereby facilitating their uptake by macrophages. The receptors involved in the recognition such as Fc and complement receptors are key players in the phagocytic capacity of macrophages (Carroll 1998; see also Ravetch 1997; Ravetch and Clynes 1998; Stahl and Ezekowitz 1998). Signaling after engagement of these receptors leads to major changes in the actin distribution and mobility of the cell, but can also lead to the production of cytokines and other factors (Wright and Silverstein 1983; Stein and Gordon 1991; Takai et al. 1994; Gerber and Mosser 2001).

2 Heterogeneity of Macrophages

The mode of activation of the macrophage that can occur after engagement of the multiple types of receptors is therefore dependent on the composition of the pathogen or particle to be internalized.

In the case of Fc receptor-mediated uptake of bacteria, strong activation of the macrophage resulting in cytoskeleton rearrangements and the release of pro-inflammatory factors such as tumor necrosis factor (TNF)-α can be seen, whereas the uptake of apoptotic cells, predominantly engaging scavenger receptors and CD14, leads to the production of anti-inflammatory factors such as interleukin (IL)-10 and transforming growth factor (TGF)-β (Savill et al. 1993; Platt et al. 1996; Savill 1998; Fadok et al. 2001). Interestingly, the way the macrophage reacts to stimuli resulting from uptake of different particles and receptor usage is also dependent on the activation state of the cell and may therefore differ considerably (Riches 1995).

In addition, it has become clear that not only the maturation stage of the macrophage in the tissue, from recently immigrated monocyte to fully differentiated macrophage, will determine the potential of the cell to react to external stimuli, but that the tissue itself, where the macrophage is maturing, is of importance for the differentiation of the cell. This leads to major differences in function of macrophages residing, e.g., in the alveoli of the lungs versus the Kupffer cells in the liver (Fathi et al. 2001; Laskin et al. 2001).

Alveolar macrophages, in addition to being efficient scavenging cells, control the homeostasis of the lungs through active downregulation of the activity of T lymphocytes and dendritic cells in the interstitial tissues of the lung (Holt et al. 1993). This capacity is unique for macrophages in the lung microenvironment and seems to be dependent on the production of inhibitory factors by the macrophage, such as nitric oxide. In addition to differences in macrophage populations between different organs, it is clear that even within tissues different subpopulations of macrophages can be discriminated, especially in lymphoid organs such as the spleen. Based on anatomical localization, receptor usage and turnover kinetics, in the spleen at least five different subpopulations of macrophages can be found (van Rooijen et al. 1989a,b; Kraal 1992).

Together with the different modes of activation that the macrophage can develop in response to stimuli, such tissue-dependent specialization adds to the complexity of the cell and forms an important factor when one attempts to manipulate macrophages by specific targeting. Furthermore, the phagocytic capacity of macrophages may lead to unwanted degradation of bioactive products before they can reach their target cells. So, although the efficient uptake of particles by macrophages makes them ideal cells to target, the rapid degradation following phagocytosis may lead to inefficient effects of bioactive materials or unwanted effects on the macrophage.

This can be a problem when macrophages themselves are the target cell, but it also creates an important drawback in attempts to reach other cell types with, e.g., encapsulated moieties. In the latter cases, such negative influences can be overcome by first killing the macrophages selectively, followed by the administration of the compound that is supposed to activate other cells (van Rooijen et al. 1997; van Rooijen and Sanders 1997).

Furthermore, because of the localization of macrophages in all tissues, often between densely packed cells or collagen fibers, it is not always easy to reach the macrophage. Administration of particulate material into the bloodstream will only reach macrophages that are located in areas where the blood has free access: the blood sinuses of liver and spleen. To reach the macrophage in other tissues, it is necessary to administer locally into the target organ, as is the case for alveolar macrophages (Pantelidis et al. 2001) or skin macrophages (Thepen et al. 2000).

3 How to Reach Macrophages in Various Tissues

The extent to which resident macrophage populations in different organs are accessible to single molecules, molecular complexes or particulate carriers depends on both the position of the macrophages in the tissues and on the properties of the molecules or particles. In general, all macrophages can be reached by small molecules when these are able to pass the capillary networks in order to penetrate into the tissues. For larger molecules, molecular complexes, or particles, the macrophage can only be reached if there is no physical barrier between the site of injection and the macrophage. Such a barrier can be formed, e.g., by endothelial cells in the wall of blood vessels, by alveolar epithelial cells in the lung, by reticular fibers or collagen fibers in the spleen, or by the presence of densely packed cells such as lymphocytes in the white pulp of the spleen or in the paracortical fields of lymph nodes. By choosing the right administration route for the materials to be injected, this barrier can be kept at a minimum.

3.1 The Use of Liposomes

Information on the comparative accessibility of various macrophages in different organs has been obtained from studies using liposomes that lead to the death of the macrophage.

Liposomes (artificially prepared phospholipid vesicles) encapsulating the bisphosphonate clodronate can be used to deplete macrophages in various organs and tissues. Such liposomes, once injected in vivo, will be ingested by macrophages. They are subsequently exposed to lysosomal phospholipases leading to intracellular degradation of the phospholipid bilayers and release of entrapped clodronate molecules within the cells. The more clodronate liposomes will be ingested and digested by the macrophages, the more clodronate will be accumulated within the cells, since it is not easy for these molecules to escape, given their poor ability to cross cell membranes. At a certain intracellular concentra-

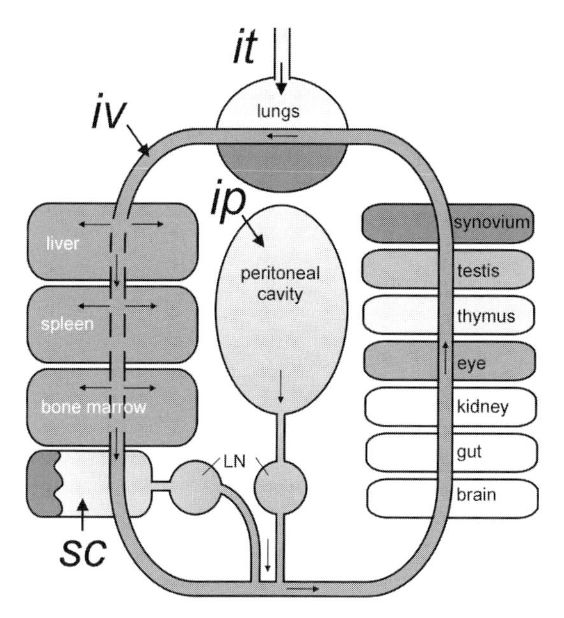

Fig. 1 Routes of administration of liposomes to reach macrophages. Intravenous (*iv*) injection of liposomes will reach all macrophages in those organs that are in direct, open contact with the bloodstream, such as liver, spleen, and bone marrow. Intraperitoneal injection (*ip*) will reach first the macrophages in the peritoneal cavity, but may ultimately also reach macrophages in the lymph nodes that drain the peritoneal cavity and may from there even reach the bloodstream. The latter will especially occur when large or repeated doses are given i.p. Subcutaneous injection of liposomes will reach local macrophages in the subcutaneous tissues and from there will be transported by lymphatics to lymph nodes (*LN*). Again, here liposomes may ultimately reach the bloodstream. Other routes of administration, such as local, intratracheal (*it*) instillation, will lead to selected access to macrophages, as seen in the lung. Intratracheally administered liposomes will readily reach the alveolar macrophages, but will not reach the macrophages in the lung interstitium. Macrophages in the organs depicted on the *right* can be reached only by local injection of liposomes, but access is often difficult and incomplete

tion of clodronate, the cell will be eliminated by apoptosis. This approach has been named the liposome-mediated macrophage suicide technique (Van Rooijen and Sanders 1994).

By comparison of the doses of clodronate liposomesliposome required for depletion of macrophages and the time interval between injection of liposomes and depletion, ample information on the in vivo accessibility of various macrophages has been obtained. The data are summarized below.

3.2 Intravenous Administration

Intravenously injected materials can reach macrophages in the liver (Kupffer cells), spleen and bone marrow (Fig. 1). Kupffer cells in the liver sinuses, as well as marginal zone macrophages and red pulp macrophages in the spleen have a strategic position with respect to large molecular aggregates and particulate materials in the circulation. Liposomes have a nearly unhindered access to these macrophages as concluded from their fast and complete depletion within 1 day after intravenous injection of clodronate liposomes in mice and rats (Van Rooijen et al. 1990). Obviously, it is a little more difficult for intravenously injected liposomes to reach the marginal metallophilic macrophages in the outer periphery of the white pulp. Depletion of the white pulp macrophages in the periarteriolar lymphocyte sheaths (PALS) is incomplete, emphasizing the barrier formed by the reticulin fiber network and/or the densely packed lymphocytes in the white pulp.

The spleen is often chosen as a model system to study macrophages in vivo, since five different subpopulations of macrophages can be found that are all restricted to their own compartment. Moreover, these macrophages are separated from the main blood flow by different barriers. Large molecules such as serum albumins and 500-kDa dextran are retained within the marginal zone and red pulp. Only small polysaccharides and proteins are able to permeate the splenic white pulp. Evidence has been presented that there exists a splenic conduit network that regulates the transport of small molecules in the white pulp (Nolte 2002).

Additionally, macrophages in the bone marrow have access to intravenously injected clodronate liposomes. However, two consecutive injections with a time interval of 2 days were required to get a nearly complete depletion of macrophages from the bone marrow (Barbe et al. 1996).

Kupffer cells in the liver play a key role in the homeostatic function of the liver. They form the largest population of macrophages in the body, make up 30% of the hepatic nonparenchymal cell population, and have easy access to particulate materials in the circulation. Consequently, a large proportion of all intravenously administered particulate carriers used for drug targeting or gene transfer will be prematurely destroyed before they reach their targets. Therefore, transient blockade of phagocytosis by Kupffer cells might be an important factor to optimize in drug targeting, gene transfer, and xenogeneic cell grafting (van Rooijen et al. 1997). Also, transient suppression of the cytokine-mediated activity of Kupffer cells might have a beneficial effect on various disorders of the liver (Schumann et al. 2000).

3.3 Subcutaneous Administration

Subcutaneously injected clodronate liposomes are able to deplete macrophages in the draining lymph nodes of mice and rats (Fig. 1). Such liposomes, when, e.g., injected in the subcutaneous tissue of the footpads of mice, led to the depletion of subcapsular sinus lining macrophages and medulla macrophages in the draining popliteal lymph nodes (Delemarre et al. 1990). Macrophages in the paracortical fields and those in the follicles of the lymph nodes were not affected, emphasizing the existence of a barrier formed by reticular fibers and/or densely packed lymphocytes in these lymph node compartments, comparable to that formed in the white pulp of the spleen.

After passing the popliteal lymph nodes, the lymph flow will be further filtered by consecutive draining lymph node stations such as the lumbar lymph nodes (in the mouse). Obviously, macrophages in these secondary lymph nodes were less efficiently depleted. Interestingly, only macrophages had been depleted in those lymph node compartments that directly drained the initial popliteal lymph nodes, indicating that different parts in the lymph nodes are corresponding each with their own draining area and have their own afferent lymph vessels. As a consequence, particles such as liposomes are not equally distributed over all macrophages in the lymph nodes.

3.4 Intraperitoneal Administration

Macrophages from the peritoneal cavity and the omentum of the rat can be depleted by two consecutive intraperitoneal injections with , given at an interval of 3 days (Biewenga et al. 1995). The peritoneal cavity is drained by the parathymic lymph nodes (in rats and mice). After passing these lymph nodes, the lymph flow reaches the blood circulation via the larger lymph vessels such as the ductus thoracicus. As a consequence, intraperitoneally injected clodronate liposomes are also able to deplete the macrophages of parathymic lymph nodes and, once they arrive in the blood circulation, they may deplete macrophages in liver and spleen (Fig. 1). Given the relatively large volume that can be administered in a single dose via the intraperitoneal route, the total number of macrophages that can be affected is even higher than that affected by intravenous injection.

3.5 Intratracheal Instillation and Intranasal Administration

Alveolar macrophages form a first line of defense against microorganisms entering the lung via the airways. In contrast to the interstitial macrophages that are separated from the alveolar space by an epithelial barrier, alveolar macrophages which are located in the alveolar space have direct access to liposomes administered via the airways, for instance by intratracheal instillation, intranasal administration, or by the application of aerosolized liposomes (Fig. 1). The direct access of clodronate liposomes to alveolar macrophages is demonstrated by the rapid elimination of these cells after intratracheal administration into mice and rats (Thepen et al. 1989), leaving the interstitial macrophage population unaffected. Alveolar macrophages make up about 80% of the total macrophage population in the lung. Given their presence in high numbers and the total mass of lung tissue, they form an important population of macrophages in the body.

3.6
Intraventricular Injection in the Central Nervous System

Stereotaxical injection of clodronate liposomes into the fourth ventricle of the central nervous system (CNS) of rats resulted in a complete depletion of perivascular and meningeal macrophages in the cerebellum, cerebrum, and spinal cord of these rats (Polfliet et al. 2001). These results also confirm that macrophages in the brain are accessible to liposomes if the latter are administered via appropriate routes.

3.7
Intra-articular Injection in the Synovial Cavity of Joints

Phagocytic synovial lining cells play a crucial role in the onset of experimental arthritis induced with immune complexes or collagen type II. A single intra-articular injection with clodronate liposomes caused the selective depletion of phagocytic synovial lining cells in mice and rats, demonstrating that this administration route allows easy access of the liposomes to the macrophages lining the synovial cavity. Recent experiments have confirmed that liposomes are also able to reach synovium lining macrophages in man (Barrera et al. 2000).

3.8
Local Injections: The Testis

Local injection of a suspension of liposomes can be performed in most organs. However, whether or not the liposomes will be able to diffuse from the injection site over the rest of the tissue will largely depend on the tissue structure. In the testis of rats, a loosely woven tissue structure allows the liposomes to reach most of the testicular macrophages, as demonstrated by the finding that at least 90% of the testicular macrophages can be depleted by clodronate liposomes (Bergh et al. 1993).

4
Specificity of Liposome-Mediated Delivery into Macrophages

During the past decade, several particulate drug-carrier devices have been developed. Among these, liposomes may be considered the most versatile and promising drug-delivery system (e.g., Gregoriadis 1995). The natural fate of liposomes, once they have been injected in vivo, is their uptake and degradation by macrophages. As a consequence, both liposome-encapsulated hydrophilic molecules and hydrophobic molecules that are associated with the phospholipid bilayers of the liposomes will be targeted into the macrophage, if no action is taken to prevent their phagocytosis. Although macrophages in general seem to prefer liposomes with an overall negative charge, e.g., achieved by incorporation of the anionic phospholipid phosphatidylserine in the bilayers, neutral and cationic liposomes are also rapidly taken up by macrophages. Several modifications of the original liposome formulations, such as the incorporation of amphipathic polyethylene glycol (PEG) conjugates in the liposomal bilayers have been proposed in order to reduce the recognition and uptake of liposomes by macrophages. Nevertheless, a large percentage of these so-called long-circulating liposomes will still be ingested by macrophages, emphasizing that macrophages form the logical target for liposomes (Litzinger et al. 1994). Liposomes of more than one micron in diameter will be internalized by non-phagocytic cells to a very low extent only. This explains why other cells such as lymphocytes and granulocytes are not depleted by clodronate liposomes. In summary, it may be concluded that liposomes form an ideal vehicle for the delivery of various compounds into macrophages.

Given the fact that macrophages will ingest all types of non-self macromolecules and particulate materials, it is difficult to achieve specific targeting to only one macrophage subset, e.g., in the spleen. In studies intended to reveal the conditions for monoclonal antibody-mediated specific targeting of enzyme molecules to marginal metallophilic macrophages in the spleen, we found that highly specific targeting of the enzyme molecules could be achieved only by using monomeric conjugates of the antibody and the enzyme. Larger conjugates lead to their uptake by all macrophage subsets in the spleen (van Rooijen et al. 1992).

5 Conclusions

Macrophages play key roles in many processes of the body. To stimulate or downregulate their functions in therapeutic ways using their extremely efficient phagocytosis capacity, one has to consider that they are often located at sites that are hard to reach. Local administration of drugs is therefore often the method of choice, whereby liposomes seem to be an ideal vehicle through their versatility and relative ease of handling, as well as their efficient uptake by macrophages.

6 References

Aderem A, Underhill DM (1999) Mechanisms of phagocytosis in macrophages. Annu Rev Immunol 17:593–623

Barbe E, Huitinga I, Dopp EA, Bauer J, Dijkstra CD (1996) A novel bone marrow frozen section assay for studying hematopoietic interactions in situ: the role of stromal bone marrow macrophages in erythroblast binding. J Cell Sci 109:2937–2945

Barrera P, Blom A, van Lent PL, van Bloois L, Beijnen JH, van Rooijen N, de Waal Malefijt MC, et al (2000) Synovial macrophage depletion with clodronate-containing liposomes in rheumatoid arthritis. Arthritis Rheum 43:1951–1959

Bergh A, Damber JE, van Rooijen N (1993) The human chorionic gonadotrophin-induced inflammation-like response is enhanced in macrophage-depleted rat testes. J Endocrinol 136:415–420

Biewenga J, van der Ende MB, Krist LF, Borst A, Ghufron M, van Rooijen N (1995) Macrophage depletion in the rat after intraperitoneal administration of liposome-encapsulated clodronate: depletion kinetics and accelerated repopulation of peritoneal and omental macrophages by administration of Freund's adjuvant. Cell Tissue Res 280:189–196

Carroll MC (1998) The role of complement and complement receptors in induction and regulation of immunity. Annu Rev Immunol 16:545–568

Delemarre FG, Kors N, Kraal G, van Rooijen N (1990) Repopulation of macrophages in popliteal lymph nodes of mice after liposome-mediated depletion. J Leukoc Biol 47:251–257

Fadok VA, Bratton DL, Henson PM (2001) Phagocyte receptors for apoptotic cells: recognition, uptake, and consequences. Journal of Clinical Investigation 108:957–962

Fathi M, Johansson A, Lundborg M, Orre L, Skold CM, Camner P (2001) Functional and morphological differences between human alveolar and interstitial macrophages. Experimental and Molecular Pathology 70:77–82

Gerber JS, Mosser DM (2001) Stimulatory and inhibitory signals originating from the macrophage Fc gamma receptors. Microbes and Infection 3:131–139

Gregoriadis G (1995) Engineering liposomes for drug delivery: progress and problems. Trends Biotechnol 13:527–537

Holt PG, Oliver J, Bilyk N, McMenamin C, McMenamin PG, Kraal G, Thepen T (1993) Downregulation of the antigen presenting cell function(s) of pulmonary dendritic cells in vivo by resident alveolar macrophages. J Exp Med 177:397–407

Janeway CA Jr (1992) The immune system evolved to discriminate infectious nonself from noninfectious self. Immunol Today 13:11–16

Kraal G (1992) Cells in the marginal zone of the spleen. Int Rev Cytol 132:31–74

Krieger M, Herz J (1994) Structures and functions of multiligand lipoprotein receptors: macrophage scavenger receptors and LDL receptor-related protein (LRP). Annu Rev Biochem 63:601–637

Laskin DL, Weinberger B, Laskin JD (2001) Functional heterogeneity in liver and lung macrophages. Journal of Leukocyte Biology 70:163–170

Litzinger DC, Buiting AM, van Rooijen N, Huang L (1994) Effect of liposome size on the circulation time and intraorgan distribution of amphipathic poly(ethylene glycol)-containing liposomes. Biochim Biophys Acta 1190:99–107

Medzhitov R, Janeway C Jr (2000) The Toll receptor family and microbial recognition. Trends Microbiol 8:452–456

Nolte M (2002) Compartments, cells and molecules in the spleen, Academic thesis. Vrije Universiteit medical center, Amsterdam

Pantelidis P, McGrath DS, Southcott AM, Black CM, du Bois RM (2001) Tumour necrosis factor-alpha production in fibrosing alveolitis is macrophage subset specific. Respiratory Research 2:365–372

Pearson AM (1996) Scavenger receptors in innate immunity. Curr Opin Immunol 8:20–28

Peiser L, Mukhopadhyay S, Gordon S (2002) Scavenger receptors in innate immunity. Curr Opin Immunol 14:123–128

Platt N, Gordon S (2001) Is the class A macrophage scavenger receptor (SR-A) multifunctional? - The mouse's tale. Journal of Clinical Investigation 108:649–654

Platt N, Suzuki H, Kurihara Y, Kodama T, Gordon S (1996) Role for the class A macrophage scavenger receptor in the phagocytosis of apoptotic thymocytes in vitro. Proceedings of the National Academy of Sciences of the United States of America 93:12456–12460

Polfliet MM, Goede PH, van Kesteren-Hendrikx EM, van Rooijen N, Dijkstra CD, van den Berg TK (2001) A method for the selective depletion of perivascular and meningeal macrophages in the central nervous system. J Neuroimmunol 116:188–195

Rabinovitch M (1995) Professional and Nonprofessional Phagocytes - an Introduction. Trends in Cell Biology 5:85–87

Ravetch JV (1997) Fc receptors. Curr Opin Immunol 9:121–125
Ravetch JV, Clynes RA (1998) Divergent roles for Fc receptors and complement in vivo. Annu Rev Immunol 16:421-432
Riches DWH (1995) Signaling Heterogeneity as a Contributing Factor in Macrophage Functional Diversity. Seminars in Cell Biology 6:377–384
Savill J (1998) Apoptosis - Phagocytic docking without shocking. Nature 392:442–443
Savill J, Fadok V, Henson P, Haslett C (1993) Phagocyte Recognition of Cells Undergoing Apoptosis. Immunology Today 14:131–136
Schumann J, Wolf D, Pahl A, Brune K, Papadopoulos T, van Rooijen N, Tiegs G (2000) Importance of Kupffer cells for T-cell-dependent liver injury in mice. Am J Pathol 157:1671–1683
Stahl PD, Ezekowitz RA (1998) The mannose receptor is a pattern recognition receptor involved in host defense. Curr Opin Immunol 10:50–55
Stein M, Gordon S (1991) Regulation of Tumor-Necrosis-Factor (Tnf) Release by Murine Peritoneal-Macrophages - Role of Cell Stimulation and Specific Phagocytic Plasma-Membrane Receptors. European Journal of Immunology 21:431–437
Takai T, Li M, Sylvestre D, Clynes R, Ravetch JV (1994) FcR gamma chain deletion results in pleiotrophic effector cell defects. Cell 76:519–529
Thepen T, Van Rooijen N, Kraal G (1989) Alveolar macrophage elimination in vivo is associated with an increase in pulmonary immune response in mice. J Exp Med 170:499–509
Thepen T, van Vuuren AJ, Kiekens RC, Damen CA, Vooijs WC, van De Winkel JG (2000) Resolution of cutaneous inflammation after local elimination of macrophages. Nat Biotechnol 18:48–51
van Rooijen N, Bakker J, Sanders A (1997) Transient suppression of macrophage functions by liposome-encapsulated drugs. Trends Biotechnol 15:178–185
van Rooijen N, Claassen E, Kraal G, Dijkstra CD (1989a) Cytological basis of immune functions of the spleen. Immunocytochemical characterization of lymphoid and non-lymphoid cells involved in the 'in situ' immune response. Prog Histochem Cytochem 19:1–71
van Rooijen N, Kors N, Kraal G (1989b) Macrophage subset repopulation in the spleen: differential kinetics after liposome-mediated elimination. J Leukoc Biol 45:97–104
Van Rooijen N, Kors N, vd Ende M, Dijkstra CD (1990) Depletion and repopulation of macrophages in spleen and liver of rat after intravenous treatment with liposome-encapsulated dichloromethylene diphosphonate. Cell Tissue Res 260:215–222
Van Rooijen N, Sanders A (1994) Liposome mediated depletion of macrophages: mechanism of action, preparation of liposomes and applications. J Immunol Methods 174:83–93
van Rooijen N, Sanders A (1997) Elimination, blocking, and activation of macrophages: three of a kind? J Leukoc Biol 62:702–709
van Rooijen N, ter Hart H, Kraal G, Kors N, Claassen E (1992) Monoclonal antibody mediated targeting of enzymes. A comparative study using the mouse spleen as a model system. J Immunol Methods 151:149–155
Wright SD, Silverstein SC (1983) Receptors for C3b and C3bi promote phagocytosis but not the release of toxic oxygen from human phagocytes. J Exp Med 158:2016–2023

Antigen-Presenting Cells and Vaccine Design

S. Y. C. Wong[1] · L. Martinez-Pomares[2]

[1] The Edward Jenner Institute for Vaccine Research,
Newbury, Berkshire, RG20 7NN, UK

[2] The Sir William Dunn School of Pathology,
Oxford University,
South Parks Road, Oxford, OX1 3RE, UK
e-mail: luisa.martinez-pomares@pathology.oxford.ac.uk

Abstract The innate immune system modulates antigen-specific adaptive immune responses in a qualitative and quantitative manner. In this chapter we propose that vaccine design would benefit from decoding this instructive function. We argue that the characterisation of cellular (particularly antigen-presenting cells) and soluble components of the innate system, and how they are interconnected, is an essential first step towards this goal.

Keywords Adaptive immune system, Adjuvants, Carbohydrate antigens, Marginal zone metallophilic macrophages, B cells, Innate immune system, T cells

Abbreviations

APC	Antigen-presenting cell
CR	Complement receptor
DC	Dendritic cell
Ig	Immunoglobulin
LPS	Lipopolysaccharide
MALT	Mucosal associated lymphoid tissue
LRR	Leucine-rich repeat
MØ	Macrophage
MR	Mannose receptor
MZ	Marginal zone
PAMP	Pathogen associated molecular pattern
Sn	Sialoadhesin
SR-A	Scavenger receptor class A
ss	Subcapsular sinus
TD	T-cell dependent
TI	T-cell independent
TLR	Toll-like receptor
TM	Transmembrane

1 Approaches to Vaccines

Historically, the aim of vaccination has been to prevent outbreaks of infectious diseases. Traditional vaccines were developed to mimic natural infections with live attenuated or inactivated bacteria or viruses. Live attenuated bacterial and viral vaccines include those for *Mycobacterium tuberculosis* (Bacille Calmette Guerin or BCG), *Salmonella typhi* Ty21a, vaccinia, polio, measles, mumps and rubella. Inactivated vaccines include those for *Vibrio cholerae*, *Bordetella pertussis*, influenza and rabies. Although their benefit-to-risk ratio is high, there is little or no knowledge of the protective antigens or immune responses responsible for immunity. Unpredictable adverse effects associated with a few conventional vaccines [eg. *B. pertussis* (Cody et al. 1981), rotavirus (Lynch et al. 2000) and respiratory syncytial virus (Kapikian et al. 1969)] are a major drawback of this empirical approach to vaccine development. Conventional vaccines with established high efficacy and safety would continue to contribute to vaccine programs, but current experimental vaccines in humans have been directed toward the use of well-defined antigens (subunit vaccines) to generate protective immunity. This rational approach has been made possible with our increased understanding of the molecular details of microbial pathogenesis, immune responses against microbial targets, and humoral and cell-mediated immune mechanisms. Subunit vaccines are either purified from bacteria or viruses or produced chemically or by recombinant gene technology. Subunit bacterial vaccines include formaldehyde-inactivated exotoxins from *Corynebacterium diphtheriae* (diphtheria toxoid) and *Clostridium tetani* (tetanus toxoid), and capsular polysaccharides from a number of encapsulated bacteria (e.g. *Haemophilus influenzae*, *Neisseria meningitidis*, *Streptococcus pneumoniae*) (Levine et al. 1997).

The present aims for vaccination are no longer restricted to the prevention of infectious diseases, but also include therapeutic intervention of chronic viral infections, cancers, alloreactivity, autoimmunity and allergy. Contraceptive vaccines are also under development (Levine et al. 1997). The types of immune responses that vaccines are required to elicit or inhibit vary with the nature of the pathogens and targeted antigens or conditions. For most extracellular pathogens, antibody responses appear to be important, whereas for most intracellular pathogens, cell-mediated responses are responsible for protection. A combination of both humoral and cell-mediated responses might be required for elimination of some pathogens. Whilst inducing neutralising antibodies against toxins or surface molecules on pathogens has been the basis for almost every past and present successful vaccine, there are many important human pathogens (e.g. malaria and HIV) and tumours that require the elicitation of strong cell-mediated immune responses. There is also a need for vaccine formulations able to induce immunological tolerance or modulate T-helper cell responses for conditions such as autoimmunity and allergy (Liu 1997).

Identification of suitable B- and T-cell epitopes (class II and class I restricted) that are targeted by the acquired immune system is necessary but insufficient for the development of effective vaccines. It has to be complemented by investigation of delivery systems and formulations (collectively referred to as adjuvants) that would compensate for the inherent lack of immunogenicity of subunit vaccines. Most adjuvants appear to act through their interactions with the innate immune system (Moingeon et al. 2001). The current view that the innate immune system not only provides short-term protection but also instructs the adaptive immune system to generate the most appropriate B- and T-cell responses following its encounter with pathogens is supported by recent evidence (Medzhitov and Janeway 2002). Thus, understanding the links between the innate and the adaptive immune systems is critically important to vaccine design.

2 Immune Responses to Antigens

Immune responses against microbes and antigens are initiated in secondary lymphoid organs, spleen, lymph nodes and mucosal-associated lymphoid tissue (MALT). In order to elicit an immune response, antigens entering at peripheral sites such as the skin, blood and gastrointestinal tract need to reach these secondary lymphoid organs where antigen-specific naïve B and T lymphocytes circulate. Not every antigen can elicit an immune response. Immunogenicity appears to vary with antigen form [soluble, cell-associated (Batista et al. 2001), or particulate (Ramachandra et al. 1999)], amount or load (Zinkernagel 2000), composition (foreignness) (Allison and Fearon 2000; Apostolopoulos et al. 2000) and location within the body (mucosal or systemic sites). It also depends on their propensity to activate complement (Carroll 1998) or engage other serum factors and natural antibodies (Sakamoto et al. 2001; Shimizu et al. 2001), and perhaps more importantly the context (Shi and Rock 2002) (homeostasis or inflammation) in which they are detected by the innate immune system.

Lack of immunogenicity to self-antigens is of obvious benefit. Tolerance to self-antigens is mostly due to the stringent process of negative selection of B and T lymphocytes in bone marrow and thymus, respectively. This central tolerance mechanism ensures the deletion of most autoreactive B and T cells. For those B and T cells that escape negative selection in the primary lymphoid organs, their full activation by self-antigens is prevented by peripheral tolerance mechanisms by which they could be killed, anergised or negatively regulated. Activation of naïve antigen-specific B and T cells to proliferate and differentiate into effector and memory cells requires antigen-presenting cells (APC) to capture, process, transport and present antigens, and provide co-stimulatory signals and cytokines.

Antigens reaching lymphoid organs are either native (cell-associated or soluble) or processed by APC. Native antigens transported with bound complement and/or antibodies or by specialised APC (Wykes et al. 1998; Berney et al. 1999) are recognised by antigen-specific B cells and are deposited on the surface of follicular dendritic cells (FDC) (Tew et al. 2001). In the case of T cell-dependent antigens (TD), B cells will acquire T-cell help and form germinal centres where immunoglobulin (Ig) class switching, affinity maturation and the production of memory cells occur. This is the case for most protein antigens and some glycolipids that can be presented to T cells by MHC class II or CD1 molecules, respectively. Humoral immune responses can also be elicited in the absence of classical B–T cognate interactions. These thymus independent (TI) responses are essential in the initial protection against pathogens and are characterised by the lack of Ig class switching and memory B cells. TI type 1 antigens [such as lipopolysaccharide (LPS)] induce a polyclonal response by activating B cells in an antigen-independent manner. TI type 2 antigens are normally polymers [carbohydrates or proteins assembled as multimers including components of viral capsules (Fehr et al. 1998)] that are able to cross-link B-cell receptors and induce mainly IgM and IgG3 production in mice and IgM and IgG2 in humans. At large antigen doses, all lymphoid organs are able to elicit TI responses. However at more realistic antigen doses, only the spleen can accumulate enough antigens to trigger TI responses (Ochsenbein et al. 2000). Therefore, the spleen has been considered as essential for the generation of TI responses.

3 Professional Antigen-Presenting Cells and Immunity: The Balance Between Clearance and Presentation

It is becoming clear that APC responses to microbes and their unique products are critical to the initiation of immune responses against pathogens. Macrophages (MØ) and dendritic cells (DC) appear to be the most important APC. Both cell types are widely distributed and located suitably throughout the body to encounter pathogens and microbial products. MØ are endowed with a number of cell surface receptors involved in opsonic and nonopsonic endocytosis and phagocytosis and upon activation produce soluble mediators that could initiate and regulate inflammatory responses. Despite having similar sets of cell surface receptors and function in antigen processing, MØ and DC have different roles in the induction and regulation of immunity, with DC being uniquely suited to activate naïve T cells in secondary lymphoid organs (Steinman et al. 2000). DC act as messengers, and instruct T cells to respond to stimuli they have received in peripheral tissues (Lanzavecchia and Sallusto 2001). The instructions would be conveyed in several ways: by the peptides presented on MHC-I and -II molecules, by the co-stimulatory and adhesion molecules displayed at the cell surface, and by the cytokines produced by the APC. The nature of the antigen and the inflammatory response elicited in the periphery would determine the subsequent B- and T-cell responses (Hawiger et al. 2001). Vaccine development will thus benefit from the identification of the mechanisms regulating APC function in situ.

Characterisation of the resident and recruited APC populations and their behaviour upon stimulation would enable the design of vaccines that will target antigens and APC to the appropriate secondary lymphoid organs and generate suitably activated T-and B-cell populations. This approach is of great interest in the case of prophylactic vaccines to be used in healthy individuals, since it will focus on delivery methods and formulations that can be produced in large scale. Therapeutic vaccines aiming at modifying an ongoing immunological process (such as allergy and chronic infection) require an understanding of the mechanisms that govern the conditions of non- or hyper-responsiveness of local APC populations, the regulation of activated T cells and the targeting systems that would act locally.

Monocytes migrate constitutively to tissues where they normally differentiate into MØ and in some cases DC (Randolph et al. 1998). MØ are heterogeneous in shape and function, depending on their local microenvironment, and can be distinguished by differences in the level of expression of certain cell surface molecules (Gordon 1999). During steady-state conditions, MØ remain in the tissue where they play an important role in normal homeostasis. Their unique endocytic and phagocytic capacity place MØ at the centre of the inflammatory process because they have the advantage of sampling the environment at the first instance (Parnaik et al. 2000; Fadok and Chimini 2001). As such, they could effectively control the availability of antigens encountered by DC. This could be relevant in the case of autoimmunity, since MØ reduce the amount of self-antigens that might be available to break tolerance. They also perform a more active role in the prevention of autoimmunity through the production of inhibitory cytokines such as interleukin (IL)-10 and transforming growth factor (TGF)-β upon apoptotic cell uptake (Fadok and Chimini 2001). These inhibitory properties have been exploited by some intracellular microbes that have developed mechanisms to gain access to the MØ intracellular environment (Ernst 1998) and are most apparent in alveolar MØ, which have also been shown to inhibit DC function (Goerdt et al. 1999; Tzachanis et al. 2002). To perform homeostatic functions, tissue MØ express endocytic receptors such as the mannose receptor (MR) and scavenger receptors, but not MHC-II and co-stimulatory molecules. This phenotype might be controlled through the interaction of the widely expressed OX2 (CD200) with its ligand (OX2L) on MØ (Hoek et al. 2000; Wright et al. 2000), or through the action of Ig-like inhibitory receptors (Dietrich et al. 2000). Finally, MØ are effector cells of the innate immune system. They are capable of destroying microbes and tumour cells and could cause tissue damage through the release or impaired clearance of inflammatory mediators, lysosomal hydrolases and reactive oxygen and nitrogen radicals. This capacity will be enhanced or modulated once activated T cells reach the tissues. In particular, T-helper (Th)1 T cells will enhance cytotoxic MØ mechanisms through the production of interferon (IFN)-γ. Th2-derived cytokines have a more subtle effect and seem to generate MØ with an alternatively activated or suppressor phenotype characterised by enhanced expression of endocytic receptors, reduced T-cell activation capacity, arginase production and synthesis of alternative macrophage-associated chemokine (AMAC)-1 (Stein et al. 1992; Goerdt et al. 1999; Tzachanis et al. 2002).

Under steady-state conditions, DC migrate through lymphatics to secondary lymphoid organs (Huang et al. 2000). This migratory DC population found in the absence of inflammation has not been well characterised, but might be involved in the maintenance of tolerance. They could transport self-antigens and present them to naïve T cells either directly, or through antigen transfer to resident DC (Steinman et al. 2000). When an inflammatory response is elicited through injury, infection or active immunisation, recruited monocytes move into the tissues and differentiate into activated MØ with increased endocytic and secretory activities. Under these conditions, enhanced cell trafficking to draining lymphatic tissue ensures antigen delivery and presentation. Though DC have been considered the likely mediators of antigen transport, several reports have identified recruited MØ as the major cells involved in antigen capture in situ. In a fluorescent microsphere-induced inflammation model, recruited monocytes with a MØ-like phenotype could be seen internalising these particles in the periphery. Antigen-bearing cells in lymphoid organs were then characterised as DC (Randolph et al. 1999). In the case of the adjuvant MF59, recruited cells at the site of injection that internalised the adjuvant were characterised as MØ (F4/80$^+$, CD11b$^+$, CD11c$^-$). When draining lymph nodes were analysed, MF59-loaded apoptotic MØ were observed at the subcapsular sinus, and a second MF59-containing cell population was detected in the T-cell area. These cells presented adjuvant-containing apoptotic bodies, indicating that antigen transfer between both cell types could have taken place (Dupuis et al. 2001).

Microbial-derived products exert a major influence in the phenotype of APC. Much research has focused in defining their capacity to modulate DC phenotype and determine the differentiation pathway (Th1 vs Th2) of Th cells upon activation (Lanzavecchia and Sallusto 2001). The use of large doses (e.g. 25 μg LPS/mouse) of antigens in most of these studies has made the assessment of the role of MØ in modulating these processes difficult (Pulendran et al. 2001). It is likely that MØ clearance/signalling systems could be overwhelmed by large doses of antigens. Under more physiological conditions, MØ could influence the extent of the inflammatory response by expressing inhibitory factors such as IL-10 following stimulation (Sutterwala et al. 1998; Gerber and Mosser 2001). IL-10 could act as a de-activator and induce uncoupling of the cytokine receptors required by DC for migration to lymphoid organs (D'Amico et al. 2000). In a model of soluble *Toxoplasma gondii* tachyzoite antigen (STAg)-induced DC migration and IL-12 production, lipoxin produced by splenic MØ induces a refractory phase in the responsiveness of DC to this stimulus (Aliberti et al. 2002).

4 Role of MØ in Secondary Lymphoid Organs in the Control of Immune Responses

There are distinct populations of MØ in secondary lymphoid organs. Blood-borne antigens are encountered by the spleen, which, as mentioned earlier, is required for TI-2 responses both in humans and mice (Amlot et al. 1985). This functional requirement in both species does not translate into structural similarity. Even though in both cases the marginal zone (MZ) is populated by B cells with an activated phenotype (CR2$^+$, IgD$^-$, IgM$^+$, CD27$^+$) distinct from the follicular recirculating pool (IgD$^+$, IgM$^+$) (Zandvoort et al. 2001), human spleen lacks a distinct marginal sinus, but contains an additional structure referred to as the perifollicular zone which surrounds B-cell follicles. However, a population of sialoadhesin (Sn)$^+$ MØ, a characteristic feature of murine MZ, has been located at these perifollicular areas (Steiniger et al. 1997). In murine spleens there are five MØ subtypes (Fig. 1). Red pulp MØ (F4/80$^+$, macrosialin$^+$, MR$^+$) do not seem to play a direct role in the regulation of immune responses except for antigen clearance and probably for their release of soluble mediators following activation. White pulp (WP) MØ are poorly characterised. They lack F4/80 expression but can be visualised by their expression of macrosialin. White pulp MØ, identified with the MOMA-2 monoclonal antibody, might act in the enhancement of immune responses through the regulated synthesis of C3 after immunisation, which could induce local opsonisation of antigens (Fischer et al. 1998). During the germinal centre reaction, tingible body macrophages (macrosialin$^+$, MR cysteine-rich domain ligands$^+$) (Martinez-Pomares et al. 1996) can be ob-

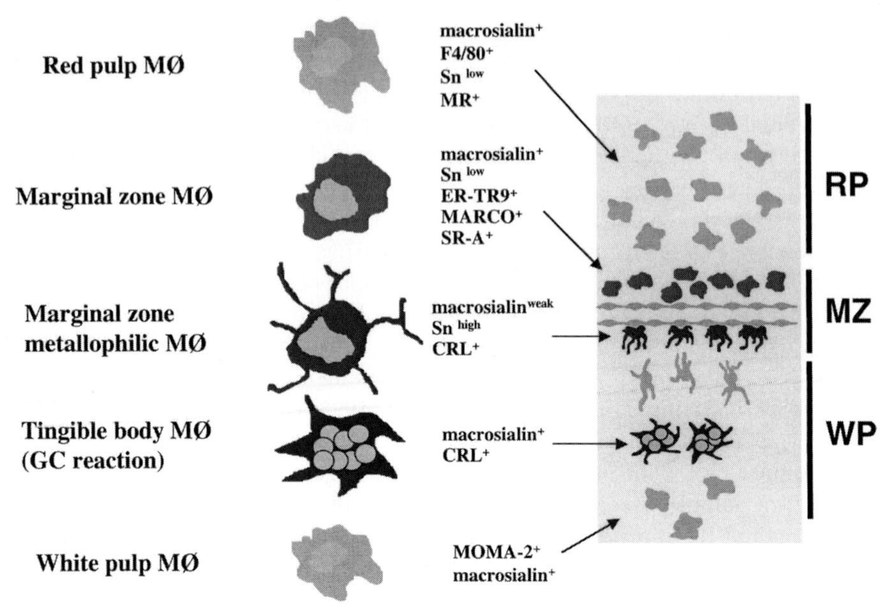

Fig. 1 In addition to their particular anatomical localisation, splenic MØ can be classified based on their differential pattern of marker expression. While red pulp MØ exclusively express the EGF-TM7 family member F4/80 and MR, they share expression of macrosialin with other MØ populations. Expression of the scavenger receptors MARCO and scavenger receptors class A, and the DC-SIGN murine homologue mSIGNR1 (recognised by the ER-TR9 antibody) is restricted to marginal zone MØ. Marginal zone metallophilic MØ are uniquely endowed with high expression of the member of the Siglec family, Sn, and the ability to interact with the CysR domain of the MR. Evenly distributed within lymphoid areas, a population of cells expressing macrosialin can be observed. During the germinal centre reaction, macrosialin expressing MØ that can bind the CysR domain of the MR (tingible body MØ) can be observed in the dark zone. These cells are probably involved in clearance of apoptotic cells

served in the dark zone. These cells are thought to be involved in the uptake of apoptotic B cells.

In murine spleens, the MZ (Kraal 1992) is a unique structure surrounding the marginal sinus where two distinct MØ populations are situated. The marginal zone metallophilic (MZM) MØ, which surround the white pulp but are more abundant around follicular areas, are strategically placed to be exposed to antigen and to control cell trafficking. These cells express large quantities of Sn, a member of the Siglec superfamily of proteins involved in cell adhesion. MZM-MØ migrate into the lymphoid area under some circumstances and could be involved in the transport of antigens to the follicles (Groeneveld et al. 1986; Yu et al. 2002). They express sulphated glycoforms of Sn and CD45 that act as counter-receptors for the cysteine-rich (CysR) domain of the MR (Martinez-Pomares et al. 1999). This interaction suggests that MR$^+$ cells or soluble MR (sMR) can be targeted to these cells. Interaction between membrane lymphotoxin expressed by B cells with its receptor on stromal cells is required for the induction of the sulphotransferases responsible for the generation of the sulphated glycoproteins that mediate CysR-domain binding (Martinez-Pomares et al. 1999; Yu et al. 2002). One day following LPS immunisation, MZMMØ are detected in the follicles. They maintain the ability to bind the CysR domain of the MR, and express MOMA-1 and Sn (Groeneveld et al. 1986; Yu et al. 2002). However, following immunisation with sheep red blood cells, the expression of the murine homologue of the human germinal centre DC marker decysin in MZMMØ was induced. Decysin$^+$ cells expressing ligands for the CysR domain were detected at 48 h in follicles (Mueller et al. 2001). These results suggest that these cells possess the flexibility to modulate their phenotype according to the type of antigen injected.

MZMMØ are of special interest in adjuvant and vaccine development for three main reasons. They are targeted by immunostimulating complexes or ISCOMs after intravenous immunisation (Claassen et al. 1995); they are required for TD responses to particulate bacterial antigens (Buiting et al. 1996); and they are major producers of IFN-α/β after viral infection (Eloranta and Alm 1999). In contrast to DC, these cells are macrophage colony-stimulating factor (M-CSF)-dependent, can be eliminated using liposomes containing clodronate (van Rooijen et al. 1989), and lack MHCII expression in situ (L. Martinez-Pomares, unpublished).

At the other side of the marginal sinus resides a population of large MØ (MZMØ) that express clearance receptors such as scavenger receptors class A (SR-A) (Hughes et al. 1994), macrophage receptor with collagenous domain (MARCO) (Elomaa et al. 1995) and an uncharacterised neutral polysaccharide receptor detected by the antibody ER-TR9 (van Vliet et al. 1985). Though it has been suggested that these cells could be involved in the generation of TI responses, depletion experiments using ER-TR9 (Kraal et al. 1989) seem to indicate that this is not the case. Indeed they might play a down-modulating role by reducing antigen load (Van den Eerwegh et al. 1992). A role in processing particulate antigens and transfer to MZ B cells cannot be ruled out. The induction of the initial TI, neutralising IgM response to vesicular stomatitis virus, poliomyelitis virus and recombinant vaccinia virus is dependent on efficient antigen trapping by CR3- and CR4-expressing MØ of the marginal zone (Ochsenbein and Zinkernagel 2000).

Material collected by lymphatics drains into lymph nodes (LN). The first cells encountered by this mixture of plasma components and cells (migratory DC among them) are the subcapsular sinus (ss) MØ. This population of M-CSF-dependent MØ express Sn and MOMA-1 and can be depleted using clodronate-loaded liposomes injected subcutaneously (van Rooijen and Sanders 1994). Their presence depends on a constant flow of lymph, since elimination of lymphatic vessels leads to their disappearance from the ss and migration into T-cell areas (Mebius et al. 1991). As in the case of MZMMØ, ssMØ express a sulphated glycoform of Sn that can be recognised by the CysR domain of the MR (Martinez-Pomares et al. 1996). After immunisation, a migratory cell population with the same phenotype but expressing CD11c and MHC-II can be found in the follicles. These cells are functionally DC and present antigen to naïve T and

B cells in vivo (Martinez-Pomares et al. 1996; Berney et al. 1999). Their origin is not clear; they could derive from the ssMØ or represent migratory cells moving from the periphery. Another major MØ populations present in LN are the medullary MØ that surround lymphatic and vascular endothelia at the medullary region. They seem to be involved in clearance and accordingly express similar markers to those found on MZMØ such as SR-A and MARCO (Elomaa et al. 1995) in addition to F4/80, MR and Sn (Linehan et al. 1999). MZMMØ and ssMØ could correspond to the antigen-transporting cells that transport immune complexes during the course of an immune response as previously described by Szakal et al. (1983). Cells expressing the MØ calcium-type lectin, found on dermal MØ but not Langerhans cells, could be transiently found in the T-cell areas of lymph nodes during the sensitisation phase of contact hypersensitivity and seemed to contribute to the efficiency of sensitisation even though they were negative for the antigen. This work opens the possibility of a modulator role for migratory MØ during the process of T-cell activation (Sato et al. 1998). Beside macrophages, immature and mature DC also reside in LN. They are able to internalise and process antigens. It has been proposed that a resident DC population in lymphoid organs could be implicated in tolerance induction (Steinman et al. 2000).

5
Adjuvants as Immunomodulators and Delivery Systems

Adjuvants are substances that enhance immune responses in general or specific to a particular antigen. Original adjuvant formulations consisting of breadcrumbs, agar, saponin, starch oil and lecithin were found to increase the levels of diphtheria or tetanus antitoxin (Ramon 1925). This empirical approach has identified a wide spectrum of substances as candidate adjuvants. These include bacterial and plant products, surface-active agents such as saponins, synthetic and biopolymers, vitamins, hormones, and aluminium and calcium salts. Their modes of action in the immune system are generally not known. To date, Alum is the only adjuvant approved for human vaccines (Edelman 2002). This is probably because it is extremely safe, rather than its effectiveness in enhancing all immune responses. Alum can only enhance antibody responses in some vaccines, and it is totally ineffective in stimulating cell-mediated immune responses. There is thus a great need for understanding the modes of action of experimental adjuvants or adjuvant formulations in order to improve current vaccines and to design novel ones. The aims for modern adjuvants or adjuvant formulations would be to boost the immunogenicity of subunit vaccines at mucosal sites or generate antigen-specific B and T cells by activating selectively Th-cell subsets (Th1 or Th2). Adjuvants could act in several potential ways. They could introduce or maintain antigen in the appropriate in vivo microenvironment. They could recruit and activate APC and lymphocytes by providing signals associated with infections, which would induce inflammation and increase Ag delivery to draining lymphoid tissues. Finally, they could deliver T-cell epitopes to

MHC class-I and -II molecules for CD8 CTL and CD4 Th cells, respectively [see O'Hagan et al. (2001) for review].

6 Endocytic Nonopsonic Receptors on Antigen-Presenting Cells as Molecular Targets for Antigen Delivery

Endocytic receptors on DC and MØ could be considered as targets for antigen delivery to MHC-II compartments and subsequent enhanced antigen presentation to T cells. In some instances, targeting to MHC-I compartment through the exogenous pathway can also take place. Since many of these receptors recognise endogenous molecules (Platt and Gordon 1998; Linehan et al. 2001; Lee et al. 2002), a balance between clearance and presentation could be established through their regulated differential expression on MØ and DC. Additionally, the requirement for inflammatory stimuli would ensure that peripheral tolerance is maintained.

6.1 The Mannose Receptor and the Scavenger Receptor Class A

In the mouse, MR and SR-A are absent from conventional DC populations in the periphery and lymphoid organs (Hughes et al. 1995; Linehan et al. 1999). This would be in accordance with their role as clearance systems for a large number of self-molecules (Table 1). Nonetheless, their presence in bone marrow-derived DC (both MR and SR-A, unpublished observations) and in cultured human DC (only shown for the MR) suggests that they might be up-regulated during DC differentiation or recruitment in vivo. Indeed, an uncharacterised population of MHC-II$^+$, MR$^+$ cells has been described in LN (Linehan et al. 1999), and MR expression has also been observed in situ in human inflamed skin (Wollenberg et al. 2002). SR-A is highly expressed by MZMØ (Hughes et al. 1995). In both cases enhanced presentation has been observed after mannosylation (Sallusto 1995; Agnes et al. 1997) or maleylation (Bansal et al. 1999) of proteins to generate MR and SR ligands respectively. MR can mediate the uptake and targeting of glycolipids for presentation by CD1b in human monocyte-derived DC (Prigozy et al. 1997). These properties were exploited in the induction of protective anti-tumoral responses by immunisation with the tumour-associated antigen MUC-1 conjugated to mannan, a ligand for the MR (Apostolopoulos et al. 1995, 2000). However, MR-mediated enhanced presentation needs to be reassessed in view of the results of several recent studies: (1) LPS-induced IL-12 production in cultured human DC can be inhibited by mannosylated lipoarabinomannans (Nigou et al. 2001); (2) The maturation state of DC is not affected by the involvement of MR in the uptake of secretory IgA (Heystek et al. 2002); and (3) MR can block the intracellular sorting and presentation of MUC-1 (Hiltbold et al. 2000). Therefore, in the absence of additional stimuli, MR ligation might not be able to enhance immunogenicity of antigens or activate DC.

The presence of MR CysR domain ligands in selected myeloid populations in secondary lymphoid organs (Martinez-Pomares et al. 1996; Berney et al. 1999; Martinez-Pomares et al. 1999; Yu et al. 2002), highlights the dual role of the MR in homeostasis and immunity. A soluble form of the MR (Martinez-Pomares et al. 1998) has been suggested as the counter-receptor for these molecules.

6.2 DEC-205 and DC-SIGN

DEC-205 (Table 1) is an endocytic receptor mostly restricted to DC populations and thymic epithelium. Even though its ligands have not been identified, using polyclonal and monoclonal antibodies it has been shown that its cytoplasmic tail is highly efficient in targeting to late endosomes and to the MHC-II compartment (Jiang et al. 1995; Mahnke et al. 2000). Recently a chimeric anti-DEC-205 monoclonal antibody bearing a hen egg lysozyme T cell epitope was shown to stimulate antigen-specific T cells in vivo. The outcome of this stimulation is unresponsiveness if this reagent was injected in the absence of adjuvant or activation and if an anti-CD40 monoclonal antibody was used (Hawiger et al. 2001). Nevertheless, these results should be carefully assessed since DEC-205 can be expressed by MØ under inflammatory conditions (Wijffels et al. 1991).

Table 1 Selected antigen-presenting cell receptors relevantin vaccine design

Receptor	Ligand	Structure	Comments
Mannose receptor	SO_4-3-Gal-SO_4-(3/4)-GalNAc-mannose, fucose *N*-acetyl-glucosamine-	CysR domain-FNII-CRD(x8)-TM-CT (endocytosis)	
Scavenger receptor Class A	Selected polyanionic compounds	Collagenous domain Type1 only SRCR domain	
DEC-205	Unknown	CysR domain-FNII-CRD (x10)-TM-CT (endocytosis)	
DC-SIGN	Mannose *N*-acetyl-glucosamine	C-type lectin-(internalisation motif)	
TLR-2	Peptidoglycan Lipoprotein Lipoarabinomannan Selected lipopolysaccharides	LRR Cytoplasmic domain related to IL-1R (TIR domain)	Signals as heterodimer in combination with TLR1 and TLR6
TLR-3	dsRNA	LRR and TIR	
TLR-4	Lipopolysaccharide Lipoteichoic acids	LRR and TIR	Requires MD2 and CD14
TLR-5	Flagellin	LRR and TIR	
TLR-7/TLR-8	Small anti-viral compounds	LRR and TIR	
TLR-9	Unmethylated CpG	LRR and TIR	
β-Glucan R	β-Glucan	C-type lectin-TM-CT (ITAM)	β-Glucan-independent binding to T cells

Another DC restricted receptor, DC-SIGN, a C-type lectin involved in T cell adhesion (through its interaction with ICAM-3), HIV-1-binding and transendothelial migration (through its interaction with ICAM-2) has been implicated in antigen internalisation and presentation to T cells (Engering et al. 2002). The requirement for antigen targeting to an endocytic receptor on DC might not be absolute since some studies indicate that targeting to other surface molecules on DC is enough to enhance presentation (Serre et al. 1998).

7 Signalling Receptors on Antigen-Presenting Cells as Sensors of Foreignness

Toll-like receptors (TLR) were thought to be genuine detectors of pathogen-associated molecular patterns (PAMP) (Medzhitov and Janeway 2000). This assessment is being challenged by the discovery of TLR-mediated signalling in responses to endogenous ligands that could be released from damaged cells (Leadbetter et al. 2002; Vabulas et al. 2002). In humans there are 10 members of this family and PAMP on specific microbes have been identified for most of these TLR (Table 1) (Akira et al. 2001; Medzhitov 2001; Hemmi et al. 2002; Kaisho and Akira 2002). Although direct binding of PAMP has not yet been demonstrated for any of the TLR, their signalling capability in response to PAMP is unequivocal. The homology of their intracellular region with the cytoplasmic tail of the IL-1R suggested that these molecules could signal through the same pathway, which involves an adaptor protein, My88D. Only TLR-4 has been shown to induce some responses through a MyD88-independent pathway mediated by another adaptor protein referred to as TIRAP or MAL (Horng et al. 2001). MyD88-deficient mice have a bias towards a Th2 response with enhanced basal levels of IgE and normal responses to ovalbumin (OVA) co-injected with Alum (a Th2-inducer) but not with complete Freund's adjuvant (CFA) (Th1-inducer). These results indicate that TLR signalling could favour the induction of a Th1 response (Schnare et al. 2001). TLR are differentially expressed in different DC populations, suggesting that there is specialisation of APC to respond to specific microbes (Kadowaki et al. 2001).

Finally dectin-1, widely expressed by myeloid cells including DC and MØ, has recently been shown to be a major β-glucan receptor on MØ (Brown and Gordon 2001; Brown et al. 2002). Since β-glucans from fungi and yeasts have a variety of immunostimulatory effects, this receptor could potentially mediate most if not all of these activities. It contains an immunoreceptor tyrosine-based activation motif (ITAM) in its cytoplasmic tail and ligand recognition is expected to activate signalling pathways of DC and MØ. In addition a role for dectin-1 as an enhancer of innate immune responses, its phagocytic activity could also lead to enhanced antigen uptake and presentation.

8 Current Challenges of Vaccine Development

Mucosal delivery of antigens is one of the most desirable routes of immunisation because it is easier to administer, has reduced toxicity and provides immunity at the sites where many pathogens establish infection. However, oral immunisation using vaccines containing pure antigens or killed/inactivated organisms has not been successful in generating protective immunity. The difficulties have been the instability of antigens in the harsh acidic and proteolytic environment of the stomach and the limited access of antigens and APC stimuli to the mucin-coated epithelial cell layer and the underlying MALT. Some promising delivery systems and potent adjuvants for oral and intranasal immunisation include genetically modified non-toxic bacterial toxins [see (O'Hagan et al. 2001) and references therein], and microparticles with entrapped protein and DNA-encoded antigens (O'Hagan et al. 2001). The mechanisms for their effectiveness have not been defined, but it is most likely due to the fact that they can enhance antigen delivery to DC and MØ through pinocytosis and/or via their interactions with specific endocytic/phagocytic receptors, and activate these APC fully by the presence of PAMP. How these signals are integrated to achieve protective immunity is being investigated at the molecular and cellular levels.

Whilst almost all vaccines are targeted to elicit adaptive immune responses toward protein antigens, there is also a need to examine non-protein components as vaccine and adjuvant candidates. Carbohydrate antigens have an important role in vaccine development because immune responses against them could prevent successful blood transfusion, organ transplantation, tumour formation and microbial infections. In addition, some microbial carbohydrates, including alginate from *Pseudomonas aeruginosa*, capsular polysaccharides from *Klebsiella pneumoniae* and *Streptococcus pneumoniae* and β-glucans from yeasts and other fungi, have immunomodulatory effects on the immune system, and as such could also be used as adjuvants to improve the immunogenicity of vaccines against infectious diseases and cancer. Their modes of action have not been clarified, but induction of macrophage functions and cytokine production (e.g. TNF-α, IL-1) would likely be their effects (Yokochi et al. 1980; Otterlei et al. 1991; Otterlei et al. 1993; Choy et al. 1996; Ho et al. 2000; Tokunaka et al. 2000; Um et al. 2000; Suzuki et al. 2001).

On the surface of most human bacterial pathogens, carbohydrate antigens are in the form of capsular polysaccharides (CPSs), lipopolysaccharides/lipooligosaccharides, teichoic acids, and lipoteichoic acid. Antibody responses to CPSs are the basis for protection against most infections due to extracellular bacteria; but the efficacy of CPS-based vaccines varies with the antibody isotype and its complement-fixing ability. Antibody responses to CPS vary with age: adults and children over 2 years are generally responsive if CPS is immunogenic, but children under 18 months (who are the most vulnerable to bacterial infections) do not respond to CPS. To achieve T-cell help in order for B cells to produce antibodies against carbohydrates, investigators have coupled CPS chemically to protein carriers in an attempt improve their immunogenicity in infants. CPS-protein conjugate vaccines for type b *Haemophilus influenzae* and serogroup C *Neisseria meningitidis* have been extremely effective in reducing disease incidence to record low levels in countries where mass vaccination programs have been implemented.

Intranasal immunisation of a pneumococcal glycoconjugate with mucosal adjuvants such as *Escherichia coli* heat-labile enterotoxin mutants has recently been shown to induce a significant increase in polysaccharide-specific and protein carrier-specific antibody responses in all IgG subclasses (Jakobsen et al. 2001). Phenotypic characterisation and determining the functions of the DCs and MØs in the nasal-associated lymphoid tissues (NALT) should enable the design of effective adjuvants and vaccines for this practical route of immunisation against respiratory pathogens. Although the successes of conjugate vaccines have encouraged further development of similar vaccines against other bacterial pathogens, the underlying mechanisms of immune responses against carbohydrate antigens remain poorly defined. MØs and DCs have been shown to contribute to antibody responses to TI antigens by responding to PAMPs and activating TLR-mediated pathways of cytokine induction. The secreted cytokines, including IL-1, IL-6, IL-12 and GM-CSF, provide the necessary second signals to B cells that are activated by multivalent mIg cross-linking polysaccharides (Vos et al. 2000). It is, however, not clear how carbohydrate antigens are transported to secondary lymphoid tissues where antigen-specific B cells are located. Further investigations on DCs and MØs would also be required to understand the immunomodulatory properties of some polysaccharides. The challenge is to understand how these types of antigens interact with the innate immune system, and how recognition and potential signalling by receptors on DCs and MØs affect immunogenicity or adjuvanticity. Insights into these mechanisms would enable the development of vaccines designed to modulate immune responses against carbohydrate antigens or other similar TI antigens in normal and immunocompromised individuals. It may also be possible to identify novel carbohydrate-based adjuvants that can achieve a fine balance between initiation of immune responses without a potentially damaging inflammatory process, because almost all CPS-based vaccines are well-tolerated in humans.

In summary, DCs and MØs play critical roles in determining antigen tolerance and immunogenicity and the generation of appropriate adaptive immune responses. Future vaccine formulations would thus not only include selected B- and T-cell epitopes, but also delivery systems and adjuvant components that specially modulate the migration, antigen presentation and cell activation of specific DC and MØ types at targeted immune sites. The main challenges in vaccine development are to identify adjuvants to generate cell-mediated immune responses for tumours and chronic viral infections, and to enhance mucosal and systemic immune responses via the oral and intranasal routes. Detailed analyses of the immunobiology of DCs and MØs would contribute greatly to the designing of vaccines to meet these challenges.

9
References

Agnes MC, Tan A, Mommaas AM, Drijfhout JW, Jordens R, Onderwater JJM, Verwoerd D, Mulder AA, van der Heiden A, Scheidegger D, et al. (1997) Mannose receptor-mediated uptake of antigens strongly enhances HLA class II-restricted antigen presentation by cultured dendritic cells. Eur J Immunol 27:2426–2435

Akira S, Takeda K, Kaisho T (2001) Toll-like receptors: critical proteins linking innate and acquired immunity. Nat Immunol 2:675–680

Aliberti J, Hieny S, Reis e Sousa C, Serhan CN, Sher A (2002) Lipoxin-mediated inhibition of IL-12 production by DCs: a mechanism for regulation of microbial immunity. Nat Immunol 3:76–82

Allison ME, Fearon DT (2000) Enhanced immunogenicity of aldehyde-bearing antigens: a possible link between innate and adaptive immunity. Eur J Immunol 30:2881–2887

Amlot PL, Grennan D, Humphrey JH (1985) Splenic dependence of the antibody response to thymus-independent (TI-2) antigens. Eur J Immunol 15:508—512

Apostolopoulos V, Pietersz GA, Gordon S, Martinez-Pomares L, McKenzie IF (2000) Aldehyde-mannan antigen complexes target the MHC class I antigen-presentation pathway. Eur J Immunol 30:1714–1723

Apostolopoulos V, Pietersz GA, Loveland BE, Sandrin MS (1995) Oxidative/reductive conjugation of mannan to antigens selects for T1 or T2 immune responses. Proc Natl Acad Sci 92:10128–10132

Bansal P, Mukherjee P, Basu SK, George A, Bal V, Rath S (1999) MHC class I-restricted presentation of maleylated protein binding to scavenger receptors. J Immunol 162:4430-4437

Batista FD, Iber D, Neuberger MS (2001) B cells acquire antigen from target cells after synapse formation. Nature 411:489–494

Berney C, Herren S, Power CA, Gordon S, Martinez-Pomares L, Kosco-Vilbois MH (1999) A member of the dendritic cell family that enters B cell follicles and stimulates primary antibody responses identified by a mannose receptor fusion protein. J Exp Med 190:851-860

Brown GD, Gordon S (2001) Immune recognition. A new receptor for beta-glucans. Nature 413:36–37

Brown GD, Taylor PR, Reid DM, Willment JA, Williams DL, Wong SYC, Martinez-Pomares L, Gordon S (2002) Dectin-1, is the principal b-glucan receptor on macrophages. J Exp Med (in press)

Buiting AM, De Rover Z, Kraal G, Van Rooijen N (1996) Humoral immune responses against particulate bacterial antigens are dependent on marginal metallophilic macrophages in the spleen. Scand J Immunol 43:398–405

Carroll MC (1998) The role of complement and complement receptors in induction and regulation of immunity. Annu Rev Immunol 16:545–568

Choy YM, Tsang SF, Kong SK, Leung KN, Parolis H, Lee CY, Fung KP (1996) K1 and K3 capsular antigens of Klebsiella induce tumor necrosis factor activities. Life Sci 58:153–158

Claassen IJ, Osterhaus AD, Claassen E (1995) Antigen detection in vivo after immunization with different presentation forms of rabies virus antigen: involvement of marginal metallophilic macrophages in the uptake of immune-stimulating complexes. Eur J Immunol 25:1446–1452

Cody CL, Baraff LJ, Cherry JD, Marcy SM, Manclark CR (1981) Nature and rates of adverse reactions associated with DTP and DT immunisations in infants and children. Pediatrics 68:650–660

D'Amico G, Frascaroli G, Bianchi G, Transidico P, Doni A, Vecchi A, Sozzani S, Allavena P, Mantovani A (2000) Uncoupling of inflammatory chemokine receptors by IL-10: generation of functional decoys. Nat Immunol 1:387–391

Dietrich J, Nakajima H, Colonna M (2000) Human inhibitory and activating Ig-like receptors which modulate the function of myeloid cells. Microbes Infect 2:323–329

Dupuis M, Denis-Mize K, LaBarbara A, Peters W, Charo IF, McDonald DM, Ott G (2001) Immunization with the adjuvant MF59 induces macrophage trafficking and apoptosis. Eur J Immunol 31:2910–2918

Edelman R (2002) The development and use of vaccine adjuvants. Mol Biotechnol 21:129–148

Elomaa O, Kangas M, Sahlberg C, Tuukkanen J, Sormunen R, Liakka A, Thesleff I, Kraal G, Tryggvason K (1995) Cloning of a novel bacteria-binding receptor structurally related to scavenger receptors and expressed in a subset of macrophages. Cell 80:603–609

Eloranta ML, Alm GV (1999) Splenic marginal metallophilic macrophages and marginal zone macrophages are the major interferon-alpha/beta producers in mice upon intravenous challenge with herpes simplex virus. Scand J Immunol 49:391–394

Engering A, Geijtenbeek TBH, van Vliet SJ, Wijers M, van Liempt E, Demaurex N, Lanzavecchia A, Fransen J, Figdor CG, Piguet V, van Kooyk Y (2002) The dendritic cell-specific adhesion receptor DC-SIGN internalizes antigen for presentation to T cells. J Immunol 168:2118–2126

Ernst JD (1998) Macrophage receptors for Mycobacterium tuberculosis. Infect Immun 66:1277–1281

Fadok VA., Chimini G (2001) The phagocytosis of apoptotic cells. Semin Immunol 13:365–372

Fehr T, Skrastina D, Pumpens P, Zinkernagel RM (1998) T cell-independent type I antibody response against B cell epitopes expressed repetitively on recombinant virus particles. Proc Natl Acad Sci U S A 95:9477–9481

Fischer MB, Ma M, Hsu NC, Carroll MC (1998) Local synthesis of C3 within the splenic lymphoid compartment can reconstitute the impaired immune response in C3-deficient mice. J Immunol 160:2619–2625

Gerber JS, Mosser DM (2001) Reversing lipopolysaccharide toxicity by ligating the macrophage Fc gamma receptors. J Immunol 166:6861–6868

Goerdt S, Politz O, Schledzewski K, Birk R, Gratchev A, Guillot P, Hakiy N, Klemke CD, Dippel E, Kodelja V, Orfanos CE (1999) Alternative versus classical activation of macrophages. Pathobiology 67:222–226

Gordon S (1999) Macrophage-restricted molecules: role in differentiation and activation. Immunol Lett 65:5–8

Groeneveld PH, Erich T, Kraal G (1986) The differential effects of bacterial lipopolysaccharide (LPS) on splenic non-lymphoid cells demonstrated by monoclonal antibodies. Immunology 58:285–290

Hawiger D, Inaba K, Dorsett Y, Guo M, Mahnke K, Rivera M, Ravetch JV, Steinman R M., Nussenzweig MC (2001) Dendritic cells induce peripheral T cell unresponsiveness under steady state conditions in vivo. J Exp Med 194:769–779

Hemmi H, Kaisho T, Takeuchi O, Sato S, Sanjo H, Hoshino K, Horiuchi T, Tomizawa H, Takeda K, Akira S (2002) Small anti-viral compounds activate immune cells via the TLR7 MyD88-dependent signaling pathway. Nat Immunol 3:196–200

Heystek HC, Moulon C, Woltman AM, Garonne P, van Kooten C (2002) Human immature dendritic cells efficiently bind and take up secretory IgA without the induction of maturation. J Immunol 168:102–107

Hiltbold EM, Vlad AM, Ciborowski P, Watkins SC, Finn OJ (2000) The mechanism of unresponsiveness to circulating tumor antigen MUC1 is a block in intracellular sorting and processing by dendritic cells. J Immunol 165:3730–3741

Ho CY, Lo TWC, Leung KN, Fung KP, Choy YM (2000) The immunostimulating activities of anti-tumor polysaccharide from K1 capsular (polysaccharide) antigen isolated from Klebsiella pneumoniae. Immunopharm 46:1–13

Hoek RM, Ruuls SR, Murphy CA, Wright GJ, Goddard R, Zurawski SM, Blom B, Homola ME, Streit WJ, Brown MH, et al. (2000) Down-regulation of the macrophage lineage through interaction with OX2 (CD200) Science 290:1768–1771

Horng T, Barton GM, Medzhitov R (2001) TIRAP: an adapter molecule in the Toll signalling pathway. Nat Immunol 2:835–841

Huang FP, Platt N, Wykes M, Major JR, Powell TJ, Jenkins CD, MacPherson GG (2000) A discrete subpopulation of dendritic cells transports apoptotic intestinal epithelial cells to T cell areas of mesenteric lymph nodes. J Exp Med 191:435–444

Hughes DA, Fraser IP, Gordon S (1994) Murine M phi scavenger receptor: adhesion function and expression. Immunol Lett 43:7–14

Hughes DA, Fraser IP, Gordon S (1995) Murine macrophage scavenger receptor: in vivo expression and function as receptor for macrophage adhesion in lymphoid and nonlymphoid organs. Eur J Immunol 25:466–473

Jakobsen H, Adarna BC, Schulz D, Rappuoli R, Jonsdottir I (2001) Characterization of the antibody response to pneumococcal glycoconjugates and the effect of heat-labile enterotoxin on IgG subclasses after intranasal immunization. J Infect Dis 183:1494–1500

Jiang W, Swiggard WJ, Heufler C, Peng M, Mirza A, Steinman RM, Nussenzweig MC (1995) The receptor DEC-205 expressed by dendritic cells and thymic epithelial cells is involved in antigen processing. Nature 375:151–155

Kadowaki N, Ho S, Antonenko S, de Vaal Malefyt R, Kastelein RA, Bazan F, Liu Y-L (2001) Subsets of human dendritic cell precursors express different Toll-like receptors and respond to different microbial antigens,. J Exp Med 194:863–869

Kaisho T, Akira S (2002) Toll-like receptors as adjuvant receptors. Biochim Biophys Acta 1589:1–13

Kapikian AZ, Mirchell RH, Chanock RM, Shvedoff RA., Stewart CE (1969) An epidemiologic study of altered clinical reactivity to respiratory syncytial (RS) virus infection in children previously vaccinated with an inactivated RS virus vaccine. Am J Epidemiol 89:405–421

Kraal G (1992) Cells in the marginal zone of the spleen. Int Rev Cytol 132:31–74

Kraal G, Ter Hart H, Meelhuizen C, Venneker G, Claassen E (1989) Marginal zone macrophages and their role in the immune response against T-independent type 2 antigens: modulation of the cells with specific antibody. Eur J Immunol 19:675–680

Lanzavecchia A., Sallusto F (2001) Regulation of T cell immunity by dendritic cells. Cell 106:263–266

Leadbetter EA, Rifkin IR, Hohlbaum AM, Beaudette BC, Shlomchik MJ, Marshak-Rothstein A (2002) Chromatin-IgG complexes activate B cells by dual engagement of IgM and Toll-like receptors. Nature 416:603–607

Lee SJ, Evers S, Roeder D, Parlow AF, Risteli J, Risteli L, Lee YC, Feizi T, Langen H, Nussenzweig MC (2002) Mannose receptor-mediated regulation of serum glycoprotein homeostasis. Science 295:1898–1901

Levine MM, Woodrow GC, Kaper JB, Cobon GS (eds) (1997) New Generation Vaccines, Second Edition edn (New York, USA, Marcel Dekker, Inc.)

Linehan SA, Martinez-Pomares L, da Silva RP, Gordon S (2001) Endogenous ligands of carbohydrate recognition domains of the mannose receptor in murine macrophages, endothelial cells and secretory cells; potential relevance to inflammation and immunity. Eur J Immunol 31:1857–1866

Linehan SA, Martinez-Pomares L, Stahl PD, Gordon S (1999) Mannose receptor and its putative ligands in normal murine lymphoid and nonlymphoid organs: In situ expression of mannose receptor by selected macrophages, endothelial cells, perivascular microglia, and mesangial cells, but not dendritic cells. J Exp Med 189:1961–1972

Liu MA (1997) The immunologist's grail: vaccines that generate cellular immunity. Proc Natl Acad Sci 94:10496–10498
Lynch M, Bresee JS, Gentsch JR, Glass RI (2000) Rotavirus vaccines. Curr Opin Infect Dis 13:495–502
Mahnke K, Guo M, Lee S, Sepulveda H, Swain SL, Nussenzweig MC, Steinman RM (2000) The dendritic cell receptor for endocytosis, DEC-205, can recycle and enhance antigen presentation via major histocompatibility complex class II-positive lysosomal compartments. J Cell Biol 151:673–683
Martinez-Pomares L, Crocker PR, Da Silva R, Holmes N, Colominas C, Rudd P, Dwek R, Gordon S (1999) Cell-specific glycoforms of sialoadhesin and CD45 are counter-receptors for the cysteine-rich domain of the mannose receptor. J Biol Chem 274:35211–35218
Martinez-Pomares L, Kosco-Vilbois M, Darley E, Tree P, Herren S, Bonnefoy JY, Gordon S (1996) Fc chimeric protein containing the cysteine-rich domain of the murine mannose receptor binds to macrophages from splenic marginal zone and lymph node subcapsular sinus and to germinal centers. J Exp Med 184:1927–1937
Martinez-Pomares L, Mahoney JA, Kaposzta R, Linehan SA, Stahl PD, Gordon S (1998) A functional soluble form of the murine mannose receptor is produced by macrophages in vitro and is present in mouse serum. J Biol Chem 273:23376–23380
Mebius RE, Streeter PR, Breve J, Duijvestijn AM, Kraal G (1991) The influence of afferent lymphatic vessel interruption on vascular addressin expression. J Cell Biol 115:85–95
Medzhitov R (2001) Toll-like receptors and innate immunity. Nature Rev Immunol 1:135–145
Medzhitov R, Janeway C Jr. (2000) Innate immune recognition: mechanisms and pathways. Immunol Rev 173:89–97
Medzhitov R, Janeway CA Jr (2002) Decoding the patterns of self and nonself by the innate immune system. Science 296:298–300
Moingeon P, Haensler J, Lindberg A (2001) Towards the rational design of Th1 adjuvants. Vaccine 19:4363–4372
Mueller CG, Cremer I, Paulet PE, Niida S, Maeda N, Lebeque S, Fridman WH, Sautes-Fridman C (2001) Mannose receptor ligand-positive cells express the metalloprotease decysin in the B cell follicle. J Immunol 167:5052–5060
Nigou J, Zelle-Rieser C, Gilleron M, Thurnher M, Puzo G (2001) Mannosylated lipoarabinomannans inhibit IL-12 production by human dendritic cells: evidence for a negative signal delivered through the mannose receptor. J Immunol 166:7477–7485
O'Hagan DT, MacKichan ML, Singh M (2001) Recent developments in adjuvants for vaccines against infectious diseases. Biomol Eng 18:69–85
Ochsenbein AF, Pinschewer DD, Odermatt B, Ciurea A, Hengartner H, Zinkernagel RM (2000) Correlation of T cell independence of antibody responses with antigen dose reaching secondary lymphoid organs: implications for splenectomized patients and vaccine design. J Immunol 164:6296–6302
Ochsenbein AF, Zinkernagel RM (2000) Natural antibodies and complement link innate and acquired immunity. Immunol Today 21:624–630
Otterlei M, Ostgaard K, Skak-Braek G, Smidsrod O, Soon-Shiong P, Espevik T (1991) Induction of cytokine production from human monocytes stimulated with alginate. J Immunother 10:286–291
Otterlei M, Sundan A, Skjak-Brae KG, Ryan L, Smidsrod O, Espevik T (1993) Similar mechanisms of action of defined polysaccharides and lipopolysaccharides: characterization of binding and tumor necrosis factor alpha induction. Infect Immun 61:1917–1925
Parnaik R, Raff MC, Scholes J (2000) Differences between the clearance of apoptotic cells by professional and non-professional phagocytes. Curr Biol 10:857–860
Platt N, Gordon S (1998) Scavenger receptors: diverse activities and promiscuous binding of polyanionic ligands. Chemistry and Biology 5:R193–R203

Prigozy TI, Sieling PA, Clemens D, Stewart PL, Behar SM, Porcelli SA, Brenner MB, Modlin RL, Kronenberg M (1997) The mannose receptor delivers lipoglycan antigens to endosomes for presentation to T cells by CD1b molecules. Immunity 6:187–197.

Pulendran B, Kumar P, Cutler CW, Mohamadzadeh M, Van Dyke T, Banchereau J (2001) Lipopolysaccharides from distinct pathogens induce different classes of immune responses in vivo. J Immunol 167:5067–5076

Ramachandra L, Chu RS, Askew D, Noss EH, Canaday DH, Potter NS, Johnsen A, Krieg AM, Nedrud JG, Boom WH, Harding CV (1999) Phagocytic antigen processing and effects of microbial products on antigen processing and T-cell responses. Immunol Rev 168:217–239.

Ramon G (1925) Sur l'augmentation anormale de l'antitoxine chez les chevaux producteurs de serum antidipherique. Bull Soc Centr Med Vet 101:227–234

Randolph GJ, Beaulieu S, Lebecque S, Steinman RM, Muller WA (1998) Differentiation of monocytes into dendritic cells in a model of transendothelial trafficking. Science 282:480–483

Randolph GJ, Inaba K, Robbiani DF, Steinman RM, Muller WA (1999) Differentiation of phagocytic monocytes into lymph node dendritic cells in vivo. Immunity 11:753–761

Sakamoto N, Shibuya K, Shimizu Y, Yotsumoto K, Miyabayashi T, Sakano S, Tsuji T, Nakayama E, Nakauchi H, Shibuya A (2001) A novel Fc receptor for IgA and IgM is expressed on both hematopoietic and non-hematopoietic tissues. Eur J Immunol 31:1310–1316

Sallusto F, Cella M, Danielli C, Lanzavecchia A (1995) Dendritic cells use macropinocytosis and the mannose receptor to concentrate macromolecules in the major histocompatibility complex class II compartment: Down regulation by cytokines and bacterial products. J Exp Med 182:389–400

Sato K, Imai Y, Irimura T (1998) Contribution of dermal macrophage trafficking in the sensitization phase of contact hypersensitivity. J Immunol 161:6835–6844

Schnare M, Barton GM, Holt AC, Takeda K, Akira S, Medzhitov R (2001) Toll-like receptors control activation of adaptive immune responses. Nat Immunol 2:947–950

Serre K, Machy P, Grivel JC, Jolly G, Brun N, Barbet J, Leserman L (1998) Efficient presentation of multivalent antigens targeted to various cell surface molecules of dendritic cells and surface Ig of antigen-specific B cells. J Immunol 161:6059–6067

Shi Y, Rock KL (2002) Cell death releases endogenous adjuvants that selectively enhance immune surveillance of particulate antigens. Eur J Immunol 32:155–162

Shimizu Y, Honda S, Yotsumoto K, Tahara-Hanaoka S, Eyre HJ, Sutherland GR, Endo Y, Shibuya K, Koyama A, Nakauchi H, Shibuya A. (2001) Fc(alpha)/mu receptor is a single gene-family member closely related to polymeric immunoglobulin receptor encoded on Chromosome 1. Immunogenetics 53:709–711

Stein M, Keshav S, Harris N, Gordon S (1992) Interleukin 4 potently enhances murine macrophage mannose receptor activity: a marker of alternative immunologic macrophage activation. J Exp Med 176:287–292

Steiniger B, Barth P, Herbst B, Hartnell A, Crocker PR (1997) The species-specific structure of microanatomical compartments in the human spleen: strongly sialoadhesin-positive macrophages occur in the perifollicular zone, but not in the marginal zone. Immunology 92:307–316

Steinman RM, Turley S, Mellman I, Inaba K (2000) The induction of tolerance by dendritic cells that have captured apoptotic cells. J Exp Med 191:411–416

Sutterwala FS, Noel GJ, Salgame P, Mosser DM (1998) Reversal of proinflammatory responses by ligating the macrophage Fcgamma receptor type I. J Exp Med 188:217-222

Suzuki Y, Adachi Y, Ohno N, Yadomae T (2001) Th1/Th2-balancing immunomodulating activity of gel-forming (1-3)-beta-glucans from fungi. Biol Pharm Bull 24:811–819

Szakal AK, Holmes KL, Tew JG (1983) Transport of immune complexes from the subcapsular sinus to lymph node follicles on the surface of nonphagocytic cells, including cells with dendritic morphology. J Immunol 131:1714–1727

Tew JG, Wu J, Fakher M, Szakal AK, Qin D (2001) Follicular dendritic cells: beyond the necessity of T-cell help. Trends Immunol 22:361–367

Tokunaka K, Ohno N, Adachi Y, Tanaka S, Tamura H, Yadomae T (2000) Immunopharmacological and immunotoxicological activities of a water-soluble (1-3)-beta-D-glucan, CSBG from Candida spp. Int J Immunopharmacol 22:383–394

Tzachanis D, Berezovskaya A, Nadler LM, Boussiotis VA (2002) Blockade of B7/CD28 in mixed lymphocyte reaction cultures results in the generation of alternatively activated macrophages, which suppress T-cell responses. Blood 99:1465–1473

Um S, Son E, Kim B, Moon E, Rhee D, Pyo S (2000) Activation of murine peritoneal macrophages by streptococcus pneumoniae type II capsular polysaccharide: involvement of CD14-dependent pathway. Scand J Immunol 52:39–45

Vabulas RM, Ahmad-Nejad P, Ghose S, Kirschning CJ, Issels RD, Wagner H (2002) HSP70 as endogenous stimulus of the Toll/interleukin-1 receptor signal pathway. J Biol Chem 277:15107–15112

Van den Eerwegh AJM, Laman JD, Schellekens MM, Boersma WJA, Claassen E (1992) Complement-mediated follicular localisation of T-independent type-2 antigens: the role of marginal zone macrophages revisited. Eur J Immunol 22:719–726

van Rooijen N, Kors N, Kraal G (1989) Macrophage subset repopulation in the spleen: differential kinetics after liposome-mediated elimination. J Leukoc Biol 45:97–104

van Rooijen N, Sanders A (1994) Liposome mediated depletion of macrophages: mechanism of action, preparation of liposomes and applications. J Immunol Methods 174:83–93

van Vliet E, Melis M, van Ewijk W (1985) Marginal zone macrophages in the mouse spleen identified by a monoclonal antibody. Anatomical correlation with a B cell subpopulation. J Histochem Cytochem 33:40–44

Vos Q, Lees A, Wu Z-Q, Snapper CM, Mond JJ (2000) B-cell activation by T-cell independent type 2 antigens as an integral part of the humoral immune response to pathogenic microorganisms. Immunological Reviews 176:154–170

Wijffels JF, De Rover Z, van Rooijen N, Kraal G, Beelen RH (1991) Chronic inflammation induces the expression of dendritic cell markers not related to functional antigen presentation on peritoneal exudate macrophages. Immunobiology 184:83–92

Wollenberg A, Mommaas M, Oppel T, Schottdorf EM, Gunther S, Moderer M (2002) Expression and function of the mannose receptor CD206 on epidermal dendritic cells in inflammatory skin diseases. J Invest Dermatol 118:327–334

Wright GJ, Puklavec MJ, Willis AC, Hoek RM, Sedgwick JD, Brown MH, Barclay AN (2000) Lymphoid/neuronal cell surface OX2 glycoprotein recognizes a novel receptor on macrophages implicated in the control of their function. Immunity 13:233–242

Wykes M, Pombo A, Jenkins C, MacPherson GG (1998) Dendritic cells interact directly with naive B lymphocytes to transfer antigen and initiate class switching in a primary T-dependent response. J Immunol 161:1313–1319

Yokochi T, Nakashima I, Kato N (1980) Adjuvant action of capsular polysaccharide of Klebsiella pneumoniae on antibody response. Microbiol Immunol 24:141–154

Yu P, Wang Y, Chin RK, Martinez-Pomares L, Gordon S, Kosco-Vibois MH, Cyster J, Fu YX (2002) B cells control the migration of a subset of dendritic cells into B cell follicles via CXC chemokine ligand 13 in a lymphotoxin-dependent fashion. J Immunol 168:5117–5123

Zandvoort A, Lodewijk ME, de Boer NK, Dammers PM, Kroese FG, Timens W (2001) CD27 expression in the human splenic marginal zone: the infant marginal zone is populated by naive B cells. Tissue Antigens 58:234–242

Zinkernagel RM (2000) Localization dose and time of antigens determine immune reactivity. Semin Immunol 12:163–171; discussion 257—344

Macrophage-Specific Gene Targeting In Vivo

D. R. Greaves · S. Gordon

Sir William Dunn School of Pathology,
South Parks Road, Oxford, OX1 3RE, UK
e-mail: david.greaves@path.ox.ac.uk

Abstract The study of macrophage biology has been hindered by our inability to efficiently direct transgene expression to cells of the mononuclear phagocyte lineage in vivo. Recent progress in understanding the transcriptional regulation of several genes expressed predominantly in macrophages has led to improved vectors for macrophage gene targeting. Vectors that can reproducibly direct high-level, macrophage-specific transgene expression may find application in gene therapy protocols that seek to correct genetic defects of metabolism such as lysosomal storage disorders. Efficient targeting of gene expression to macrophages and dendritic cells may allow us to modulate immune responses and to develop more efficient genetic vaccines.

Keywords Phagocytes, Transgenes, Vaccines

1 Background

1.1 Introduction to the Cells of the Mononuclear Phagocyte Lineage

All blood cells develop from a series of haematopoietic progenitor cells, which ultimately originate from haematopoietic stem cells. Blood monocytes develop from monoblasts within the adult bone marrow and are released into the circulation. Monocytes are recruited via constitutive and inflammatory mechanisms to virtually all the tissues of the body where they terminally differentiate into the various cell types of the mononuclear phagocyte lineage. Mononuclear phagocytes in adult tissues include Kupffer cells in the liver, microglia in the brain, Langerhans cells in the skin and mucosa and tissue-resident macrophages in tissues such as the lamina propria detectable with specific monoclonal antibodies including F4/80 and FA/11, which recognise the gene products of the Emr1 and CD68 genes respectively. These tissue resident macrophages play important roles in tissue homeostasis and are key players in innate and adaptive immune responses. Other cell types of the mononuclear phagocyte family include osteoclasts which play an essential role in bone morphogenesis, bone remodelling and bone homeostasis. Monocytes can also differentiate into myeloid dendritic cells (DCs), which capture and process antigens in the tissues before presenting antigenic peptides to naïve T cells in the context of MHC class II within lymph nodes. While protocols for dendritic-cell differentiation from human monocytes in culture systems are well established, the exact linage relationships between macrophages and DCs in vivo are less well defined.

With the advent of new genomic technologies such as gene arrays, we have access to extensive databases that list the most highly expressed genes in macrophages, myeloid DCs, macrophages treated with oxidised low-density lipoprotein (LDL) and DCs challenged with different pathogens [to list only four recent papers: Hashimoto et al. (1999, 2000); Shiffman et al. 2000; Huang et al. 2001)]. One striking feature of this type of expression analysis is how few genes in these collections are solely restricted to mononuclear phagocytes. Among the "macrophage-restricted" genes of known function, we can count a number of pattern-recognition receptors including the scavenger receptors SR-A and MARCO as well as the macrophage mannose receptor (MMR) (McKnight and Gordon 1998). Other macrophage-restricted genes include the murine Emr1 gene, which encodes the antigen detected by the F4/80 monoclonal antibody (McKnight et al. 1996). Other members of this EGF TM7 family (Stacey et al. 2000) include the recently described membrane protein FIRE (Emr4), expressed by macrophages and DCs (Caminschi et al. 2001); Emr2, expressed by neutrophils and activated macrophages (Lin et al. 2000) and Emr3, expressed by neutrophils, monocytes and macrophages (Stacey et al. 2001). Other cell surface receptors that are restricted to mononuclear phagocytes include the M-CSF receptor, which is the product of the *c-fms* proto-oncogene and Siglec-1 a membrane lectin expressed by a subset of macrophages that recognises carbohydrate ligands containing sialic acid (Munday et al. 1999). Recently we have cloned and characterised a number of macrophage-restricted genes including two cell surface-expressed lectin molecules, a macrophage-restricted C-type lectin, MARVIN (Balch et al. 1998) and a functional β-glucan receptor previously reported as a DC-restricted molecule, Dectin-1 (Brown and Gordon 2001; Willment et al. 2001). Currently the best candidate that we have for a pan macrophage marker is CD68, a member of the lysosomal associated membrane protein (LAMP) family (Holness and Simmons 1993). Anti CD68 monoclonal antibodies recognise tissue-resident macrophages in a wide range of normal human tissues and reveal the presence of activated macrophages in human pathologies such as atherosclerosis and arthritis (Pulford et al. 1990). The FA/11 monoclonal antibody recognises the mouse homologue of human CD68, macrosialin (the product of the murine *Cd68* gene). FA/11 detects a similarly broad range of tissue resident macrophages in murine tissues (Rabinowitz and Gordon 1991). For all of the candidate macrophage-restricted markers that we have mentioned in this section, mRNAs and proteins expressed by these genes can be found in some non-myeloid cell types in certain tissues, e.g. SR-A expression in hepatic endothelium or MMR expression in placenta. The term "macrophage-restricted" should be seen as a relative rather than an absolute description for the pattern of expression of these genes. Strikingly, there have been no reports of any macrophage-restricted or even myeloid-restricted transcription factors to date. Many macrophage-expressed gene promoters contain binding sites for the ETS family transcription factor PU.1, but this helix-loop-helix transcription factor is also expressed in B lymphocytes and granulocytes and has been implicated in eosinophil gene expression (Chen et al. 1995; van Dijk et al. 1998).

1.2 Potential Applications of Macrophage-Specific Gene Targeting

If we could develop gene expression systems that would allow us to reproducibly direct high-level transgene expression to macrophages in vivo, this genetic technology would find multiple applications both in basic science and, potentially, in clinical medicine.

Successful macrophage gene targeting would allow us to test the effect of specific transgenes on macrophage function in normal animals. These studies could be extended to study the role of macrophages in animal models of important human pathologies such as atherosclerosis, arthritis and neuroinflammation as well as infectious disease models. One potential application for lineage-specific gene targeting would be the development of lineage-specific gene knockout animals. Macrophage-restricted expression of the Cre recombinase would allow the excision of a "phloxed" allele from the genome only in macrophages. The first reports of such a myeloid-restricted excision strategy used mice in which the Cre recombinase was recombined into the murine lysozyme gene. When crossed with two different strains of mice carrying loxP-flanked genes, highly efficient excision was seen in both granulocytes and mature macrophages (Clausen et al. 1999). Clearly using the lysozyme locus does not direct gene excision targeted only to macrophages.

A recently developed technology for interfering with gene expression is post-transcriptional gene silencing induced by RNA interference or RNAi. This technique has been successfully used to study gene function in invertebrates and plants and recent technical developments may make this system more useful for interfering with mammalian gene expression (Caplen et al. 2002; Paddison et al. 2002). The RNAi technique relies on expressing specific double-stranded RNAs in specific cell types. Obviously, provision of macrophage-specific expression could allow for macrophage-specific ablation of selected genes.

One potential clinical application for high-level macrophage-specific gene expression would be somatic gene therapy for inherited metabolic diseases. One obvious candidate for macrophage gene therapy would be amelioration of the debilitating effects of inherited lysosomal storage disorders including Gaucher disease, Fabry disease and Niemann-Pick disease. This very heterogeneous group of metabolic diseases has been an attractive target for the development of somatic gene therapy protocols because of the paucity of good therapeutic options for afflicted individuals (currently limited to bone marrow transplantation and costly enzyme-replacement therapy). In Gaucher disease, the enzyme glucocerebrosidase is absent from all cells leading to an intracellular accumulation of undigested glucosylceramide, particularly affecting macrophages. Retroviral gene transfer of the glucocerebrosidase gene can correct the enzyme deficiency in Gaucher disease cell lines (Aran et al. 1996; Schuening et al. 1997) and significant levels of glucocerebrosidase have been obtained in transgenic animals (Guy et al. 1999). Retroviral gene delivery via $CD34^+$ bone marrow cells transduced ex vivo has been attempted in a pilot study of three patients. After reintroduction into patients, retrovirus-infected cells could be detected for up to 3 months, but the very low efficiency of gene correction (~0.2% of cells) and the low levels of glucocerebrosidase expression were insufficient to make any impact on the course of the disease (Dunbar et al. 1998). The macrophage could also be a useful vehicle for production of secreted proteins absent in the plasma of inherited diseases such as factor VIII in haemophilia A or factor IX in haemophilia B. Alternatively, targeting therapeutic gene expression to the alveolar macrophage could be an attractive therapeutic modality for respiratory diseases including emphysema and cystic fibrosis.

Given the central role of macrophages in innate immunity and their ability to present antigen in the context of MHC class II, efficient macrophage and DC gene targeting would allow the development of new protocols for immunomodulation. Such protocols could involve either in vivo delivery via a "gene gun" apparatus or ex vivo transfection of macrophages or DCs with a view to eliciting a strong adaptive immune response against tumour-specific antigens. Several recent reports have shown successful vaccination against HIV and malaria using a combination of DNA vaccine priming followed by recombinant virus booster. Nearly all DNA vaccination has been performed using bacterial plasmids that use the human cytomegalovirus (CMV) immediate early promoter/enhancer. To date there have been very few reports of using promoters that would direct high-level gene expression specifically in macrophages or DCs. While broad cell-type specificity might be suitable or preferable for local antigen delivery, the ability to direct gene expression specifically to Langerhans cells in vivo might allow for the manipulation of antigen-presenting cell biology through expression of specific cytokines, co-receptors or chemokine receptors. Other potential approaches to immunotherapy might be to use macrophage-specific gene expression to deliver therapeutic doses of anti-inflammatory cytokines such as interleukin (IL)-10 (Lang et al. 2002).

1.3
A Very Brief Introduction to Mammalian Gene Regulation

Before discussing how macrophage-specific gene expression programmes are established, it is worth briefly reviewing current ideas about mammalian gene regulation. Differential gene expression in mammalian cells appears to be regulated primarily at the level of transcription initiation. For this reason, studies of gene expression have focussed on DNA–protein interactions that lead to the productive engagement of RNA polymerase II with the transcription initiation sites of mammalian genes. The DNA sequences immediately 5′ of the major transcription initiation site constitute the gene's promoter and contain a number of recognisable sequence motifs including CCAAT boxes, GC sequences and TATA boxes. Nearly all the macrophage-restricted genes studied to date do not conform to the typical mammalian promoter organisation in that they do not contain recognisable TATA-box sequences within their promoters. Instead they contain one or several purine-rich DNA sequences containing the motif GGAA around the transcription initiation site (Tenen et al. 1997; Clarke and Gordon 1998). For many macrophage-expressed genes, these GGAA sites have been shown to bind the ETS family transcription factor PU.1, and mutations that abolish PU.1 binding markedly reduce promoter activity. PU.1 is not a macrophage-specific transcription factor, however, as PU.1 binding sites have been shown to be important for expression of genes in B lymphocytes and eosinophils, and PU.1 is also expressed in neutrophils (Chen et al. 1995). PU.1 has been shown to interact with a number of transcription factors including AML-1, CCAAT enhancer-binding proteins (C/EBPs), interferon regulatory factor (IRF)-4 and microphthalmia transcription factor (MITF). These combinatorial interactions with other classes of transcription factor may act to restrict PU.1's transcription enhancing activity to cells of the macrophage lineage. Transcription factors bound to the promoter and more distant enhancer elements interact with a cohort of other nuclear proteins to form a stable pre-initiation complex containing RNA polymerase II. These large multi-protein complexes contain classes of proteins called coactivators and corepressors that play a key role in mammalian gene regulation (Rosenfeld and Glass 2001).

For several genes that are expressed in other haematopoietic cell types, compelling evidence has been presented that sequences residing up to 50 kbp away from the gene can exert a powerful influence on gene expression. These sequence elements, termed locus control regions (LCRs), are especially important in allowing genes to escape the repressive effects of assembly into inactive chromatin. The first LCR to be identified was associated with a series of DNase I hypersensitive sites found in the chromatin of the human β-globin locus (Grosveld et al. 1987). When these LCR sequences were incorporated into DNA used for microinjection of mouse embryos, the resultant transgenic mice were shown to express human β-globin transgenes at a high level only in cells of the erythroid lineage. Importantly, the human β-globin transgenes were expressed at the same level as the endogenous murine β-globin genes (Blom van Assendelft et al. 1989; Talbot et al. 1989). The functional definition of an LCR sequence is that it can direct the position independent expression of a transgene and that the observed levels of transgene expression are directly proportional to copy number (Orkin 1990). LCRs were subsequently described in the human T-cell gene CD2, the human T-cell receptor gene locus and at least two gene loci that direct expression in B cells (Greaves et al. 1989; Diaz et al. 1994; Madisen and Groudine 1994; Sabbattini et al. 1999; reviewed in Grosveld 1999; Dillon and Sabbattini 2000; Festenstein and Kioussis 2000).

2
Macrophage-Specific Gene Targeting in Transgenic Animals

In Table 1 we have listed published reports of successful targeting of heterologous transgenes to macrophages in vivo. In the following section of our review we will briefly discuss published reports of macrophage gene targeting in transgenic mouse experiments.

Table 1 Examples of successful targetingof transgene expression to macrophages in transgenic animals

Transgene	Targeting sequence	Reference
Scavenger Receptor SR-AIII	CD68 promoter and IVS1	Gough et al. 2001
Interleukin-10	CD68 promoter and IVS1	Lang et al. 2002
Murine Abq MHC class II	CD68 promoter and IVS1	Unpublished data
Matrix metalloproteinase-1	Human SR-A promoter	Lemaitre et al. 2001
Hormone-sensitive lipase	Human SR-A promoter	Escary et al. 1999
Lipoprotein lipase	Human SR-A promoter	Wilson et al. 2001
Human apoA1	Human SR-A promoter	Major et al. ApoAI
Human growth hormone	Human SR-A promoter	Horvai et al. 1995
Bovine scavenger receptor SR-A	Chicken lysozyme minigene	Daugherty et al. 2001
Interferon-γ receptor dominant negative mutant	Human lysozyme gene	Dighe et al. 1995
Glucocerebrosidase	Murine MHC Class II LCR	Guy et al. 1999
PML/RARalpha	CD11b promoter	Early et al. 1996

2.1
The Human CD11b Promoter

CD11b is an integrin subunit expressed by monocytes, most macrophage populations, granulocytes, natural killer cells and a subset of $CD5^+$ B lymphocytes. Daniel Tenen and colleagues published a series of reports in which they characterised the promoter of the human CD11b gene and identified sequences that directed reporter gene expression in transiently transfected myeloid cell lines. Key elements within the human CD11b promoter included a PU.1 binding site at position 20 and an Sp1 binding site at position 60 (Pahl et al. 1992; Chen et al. 1993; Pahl et al. 1993).

A 1.8 kbp fragment of the human CD11b promoter was tested for its ability to direct macrophage-restricted expression of two different reporter genes, the *Escherichia coli* β-galactosidase gene and a murine Thy1.1 CDNA with human growth hormone introns and polA addition sequence (Dziennis et al. 1995). One line of transgenic mice was established carrying the Thy 1.1 transgene, and two lines of transgenic mice were established carrying the β-galactosidase transgene. Northern blot and RNase protection analysis demonstrated transgene expression in elicited peritoneal macrophages and fluorescence-activated cell sorter (FACS) analysis with a Thy 1.1-specific monoclonal antibody showed that ~50% of peritoneal macrophages expressed the Thy 1.1 transgene. Two-colour FACS analysis using the B-cell specific marker B220 and the granulocyte-specific reagent Gr1 showed that the Thy1.1 transgene was expressed in granulocytes and B cells as well as a subpopulation of $Mac1^+$ macrophages (Dziennis et al. 1995).

In a separate study, Back et al. used a 1.5-kbp fragment of the CD11b promoter to direct expression of a human CD4 reporter gene in transgenic mice (Back et al. 1995). In three independent lines of transgenic mice Back et al. saw no expression in macrophages and detected human CD4 transgene expression in granulocytes and B lymphocytes in two of the three lines of mice (Back et al. 1995). The discrepancy in the expression pattern reported by the two groups could be caused by the relatively small number of transgenic lines studied or differences in the design of the transgenic targeting construct used, notably the provision of human growth hormone gene sequences in the CD11b Thy 1.1 construct.

In the only other published study in which the human CD11b promoter was used to direct transgene expression, Early et al. reported the phenotype of transgenic mice expressing a PML/RARα transgene using the CD11b expression cassette developed by the Tenen laboratory. In one line of founder mice, transgene expression was detected in bone marrow by RT-PCR, but no experiments were performed to demonstrate which haematopoietic cell type was expressing the transgene (Early et al. 1996). In summary, published reports show that the CD11b promoter can direct transgene expression to myeloid cell types, but the efficiency and specificity of targeting to macrophages in vivo leave much to be desired.

2.2
Lysozyme-Directed Transgene Expression

The enzyme lysozyme degrades components of the bacterial cell wall. In mammalian species, lysozyme is expressed by cells of the innate immune system especially neutrophils and activated macrophages and by the Paneth cells of the intestine. In the mouse genome there are two lysozyme genes, the M lysozyme gene that directs lysozyme synthesis in neutrophils and macrophages while the linked P lysozyme gene is expressed specifically in the Paneth cells of the intestine (Cross et al. 1988). In contrast, there is only one lysozyme gene in the human genome. The transcriptional regulatory sequences of the murine M lysozyme gene have been studied in transient transfection assays. Sequences capable of directing expression in myeloid cells have been identified in the promoter and an enhancer element identified in the 3′ flanking sequences of the M lysozyme gene (Mollers et al. 1992). The only report of using murine lysozyme sequences to target myeloid cell expression in vivo is the report of Clausen et al. who integrated a copy of the Cre recombinase into the murine M lysozyme gene by homologous recombination and demonstrated myeloid restricted excision of P lox flanked target genes in two different strains of transgenic mice (Clausen et al. 1999). Lysozyme Cre mice have recently been used to generate transgenic mice that have macrophage-specific ablation of the IL-4 receptor α chain (F. Brombacher, personal communication).

Work from our own laboratory showed that a 3.5-kbp fragment of the human lysozyme gene was able to direct expression of a bacterial reporter gene, chloramphenicol acetyltransferase (CAT), to myeloid cells in three independent lines of transgenic mice. Transgene activity was highest in bone marrow, spleen, lung and thymus, and CAT enzyme activity was detectable in elicited granulocytes and macrophages (Clarke et al. 1996). The same 3.5-kbp human lysozyme promoter fragment was used to direct myeloid-restricted expression of a dominant negative mutant of the interferon-γ receptor alpha chain (Dighe et al. 1995). We have demonstrated that human lysozyme promoter sequences can target gene expression to immature and mature myeloid cells, but our experience is that the levels of transgene mRNA expression are low and expression is confined to a subset of tissue macrophages in vivo.

The chicken lysozyme gene has been used as a model system for studying changes in chromatin organisation related to gene expression. Additionally, the ability of different fragments of the chicken lysozyme gene locus to direct position-independent transgene expression has been tested in stably transfected cell lines and transgenic animals (reviewed in Bonifer et al. 1997). The initial experiments of Stief and Sippel showed that inclusion of chicken lysozyme gene sequences with the properties of matrix attachment regions (MARs) in reporter gene plasmids gave position-independent expression in stably transfected mammalian cell lines (Stief et al. 1989). This property of chicken lysozyme gene sequences was confirmed in the experiments of Bonifer et al., who used a 21-kbp DNA fragment encompassing the chicken lysozyme gene locus and its associated DNase I hypersensitive sites and MARs to generate transgenic mice. The chicken lysozyme gene was expressed in mature myeloid cells, including macrophages, in a copy-number dependent manner regardless of the site of integration into the mouse genome (Bonifer et al. 1990, 1997; Bonifer et al. 1994). Chicken lysozyme MAR sequences have been used as a component of several eukaryotic gene expression vectors used for stable transfection experiments in mammalian cells (Zahn et al. 2001) and even in zebrafish transgenesis (Caldovic et al. 1999).

We are aware of only one report in which chicken lysozyme gene sequences have been used to express a heterologous gene in the macrophages of transgenic mice. Daugherty et al. recently reported the generation of transgenic mice in which a bovine scavenger receptor bSR-A was expressed under the control of the chicken lysozyme promoter. The authors detected transgene mRNA in elicited macrophages but they were unable to detect expression of bSR-A protein due to a lack of specific antibodies. The bSR-A mice were shown to have altered peritoneal macrophage adhesion in vitro and enhanced granuloma formation in vivo, both properties that would be consistent with enhanced scavenger receptor expression. In the absence of transgene protein expression data, it is hard to evaluate the usefulness of the chicken lysozyme cassette for directing macrophage-specific expression in transgenic mice.

2.3
Regulatory Elements of the Human *c-fes* Gene

The *c-fes* proto-oncogene encodes a 92-kDa protein tyrosine kinase that is associated with the common β-chain subunit of the granulocyte-macrophage (GM)-CSF and IL-3 receptors. A 13.2-kbp fragment of human genomic DNA containing all 19 exons and 18 introns of the *c-fes* gene together with 446 bp of 5′ flanking and 1.5 kb of 3′ flanking sequence was used to generate transgenic mice (Greer et al. 1990). The human *c-fes* transgene mRNA showed the same pattern of expression as the endogenous mouse *c-fes* gene, being expressed at high level in bone marrow-derived macrophages. The authors reported that the 13.2-kbp fragment of DNA containing the human *c-fes* gene was able to confer position-independent expression that was proportional to copy number and that the human *c-fes* transgene appeared to be expressed as efficiently as the endogenous *c-fes* gene (Greer et al. 1990).

Heydemann et al. made a series of plasmid constructs in which different fragments of the 13.2-kb *c-fes* gene fragment were tested for their ability to direct myeloid-specific gene expression in transgenic mice. Sequences important for transgenic expression were shown to reside in introns 1 and 3 of the human *c-fes* gene (Heydemann et al. 2000). On the basis of these observations, the authors concluded that sequences within introns 1 and 3 constitute a myeloid-specific locus control region (LCR). In the same paper, the authors went on to explore the ability of the human *c-fes* LCR to direct myeloid expression of a heterologous enhanced green fluorescent protein (EGFP) transgene. EGFP transgene expression was only detected in 3 of the 8 transgenic lines analysed. Transgene expression was analysed by flow cytometry and EGFP fluorescence was detected in 50% of $Gr1^+$ granulocytes and 50% of $Mac1^+$ myeloid cells but absent in $B220^+$ B lymphocytes and absent in thymocytes (Heydemann et al. 2000). The failure of the *c-fes* LCR to give a higher efficiency of heterologous transgene expression in myeloid cells is surprising. Given that *c-fes* LCR sequences are completely contained within the *c-fes* gene it is possible that these sequences need to be included in the primary transcription unit for maximal effect.

2.4
Use of the Human SR-A Gene Promoter to Target Gene Expression in Transgenic Mice

Macrophage scavenger receptors mediate the uptake of modified forms of LDL. The first macrophage scavenger receptor gene to be cloned was SR-A, which gives rise to at least two different functional isoforms of the receptor, SR-AI and SR-AII, by alternative splicing (Emi et al. 1993). SR-A is expressed by a subset of tissue-resident macrophages in vivo, such as alveolar macrophages and Kupffer cells, as well as macrophages in inflammatory pathologies, such as atherosclerosis (Gough et al. 1999). The promoter of the human SR-A gene has been analysed by a number of laboratories in transient transfection assays. The laboratory of Christopher Glass identified SR-A promoter sequences between positions 245 and +46 as being important for expression in the myeloid leukaemia cell line THP-1 and implicated the transcription factors AP1, PU.1 and ets-2 as being important for SR-A promoter activity (Wu et al. 1994). Further analysis of the SR-A promoter by the Glass laboratory identified an enhancer sequence between positions 4500 and 4100 (Moulton et al. 1994). A human SR-A promoter/enhancer cassette containing human SR-A promoter sequences between 245 and +46 fused to the upstream enhancer sequence was used in transgenic experiments. Horvai et al. first demonstrated the ability of this SR-A enhancer/promoter cassette to direct macrophage-restricted expression of a human growth hormone reporter gene in macrophages of transgenic mice and demonstrated transgene expression by bone marrow-derived macrophages cells differentiated in the presence of M-CSF (Horvai et al. 1995) The current literature furnishes several examples of successful targeting to macrophages using the SR-A enhancer/promoter cassette developed by the Glass laboratory (Table 1).

2.5
Regulatory Sequences of the Human CD68 Gene

CD68 is probably the best pan macrophage marker in immunohistochemistry studies of human tissues (Pulford et al. 1990). The protein recognised by CD68 monoclonal antibodies is a member of the lysosome-associated membrane protein (LAMP) family (Holness and Simmons 1993) and human CD68 shows 72% amino acid identity with the murine macrosialin protein (Holness et al. 1993).

Li et al. have studied the elements of the murine macrosialin promoter required for macrophage-specific expression (Li et al. 1998) and our laboratory has characterised the sequences responsible for transcriptional regulation of the human CD68 gene (Greaves et al. 1998). The draft human genome sequence shows that the human CD68 gene is one of 9 genes in a ~140 kbp region of chromosome 17p13 and the CD68 gene ATG initiation codon lies 669 bp 3′ of the EIF4A1gene, which encodes eukaryotic initiation factor 4AI (Jones et al. 1998).

The 666-bp EIF4A1 CD68 intergenic region has significant promoter activity in murine macrophage cell lines and a number of other cell types (Greaves et al. 1998). A 5′ deletion analysis of the CD68 5′ flanking sequence showed that a promoter fragment of only 150 bp directs reporter gene expression in transiently transfected macrophage cell lines to a level that is twice that seen using the SV40 enhancer in the same reporter plasmid. We cloned each of the five CD68 gene introns 3′ of a minimal CD68 promoter and showed that only intron 1 was able to contribute to macrophage-specific expression (Greaves et al. 1998). Taken together, these observations led us to conclude that the combination of the human CD68 promoter and the 83-bp first intron of the human CD68 gene cooperate to direct high-level reporter gene expression in transiently transfected macrophage cell lines.

To test the utility of human CD68 gene sequences in macrophage gene-targeting experiments, we have generated a series of CD68 expression vectors in which cDNA fragments of different transgenes have been cloned downstream of a 2940 CD68 promoter fragment and the 83-bp first intron of the CD68 gene. Initial experiments using splice variants of the human macrophage scavenger receptor SR-A showed that human CD68 gene sequences were able to direct high-level expression of SR-A transgenes in stably transfected RAW cells (Gough et al. 2001). In two lines of transgenic mice, we were able to demonstrate high levels of a SR-AIII transgene in bone marrow and elicited peritoneal macrophages (Gough et al. 2001). More recently we have shown the utility of the CD68 promoter intron 1 cassette to direct macrophage-specific expression of an IL-10 transgene in transgenic mice. Macrophages of these transgenic mice constitutively express high levels of the deactivating cytokine IL-10 and display profound changes in their response to bacterial pathogens. No transgene expression is detected in neutrophils, B cells or T cells (Lang et al. 2002). These initial experiments suggest that CD68 transcriptional regulatory sequences may be useful for directing transgene expression in cells of the mononuclear phagocyte lineage. One reason for the success of the CD68 vector may be the inclusion of the CD68 intron close to the 5′ end of the transgene.

3 Gene Delivery to Macrophages Using Viral and Non-viral Vectors

3.1 Retroviral Vectors

Retroviruses have been used widely in gene therapy protocols and as a vehicle for gene delivery to primary cells, including bone marrow progenitor cells. First generation retroviral vectors were of limited usefulness in macrophage transduction because they were only able to deliver transgene expression in dividing cells. More recently, developed recombinant lentiviral vectors have shown great promise in gene delivery to a wide range of cell types in vivo (Somia and Verma 2000). The most spectacular example of the power of this technology is shown by expression of GFP in every tissue of an adult mouse following transduction of single-cell mouse embryos with a recombinant lentiviral vector (Lois et al. 2002). The majority of retroviral vectors rely on viral or housekeeping gene promoters to direct transgene expression. However, a recent paper by Cui et al. compared the efficiency of self-inactivating lentiviral vectors containing the promoter of the housekeeping gene EF-1α or the class II gene HLA-DRα to drive GFP expression in antigen-presenting cells following infection of human CD34^{+} bone marrow-derived progenitor cells. The HLA-DRα promoter efficiently targeted GFP expression to differentiated DCs in non-obese diabetic, severe combined immunodeficiency (NOD/SCID) mice engrafted with human haematopoietic stem cells (Cui et al. 2002). This report shows the feasibility of developing myeloid-restricted transgene expression in vivo through development of retroviral vectors containing macrophage or DC-specific promoters.

3.2 Adenovirus-Mediated Gene Transfer

Recombinant adenoviruses offer several significant advantages over retroviral vectors for gene delivery, notably the ability to routinely prepare high titres of recombinant virus and the ability to efficiently infect non-dividing cells. Recombinant adenoviruses have been used to drive heterologous gene expression in human macrophages and DCs cultured ex vivo. Examples include the restoration of respiratory burst in monocyte-derived macrophages from patients with X-linked chronic granulomatous disease using a recombinant adenovirus encoding the gp91 phox subunit (Schneider et al. 1997), MUC1 gene transduction of human blood-derived DCs (Maruyama et al. 2001) and modification of murine DCs to secrete the CC chemokine macrophage-derived chemokine (MDC) (Kikuchi and Crystal 2001). There are also reports of recombinant adenovirus transduction of tissue resident macrophages in vivo, for instance the study of Wheeler et al., which demonstrated transgene expression in Kupffer cells in mice infected with a recombinant adenovirus (Wheeler et al. 2001). Not all tissue-resident macrophages are equally susceptible to adenovirus infection due to differences in the level of expression of the cell-surface integrins that mediate adenoviral gene entry into cells. It has been reported that alveolar macrophages are refractory to adenovirus infection and that this block to infection is not alleviated by treatment with M-CSF (Kaner et al. 1999; Conron et al. 2001). One approach that might be adopted to circumvent this obstacle to in vivo gene delivery would be the development of recombinant adenoviral vectors with altered viral coat proteins (Wickham 2000).

For experiments where primary cells are manipulated ex vivo there is no strict requirement for including a macrophage or DC-specific promoter in the adenovirus vector. However, the utility of recombinant adenoviruses or adeno-associated adenoviruses for in vivo gene delivery could be greatly enhanced by developing adenoviral vectors containing macrophage-specific promoters. We have made a recombinant adenovirus that uses the human CD68 promoter to drive expression of a soluble form of the human scavenger receptor (Wickham 2000) and we are developing recombinant viral vectors that include both the CD68 promoter and the macrophage-specific enhancer element within the first intron of the CD68 gene.

3.3 Non-Viral Vectors

A number of different compounds have been developed that facilitate DNA entry into cultured cells in vitro, and some of these compounds have been shown to mediate DNA delivery in vivo. Peritoneal macrophages have been shown to take up naked plasmid DNA via a non-scavenger receptor-mediated mechanism, although this process is very inefficient (Takakura et al. 1999). One approach to directing DNA delivery specifically to macrophages uses potential ligands for known macrophage receptors. The first example of receptor-mediated gene transfer to macrophages used DNA coupled to a ligand for the macrophage mannose receptor (Ferkol et al. 1996). Attempts to use this technology to deliver an alpha1 antitrypsin gene to alveolar macrophages in vivo were very inefficient (Ferkol et al. 1998). Recent papers have used mannose coupled to polyethylenimine (PEI) to deliver DNA to DCs (Diebold et al. 1999; Diebold et al. 2001) and Kawakami et al. have explored the possibility of using mannosylated, fucosylated and galactosylated liposome-DNA complexes for macrophage-specific gene delivery (Kawakami et al. 2000a,b).

One interesting approach to in vivo protein, drug or DNA delivery to selected cell populations has been developed by the laboratory of Seppo Ylä-Herttuala. Lehtolainen et al. constructed a novel endocytic receptor in which the C-terminal ligand-binding domain of the bovine scavenger receptor has been replaced with avidin. This "scavidin" receptor behaves as a novel endocytic receptor that binds and internalises biotinylated molecules (Lehtolainen et al. 2001). Macrophage-restricted expression of this or similar novel receptors could allow for highly selective delivery of therapeutic compounds to macrophages in vivo.

4
Future Prospects

Significant progress has been made towards identifying important regulatory elements for macrophage-specific gene expression but we are still some way from having a macrophage expression vector that will reproducibly yield high-level expression in macrophage populations in vivo. The identification of potential locus control regions in the human *c-fes* and the murine *spi-1* genes is an exciting development in the field and analysis of these regions may reveal important molecular mechanisms that underlie the development of the mononuclear phagocyte lineage. It will be very interesting to see if these sequences can be used to develop macrophage-specific expression vectors.

Table 1 shows that the human SR-A promoter has found the most widespread application in directing macrophage-specific expression in transgenic mice so far, and the usefulness of the SR-A promoter may be augmented by changes in vector design such as the introduction of heterologous introns into the primary transcription unit and careful selection of poly A addition sequences (C. Glass, personal communication). Most of the experiments we have discussed have used pronuclear microinjection of naked DNA to transfer transgene expression constructs to the mouse germline. So far there have been few reports of attempts to incorporate macrophage-specific promoters into viral vectors that would allow for macrophage-specific expression following transduction of bone marrow-derived stem cells or haematopoietic progenitor cells.

All the examples of macrophage gene targeting that we have discussed in this review have been aimed at directing constitutive transgene expression in macrophages. Recently we have used human CD68 sequences to drive IL-10 expression in transgenic mice, and mice with more than two copies of the transgene had very high serum levels of IL-10 and died from opportunistic bacterial infections (Lang et al. 2002). This observation suggests that for some transgenes it will be important to start designing vectors that allow for inducible macrophage gene expression. Currently, inducible gene expression systems based on the bacterial Tet repressor and the tetracycline analogue doxycycline seem to offer the best prospects for mediating regulated gene expression in transgenic mice (Lewandoski et al. 2001). The development of an inducible macrophage expression system would allow a detailed appreciation of macrophage function in a whole range of pathological settings, ranging from endotoxic shock to atherosclerosis.

Impressive preliminary results have been obtained using naked DNA to induce humoral and cellular immune responses to test antigens in animal systems. The success obtained using DNA vaccine prime, recombinant virus boost vaccination protocols has focused attention on generating specific immune responses to candidate antigens derived from infectious disease agents or tumours (Hanke et al. 1999). The ability to target macrophage and DC populations in vivo could have very important consequences for our ability to develop immunomodulation protocols for therapeutic benefit. Of especial interest would be switching a predominantly Th2-type immune response to a more Thl-type immune response in allergic asthma or the induction of immune tolerance in transplantation or autoimmunity.

It is our belief that progress in understanding the basic biology of macrophages will yield benefits in developing new treatments for pathologies characterised by monocyte recruitment, macrophage differentiation or macrophage dysfunction such as atherosclerosis, arthritis and lysosomal storage disorders.

Acknowledgements. We thank Peter Gough and Chris Glass for helpful discussions and Gordon Brown for comments on the manuscript. Work in our laboratories is supported by the Wellcome Trust, the Medical Research Council and the British Heart Foundation. D.R.G. is a British Heart Foundation Basic Science Lecturer.

5 References

Aran, J.M., et al., Complete restoration of glucocerebrosidase deficiency in Gaucher fibroblasts using a bicistronic MDR retrovirus and a new selection strategy. Hum Gene Ther, 1996. 7(17): p. 2165–75

Back, A., K. East, and D. Hickstein, Leukocyte integrin CD11b promoter directs expression in lymphocytes and granulocytes in transgenic mice. Blood, 1995. 85(4): p. 1017–24

Balch, S.G., et al., Cloning of a novel C-type lectin expressed by murine macrophages. J Biol Chem, 1998. 273(29): p. 18656–64

Blom van Assendelft, M., et al., The beta-globin dominant control region activates homologous and heterologous promoters in a tissue-specific manner. Cell, 1989. 56: p. 969–977

Bonifer, C., et al., Dissection of the locus control function located on the chicken lysozyme gene domain in transgenic mice. Nucleic Acids Res, 1994. 22(20): p. 4202–10

Bonifer, C., et al., Tissue specific and position independent expression of the complete gene domain for chicken lysozyme in transgenic mice. Embo J, 1990. 9(9): p. 2843–8

Bonifer, C., U. Jagle, and M.C. Huber, The chicken lysozyme locus as a paradigm for the complex developmental regulation of eukaryotic gene loci. J Biol Chem, 1997. 272(42): p. 26075–8

Brown, G.D. and S. Gordon, Immune recognition. A new receptor for beta-glucans. Nature, 2001. 413(6851): p. 36–7

Caldovic, L., D. Agalliu, and P.B. Hackett, Position-independent expression of transgenes in zebrafish. Transgenic Res, 1999. 8(5): p. 321–34

Caminschi, I., et al., Molecular cloning of F4/80-like-receptor, a seven-span membrane protein expressed differentially by dendritic cell and monocyte-macrophage subpopulations. J Immunol, 2001. 167(7): p. 3570–6

Caplen, N.J., et al., Rescue of polyglutamine-mediated cytotoxicity by double-stranded RNA-mediated RNA interference. Hum Mol Genet, 2002. 11(2): p. 175–84

Chen, H.M., et al., Neutrophils and monocytes express high levels of PU.1 (Spi-1) but not Spi-B. Blood, 1995. 85(10): p. 2918–28

Chen, H.M., et al., The Sp1 transcription factor binds the CD11b promoter specifically in myeloid cells in vivo and is essential for myeloid-specific promoter activity. J Biol Chem, 1993. 268(11): p. 8230–9

Clarke, S. and S. Gordon, Myeloid-specific gene expression. J Leukoc Biol, 1998. 63(2): p. 153–68

Clarke, S., et al., The human lysozyme promoter directs reporter gene expression to activated myelomonocytic cells in transgenic mice. Proc Natl Acad Sci USA, 1996. 93(4): p. 1434–8
Clausen, B.E., et al., Conditional gene targeting in macrophages and granulocytes using LysMcre mice. Transgenic Res, 1999. 8(4): p. 265–77
Conron, M., et al., Alveolar macrophages and T cells from sarcoid, but not normal lung, are permissive to adenovirus infection and allow analysis of NF-kappa b-dependent signaling pathways. Am J Respir Cell Mol Biol, 2001. 25(2): p. 141–9
Cross, M., et al., Mouse lysozyme M gene: isolation, characterization, and expression studies. Proc Natl Acad Sci U S A, 1988. 85(17): p. 6232–6
Cui, Y., et al., Targeting transgene expression to antigen-presenting cells derived from lentivirus-transduced engrafting human hematopoietic stem/progenitor cells. Blood, 2002. 99(2): p. 399–408
Daugherty, A., et al., Macrophage-specific expression of class A scavenger receptors enhances granuloma formation in the absence of increased lipid deposition. J Lipid Res, 2001. 42(7): p. 1049–1055
Diaz, P., D. Cado, and A. Winoto, A locus control region in the T cell receptor alpha/delta locus. Immunity, 1994. 1(3): p. 207–17
Diebold, S.S., et al., Mannose polyethylenimine conjugates for targeted DNA delivery into dendritic cells. J Biol Chem, 1999. 274(27): p. 19087–94
Diebold, S.S., et al., MHC class II presentation of endogenously expressed antigens by transfected dendritic cells. Gene Ther, 2001. 8(6): p. 487–93
Dighe, A.S., et al., Tissue-specific targeting of cytokine unresponsiveness in transgenic mice. Immunity, 1995. 3(5): p. 657–66
Dillon, N. and P. Sabbattini, Functional gene expression domains: defining the functional unit of eukaryotic gene regulation. Bioessays, 2000. 22(7): p. 657–665
Dunbar, C.E., et al., Retroviral transfer of the glucocerebrosidase gene into CD34+ cells from patients with Gaucher disease: in vivo detection of transduced cells without myeloablation. Hum Gene Ther, 1998. 9(17): p. 2629–40
Dziennis, S., et al., The CD11b promoter directs high-level expression of reporter genes in macrophages in transgenic mice. Blood, 1995. 85(2): p. 319–29
Early, E., et al., Transgenic expression of PML/RARalpha impairs myelopoiesis. Proc Natl Acad Sci U S A. 1996 Jul 23;93(15):7900–4., 1996. 93(15): p. 7900–7904
Emi, M., et al., Structure, organization, and chromosomal mapping of the human macrophage scavenger receptor gene. J Biol Chem, 1993. 268(3): p. 2120–5
Escary, J.L., et al., Paradoxical effect on atherosclerosis of hormone-sensitive lipase overexpression in macrophages. J Lipid Res, 1999. 40(3): p. 397–404
Ferkol, T., et al., Receptor-mediated gene transfer into macrophages. Proc Natl Acad Sci U S A, 1996. 93(1): p. 101–5
Ferkol, T., et al., Transfer of the human Alpha1-antitrypsin gene into pulmonary macrophages in vivo. Am J Respir Cell Mol Biol, 1998. 18(5): p. 591–601
Festenstein, R. and D. Kioussis, Locus control regions and epigenetic chromatin modifiers. Curr Opin Genet Dev, 2000. 10(2): p. 199–203
Gough, P.J., et al., Analysis of macrophage scavenger receptor (SR-A) expression in human aortic atherosclerotic lesions. Arterioscler Thromb Vasc Biol, 1999. 19(3): p. 461–71
Gough, P.J., S. Gordon, and D.R. Greaves, The use of human CD68 transcriptional regulatory sequences to direct high level expression of scavenger receptor SR-A in macrophages in vitro and in vivo. Immunology, 2001. 103: p. 351–361
Greaves, D.R., et al., Functional comparison of the murine macrosialin and human CD68 promoters in macrophage and non macrophage cell lines. Genomics, 1998. 54: p. 165–168
Greaves, D.R., et al., Human CD2 3' flanking sequences confer high-level, T- cell specific, position independent gene expression in transgenic mice. Cell, 1989. 56: p. 979–986

Greer, P., et al., Myeloid expression of the human c-fps/c-fes proto-oncogene in transgenic mice. Mol Cell Biol, 1990. 10(6): p. 2521–2527
Grosveld, F., Activation by locus control regions? Curr Opin Genet Dev, 1999. 9(2): p. 152–7
Grosveld, F., et al., Position-independent, high-level expression of the human beta globin gene in transgenic mice. Cell, 1987. 51: p. 975–985
Guy, J., et al., Murine MHC class II locus control region drives expression of human beta-glucocerebrosidase in antigen presenting cells of transgenic mice. Gene Ther, 1999. 6(4): p. 498–507
Hanke, T., et al., Effective induction of simian immunodeficiency virus-specific cytotoxic T lymphocytes in macaques by using a multiepitope gene and DNA prime-modified vaccinia virus Ankara boost vaccination regimen. J Virol, 1999. 73(9): p. 7524–32
Hashimoto, S., et al., Serial analysis of gene expression in human monocytes and macrophages. Blood, 1999. 94(3): p. 837–44
Hashimoto, S.I., et al., Identification of genes specifically expressed in human activated and mature dendritic cells through serial analysis of gene expression. Blood, 2000. 96(6): p. 2206–14
Heydemann, A., et al., A minimal c-fes cassette directs myeloid-specific expression in transgenic mice. Blood, 2000. 96(9): p. 3040–3048
Holness, C.L. and D.L. Simmons, Molecular cloning of CD68, a human macrophage marker related to lysosomal glycoproteins. Blood, 1993. 81(6): p. 1607–13
Holness, C.L., et al., Macrosialin, a mouse macrophage-restricted glycoprotein, is a member of the lamp/lgp family. J Biol Chem, 1993. 268(13): p. 9661–6
Horvai, A., et al., Scavenger receptor A gene regulatory elements target gene expression to macrophages and to foam cells of atherosclerotic lesions. Proc Natl Acad Sci USA, 1995. 92(12): p. 5391–5
Huang, Q., et al., The plasticity of dendritic cell responses to pathogens and their components. Science, 2001. 294(5543): p. 870–5
Jones, E., et al., The linked human elongation Initiation Factor 4A1 (EIF4A1) and CD68 genes map to chromosome 17p13. Genomics, 1998. 53: p. 248–250
Kaner, R.J., et al., Modification of the genetic program of human alveolar macrophages by adenovirus vectors in vitro is feasible but inefficient, limited in part by the low level of expression of the coxsackie/adenovirus receptor. Am J Respir Cell Mol Biol, 1999. 20(3): p. 361–70
Kawakami, S., et al., Biodistribution characteristics of mannosylated, fucosylated, and galactosylated liposomes in mice. Biochim Biophys Acta, 2000a. 1524(2–3): p. 258–65
Kawakami, S., et al., Mannose receptor-mediated gene transfer into macrophages using novel mannosylated cationic liposomes. Gene Ther, 2000b. 7(4): p. 292–9
Kikuchi, T. and R.G. Crystal, Antigen-pulsed dendritic cells expressing macrophage-derived chemokine elicit Th2 responses and promote specific humoral immunity. J Clin Invest, 2001. 108(6): p. 917–27
Lang, R., et al., Autocrine deactivation of macrophages in transgenic mice constitutively overexpressing IL-10 under control of the human CD68 promoter. Journal of Immunology, 2002. in press
Lehtolainen, P., et al., Cloning and characterisation of scavidin, a fusion protein for the targeted delivery of biotinylated molecules. J Biol Chem, 2001
Lemaitre, V., et al., ApoE knockout mice expressing human matrix metalloproteinase-1 in macrophages have less advanced atherosclerosis. J Clin Invest, 2001. 107(10): p. 1227–34
Lewandoski, M., Conditional control of gene expression in the mouse. Nat Rev Genet, 2001. 2(10): p. 743–55
Li, A.C., et al., The macrosialin promoter directs high levels of transcriptional activity in macrophages dependent on combinatorial interactions between PU.1 and c-Jun. J Biol Chem, 1998. 273(9): p. 5389–99

Lin, H.H., et al., Human EMR2, a novel EGF-TM7 molecule on chromosome 19p13.1, is closely related to CD97. Genomics, 2000. 67(2): p. 188–200
Lois, C., et al., Germline transmission and tissue-specific expression of transgenes delivered by lentiviral vectors. Science, 2002. 295(5556): p. 868–72
Madisen, L. and M. Groudine, Identification of a locus control region in the immunoglobulin heavy-chain locus that deregulates c-myc expression in plasmacytoma and Burkitt's lymphoma cells. Genes Dev, 1994. 8(18): p. 2212–26
Major, A.S., et al., Increased cholesterol efflux in apolipoprotein AI (ApoAI)-producing macrophages as a mechanism for reduced atherosclerosis in ApoAI((-/-)) mice. Arterioscler Thromb Vasc Biol, 2001. 21(11): p. 1790–5
Maruyama, K., et al., Adenovirus-Mediated MUC1 gene transduction into human blood-derived dendritic cells. J Immunother, 2001. 24(4): p. 345–53
McKnight, A.J. and S. Gordon, Membrane molecules as differentiation antigens of murine macrophages. Adv Immunol, 1998. 68: p. 271–314
McKnight, A.J., et al., Molecular cloning of F4/80, a murine macrophage-restricted cell surface glycoprotein with homology to the G-protein-linked transmembrane 7 hormone receptor family. J Biol Chem, 1996. 271(1): p. 486–9
Mollers, B., et al., The mouse M-lysozyme gene domain: identification of myeloid and differentiation specific DNaseI hypersensitive sites and of a 3'-cis acting regulatory element. Nucleic Acids Res, 1992. 20(8): p. 1917–24
Moulton, K.S., et al., Cell-specific expression of the macrophage scavenger receptor gene is dependent on PU.1 and a composite AP-1/ets motif. Mol Cell Biol, 1994. 14(7): p. 4408–18
Munday, J., H. Floyd, and P.R. Crocker, Sialic acid binding receptors (siglecs) expressed by macrophages. J Leukoc Biol, 1999. 66(5): p. 705–11
Orkin, S.H., Globin gene regulation and switching: circa 1990. Cell, 1990. 63(4): p. 665–72
Paddison, P.J., A.A. Caudy, and G.J. Hannon, Stable suppression of gene expression by RNAi in mammalian cells. Proc Natl Acad Sci U S A, 2002. 99(3): p. 1443–8
Pahl, H.L., A.G. Rosmarin, and D.G. Tenen, Characterization of the myeloid-specific CD11b promoter. Blood, 1992. 79(4): p. 865–70
Pahl, H.L., et al., The proto-oncogene PU.1 regulates expression of the myeloid-specific CD11b promoter. J Biol Chem, 1993. 268(7): p. 5014–20
Pulford, K.A., et al., Distribution of the CD68 macrophage/myeloid associated antigen. Int Immunol, 1990. 2(10): p. 973–80
Rabinowitz, S.S. and S. Gordon, Macrosialin, a macrophage-restricted membrane sialoprotein differentially glycosylated in response to inflammatory stimuli. J Exp Med, 1991. 174(4): p. 827–36
Rosenfeld, M.G. and C.K. Glass, Coregulator codes of transcriptional regulation by nuclear receptors. J Biol Chem, 2001. 276(40): p. 36865–8
Sabbattini, P., et al., Analysis of mice with single and multiple copies of transgenes reveals a novel arrangement for the lambda5-VpreB1 locus control region. Mol Cell Biol, 1999. 19(1): p. 671–9
Schneider, S.D., et al., Adenovirus-mediated gene transfer into monocyte-derived macrophages of patients with X-linked chronic granulomatous disease: ex vivo correction of deficient respiratory burst. Gene Ther, 1997. 4(6): p. 524–32
Schuening, F., et al., Retrovirus-mediated transfer of the cDNA for human glucocerebrosidase into peripheral blood repopulating cells of patients with Gaucher's disease. Hum Gene Ther, 1997. 8(17): p. 2143–60
Shiffman, D., et al., Large scale gene expression analysis of cholesterol-loaded macrophages. J Biol Chem, 2000. 275(48): p. 37324–32
Somia, N. and I.M. Verma, Gene therapy: trials and tribulations. Nat Rev Genet, 2000. 1(2): p. 91–9
Stacey, M., et al., Human epidermal growth factor (EGF) module-containing mucin-like hormone receptor 3 is a new member of the EGF-TM7 family that recognizes a li-

gand on human macrophages and activated neutrophils. J Biol Chem, 2001. 276(22): p. 18863–70

Stacey, M., et al., LNB-TM7, a group of seven-transmembrane proteins related to family-B G-protein-coupled receptors. Trends Biochem Sci, 2000. 25(6): p. 284–9

Stief, A., et al., A nuclear DNA attachment element mediates elevated and position-independent gene activity. Nature, 1989. 341(6240): p. 343–5

Takakura, Y., et al., Characterization of plasmid DNA binding and uptake by peritoneal macrophages from class A scavenger receptor knockout mice. Pharm Res, 1999. 16(4): p. 503–8

Talbot, D.J., et al., A dominant control region from the human beta- globin locus conferring integration site- independent gene expression. Nature, 1989. 338: p. 352–355

Tenen, D.G., et al., Transcription factors, normal myeloid development, and leukemia. Blood, 1997. 90(2): p. 489–519

van Dijk, T.B., et al., The role of transcription factor PU.1 in the activity of the intronic enhancer of the eosinophil-derived neurotoxin (RNS2) gene. Blood, 1998. 91(6): p. 2126–32

Wheeler, M.D., et al., Adenoviral gene delivery can inactivate Kupffer cells: role of oxidants in NF-kappaB activation and cytokine production. J Leukoc Biol, 2001. 69(4): p. 622–30

Wickham, T.J., Targeting adenovirus. Gene Ther, 2000. 7(2): p. 110–4

Willment, J.A., S. Gordon, and G.D. Brown, Characterization of the human beta -glucan receptor and its alternatively spliced isoforms. J Biol Chem, 2001. 276(47): p. 43818–23

Wilson, K., et al., Macrophage-specific expression of human lipoprotein lipase accelerates atherosclerosis in transgenic apolipoprotein e knockout mice but not in C57BL/6 mice. Arterioscler Thromb Vasc Biol, 2001. 21(11): p. 1809–1815

Wu, H., et al., Combinatorial interactions between AP-1 and ets domain proteins contribute to the developmental regulation of the macrophage scavenger receptor gene. Mol Cell Biol, 1994. 14(3): p. 2129–39

Zahn-Zabal, M., et al., Development of stable cell lines for production or regulated expression using matrix attachment regions. J Biotechnol, 2001. 87(1): p. 29–42

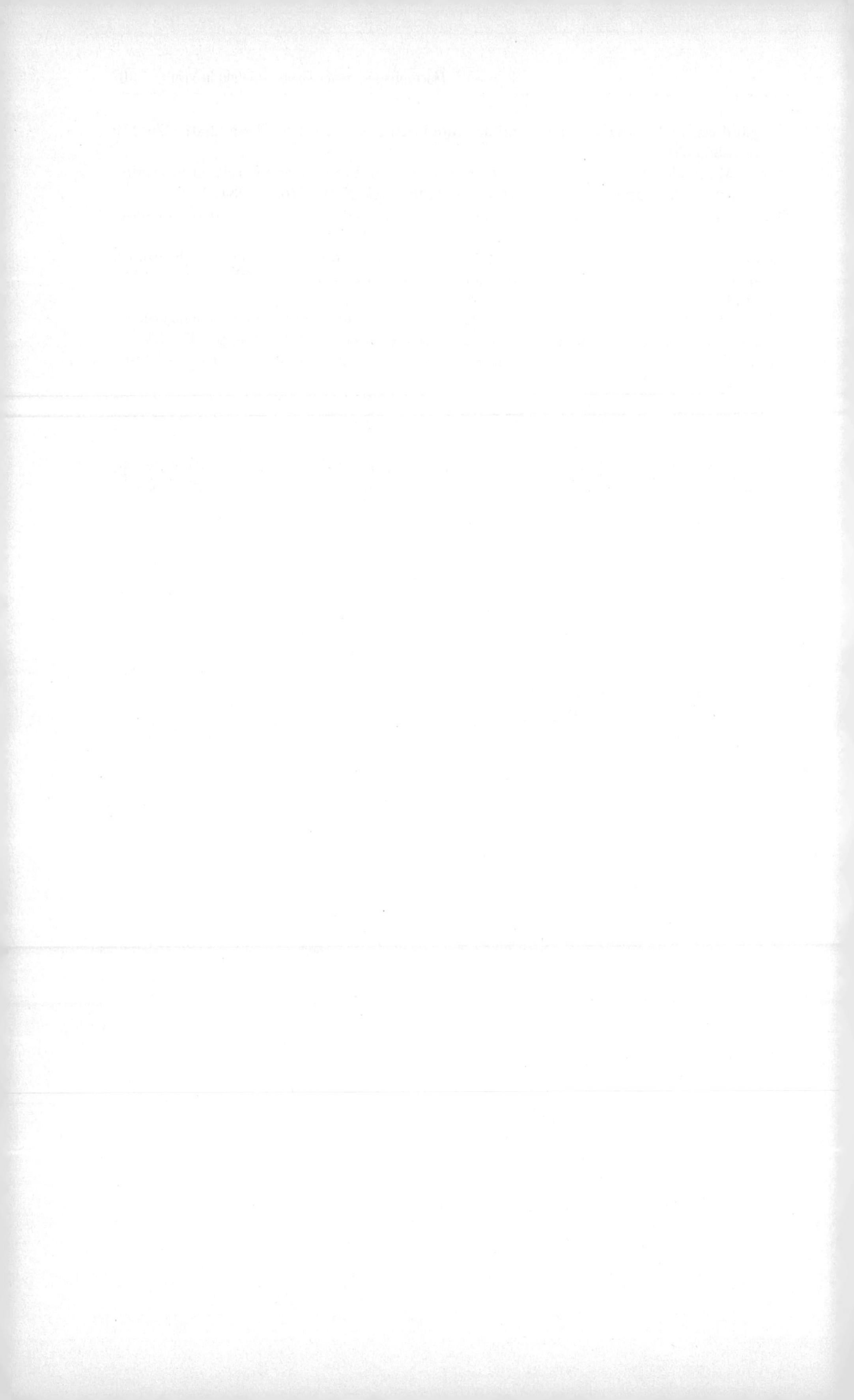

Part 2
Macrophage Targets in Inflammation

Integrins of Macrophages and Macrophage-Like Cells

E. J. Brown

Program in Host-Pathogen Interactions,
University of California,
513 Parnassus Avenue, Campus Box 0654, San Francisco, CA 94143, USA
e-mail: ebrown@medicine.ucsf.edu

Abstract The integrins are a family of heterodimeric adhesion receptors present on virtually every cell in metazoan organisms. Macrophage integrins are involved in adhesion to extracellular matrix and to other cells, in phagocytosis, and in cell migration and spreading. Macrophage integrins also transduce signals from the extracellular environment, both through activation of specific kinase cascades and through modulation of cytoskeletal elements. The ligand-binding ability of macrophage integrins can be regulated by environmental cues including growth factors, lipid mediators, bacterial peptides, and fragments of complement, clotting, and other proteins that may accumulate at sites of inflammation. In addition, integrins can function as components of multimolecular plasma membrane complexes that include tetraspanins, proteases, and other receptors. These multimolecular complexes can be endowed with functional properties not inherent in the isolated components. This review summarizes current understanding of the complex biology of macrophage integrins. The involvement of integrins in many macrophage processes fundamental to the function of these cells for inflammation and host defense makes them double-edged swords as candidates for therapeutic intervention. Beneficial effects of broad integrin blockade in ameliorating idiopathic inflammation may be accompanied by unacceptable susceptibility to infection or other deleterious side effects. However, more specific blockade of less widely expressed integrins holds the promise of a better therapeutic effect. For example, recent studies in animals suggest potential therapeutic use for blockade of osteoclast $\alpha V\beta 3$ in osteoporosis and of T-cell and monocyte $\alpha 1\beta 1$ and $\alpha 2\beta 1$ in arthritis. Further understanding of the molecular mechanisms through which macrophage integrins control key events in a variety of diseases may lead to the development of inhibitors of even greater specificity. The integrins remain appealing therapeutic targets because of their central role in macrophage biology.

Keywords Adhesion, Affinity modulation, Cytoskeleton, Dietary fatty acids, Extracellular matrix, ICAM-1, Macrophages, Migration, Phagocytosis, Signal transduction, VCAM-1

The integrins are a family of 22 adhesion receptors that recognize a wide variety of ligands, including: various proteins and proteoglycans of basement membranes and extracellular matrix; members of the immunoglobulin superfamily on cell surfaces; and potentially some anionic phospholipids. Integrins are present on all nucleated cells and on platelets; only erythrocytes are apparently devoid of integrins, although relatively mature erythrocyte precursors do express these receptors. Integrins are involved in leukocyte proliferation, maturation, migration, activation, and multiple effector functions including phagocytosis, cytotoxicity, synthesis of cytokines, and activation of the nicotinamide adenine dinucleotide phosphate (NADPH) oxidase. Based on the importance of integrins in leukocyte migration to and activation at sites of inflammation, these receptors have for some time been considered excellent targets for therapy of inflammatory diseases. Recently, understanding of the critical role of osteoclast integrins has led to great enthusiasm for targeting these integrins to prevent or treat osteoporosis. After a brief review of general aspects of integrin structure and function, this review will consider specifically macrophage integrins and the physiologic and pathologic processes in which they are involved, and ultimately the potential therapeutic applications of anti-integrin therapy to treat diseases in which macrophages and related cells may have an important role.

1 Integrin Structure

All integrins are heterodimers of two type I membrane proteins (N terminal on the extracytoplasmic face of the membrane) that function primarily as adhesion receptors. There are 17 known α chain proteins ($\alpha 1$–$\alpha 11$, αM, αL, αX, αD, αV, αE) and 8 known β chain proteins ($\beta 1$–$\beta 8$); various α and β chains combine in the endoplasmic reticulum to create a family of 22 cell surface receptors. The interactions between the two chains are complex, and failure to associate appropriately leads to failure of secretion. Recently, the crystal structure of the inte-

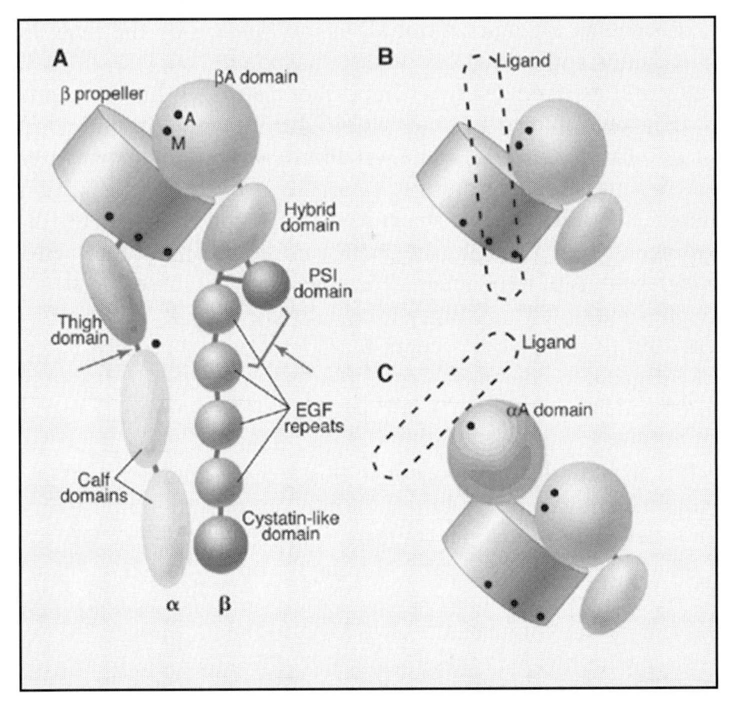

Fig. 1A–C Integrin structure. **A** Structure of integrin $\alpha V\beta 3$ based on the crystallographic data from Xiong et al. (2001). **B** The extensive contact between the β propeller domain of the α chain and the βA domain of the β chain provides the interface for ligand binding. **C** For integrins that contain an I domain (αA domain, **C**) in the α chain, this domain likely interacts with the α/β interface created by the β propeller and the βA domain and provides an independent ligand binding face. (Reproduced from Humphries and Mould 2001, with permission)

grin $\alpha V\beta 3$ was solved (Xiong et al. 2001). A key feature of the structure is a previously predicted "beta-propeller" domain in the α chain with heptad symmetry, reminiscent of the structure of the beta chain of heterotrimeric G proteins (Springer 1997). The aminoterminal domain of the β chain (the so-called βA domain) interacts with this beta propeller structure and has a fold similar to the alpha chain of heterotrimeric G proteins, as well as to domains in a variety of extracellular molecules, including von Willebrand factor. The interface between the beta propeller and the βA domain closely resembles the interface between alpha and beta chains in heterotrimeric G proteins. Ligands for $\alpha V\beta 3$ are thought to bind at this interface between α and β chains (Fig. 1); this has been confirmed very recently for the peptide ligand Arg-Gly-Asp (Xiong et al. 2002). Carboxyterminal to the βA domain are four highly disulfide bonded epithelial growth factor (EGF)-like domains which also participate in interaction with the α chain (Fig. 1). Some α chains have an additional domain with high homology to the βA domain, also called the A or I (for inserted) domain, which is inserted between spokes 2 and 3 of the beta-propeller structure. When this α chain I domain is present, it accounts for most if not all ligand binding by the integrin (Fig. 1). Integrin α chains containing this domain include $\alpha1$, $\alpha2$, αL, αM, αX, and αD. Historically and experimentally, an important feature of this domain is its ability to fold on its own, allowing important insights into integrin function.

When the first I domain (from αM) was crystallized, two different structures were noted which depended on the divalent cation used in the crystallization process. The difference between the two structures was a small change in the divalent cation binding site and a much greater shift in the position of an alpha helix at the carboxyterminus of the domain. It is now clear that these two structures represent conformations which differ in affinity for ligand. This conformation-dependent affinity change can account for at least part of the well-documented ability of integrins to alter their ligand binding properties in response to cell activation ("inside-out signaling"). A caveat to this hypothesis is that it would apply only to those few integrins with α chain I domains. However, if the βA domain, which is structurally homologous to this α chain I domain, can undergo a similar conformational change in response to cell activation, the ability of the βA domain to assume two distinct conformations could go a long way toward explaining cellular regulation of ligand binding for non-I domain-containing integrins. The recent crystal structure of $\alpha V\beta3$ with its peptide ligand shows that ligand indeed can induce conformational changes in the βA domain (Xiong et al. 2002), consistent with this hypothesis about integrin activation. In this respect, the homology of integrins with heterotrimeric G proteins is quite intriguing; it appears that nature has conserved a specific molecular mechanism for communicating cell activation in two quite disparate pathways.

Understanding of the mechanism for regulation of ligand binding by I domain-containing integrins has led to the generation of a unique reagent for inhibition of lymphocyte function associated antigen (LFA)-1 ($\alpha L\beta2$) function. Shimaoka et al. (2001) created a soluble I domain that was locked into the high-affinity conformation by introduction of a disulfide bond at relevant sites in the domain. This constitutively high-affinity receptor had almost 10,000-fold increased affinity for intercellular adhesion molecule (ICAM)-1 compared to wild-type I domain or intact $\alpha L\beta2$ and acted as an inhibitor of $\alpha L\beta2$ function both in vitro and in vivo. To this point, only monoclonal antibodies have been available to inhibit $\alpha L\beta2$ function; the new inhibitor has the advantage that it is smaller and, at least theoretically, more readily diffusible to sites where leukocyte $\alpha L\beta2$ might be engaged outside the vasculature.

2
Integrin Expression in Macrophages and Macrophage-Like Cells

Monocytes and macrophages constitutively express integrins of the $\beta1$, $\beta2$, and $\beta5$ families (Table 1). Of the 12 known $\beta1$ integrins, monocytes and macrophages have been shown to express $\alpha1$, $\alpha2$, $\alpha3$, $\alpha4$, $\alpha5$, and $\alpha6$ associated with $\beta1$. These cells also constitutively express αL and αM associated with $\beta2$. In humans, numerous macrophage populations also express $\alpha X\beta2$ (CDllc/CD18), and

Table 1 Monocyte and macrophage integrins

β chain family	Associated α	Ligands
$\beta1$	$\alpha1$	Collagen
	$\alpha2$	Collagen, laminin
	$\alpha3$	Laminin
	$\alpha4$	Fibronectin, VCAM-1
	$\alpha5$	Fibronectin
	$\alpha6$	Laminin
$\beta2$	αM	ICAM-1, fibrinogen, others[b]
	αL	ICAM-1, ICAM-2
	αX	Fibrinogen
	αD	ICAM-3
$\beta3$	αV	Many RGD-containing proteins
$\beta5$	αV	Similar to $\alpha V\beta3$
$\beta7$[a]	$\alpha4$	MADCAM-1, fibronectin
	αE	E-cadherin

[a] For a discussion of ligands for $\alpha M\beta2$, see Brown (1991), Ross and Vetvicka (1993), Ehlers (2000), and Humphries and Mould (2001)

[b] There is a single report of $\beta7$ expression on macrophages

expression of this receptor increases during monocyte differentiation to macrophages, but in mouse this ICAM-3 receptor seems to be restricted to dendritic cells. A fourth $\beta2$-associated a chain, αD, also seems to be expressed to a greater extent on tissue macrophages than monocytes. Coordinate expression of αX and αD is not surprising in view of their tight linkage (11.5 kbp apart) on chromosome 22. αV is the integrin α chain with the most promiscuous β chain pairings, since it can interact with $\beta1$, $\beta3$, $\beta5$, $\beta6$, and $\beta8$. On human monocytes and murine bone marrow macrophages, the major αV-containing integrin is $\alpha V\beta5$. Of importance, during the maturation of monocytic cells to osteoclasts, $\alpha V\beta3$ expression increases markedly, and $\alpha V\beta5$ expression diminishes. Indeed $\alpha V\beta3$ is the major osteoclast integrin, and it appears to be important both for migration of osteoclasts to and on bone and for regulation of osteoclast survival (McHugh et al. 2000). Unligated $\alpha V\beta3$ leads to rapid apoptosis (Stupack et al. 2001). $\beta7$ integrins ($\alpha E\beta7$ and $\alpha4\beta7$), expressed on gut-homing lymphocytes, are not present on blood monocytes, but can be induced during inteferon (IFN)-γ-mediated differentiation (Tiisala et al. 1995). These integrins are present on macrophages in liver, lung, and lymph node sinuses, suggesting that they may be markers of differentiation from monocyte to tissue macrophage.

3
Integrins in Macrophage Biology

There have been numerous reviews of leukocyte integrin function (e.g., Madri and Graesser 2000; Ley 2001; Woods and Shimizu 2001; Bunting et al. 2002), so this section will only highlight briefly the major areas of relevance to macrophage biology.

3.1 Integrins in Monocyte/Macrophage Extravasation

Integrins can mediate both cell–cell and cell–extracellular matrix adhesions. Integrins are necessary for monocyte migration from blood to tissue, which is necessary for macrophage accumulation at sites of inflammation. Whether monocyte extravasation is necessary for replenishing tissue macrophages during normal homeostasis is less certain. There is evidence both for self-renewal of tissue macrophages and for repletion of this population from the bone marrow (Naito et al. 1996). Essentially all studies of integrin involvement in monocyte extravasation have been done in the context of inflammation. The first point of integrin engagement is during monocyte-endothelial interaction. Firm integrin-mediated adhesion of monocyte to endothelium is required for successful extravasation; the major integrins involved in this appear to be $\alpha L\beta 2$ and $\alpha 4\beta 1$. These integrins appear to have overlapping functions in mediating firm adhesion. $\alpha 4\beta 1$ recognizes vascular cell adhesion molecule (VCAM)-1, an endothelial ligand induced at sites of inflammation. $\alpha L\beta 2$ is known to interact with both ICAM-1 and ICAM-2 expressed on endothelial cells; it has also recently been shown to recognize the endothelial tight junction protein junctional adhesion molecule (JAM)-1 may be important in initiating leukocyte migration through the interendothelial junctions (Ostermann et al. 2002). Presumably, mechanisms of transendothelial migration are similar in maintenance of the tissue macrophage pool in homeostatic conditions; since $\alpha L\beta 2$ can recognize endothelial ligands that do not require inflammation for expression (ICAM-2 and JAM-1), it would be a leading candidate to mediate monocyte-endothelial interaction in this circumstance. Consistent with this, unperturbed $\beta 2$ integrin-deficient mice have decreased numbers of pulmonary dendritic cells and macrophages (Schneeberger et al. 2000). However, the defect is only partial, suggesting that there are alternative pathways for maintenance of the tissue macrophage pool.

Transendothelial migration also is important in dendritic cell migration from tissues to lymph nodes after antigen exposure. In this case, cells must cross the lymphatic endothelium to gain entry into the lymph nodes where they can interact with T cells to initiate an immune response. There is some evidence that this migration may be mediated by $\alpha 4\beta 1$ (Zou et al. 2001), although other integrins may contribute as well (Weiss et al. 2001). $\alpha 4\beta 1$ expression is increased during the dendritic cell differentiation that occurs concomitant with migration back to lymph nodes (Puig-Kroger et al. 2000). "Reverse transmigration" (movement across endothelium from the abluminal to the lumenal side) may be important in dendritic cell entry into the afferent lymphatic circulation in order to migrate to the lymph nodes. In vitro, monocyte maturation to dendritic cells is required for this migration, and the maturation process requires both phagocytosis of antigen and interaction of monocytes with collagen (Randolph et al. 1998). This may implicate integrins not only in the transmigration process but also in differentiation to dendritic cells. However, no investigation of the roles of specific integrins in either reverse transmigration or dendritic differentiation of monocytes has been reported.

3.2 Macrophage Interaction with Extracellular Matrix

In addition to transendothelial migration, macrophage integrins mediate both cell migration through tissue and stable adhesion at sites of inflammation, wound repair, and other perturbations of homeostasis. Macrophages have not to date been a popular model system for study of the mechanisms of cell migration, presumably because they are neither as motile as neutrophils or lymphocytes nor as easy to transfect as fibroblasts. Progress in understanding integrins and cell migration in these other model systems has been rapid and reviewed extensively in recent years (e.g., Eliceiri 2001; van der Flier and Sonnenberg 2001; Woods and Shimizu 2001) and will not be reviewed in detail here. Among the key enzymes involved in integrin-mediated adhesion to and/or migration on extracellular matrix are focal adhesion kinase (FAK), PI 3-kinase, and src family kinases. In fibroblasts, endothelial cells, and T cells, FAK can act as a scaffolding molecule for assembly of the other kinases together with certain key structural molecules, such as paxillin, upon integrin ligation. However, FAK itself is only minimally expressed in macrophages which instead express a homologue called pyk2 (Duong and Rodan 2000). While there is good evidence that pyk2 phosphorylation can be activated by adhesion, whether it serves an analogous role to FAK in macrophage migration remains debated. Despite this caveat, there is excellent evidence that phosphatidyl inositol (PI) 3-kinase and src family kinases are involved in integrin mediated adhesion and migration (Suen et al. 1999; Ridley 2001). This suggests that, despite differences in detail, the basic mechanisms of regulation of integrin function are similar in macrophages and fibroblasts.

A potentially important role for integrin–extracellular matrix interactions in macrophages is to regulate phagocytic function. Interaction with several different extracellular matrix proteins, including fibronectin, collagen, and laminin stimulates phagocytosis via both Fcγ and complement receptors (Pommier et al. 1983; Bohnsack et al. 1985; Newman and Tucci 1990). This activation of phagocytic function occurs in polymorphonuclear leukocytes as well, but may have a more complex regulation in that cell type. The $\alpha V\beta 3$ integrin has a major role in this activation and requires association with a second plasma membrane protein, CD47, to initiate the signal. Interferon-γ treatment can downregulate extracellular matrix-mediated activation of macrophages (Wright et al. 1986), suggesting that immunologically activated macrophages may lose some integrin signaling mechanisms; the molecular basis for this phenomenon is unknown.

3.3 Macrophage Integrins in Macrophage Cell–Cell Interactions

Macrophages and dendritic cells participate in a variety of cell–cell interactions outside the vasculature as an essential aspect of their biological functions. For example, they must interact with other cells of the immune system (T cells and B cells), with epithelial cells in the lung and gut, with virally infected cells as part of the immune response, with tumor cells as part of the innate anti-tumor response, and with bone marrow stromal cells during development. Furthermore, macrophages are important in removal of apoptotic cells during development and in adult life.

Interaction with bone marrow stroma seems to depend on interactions between $\alpha4\beta1$ on monocyte precursors and VCAM-1 expressed on stromal cells (Teixido et al. 1992; Harada et al. 1998; Berrios et al. 2001; Hidalgo et al. 2001). $\beta2$ integrins may contribute to this interaction as well; they are clearly important in interactions with tumor cells, infected fibroblasts, and possibly some epithelial cells that express ICAM-1 (Shang and Issekutz 1998; Rosseau et al. 2000; Chang et al. 2002). However, in the case of macrophage interactions with B and T cells, most of the emphasis in studies to date has been on lymphocyte integrin interactions with macrophage immunoglobulin (Ig) superfamily ligands, both for formation of the immunologic synapse (which involves interaction of T cell $\alpha L\beta2$ with ICAM-1 on dendritic cells or other antigen-presenting cells) and for T-cell homing to thymus or B-cell homing to lymph nodes. Nonetheless, both B and T cells can express ICAM-1 and ICAM-3, so it seems likely that some contribution to cell–cell interactions may arise from recognition of these lymphocyte Ig ligands by macrophage integrins.

In contrast to the almost exclusive use of $\alpha4\beta1$ and $\beta2$ integrins in these cell–cell interactions, several groups have described a role for αV integrins in macrophage recognition of apoptotic cells (Leverrier and Ridley 2001; Ren et al. 2001). Because of the large number of receptors involved in recognition of apoptotic cells by macrophages (Fadok and Henson 1998), inhibition of αV integrins, either with antibody or arginine-glycine-aspartic acid (RGD) peptides, only leads to about 50% diminution of phagocytosis of apoptotic cells. Nonetheless, transfection of $\alpha V\beta5$ can confer ability to recognize and ingest apoptotic cells on 293 epithelial cells (Albert et al. 2000), and ingestion is initiated by a signaling pathway reminiscent of the pathway defined by studies of apoptotic cell phagocytosis in *Caenorhabditis elegans* (Henson et al. 2001). Phagocytosis of apoptotic cells is anti-inflammatory because these targets induce macrophage production of transforming growth factor (TGF)-β. It has been reported that ligation of αV integrins leads to TGF-β synthesis (Freire-de-Lima et al. 2000), which would suggest that these integrins are critical to the macrophage response to apoptotic cells, but this view has been challenged recently (Huynh et al. 2002).

4
Integrins and the Macrophage Cytoskeleton

Integrins are named because they integrate the extracellular matrix with the intracellular matrix, or cytoskeleton (Tamkun et al. 1986). There are direct connections between integrin cytoplasmic domains and the actin cytoskeleton, mediated via talin, vinculin, and α-actinin. While no direct links are known between integrins and other components of the cytoskeleton, there is evidence that indirect interactions between microtubules and integrins may regulate integrin avidity in macrophages (Zhou et al. 2001). Perhaps this is unsurprising, given the extensive crosstalk among various structural elements of the cytoskeleton, but the involvement of microtubules in regulating integrin adhesion points to potential new mechanisms of cellular control and new targets for pharmacologic regulation. Moreover, the connection between microtubules and integrins may be more fundamental than initially apparent, since the microtubule polymerization inhibitor colchicine blocks integrin-mediated but not FcγR-mediated phagocytosis (Munthe-Kaas et al. 1976). Possibly, microtubule-dependent delivery of intracellular vesicle membranes is important for integrin regulation.

5
Integrin Signaling in Migration and Phagocytosis

Macrophage migration and spreading on extracellular matrix proteins are dependent on integrin contacts with cytoskeleton. The basic paradigm developed from work in fibroblasts—that different rho family GTPases regulate distinct aspects of cytoskeletal assembly—seems to operate in macrophages as well. cdc42 and rac are apparently important in filapodial and lamellar extension, while rho is involved in the generation of contractility. In macrophage migration, both rac and rho are required for migration, and rac seems to be activated downstream of PI 3-kinase, at least in response to CSF-1 (Ridley 2001). In contrast, the requirement for cdc42 is not for migration but for appropriate sensing of a chemotactic gradient; when cdc42 is inhibited, macrophages migrate randomly (Ridley et al. 1999). One difference between fibroblasts and macrophages is that actin stress fibers are less prominent in macrophages than fibroblasts, and "focal contacts"—the linear arrangement of integrins at adhesion sites in response to the contractile force of the stress fibers—are very much more frequent in fibroblasts. Instead, rhoA—acting through Rho kinase and perhaps other effectors—is very important for the contractile force on the uropod required for macrophage migration. Inhibition of rhoA or rho kinase leads to macrophages that have lengthy, extended uropods, and fail to migrate (Worthylake et al. 2001).

Recent work has emphasized that, in contrast to migration, integrin-mediated phagocytosis requires rho, but is independent of cdc42 and rac (Caron and Hall 1998). This is quite distinct from FcγR-mediated ingestion, which is dependent on cdc42 and rac, but independent of rho. The exclusive dependence of integrin-mediated phagocytosis on rho may correlate with the morphology of this sort of ingestion. In FcγR-mediated phagocytosis, macrophage membrane protrudes around the target; in contrast, in integrin-mediated phagocytosis, target particles sink into the cytoplasm, and there is little or no protrusion of membrane around the particle, emphasizing the contractile rather than protrusive functions of cytoskeleton. This dependence on contractility is consistent with the requirement for rho guanosine triphosphate (GTP)ase, which seems most closely associated with this cytoskeletal function. Almost all studies of integrin-mediated phagocytosis have examined $\alpha M\beta 2$; it remains to be determined whether the rho dependence of the process is true for integrins in general. There are many examples of macrophage phagocytosis via other integrins, including $\alpha V\beta 3$, $\alpha V\beta 5$, and $\alpha 5\beta 1$.

There are other distinctions between the signals required for integrin-mediated spreading and phagocytosis. While syk is required for integrin $\beta 2$-dependent cell spreading in macrophages (Vines et al. 2001), it is not necessary for $\alpha M\beta 2$-mediated phagocytosis (Kiefer et al. 1998). Indeed, even quite broadly reactive tyrosine kinase inhibitors do not block $\alpha M\beta 2$-mediated phagocytosis (Allen and Aderem 1996). Consistent with a different requirement of tyrosine kinases in integrin-dependent adhesion and phagocytosis, macrophages that fail to express the transmembrane tyrosine phosphatase CD45 show abnormal adhesion and spreading, but ingest normally (Roach et al. 1997). Together, these studies suggest that integrin ligation can activate a tyrosine kinase-initiated pathway in macrophages involving syk that likely leads to cdc42 and rac activation downstream from syk activation of PI 3-kinase. This pathway is required for events at the leading edge of an orientated macrophage, including lamellar protrusion, that are required for migration and spreading. In contrast, this pathway is not required for integrin-mediated ingestion. The data imply that there must be a pathway for rho activation during phagocytosis—independent of tyrosine kinases—that is sufficient to support integrin-mediated ingestion. Rho can be activated by certain G protein-coupled receptors, and there is evidence that integrin ligation can activate heterotrimeric G protein signaling (Erb et al. 2001; Wei et al. 2001). Thus, it is intriguing to speculate that integrin-mediated ingestion may be initiated or sustained by a pathway involving heterotrimeric G protein activation of rho. It is possible that protein kinase C, which also is required for integrin-mediated ingestion (Allen and Aderem 1995), synergizes with rho to activate phospholipase D (Du et al. 2000). Phospholipase D activity is closely correlated with integrin-mediated phagocytosis (Kusner et al. 1996; Serrander et al. 1996). Given these distinctions, it might be possible to target pharmacologically either integrin-mediated phagocytosis or integrin-mediated adhesion and migration without affecting the other function. This could be useful in modulating, e.g., presentation of particulate antigens, without interfering with other host defense functions of the macrophages.

6
Integrins as Components of Multimolecular Complexes on Macrophages

An emerging area of understanding is that integrins can function as components of multimolecular complexes at the plasma membrane that have unique functions that are more than the sum of the individual receptors. Although not the first to be discovered, complexes of growth factor receptors and integrins have received the most intense scrutiny because it is clear that these complexes are responsible for many of the survival or proliferative signals initially attributed to growth factor signaling in anchorage-dependent cells. For many years, it has been known that most nontransformed cells require adhesion for survival or proliferation, even in the presence of adequate growth factor. This role for adhesion is now clearly attributable to integrin signaling. While fibroblasts undergo only growth arrest on loss of adhesion, endothelial cells and epithelial cells die from a form of apoptosis termed anoikis ("homelessness"). Even fibroblasts require adhesion for survival in the absence of growth factors (Ilic et al. 1998). In fibroblasts, endothelial cells, and epithelial cells, complexes among integrins and EGF or platelet-derived growth factor (PDGF) receptors have been identified. The growth factor receptors in these complexes are phosphorylated to a greater extent and more efficiently signal microtubule-associated protein (MAP) kinase activation than receptors not associated with integrins (DeMali et al. 1999; Yu et al. 2000). In addition to survival or proliferation, these complexes may be important in cell migration in response to growth factors. Although these phenomena have not been studied extensively in macrophages, they may have relevance to macrophage biology. Colony-stimulating factor (CSF)-1 withdrawal leads to both cell rounding and apoptosis of murine macrophages (Pixley et al. 2001); both processes require the tyrosine phosphatase Src-homology-containing phosphatase (SHP)-1 (Roach et al. 1998; Berg et al. 1999), suggesting a possible functional link between the two phenomena. Moreover, deliberate inhibition of adhesion induces apoptosis in bone marrow-derived macrophages, suggesting that these cells may undergo anoikis. Finally, genetic defects that lead to diminished adhesion (e.g., absence of the transmembrane tyrosine phosphatase CD45 or the integrin CD18) lead to rapid apoptosis on growth factor withdrawal. These data are all consistent with an important role for cooperation between integrins and c-fms, the CSF-1 receptor, in macrophage biology. To date, no evidence about whether or not macrophage integrins form a complex with the CSF-1 receptor has been published.

A second multimolecular complex at the plasma membrane involves integrins and a family of proteins known as tetraspanins (Berditchevski 2001). As their name suggests, the tetraspanin family all have four transmembrane segments, with a large extracellular loop between the third and fourth segment and short aminoterminal and carboxyterminal cytoplasmic domains. There are at least 26 members of the tetraspanin family in higher eukaryotes, many of which have broad cellular distribution. There has been no systematic study of tetraspanin expression in macrophages, but several, including CD9 and CD81, which have important integrin associations, are known to be expressed on macrophages. Tetraspanin–integrin complexes have been implicated both in cell adhesion and cell motility. Tetraspanins may contribute to these integrin-dependent phenomena by recruiting important signaling enzymes, such as phosphatidylinositol 4-kinase and protein kinase C, to sites of adhesion. Tetraspanins may also contribute to regulation of integrin expression at the plasma membrane or recycling from coated pits. Very little investigation has been done to date on the role of these undoubtedly physiologically important complexes in macrophage integrin function.

The association of specific integrins with particular proteases is important in cell migration through extracellular matrix, presumably because the interaction allows the adhesive integrin to direct the matrix degrading activity to specific sites on the cell surface. The association of matrix metalloproteinase (MMP)-2 with $\alpha V\beta 3$ has been shown to be involved in angiogenesis (Brooks et al. 1998), and the protease and integrin seem to act synergistically, since while intact collagen is not a ligand for $\alpha V\beta 3$, the integrin can recognize collagen that has been degraded by the MMP. Association of $\alpha V\beta 6$ with MMP-9 is implicated in metastasis of colorectal cancer (Niu et al. 1998), and $\alpha 2\beta 1$ ligation by collagen can induce synthesis of MMP-1 (Jones and Walker 1997). Although macrophage MMPs have been implicated in a variety of chronic inflammatory and destructive diseases, such as arthritis, emphysema, and atherosclerosis, there has not been a systematic examination of the significance of the MMP–integrin interaction for these diseases.

A second example of protease–integrin association involves a multimolecular complex involving integrins and the urokinase receptor (uPAR). UPAR is a glycan phosphoinositol (GPI)-linked receptor that has been coprecipitated with $\alpha V\beta 3$, $\alpha M\beta 2$, $\alpha 3\beta 1$, and $\alpha 5\beta 1$ in a variety of cells. Binding of the enzyme ligand plasminogen activator to uPAR enhances integrin-mediated adhesion and migration. The site of interaction of $\alpha M\beta 2$ with uPAR has been mapped to a region of the αM chain in the beta propeller domain. A peptide based on this αM sequence also inhibited uPAR association with $\alpha 3\beta 1$ and $\alpha 5\beta 1$, suggesting that a common site on uPAR interacts with diverse integrins. Other GPI-linked receptors can be physically associated with the $\alpha M\beta 2$ integrin, including FcγRIIIB and CD14 (Poo et al. 1995; Pfeiffer et al. 2001). FcγRIIIB function as a signaling or phagocytic receptor may depend on its association with $\alpha M\beta 2$. Finally, some transmembrane receptors (e.g., FcgRIIA; Petty and Todd 1996) physically associate with specific integrins; in these circumstances, integrins and associated receptors reciprocally influence each other's function.

The best-studied example of a plasma membrane multimolecular complex involving association of integrins with other adhesion receptors is the interaction of a subset of integrins with CD47 (Brown and Frazier 2001). CD47 is an Ig superfamily member that has five transmembrane segments that can be coprecipitated with $\beta 3$ integrins and $\alpha 2\beta 1$. It also may associate with other integrins, perhaps with affinity too low to be detected by coprecipitation. The known ligands for CD47 include thrombospondins, through which CD47 may participate in recognition of the extracellular matrix, and signal regulatory protein (SIRP)α, a plasma membrane protein expressed on a variety of cells, through which CD47 may participate in cell–cell interactions. The unusual structure of CD47 suggests that it can act as a signaling molecule, and there is abundant experimental evidence that supports this hypothesis. In association with integrins, CD47 can activate heterotrimeric G proteins, especially Gi (Brown 2001); this complex may be important in the integrin-mediated functions that appear to depend on heterotrimeric G protein signaling. However, it is clear that not all signaling by CD47 is inhibited by pertussis toxin; in these cases, the proximal signals from CD47 ligation are uncertain. Ligation of the integrin-CD47 complex can lead to cell migration, activation, and aggregation, and mice lacking CD47 have a defect in integrin-mediated cell activation that leads to a deficiency in host defense. The significance of the complex for macrophage biology specifically has not been studied in detail.

7 Therapeutic Inhibition of Macrophage Integrin Function

The previous discussion makes it clear that there are a number of integrins expressed on macrophages that might be rational targets for therapeutic intervention, but several conceptual problems exist that may limit the utility of therapeutic integrin blockade. The first problem is that integrins present on macrophages are usually expressed on other cells as well. Therefore, integrin inhibition might lead to undesired side effects because of blockade of essential functions on these other cells. The second problem is redundancy in integrin function. For example, multiple $\beta1$ integrins recognize the extracellular matrix fibronectin, so that inhibition of cell binding to this ligand could require inhibition of as many as six different integrins. At the same time, too broad an inhibition of integrins could lead to unwanted side effects. For example, inhibition of $\beta2$ integrin function with monoclonal antibodies, although it has beneficial effects on acute survival from reperfusion injury (Winn et al. 1997), markedly increases susceptibility to bacterial infection (Mileski et al. 1993). Nonetheless, there are some diseases in which anti-integrin therapy has had clear utility. The greatest success to date has been the use of inhibitors of $\alpha IIb\beta3$, the major platelet integrin required for platelet aggregation, as acute anti-thrombotic therapy following angioplasty (Coller 2001).

There are several chronic inflammatory diseases that seem to be natural candidates for anti-macrophage integrin therapy because they are thought to play an important role in pathogenesis. These include multiple sclerosis, inflammatory bowel disease, various arthritides, and atherosclerosis. At this point, the data for involvement of integrins comes primarily from in vitro studies or experiments in animal models. Much of the work has focused on the integrin $\alpha4\beta1$ because of its predominant expression on leukocytes and its importance for transendothelial migration of both lymphocytes and macrophages. Thus, inhibition of $\alpha4\beta1$ has the possibility to block trafficking of these two cell types without significant effect on other biologic processes. Moreover, since polymorphonuclear neutrophil (PMN) express little $\alpha 4\beta 1$, blockade of this integrin has less potential to increase susceptibility to infection than blockade of $\beta 2$ integrins, which are highly expressed on PMN. A humanized monoclonal antibody that blocks $\alpha 4\beta 1$ function is now in clinical trials for both multiple sclerosis and Crohn's disease. Other potentially important targets are the collagen-binding integrins $\alpha 1\beta 1$ and $\alpha 2\beta 1$ that are expressed more widely than $\alpha 4\beta 1$. These two integrins are expressed on macrophages and on lymphocytes at sites of chronic inflammation, and inhibition of these integrins has a preventive or therapeutic effect in several murine models of chronic inflammation (de Fougerolles 2000). While most therapeutic effects of anti-$\alpha 4\beta 1$ may be attributable to effects on lymphocyte trafficking, inhibition of $\alpha 1$ and $\alpha 2$ blocks inflammation in diseases that do not have a lymphocyte component, implicating a therapeutic effect of blockade of these molecules on macrophages.

Because of the important role of macrophages in the pathogenesis of atherosclerosis and the early expression of VCAM-1 (an $\alpha 4\beta 1$ ligand) on endothelium overlying sites of intimal proliferation, atherosclerosis seems to be a good target for anti-integrin therapy (Li et al. 1993). Furthermore, ICAM-1 (a ligand for $\alpha L\beta 2$ and $\alpha M\beta 2$) likely contributes to the atherogenic process, which is modestly delayed in ICAM-1-deficient mice (Collins et al. 2000). While CD11b deficiency does not have the same protective effect against atherogenesis (Kubo et al. 2000), combined deficiency of $\alpha L\beta 2$ and $\alpha M\beta 2$ has not been investigated. Because $\alpha 4$ deficiency leads to embryonic lethality, atherosclerosis in $\alpha 4$-deficient mice has not been investigated. However, in some models, short-term treatment with anti-$\alpha 4$ antibodies or $\alpha 4\beta 1$ blocking peptides reduces both monocyte and lipid accumulation in plaques (Huo and Ley 2001), suggesting the possibility that successful targeting of monocyte $\alpha 4\beta 1$ could have a significant impact on generation of the atheromatous lesion.

Finally, a very promising target for anti-integrin therapy is in treatment of osteoporosis, a disease caused by increased bone resorption by osteoclast, a blood monocyte-derived cell, relative to bone formation by osteoblasts. The multinucleate osteoclast develops a specialized bone-apposed plasma membrane that is rich in the H^+ ATPase proton pump, allowing secretion of acid into a space between the cell and the mineralized surface of the bone, that is thought to be important in the process of bone resorption. The fact that the pH of this compartment is quite different from the surrounding medium implies that there must be a tight barrier between this compartment and the rest of the milieu of the osteoclast. It is thought that integrin $\alpha V\beta 3$-mediated adhesion is required for this tight seal. During differentiation from monocyte to osteoclasts, cells begin to express the integrin $\alpha V\beta 3$ at high levels. Moreover, $\alpha V\beta 3$ is present at the circumference of the acid bone-resorbing compartment. In mice, $\beta 3$ deficiency leads to enhanced bone mineralization (osteosclerosis) (McHugh et al. 2000), and inhibitors of $\alpha V\beta 3$ significantly block bone resorption in several animal models of osteoporosis (Crippes et al. 1996; Engleman et al. 1997). Blockade of $\alpha v\beta 3$ also inhibits tumor-induced angiogenesis and neovascularization at inflammatory sites (Eliceiri and Cheresh 1999). While this anti-angiogenic effect is a potential problem, these antagonists apparently have no effect on already established vasculature. Indeed, the additional effect of antagonizing $\alpha V\beta 3$ may make this integrin a target in inflammation and cancer as well as osteoporosis.

Because of the side effects associated with general inhibition of $\beta 2$ integrins, these have been considered poor targets for therapeutic intervention. However, as we learn more about the signaling cascades associated with integrin ligation, it may be possible to develop more specific inhibitors that can block some aspects of integrin signaling without disturbing integrin-mediated adhesion and migration. Such a strategy might lead to blockade of consequences of chronic adhesion—for example, macrophage synthesis of growth factors and chemoattractants for smooth muscle cells at atherosclerotic plaques—while allowing normal migration of PMN and macrophages to sites of acute infection. However, neither the molecular mechanisms required for activation of adhesion and migration—"inside out signaling"—or the signaling pathways impacted by integrin engagement—"outside in signaling"—are known in sufficient detail to exploit this strategy for therapeutic purposes. This both justifies further basic research into understanding the molecular mechanisms of regulation of integrin function and holds hope that the future will see important advances in manipulation of integrin targets for therapeutic purposes.

8 References

Albert,M.L., Kim,J.I., and Birge,R.B. 2000. alphavbeta5 integrin recruits the CrkII-Dock180-rac1 complex for phagocytosis of apoptotic cells. Nat.Cell Biol. 2:899–905

Allen,L.A.H. and Aderem,A. 1996. Molecular definition of distinct cytoskeletal structures involved in complement- and Fc receptor-mediated phagocytosis in macrophages. J.Exp.Med. 184:627–637

Allen,L.H. and Aderem,A. 1995. A role for MARCKS, the alpha isozyme of protein kinase C and myosin I in zymosan phagocytosis by macrophages. J.Exp.Med. 182:829–840

Berditchevski,F. 2001. Complexes of tetraspanins with integrins: more than meets the eye. J.Cell Sci. 114:4143–4151

Berg,K.L., Siminovitch,K.A., and Stanley,E.R. 1999. SHP-1 regulation of p62(DOK) tyrosine phosphorylation in macrophages. J.Biol.Chem. 274:35855–35865

Berrios,V.M., Dooner,G.J., Nowakowski,G., Frimberger,A., Valinski,H., Quesenberry,P.J., and Becker,P.S. 2001. The molecular basis for the cytokine-induced defect in homing and engraftment of hematopoietic stem cells. Exp.Hematol. 29:1326–1335

Bohnsack,J.F., Kleinman,H., Takahashi,T., O'Shea,J.J., and Brown,E.J. 1985. Connective tissue proteins and phagocytic cell function: laminin enhances complement and Fc-mediated phagocytosis by cultured human macrophages. J.Exp.Med. 161:912–923

Brooks,P.C., Silletti,S., Von Schalscha,T.L., Friedlander,M., and Cheresh,D.A. 1998. Disruption of angiogenesis by PEX, a noncatalytic metalloproteinase fragment with integrin binding activity. Cell 92:391–400

Brown,E. 2001. Integrin-associated protein (CD47): an unusual activator of G protein signaling. J.Clin.Invest 107:1499–1500

Brown,E.J. 1991. Complement receptors and phagocytosis. Curr.Opin.Immunol. 3:76–82

Brown,E.J. and Frazier,W.A. 2001. Integrin-associated protein (CD47) and its ligands. Trends Cell Biol. 11:130–135

Bunting,M., Harris,E.S., McIntyre,T.M., Prescott,S.M., and Zimmerman,G.A. 2002. Leukocyte adhesion deficiency syndromes: adhesion and tethering defects involving beta 2 integrins and selectin ligands. Curr.Opin.Hematol. 9:30–35

Caron,E. and Hall,A. 1998. Identification of two distinct mechanisms of phagocytosis controlled by different Rho GTPases. Science 282:1717–1721

Chang,Y.J., Holtzman,M.J., and Chen,C.C. 2002. Interferon-gamma-induced epithelial ICAM-1 expression and monocyte adhesion. Involvement of protein kinase C-dependent c-Src tyrosine kinase activation pathway. J.Biol.Chem. 277:7118–7126

Coller,B.S. 2001. Anti-GPIIb/IIIa drugs: current strategies and future directions. Thromb.-Haemost. 86:427–443

Collins,R.G., Velji,R., Guevara,N.V., Hicks,M.J., Chan,L., and Beaudet,A.L. 2000. P-Selectin or intercellular adhesion molecule (ICAM)-1 deficiency substantially protects against atherosclerosis in apolipoprotein E-deficient mice. J.Exp.Med. 191:189–194

Crippes,B.A., Engleman,V.W., Settle,S.L., Delarco,J., Ornberg,R.L., Helfrich,M.H., Horton, M.A., and Nickols,G.A. 1996. Antibody to beta3 integrin inhibits osteoclast-mediated bone resorption in the thyroparathyroidectomized rat. Endocrinology 137:918–924

de Fougerolles,A.R., Sprague,A.G., Nickerson-Nutter,C.L., Chi-Rosso,G., Rennert,P.D., Gardner,H., Gotwals,P.J., Lobb,R.R., and Koteliansky,V.E. 2000. Regulation of inflammation by collagen-binding integrins alpha1beta1 and alpha2beta1 in models of hypersensitivity and arthritis. J.Clin.Invest 105:721–729

DeMali,K.A., Balciunaite,E., and Kazlauskas,A. 1999. Integrins enhance platelet-derived growth factor (PDGF)-dependent responses by altering the signal relay enzymes that are recruited to the PDGF beta receptor. J.Biol.Chem. 274:19551–19558

Du,G., Altshuller,Y.M., Kim,Y., Han,J.M., Ryu,S.H., Morris,A.J., and Frohman,M.A. 2000. Dual requirement for rho and protein kinase C in direct activation of phospholipase D1 through G protein-coupled receptor signaling. Mol.Biol.Cell 11:4359–4368

Duong,L.T. and Rodan,G.A. 2000. PYK2 is an adhesion kinase in macrophages, localized in podosomes and activated by beta(2)-integrin ligation. Cell Motil.Cytoskeleton 47:174–188

Ehlers,M.R. 2000. CR3: a general purpose adhesion-recognition receptor essential for innate immunity. Microbes.Infect. 2:289–294

Eliceiri,B.P. 2001. Integrin and growth factor receptor crosstalk. Circ.Res. 89:1104–1110

Eliceiri,B.P. and Cheresh,D.A. 1999. The role of alphav integrins during angiogenesis: insights into potential mechanisms of action and clinical development. J.Clin.Invest 103:1227–1230

Engleman, V.W., Nickols,G.A., Ross,F.P., Horton,M.A., Griggs,D.W., Settle,S.L., Ruminski, P.G., and Teitelbaum,S.L. 1997. A peptidomimetic antagonist of the alpha(v)beta3 integrin inhibits bone resorption in vitro and prevents osteoporosis in vivo. J.Clin.Invest 99:2284–2292

Erb,L., Liu,J., Ockerhausen,J., Kong,Q., Garrad,R.C., Griffin,K., Neal,C., Krugh,B., Santiago-Perez,L.I., Gonzalez,F.A. et al. 2001. An RGD sequence in the P2Y(2) receptor interacts with alpha(V)beta(3) integrins and is required for G(o)-mediated signal transduction. J.Cell Biol. 153:491–501

Fadok,V.A. and Henson,P.M. 1998. Apoptosis: getting rid of the bodies. Curr.Biol. 8:R693-R695

Freire-de-Lima,C.G., Nascimento,D.O., Soares,M.B., Bozza,P.T., Castro-Faria-Neto,H.C., de Mello,F.G., DosReis,G.A., and Lopes,M.F. 2000. Uptake of apoptotic cells drives the growth of a pathogenic trypanosome in macrophages. Nature 403:199–203

Harada,H., Kukita,T., Kukita,A., Iwamoto,Y., and Iijima,T. 1998. Involvement of lymphocyte function-associated antigen-1 and intercellular adhesion molecule-1 in osteoclastogenesis: a possible role in direct interaction between osteoclast precursors. Endocrinology 139:3967–3975

Henson,P.M., Bratton,D.L., and Fadok,V.A. 2001. Apoptotic cell removal. Curr.Biol. 11:R795-R805

Hidalgo,A., Sanz-Rodriguez,F., Rodriguez-Fernandez,J.L., Albella,B., Blaya,C., Wright,N., Cabanas,C., Prosper,F., Gutierrez-Ramos,J.C., and Teixido,J. 2001. Chemokine stromal cell-derived factor-1alpha modulates VLA-4 integrin-dependent adhesion to fibronectin and VCAM-1 on bone marrow hematopoietic progenitor cells. Exp.Hematol. 29:345–355

Humphries,M.J. and Mould,A.P. 2001. Structure. An anthropomorphic integrin. Science 294:316–317

Huo,Y. and Ley,K. 2001. Adhesion molecules and atherogenesis. Acta Physiol Scand. 173:35–43

Huynh,M.L., Fadok,V.A., and Henson,P.M. 2002. Phosphatidylserine-dependent ingestion of apoptotic cells promotes TGF-beta1 secretion and the resolution of inflammation. J.Clin.Invest 109:41–50

Ilic,D., Almeida,E.A., Schlaepfer,D.D., Dazin,P., Aizawa,S., and Damsky,C.H. 1998. Extracellular matrix survival signals transduced by focal adhesion kinase suppress p53-mediated apoptosis. J.Cell Biol. 143:547–560

Jones,J.L. and Walker,R.A. 1997. Control of matrix metalloproteinase activity in cancer. J.Pathol. 183:377–379

Kiefer,F., Brumell,J., Al Alawi,N., Latour,S., Cheng,A., Veillette,A., Grinstein,S., and Pawson,T. 1998. The Syk protein tyrosine kinase is essential for Fcgamma receptor signaling in macrophages and neutrophils. Mol.Cell Biol. 18:4209–4220

Kubo,N., Boisvert,W.A., Ballantyne,C.M., and Curtiss,L.K. 2000. Leukocyte CD11b expression is not essential for the development of atherosclerosis in mice. J.Lipid Res. 41:1060–1066

Kusner,D.J., Hall,C.F., and Schlesinger,L.S. 1996. Activation of phospholipase D is tightly coupled to the phagocytosis of Mycobacterium tuberculosis or opsonized zymosan by human macrophages. J.Exp.Med. 184:585–595

Leverrier,Y. and Ridley,A.J. 2001. Requirement for Rho GTPases and PI 3-kinases during apoptotic cell phagocytosis by macrophages. Curr.Biol. 11:195–199

Ley,K. 2001. Pathways and bottlenecks in the web of inflammatory adhesion molecules and chemoattractants. Immunol.Res. 24:87–95

Li,H., Cybulsky,M.I., Gimbrone,M.A., Jr., and Libby,P. 1993. An atherogenic diet rapidly induces VCAM-1, a cytokine-regulatable mononuclear leukocyte adhesion molecule, in rabbit aortic endothelium. Arterioscler.Thromb. 13:197–204

Madri,J.A. and Graesser,D. 2000. Cell migration in the immune system: the evolving inter-related roles of adhesion molecules and proteinases. Dev.Immunol. 7:103–116

McHugh,K.P., Hodivala-Dilke,K., Zheng,M.H., Namba,N., Lam,J., Novack,D., Feng,X., Ross,F.P., Hynes,R.O., and Teitelbaum,S.L. 2000. Mice lacking beta3 integrins are osteosclerotic because of dysfunctional osteoclasts. J.Clin.Invest 105:433–440

Mileski,W.J., Sikes,P., Atiles,L., Lightfoot,E., Lipsky,P., and Baxter,C. 1993. Inhibition of leukocyte adherence and susceptibility to infection. J.Surg.Res. 54:349–354

Munthe-Kaas,A.C., Kaplan,G., and Seljelid,R. 1976. On the mechanism of internalization of opsonized particles by rat Kupffer cells in vitro. Exp.Cell Res. 103:201–212

Naito,M., Umeda,S., Yamamoto,T., Moriyama,H., Umezu,H., Hasegawa,G., Usuda,H., Shultz,L.D., and Takahashi,K. 1996. Development, differentiation, and phenotypic heterogeneity of murine tissue macrophages. J.Leukoc.Biol. 59:133–138

Newman,S.L. and Tucci,M.A. 1990. Regulation of human monocyte/macrophage function by extracellular matrix. Adherence of monocytes to collagen matrices enhances phagocytosis of opsonized bacteria by activation of complement receptors and enhancement of Fc receptor function. J.Clin.Invest. 86:703–714

Niu,J., Gu,X., Turton,J., Meldrum,C., Howard,E.W., and Agrez,M. 1998. Integrin-mediated signalling of gelatinase B secretion in colon cancer cells. Biochem.Biophys.Res.Commun. 249:287–291

Ostermann,G., Weber,K.S., Zernecke,A., Schroder,A., and Weber,C. 2002. JAM-1 is a ligand of the beta(2) integrin LFA-1 involved in transendothelial migration of leukocytes. Nat.Immunol. 3:151–158

Petty,H.R. and Todd,R.F. 1996. Integrins as promiscuous signal transduction devices. Immunology Today 17:209–212

Pfeiffer,A., Bottcher,A., Orso,E., Kapinsky,M., Nagy,P., Bodnar,A., Spreitzer,I., Liebisch,G., Drobnik,W., Gempel,K. et al. 2001. Lipopolysaccharide and ceramide docking to CD14 provokes ligand-specific receptor clustering in rafts. Eur.J.Immunol. 31:3153–3164

Pixley,F.J., Lee,P.S., Condeelis,J.S., and Stanley,E.R. 2001. Protein tyrosine phosphatase phi regulates paxillin tyrosine phosphorylation and mediates colony-stimulating factor 1-induced morphological changes in macrophages. Mol.Cell Biol. 21:1795–1809

Pommier,C.G., Inada,S., Fries,L.F., Takahashi,T., Frank,M.M., and Brown,E.J. 1983. Plasma fibronectin enhances phagocytosis of opsonized particles by human peripheral blood monocytes. J.Exp.Med. 157:1844–1854

Poo,H., Krauss,J.C., Mayobond,L., Todd,R.F., and Petty,H.R. 1995. Interaction of Fc-gamma receptor type IIIB with complement receptor type 3 in fibroblast transfectants—evidence from lateral diffusion and resonance energy transfer studies. J.Mol.Biol. 247:597–603

Puig-Kroger,A., Sanz-Rodriguez,F., Longo,N., Sanchez-Mateos,P., Botella,L., Teixido,J., Bernabeu,C., and Corbi,A.L. 2000. Maturation-dependent expression and function of the CD49d integrin on monocyte-derived human dendritic cells. J.Immunol. 165:4338–4345

Randolph,G.J., Beaulieu,S., Lebecque,S., Steinman,R.M., and Muller,W.A. 1998. Differentiation of monocytes into dendritic cells in a model of transendothelial trafficking. Science 282:480–483

Ren,Y., Stuart,L., Lindberg,F.P., Rosenkranz,A.R., Chen,Y., Mayadas,T.N., and Savill,J. 2001. Nonphlogistic clearance of late apoptotic neutrophils by macrophages: efficient phagocytosis independent of beta 2 integrins. J.Immunol. 166:4743–4750

Ridley,A.J. 2001. Rho proteins, PI 3-kinases, and monocyte/macrophage motility. FEBS Lett. 498:168–171

Ridley,A.J., Allen,W.E., Peppelenbosch,M., and Jones,G.E. 1999. Rho family proteins and cell migration. Biochem.Soc.Symp. 65:111-23:111–123

Roach,T., Slater,S., Koval,M., White,L., McFarland,E., Okumura,M., Thomas,M., and Brown,E. 1997. CD45 regulates src family member kinase activity associated with macrophage integrin-mediated adhesion. Curr.Biol. 7:408–417

Roach,T.I., Slater,S.E., White,L.S., Zhang,X., Majerus,P.W., Brown,E.J., and Thomas,M.L. 1998. The protein tyrosine phosphatase SHP-1 regulates integrin-mediated adhesion of macrophages. Curr.Biol. 8:1035–1038

Ross,G.D. and Vetvicka,V. 1993. CR3 (CD11b, CD18): a phagocyte and NK cell membrane receptor with multiple ligand specificities and functions. Clin.Exp.Immunol. 92:181–184

Rosseau,S., Selhorst,J., Wiechmann,K., Leissner,K., Maus,U., Mayer,K., Grimminger,F., Seeger,W., and Lohmeyer,J. 2000. Monocyte migration through the alveolar epithelial barrier: adhesion molecule mechanisms and impact of chemokines. J.Immunol. 164:427–435

Schneeberger,E.E., Vu,Q., LeBlanc,B.W., and Doerschuk,C.M. 2000. The accumulation of dendritic cells in the lung is impaired in CD18-/- but not in ICAM-1-/- mutant mice. J.Immunol. 164:2472–2478

Serrander,L., Fallman,M., and Stendahl,O. 1996. Activation of phospholipase D is an early event in integrin-mediated signalling leading to phagocytosis in human neutrophils. Inflammation 20:439–450

Shang,X.Z. and Issekutz,A.C. 1998. Contribution of CD11a/CD18, CD11b/CD18, ICAM-1 (CD54) and -2 (CD102) to human monocyte migration through endothelium and connective tissue fibroblast barriers. Eur.J.Immunol. 28:1970–1979

Shimaoka,M., Lu,C., Palframan,R.T., Von Andrian,U.H., McCormack,A., Takagi,J., and Springer,T.A. 2001. Reversibly locking a protein fold in an active conformation with a disulfide bond: integrin alphaL I domains with high affinity and antagonist activity in vivo. Proc.Natl.Acad.Sci.U.S.A 98:6009–6014

Springer,T.A. 1997. Folding of the N-terminal, ligand-binding region of integrin alpha-subunits into a beta-propeller domain. Proc.Natl.Acad.Sci.U.S.A 94:65–72

Stupack,D.G., Puente,X.S., Boutsaboualoy,S., Storgard,C.M., and Cheresh,D.A. 2001. Apoptosis of adherent cells by recruitment of caspase-8 to unligated integrins. J.Cell Biol. 155:459–470

Suen,P.W., Ilic,D., Caveggion,E., Berton,G., Damsky,C.H., and Lowell,C.A. 1999. Impaired integrin-mediated signal transduction, altered cytoskeletal structure and reduced motility in Hck/Fgr deficient macrophages. J.Cell Sci. 112:4067–4078

Tamkun,J.W., Desimone,D.W., Fonda,D., Patel,R.S., Buck,C., Horowitz,A.F., and Hynes,-R.O. 1986. Structure of integrin,a glycoprotein involved in the transmembrane linkage between fibronectin and actin. Cell 46:271–282

Teixido,J., Hemler,M.E., Greenberger,J.S., and Anklesaria,P. 1992. Role of beta 1 and beta 2 integrins in the adhesion of human CD34hi stem cells to bone marrow stroma. J.Clin.Invest 90:358–367

Tiisala,S., Paavonen,T., and Renkonen,R. 1995. Alpha E beta 7 and alpha 4 beta 7 integrins associated with intraepithelial and mucosal homing, are expressed on macrophages. Eur.J.Immunol. 25:411–417

van der Flier, A. and Sonnenberg,A. 2001. Function and interactions of integrins. Cell Tissue Res. 305:285–298

Vines,C.M., Potter,J.W., Xu,Y., Geahlen,R.L., Costello,P.S., Tybulewicz,V.L., Lowell,C.A., Chang,P.W., Gresham,H.D., and Willman,C.L. 2001. Inhibition of beta2 Integrin Receptor and Syk Kinase Signaling in Monocytes by the Src Family Kinase Fgr. Immunity. 15:507–519

Wei,Y., Eble,J.A., Wang,Z., Kreidberg,J.A., and Chapman,H.A. 2001. Urokinase receptors promote beta1 integrin function through interactions with integrin alpha3beta1. Mol.Biol.Cell 12:2975–2986

Weiss,J.M., Renkl,A.C., Maier,C.S., Kimmig,M., Liaw,L., Ahrens,T., Kon,S., Maeda,M., Hotta,H., Uede,T. et al. 2001. Osteopontin is involved in the initiation of cutaneous contact hypersensitivity by inducing Langerhans and dendritic cell migration to lymph nodes. J.Exp.Med. 194:1219–1229

Winn,R.K., Ramamoorthy,C., Vedder,N.B., Sharar,S.R., and Harlan,J.M. 1997. Leukocyte-endothelial cell interactions in ischemia-reperfusion injury. Ann.N.Y.Acad.Sci. 832:311–21:311–321

Woods,M.L. and Shimizu,Y. 2001. Signaling networks regulating beta1 integrin-mediated adhesion of T lymphocytes to extracellular matrix. J.Leukoc.Biol. 69:874–880

Worthylake,R.A., Lemoine,S., Watson,J.M., and Burridge,K. 2001. RhoA is required for monocyte tail retraction during transendothelial migration. J.Cell Biol. 154:147–160

Wright,S.D., Detmers,P.A., Jong,M.T., and Meyer,B.C. 1986. Interferon-gamma depresses binding of ligand by C3b and C3bi receptors on cultured human monocytes, an effect reversed by fibronectin. J.Exp.Med. 163:1245–1259

Xiong,J.P., Stehle,T., Diefenbach,B., Zhang,R., Dunker,R., Scott,D.L., Joachimiak,A., Goodman,S.L., and Arnaout,M.A. 2001. Crystal structure of the extracellular segment of integrin alpha Vbeta3. Science 294:339–345

Xiong,J.P., Stehle,T., Zhang,R., Joachimiak,A., Frech,M., Goodman,S.L., and Arnaout,M.A. 2002. Crystal Structure of the Extracellular Segment of Integrin {alpha}V{beta}3 in Complex with an Arg-Gly-Asp Ligand. Science In press

Yu,X., Miyamoto,S., and Mekada,E. 2000. Integrin alpha 2 beta 1-dependent EGF receptor activation at cell-cell contact sites. J.Cell Sci. 113:2139–2147

Zhou,X., Li,J., and Kucik,D.F. 2001. The microtubule cytoskeleton participates in control of beta2 integrin avidity. J.Biol.Chem. 276:44762–44769

Zou,W., Machelon,V., Coulomb-L'Hermin,A., Borvak,J., Nome,F., Isaeva,T., Wei,S., Krzysiek,R., Durand-Gasselin,I., Gordon,A. et al. 2001. Stromal-derived factor-1 in human tumors recruits and alters the function of plasmacytoid precursor dendritic cells. Nat.Med. 7:1339–1346

Macrophage Targets in Inflammation: Purinergic Receptors

F. Di Virgilio · D. Ferrari

Department of Experimental and Diagnostic Medicine, Section of General Pathology and Interdisciplinary Center for the Study of Inflammation (ICSI), University of Ferrara, Via Borsari 46, 44100 Ferrara, Italy
e-mail: fdv@unife.it

Abstract Extracellular adenosine and adenine nucleotides are potent endogenous modulators of inflammation that accumulate in the pericellular space under various physiological and pathological conditions. Their action is mediated by plasma membrane receptors named P1 (adenosine) or P2 (adenine nucleotides). Adenosine is generated extracellularly by ubiquitous ecto-ATPases/ecto-nucleotidases that dephosphorylate adenosine 5′-triphosphate (ATP). Adenosine is then either finally degraded by adenosine deaminase or taken up by the cells. Activation of P1 receptors by adenosine mainly inhibits, and activation of P2 receptors by adenine nucleotides mainly stimulates inflammatory cell responses. Differential expression by inflammatory cells and modulation during differentiation make P1 and P2 receptors potential targets for development of novel anti-inflammatory drugs.

Keywords Adenosine, Adenine, Cytokines, Ecto-ATPases, Adenosine 5′-triphosphate, Intracellular parasites

1 Introduction

It has been known for some years that adenosine inhibits in vitro inflammatory cell responses, and there are grounds to believe that locally released adenosine downmodulates inflammation in vivo by acting at A_{2A} receptors and may mediate the in vivo therapeutical effect of anti-inflammatory drugs (Cronstein et al. 1983; Cronstein et al. 1993; Ohta and Sitkovsky 2001). It was less appreciated until recently that adenosine precursors, the adenine nucleotides, may also modulate responses of inflammatory cells, among which, most notably, are the macrophages. At inflammatory sites, cell injury or activation causes release of substantial amounts of intracellular nucleotides that may dramatically affect responses of resident or infiltrating inflammatory cells (Di Virgilio 1995; Morabito et al. 1998) (the cytoplasmic ATP concentration ranges from 5 to 10 mM in most cell types, while that of other nucleotides, e.g., uridine triphosphate (UTP), can reach about one third of that of ATP) (Lazarowski and Boucher 2001). Extracellular nucleotides trigger a complex network of autocrine/paracrine interactions that, depending on the relative amounts of nucleotide or nucleoside present, the type and level of specific plasma membrane receptors expressed, and the state of priming of neighbouring cells by other inflammatory mediators, will eventually lead to the amplification or inhibition of the inflammatory response. In addition, the ubiquitous ecto-ATPases/ecto-nucleotidases, that rapidly hydrolyse extracellular nucleotides to adenosine, are powerful modulatory elements. Adenosine is then either finally degraded by adenosine deaminase (ADA), or taken up by the cells to be further utilized in intracellular synthetic pathways. While a number of agents able to modulate the effects of adenosine have been available for several years and are currently tested for clinical use, drugs that interfere with inflammatory cell stimulation by extracellular nucleotides have attracted interest only very recently, and none of them has entered clinical trials.

2 Purinergic Receptors: What Are They?

The name "purinergic" receptors was initially coined on a purely functional and pharmacological basis, indicating those plasma membrane receptors mediating the action of extracellular adenine nucleotides or nucleosides, i.e. ATP or adenosine. As it has recently become clear that some P2 receptor subtypes are highly specific for UTP or uridine diphosphate (UDP), it has been agreed to drop the word "purinergic" (UTP and UDP are pyrimidines) and simply name receptors for extracellular nucleotides or nucleosides as P2 or P1, respectively. The original classification based on the pharmacological profile has been replaced by that based on the molecular structure; accordingly, P2 receptors are grouped into two subfamilies (P2Y and P2X), numbering eight and seven members, respectively, and P1 receptors into four subtypes. Very recently, the receptor for uri-

Table 1 Classification of mammalian adenosine/P1 receptors

Subtype	Amino acid number	Signal transduction	
A_1	326–328	↑IP_3	↓cAMP
A_{2A}	409–411		↑cAMP
A_{2B}	328–332	↑IP_3	↑cAMP
A_3	318–320	↑IP_3	↓cAMP

Size of P1 receptororthologues depends on the speciesof origin.

Table 2 Classification of mammalian P2Y and P2X receptors

Subtype	Amino acid number	Preferred naturally occurring agonist	Signal transduction	
$P2Y_1$	362	ADP	↑IP_3	↓cAMP
$P2Y_2$	373	UTP, ATP	↑IP_3	
$P2Y_4$	352	UTP	↑IP_3	
$P2Y_6$	379	UDP	↑IP_3	
$P2Y_{11}$	371	ATP	↑IP_3	↑cAMP
$P2Y_{12}$	342	ADP		↓cAMP
$P2Y_{13}$	334	ADP	↑IP_3	↑↓cAMP
PSY_{14}	338	UDP-glucose	↑IP_3	
$P2X_1$	399	ATP	Ion currents	
$P2X_2$ [a]	472	ATP	Ion currents	
$P2X_3$	397	ATP	Ion currents	
$P2X_4$ [a]	388	ATP	Ion currents	
$P2X_5$	455	ATP	Ion currents	
$P2X_6$	379	ATP	Ion currents	
$P2X_7$	595	ATP	Ion currents Protein–protein interaction	
$P2X_2/P2X_3$		ATP	Ion currents	
$P2X_1/P2X_5$		ATP	Ion currents	
$P2X_4/P2X_6$		ATP	Ion currents	

[a] Splice variantsof $P2X_2$ and $P2X_4$ have been identified. P2X receptors may assemble as heteroligomers. At $P2Y_1$ and $P2Y_{12}$ ATP may act as an antagonist. $P2Y_{13}$ mediatesinhibition of adenylatecyclase at low and stimulation at high concentrations.

dine 5′-diphosphoglucose (UDP-glucose) has been included in the P2 family (Tables 1 and 2).

2.1 P2 Receptors

P2 receptors are divided into two subfamilies: P2X (ligand-gated ion channels, ionotropic, receptors) and P2Y (G protein-coupled, metabotropic, receptors). P2X receptors are activated by extracellular ATP (the only known physiological ligand), and mediate fast mono- and di-valent cation (Na^+, K^+ and Ca^{2+}) fluxes across the plasma membrane (North 2002). Intriguingly, permeability of some

P2X receptors (especially $P2X_7$, but to a lesser extent also $P2X_2$ and $P2X_4$) increases during continuous ATP stimulation or upon application of repeated ATP pulses (Di Virgilio 1995; Falzoni et al. 1995; Virginio et al. 1999; Khakh et al. 1999). The molecular basis of this odd phenomenon is as yet unknown, but it is hypothesized that it might depend on the recruitment of additional receptor subunits, thus increasing the size of the channel pore. All P2X receptors are multimeric structures of which seven basic subunits, and some splice variants, have been cloned ($P2X_{1-7}$).

An open problem is P2X receptor subunit stoichiometry and composition. Growing consensus supports the view that P2X receptors assemble as trimers or hexamers (Nicke et al. 1998), but assembly into tetramers has also been reported (Kim et al. 1997). Functional expression studies show that heterologously expressed P2X subunits may assemble to form heteromultimeric structures, and patch-clamp data suggests that native P2X receptors may also assemble as heteromultimers (Lewis et al. 1995; Torres et al. 1999). A relevant exception is $P2X_7$, the P2X receptor mainly expressed by macrophages. While there is good evidence to support a tri/hexameric stoichiometry for this receptor, there is likewise strong evidence that the $P2X_7$ subunit does not assemble with any other P2X subunit (Torres et al. 1999; Kim et al. 2001b). Size of P2X subunits is comprised between 379 ($P2X_6$) and 595 ($P2X_7$) amino acids, for a predicted molecular weight of 42 to 65 kDa. Hydropathy plots and absence of a leader sequence predict a membrane topology with only two transmembrane stretches separated by a bulky extracellular region with both the N- and C-termini on the cytoplasmic side (Brake et al. 1994; Surprenant et al. 1996). The extracellular domain contains 10 cysteines and 2–6 N-linked glycosylation sites.

Within the P2X subfamily, a special place is occupied by $P2X_7$ for several reasons: (1) it has a long (242 residues) cytoplasmic carboxy-terminal tail compared to the other members of the family; (2) the COOH tail contains a hydrophobic stretch that might insert into the plasma membrane or interact with cytoplasmic vesicles; (3) it does not heteropolymerize with other P2X subunits; (4) its activation needs ATP concentrations that are 10- to 100-fold higher than those required to activate other P2X receptors; (5) it is the P2X receptor that more readily undergoes a large increment in conductance (channel-to-pore transition) upon sustained stimulation with ATP (in fact it is the only P2X receptor for which this phenomenon has been reproducibly demonstrated and extensively characterized); (6) it is mainly, although not exclusively, localized to immune cells; (7) it is the only P2X receptor whose activation undisputedly triggers cell death.

Of particular relevance for the participation of $P2X_7$ in the modulation of macrophage responses is the finding that its COOH tail harbours an amino acid stretch (573–590) highly homologous (about 90%) with the lipopolysaccharide (LPS)-binding site of LPS-binding protein (Denlinger et al. 2001). Additional studies have provided evidence that $P2X_7$ interacts with several extracellular, transmembrane or cytoskeletal proteins to form a highly complex and versatile signalling complex (receptosome) (Denlinger et al. 2001; Kim et al. 2001a). At least eleven $P2X_7$-associated proteins have been conclusively identified: laminin $\alpha3$, integrin $\beta2$, receptor protein tyrosine phosphatase β (RPTPβ), α-actinin, β-actin, supervillin, heat shock protein (Hsp) 90, heat shock cognate (Hsc) protein 71, Hsp70, phosphatidylinositol 4-kinase (PI4 K) 230 and membrane-associated guanylate kinase (MAGuK) P55 (Kim et al. 2001a). Finally, all P2X receptors contain in the N-terminus a consensus sequence for protein kinase C (Thr-X-Lys/Arg).

P2Y receptors are members of the rhodopsin-like G protein superfamily with seven transmembrane domains (North and Barnard 1997; von Kugelgen and Wetter 2000). They number from 333 ($P2Y_{13}$) to 379 ($P2Y_6$) amino acids, for a predicted molecular mass of 36–42 kDa. Agonist selectivity is a discriminant feature of P2Y receptors: $P2Y_1$, $P2Y_{11}$, $P2Y_{12}$ and $P2Y_{13}$ are selective for adenine nucleotides, whereas other members of the P2Y family can also be activated by uracil nucleotides. Furthermore, $P2Y_1$, $P2Y_6$, $P2Y_{12}$ and $P2Y_{13}$ are selectively activated by nucleoside diphosphates, while the other P2Y receptors are preferentially activated by nucleoside triphosphates. All P2Y receptors, with the exception of $P2Y_{12}$, couple to phospholipase C, thus causing inositol triphosphate formation and Ca^{2+} release from intracellular stores. Coupling to phospholipase D has also been described (Purkiss and Boarder 1992). $P2Y_1$ and $P2Y_2$ mediate inhibition of adenylate cyclase, while $P2Y_{11}$ mediates stimulation. $P2Y_{13}$ mediates inhibition of adenylate cyclase at low and stimulation at high concentration. The receptor for UDP-glucose has been included in the P2Y family as $P2Y_{14}$. This receptor, expressed in a wide variety of human tissues, couples to Ca^{2+} release from intracellular stores (Chambers et al. 2000).

2.2 P1 Receptors

P1 receptors are defined as those plasma membrane receptors that bind adenosine. This family comprises four members: A_1, A_{2A}, A_{2B} and A_3. All adenosine receptors are seven membrane-spanning and couple via G proteins to phospholipase C and/or adenylate cyclase. Activation of the A_1 subtype causes phospholipase C stimulation and a decrease in cyclic adenosine monophosphate (cAMP), while activation of A_{2A} causes only stimulation of adenylate cyclase. Activation of A_{2B} causes an elevation of both IP_3 and cAMP. Finally, activation of A3 causes a decrease in cAMP and an increase in IP_3 concentration (Ralevic and Burnstock 1998). Like all the other members of the serpentine receptors superfamily, P1 receptors were until recently thought of as monomeric molecules, but a recent paper by Yoshioka et al. (2001) suggests that A_1 and $P2Y_1$ receptors may form heteromeric structures with $P2Y_1$-like agonistic pharmacology.

3 Sources and Mechanism for ATP Release into the Extracellular Space

There is no doubt that ATP is a major neurotransmitter in the central and peripheral nervous system, often in combination with other mediators. Less appreciated is the occurrence of ATP release outside the nervous system, but proof that this is a much more frequent event than commonly thought is being provided by an increasing number of laboratories (Dubyak et al. 2002; Schwiebert et al. 2002). Concentration of ATP in the blood or in the interstitial fluid under resting conditions is thought to be in the low nanomolar range, although this is likely to be a large underestimate of the actual ATP concentration in the pericellular space. Most cell types have been shown to continuously release ATP thus generating an autocrine tonic stimulation of P2 receptor (Ostrom et al. 2000), that may even lead to P2-receptor desensitisation. Several basal cellular parameters (e.g. the resting cytoplasmic Ca^{2+} and cAMP concentration, and the basal arachidonic acid release) appear to be modulated by the tonic stimulation of P2 receptors (Grierson et al. 1995; Ostrom et al. 2000). More interestingly, several agents are capable of inducing ATP release thus raising several fold the concentration of this nucleotide in the proximity of the plasma membrane (Ferrari et al. 1997b; Sperlagh et al. 1998; Warny et al. 2001).

One of the most important sources of ATP release are platelets. These cells store ATP up to a several millimolar concentration within their dense granules, thus massive ATP release, up to 20–50 μM, occurs during platelet aggregation (Beigi et al. 1999). Since activated platelets adhere to leukocytes and can establish protected compartments at the site of interaction with adjacent cells, it has to be expected that much higher ATP concentrations can be reached within these secluded environments, in a way reminiscent of the high concentrations achieved by neurotransmitters secreted into the synaptic cleft. Endothelial cells are another important source of extracellular ATP in response to shear stress (Bodin et al. 1991), swelling (Oike et al. 2000), or stimulation of plasma membrane receptors (Yang et al. 1994). Furthermore, leukocytes stimulated with bacterial endotoxin (LPS) also release ATP (Ferrari et al. 1997b; Sperlagh et al. 1998; Sikora et al. 1999; Imai et al. 2000; Warny et al. 2001; but see also Grahames et al. 1999 and Beigi and Dubyak 2000). There is hint that some well-known anti-inflammatory drugs (e.g. sulphasalazine and methotrexate) owe their action to their ability to promote ATP release from macrophages and conversion of this nucleotide to adenosine (Morabito et al. 1998).

While it is obvious that cell lysis may well be responsible for the discharge of cellular ATP, how this nucleotide is extruded from intact cells is still an open question. As mentioned above, platelets and a few other cell types certainly release ATP by exocytosis of their cytoplasmic granules (Meyers et al. 1982; Lages and Weiss 1999), but secretion of ATP from many other different cell types does not appear to occur via stimulated exocytosis. It has been suggested that ATP might be stored within vesicles originating from the Golgi and released by constitutive exocytosis (Maroto and Hamill 2001). Plasma membrane transporters such as the multidrug resistance protein (Abraham et al. 1993) or the cystic fibrosis protein (CFTR) (Jiang et al. 1998) have also been candidated to the role of ATP transporters. Connexin 43 hemichannels have been reported to mediate ATP release (Cotrina et al. 1998), but this evidence has been recently questioned (Romanello et al. 2001). Rather intriguingly, it has been shown that transfection with the differentiation antigen CD39 (which is a member of plasma membrane ecto-nucleoside triphosphate diphosphohydrolase family) enhances ATP transport across the plasma membrane of *Xenopus* oocytes (Bodas et al. 2000). Finally, data by Baricordi et al. (1999) show that transduction of human lymphoblastoid cell lines with the $P2X_7$ receptor increases their ability to release ATP, thus suggesting that $P2X_7$ might participate in ATP secretion.

4
Systems Involved in Degradation of Extracellular ATP

Once in the extracellular space, ATP is quickly degraded by very active ecto-enzymes grouped into four families: ecto-nucleotide triphosphate diphosphohydrolase (E-NTPDase), ecto-nucleotide pyrophosphatase/phosphodiesterase (E-NPP), alkaline phosphatase and ecto-5′-nucleotidase (Zimmermann 2000, 2001). E-NTPDases cleave ATP, adenosine diphosphate (ADP) and several other purine and pyrimidine nucleotides to AMP and Pi. An important member of this family is the lymphocyte differentiation marker CD39. E-NPPases hydrolyse ATP to AMP and Ppi, ADP to AMP and Pi or NAD^+ to nicotinamide mononucleotide. E-NPPases are known as differentiation antigens for plasma cells, motility-stimulating proteins (autotaxins) or neural differentiation and tumour surface markers. Alkaline phosphatases are non-specific ecto-phosphomonoesterases with a broad substrate specificity. Finally, ecto-5′-nucleotidases catalyse the conversion of nucleoside 5′-monophosphates to the respective nucleoside and Pi. Ecto-5′-nucleotidase is the main enzyme responsible for the extracellular generation of adenosine.

5
P1 and P2 Receptor Subtypes Expressed by Macrophages

An extensive characterization of P1 receptor expression by macrophages has not yet been carried out. However, broadly converging biochemical and pharmacological data from human monocyte/macrophages or mouse macrophages and human and mouse macrophage cell lines suggest that all the four P1 subtypes are expressed by these cells (McWhinney et al. 1996; Xaus et al. 1999; Montesinos and Cronstein 2001; Johnston et al. 2001). As regards P2 receptors, mRNA for $P2Y_1$, $P2Y_2$, $P2Y_4$ and $P2Y_6$ has been amplified from monocytes and macrophages (Di Virgilio et al. 2001a; Dubyak 2001). Expression of $P2Y_{11}$ has been demonstrated in human dendritic cells (Wilkin et al. 2001), but as yet there is no published evidence that it may also be present in macrophages. Expression of members of the P2X subfamily is much less characterized, with the exception

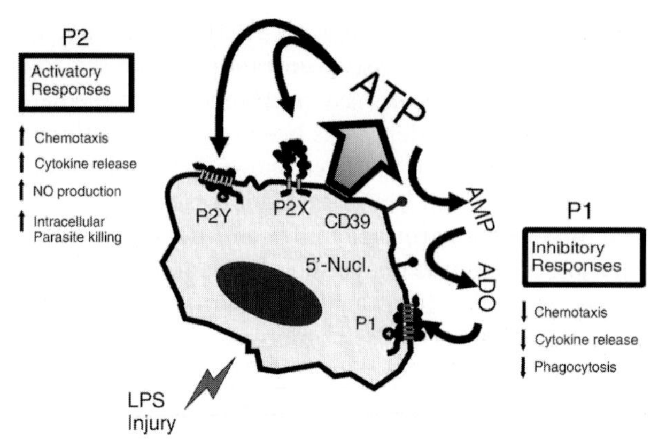

Fig. 1 Autocrine/paracrine loop based on ATP release from the macrophage and P2/P1 receptor activation

of $P2X_7$, that has been the focus of intense investigation over recent years (Di Virgilio et al. 2001a,b). While there is no doubt that $P2X_7$ is the main P2X receptor subtype expressed by macrophages (and very likely also by monocytes, despite previous evidence of the contrary) (Falzoni et al. 2000; Gu et al. 2000; Gudipaty et al. 2001), scattered evidence suggests that $P2X_1$ and $P2X_4$ might also be expressed, at least at certain developmental stages (Buell et al. 1996; Soto et al. 1996).

6 Role of P1 and P2 Receptors in Macrophage Physiology

While it is generally appreciated that extracellular ATP has an important role as a neuromediator, the notion that this nucleotide may also be a mediator of inflammation has received attention only very recently. Macrophages are fully equipped to exploit the potential of nucleotides and nucleosides as extracellular messengers as (1) they express P2Y, P2X and P1 receptors, (2) express ecto-ATPase/nucleotidase that convert nucleotides to nucleosides, and (3) can release ATP. Thus, macrophages are not only the target of ATP released by neighbour cells, but also the centre of an ATP-based autocrine/paracrine loop (Fig. 1). However, our understanding of signalling via extracellular nucleotides in the immune system is still very rudimentary. Some authors have suggested that extracellular nucleotides might be early signals for alerting the immune system of an impendent danger, and thus drive recruitment and activation of inflammatory cells (Greenberg et al. 1988; Di Virgilio et al. 1995; Gallucci and Matzinger 2001). Nucleotides have a strong macrophage chemotactic activity (McCloskey et al. 1999; Oshimi et al. 1999), activate the NADPH (nicotinamide adenine dinucleotide phosphate, reduced) oxidase, especially in the presence of phagocytic particles (Schmid-Antomarchi et al. 1997), stimulate release of interleukin (IL)-1α, IL-1β, IL-8 and IL-18 (Perregaux and Gabel 1994; Ferrari et al. 1997a; Perregaux and Gabel 1998; Perregaux et al. 2000; Warny et al. 2001), drive the externalisation of active caspase-1 (Laliberte et al. 1999), trigger nitric oxide generation (Denlinger et al. 1996), increase expression of tumour necrosis factor (TNF)-α mRNA (Tonetti et al. 1995), and finally ATP at high doses has a potent cytotoxic effect (Steinberg and Silverstein 1987; Murgia et al. 1992). The ATP-dependent cytotoxic activity is of particular interest because a side-effect, not shared by other cytotoxic stimuli, is the killing of intracellular parasites (i.e. *Mycobacterium tuberculosis*) that are normally able to survive within phagocytic vacuoles. A number of different laboratories have shown that stimulation with ATP of macrophages that have ingested *Mycobacteria* causes an increased rate of phagosome-lysosome fusion that greatly enhances the microbicidal efficacy of the phagocytes (Molloy et al. 1994; Lammas et al. 1997; Fairbairn et al. 2001; Mancino et al. 2001). In the context of granulomatous diseases such as tuberculosis or sarcoidosis, it is of interest that one of the P2X subtypes, $P2X_7$, appears to be necessary for the formation of multinucleated giant cells (MGCs) (Chiozzi et al. 1997; Falzoni et al. 2000). A recent report shows that this same receptor is overexpressed in macrophages from sarcoidosis patients (Mizuno et al. 2001). The $P2X_7$ receptor has also an important role in osteoclast differentiation and in communication between osteoblasts and osteoclasts (Morrison et al. 1998; Jorgensen et al. 2002).

Most attention has so far concentrated on the involvement of $P2X_7$ in inflammation; however, other P2 receptor are increasingly implicated in macrophage responses to nucleotides. $P2Y_1$ and/or $P2Y_2$ are likely to be the main receptors mediating monocyte/macrophage chemotaxis in response to nucleotides, while $P2Y_6$ is coupled to release of the granulocyte selective chemokine IL-8. An intriguing, and as yet little studied subtype is $P2Y_{11}$. This receptor appears to have an important role in the maturation of dendritic cells, alone or in association with other P2 receptors (Wilkin et al. 2001; La Sala et al. 2001). Whether it is also involved in monocyte/macrophage differentiation is not yet known.

Participation of P1 and P2 receptors in the modulation of so many different macrophage responses makes these molecules obvious targets for the development of innovative anti-inflammatory drugs. The clinical use of P1 or P2 agonists/antagonists is complicated by the several side-effects due to the widespread expression of their receptors in many different organs. Thus, one should either concentrate on the development of agents able to affect the local nucleoside or nucleotide concentration, or try to exploit the tissue-specific expression of P1and P2 subtypes. A simpleminded, but probably not as naïve as it might appear, approach would be the attempt to modulate the local concentration of the P1 and/or P2 agonists, i.e. adenosine and the nucleosides di- and tri-phosphates (ATP, ADP, UTP and UDP). This strategy is supported by two observations: (1) the mechanism of action of some anti-inflammatory drugs is understood to be mediated by the local accumulation of adenosine (Montesinos et al. 1998), and (2) an excessive accumulation of extracellular ATP as a consequence of CD39 (ecto-ATPase) reduced activity severely impairs inflammatory cell migration (Goepfert et al. 2001). These observations make CD39 an additional appealing target for pharmacological modulation of inflammation.

Design and synthesis of selective P1 agonists or antagonists has been the field of extensive investigation over recent years, and some compounds are being evaluated for clinical use (see Baraldi and Borea 2000, and Linden 2001 for recent reviews). In contrast, comparatively fewer agonist or antagonists selective for P2Y or P2X subtypes are available. At the moment, $P2Y_{12}$ is the only receptor targeted by antagonists widely used in clinical trials (Storey 2001). But $P2Y_{12}$ is not expressed by macrophages. An interesting selective antagonist of $P2Y_1$ (MRS 2179) has been recently described, but no clinical data are as yet available (Camaioni et al. 1998). Stable agonists for the $P2Y_2$ receptor are available for the local treatment of respiratory and ophthalmic diseases (cystic fibrosis, dry eye syndrome) (Boeynaems et al. 2001). Some stable ATP analogues (α, β-methylene ATP and β,γ-methylene ATP) have a strong preference for P2X over P2Y receptors (Ralevic and Burnstock 1998) and might be used to selectively stimulate P2X responses. A few selective antagonists are available for P2X receptors. The most widely used is pyridoxalphosphate-6-azophenyl 2′-4′-disulfonic acid (PPADS), a pyridoxal phosphate derivative that, however, may also inhibit $P2Y_1$. PPADS is a non-competitive antagonist that is thought to form Schiff's bases with lysines (possibly K64 and K311) that are key constituents of the ATP binding site (Thompson 2002). The same mechanism of action is shared by another irreversible blocker: oxidized ATP. This compound too, thanks to its aldehyde groups, covalently binds the same lysines and irreversibly blocks P2X receptors (Murgia et al. 1993). Oxidized ATP was initially thought to be selective for $P2X_7$, but it is now clear that it blocks other P2X subtypes. Another compound, the quinoline derivative 1-[*N*,*O*-bis(5-isoquinolinesulphonyl)-*N*-methyl-l-tyrosyl]-4-phenylpiperazine (KN-62) originally introduced as a P2 reagent by Gargett and Wiley (1997), is proving very useful as a lead compound for the synthesis of novel $P2X_7$ blockers (Baraldi et al. 2000; Ravi et al. 2001). As of now, none of these compounds has entered clinical trials.

7 Conclusions

Receptors for extracellular nucleotides and nucleosides are emerging as novel and important modulators of inflammation. Expression of specific P1 and P2 subtypes by different white blood cell types at various developmental stages confers to these receptors an intriguing plasticity that on the one hand underlines their importance in leukocyte function and differentiation, and on the other suggests new avenues for the development of innovative anti-inflammatory drugs. The relevance of P2 receptors in the overall inflammatory process is further underscored by their increasingly recognized role in nociception (Burnstock 2000; Cook and McCleskey 2002). These studies suggest that modulation of signal transduction based on extracellular ATP and its metabolites may allow us to design a new generation of drugs targeted at many manifestations of inflammation, ranging from pain sensation to leukocyte recruitment and pathogen killing.

8 References

Abraham EH, Prat AG, Gerweck L, Seneveratne T, Arceci RJ, Kramer R, Guidotti G, Cantiello HF (1993) The multidrug resistance (mdr1) gene product functions as an ATP channel. Proc Natl Acad Sci USA 90:312–316

Baricordi OR, Melchiorri L, Adinolfi E, Falzoni S, Chiozzi P, Buell G, Di Virgilio F (1999) Increased proliferation rate of lymphoid cells transfected with the $P2X_7$ receptor. J Biol Chem 274:33206–33208

Baraldi PG and Borea PA (2000) New potent and selective human adenosine (A3) receptor antagonists. Trends Pharmacol Sci 21:456–459

Baraldi PG, Romagnoli R, Tabrizi MA, Falzoni S, Di Virgilio F (2000) Synthesis of conformationally constrained analogues of KN62, a potent antagonist of the $P2X_7$ receptor. Bioorg Med Chem Lett 10:681–684

Beigi R, Kobatake E, Aizawa M, Dubyak GR (1999) Detection of local ATP release from activated platelets using cell surface-attached firefly luciferase. Am J Physiol 276:C267-C278

Beigi RD, Dubyak GR (2000) Endotoxin activation of macrophages does not induce ATP release and autocrine stimulation of P2 nucleotide receptors. J Immunol 165:7189–7198

Bodas E, Aleu J, Pujol G, Martin-Satuè M, Marsal J, Solsona C (2000) ATP crossing the cell plasma membrane generates an ionic current in *Xenopous* oocytes. J Biol Chem 275:20268–20273

Bodin P, Bailey D, Burnstock G (1991) Increased flow-induced ATP release from isolated vascular endothelial cells but not smooth muscle cells. Br J Pharmacol 103:1203–1205

Boeynaems J-M, Robaye B, Janssens R, Suarez-Huerta N, Communi D (2001) Overview of P2Y receptors as therapeutic targets. Drug Dev Res 52:187–189

Brake AJ, Wagenbach MJ, Julius D (1994) New structural motif for ligand-gated ion channels defined by an ionotropic ATP receptor. Nature 371:519–523

Buell G, Michel AD, Lewis C, Collo G, Humphrey PPA, Surprenant A (1996) $P2X_1$ receptor activation in HL-60 cells. Blood 87:2659–2663

Burnstock G (2000) P2X receptots in sensory neurons. Br J Anesth 84:476–488

Camaioni E, Boyer JL, Mohanram A, Harden TK, Jacobson KA (1998) Deoxyadenosine-bisphosphate derivatives as potent antagonists at $P2Y_1$ receptors. J Med Chem 41:183–190

Chambers JK, MacDonald LE, Sarau HM, Ames RS, Freeman K, Foley JJ, Zhu Y, McLaughlin MM, Murdock P, McMillan L, Trill J, Swift A, Aiyar N, Taylor P, Vawter L, Naheed S, Szekeres P, Hervieu G, Scott C, Watson JM, Murphy AJ, Duzic E, Klein C, Bergsma DJ, Wilson S, Livi GP (2000) A G protein-coupled receptor for UDP-glucose. J Biol Chem 275:10767–10771

Chiozzi P, Sanz JM, Ferrari D, Falzoni S, Aleotti A, Buell GN, Collo G, Di Virgilio F (1997) Spontaneous cell fusion in macrophage cultures expressing high levels of the P2Z/$P2X_7$ receptor. J Cell Biol 138:697–706

Cook SP, and McCleskey EW (2002) Cell damage excites nociceptors through release of cytosolic ATP. Pain 95:41–47

Cotrina ML, Lin J.H-C, Alves-Rodrigues A, Liu S, Li J, Azmi-Ghadimi H, Kang J, Naus CCG, Nedergaard M (1998) Connexins regulate calcium signalling by controlling ATP release. Proc Natl Acad Sci USA 95:15735–15740

Cronstein BN, Kramer SB, Weismann G, Hirschhorn R (1983) Adenosine: a physiological modulator of superoxide anion generation by human neutrophils. J Exp Med 158:1160–1177

Cronstein BN, Naime D, Ostad E (1993) The anti-inflammatory mechanism of methotrexate: increased adenosine release at inflamed sites diminishes leukocyte accumulation in an in vivo model of inflammation. J Clin Invest 92:2675–2682

Denlinger LC, Fisette PL, Garis KA, Kwon G, Vasquez-Torres A, Simon AD, Nguyen B, Proctor RA, Bertics PJ, Corbett JA (1996) Regulation of inducible nitric oxide synthase expression by macrophage purinoceptor and calcium. J Biol Chem 271:337–342

Denlinger LC, Fisette PL, Sommer JA, Watters JJ, Prabhu U, Dubyak GR, Proctor RA, Bertics PJ (2001) Cutting edge: The nucleotide receptor $P2X_7$ contains multiple protein and lipid-interaction motifs including a potential binding site for bacterial lipopolysaccharide. J Immunol 167:1871–1876

Di Virgilio F (1995) The P2Z purinoceptor: an intriguing role in immunity, inflammation and cell death. Immunol Today 16:524–528

Di Virgilio F, Chiozzi P, Ferrari D, Falzoni S, Sanz JM, Morelli A, Torboli M, Bolognesi G, Baricordi OR (2001a) Nucleotide receptors: an emergine family of regulatory molecules in blood cells. Blood 97:587–600

Di Virgilio F, Vishwanath V, Ferrari D (2001b) On the role of the $P2X_7$ receptor in the immune system. Handbook of Experimental Pharmacology 151/II:356–374

Dubyak GR (2001) Role of P2 receptors in the immune system. Handbook of Experimental Pharmacology 151/II:324–354

Dubyak GR (2002) Focus on “Extracellular ATP signalling and P2X nucleotide receptors in monolayers of primary human vascular endothelial cells.”. Am J Physiol Cell Physiol 282:C242–C244

Fairbairn IP, Stober CB, Kumararatne DS, Lammas DA (2001) ATP-mediated killing of intracellular mycobacteria by macrophages is a $P2X_7$-dependent process inducing bacterial death by phagosome-lysosome fusion. J Immunol 167:3300–3307

Falzoni S, Chiozzi P, Ferrari D, Buell G, Di Virgilio F (2000) $P2X_7$ receptor and polykarion formation. Mol Bio Cell 11:3169–3176

Falzoni S, Munerati M, Ferrarai D, Spisani S, Moretti S, Di Virgilio F (1995) The purinergic P2Z receptor of human macrophage cells. Characterization and possibile physiological role. J Clin Invest 95:1207–1216

Ferrari D, Chiozzi P, Falzoni S, Dal Susino M, Melchiorri L, Baricordi OR, Di Virgilio F (1997a) Extracellular ATP triggers IL-1β release by activating the purinergic P2Z receptor of human macrophages. J Immunol 159:1451–1458

Ferrari D, Chiozzi P, Falzoni S, Hanau S, Di Virgilio F (1997b) Purinergic modulation of interleukin-1b release from microglial cells stimulated with bacterial endotoxin. J Exp Med 185:579–582

Gallucci S and Matzinger P (2001) Danger signals: SOS to the immune system. Curr Opin Immunol 13:114–119

Gargett CE, Wiley JS (1997) The isoquinoline derivative KN-62: a potent antagonist of the P2Z-receptor of human lymphocytes. Br J Pharmacol 120:1483–1490

Goepfert C, Sundberg C, Sevigny J, Enjyoji K, Hoshi T, Csizmadia E, Robson S (2001) Disordered cellular migration and angiogenesis in cd39-null mice. Circulation 104:3109–3115

Grahames CB, Michel AD, Chessell IP, Humphrey PP (1999) Pharmacological characterization of ATP-and LPS-induced IL-1β release in human monocytes. Br J Pharmacol 127:1915–1921

Greenberg S, Di Virgilio F, Steinberg TH, Silverstein SC (1988) Extracellular nucleotides mediate Ca^{2+} fluxes in J774 macrophages by two distinct mechanisms. J Biol Chem 263:10337–10343

Grierson JP and Meldolesi J (1995) Shear stress-induced [Ca^{2+}]i transients and oscillations in mouse fibroblasts are mediated by endogenously released ATP. J Biol Chem 270:4451–4456

Gudipaty L, Humphreys BD, Buell G, Dubyak GR (2001) Regulation of $P2X_7$ nucleotide receptor function in human monocytes by extracellular ions and receptor density. Am J Physiol Cell Physiol 280:C943-C953

Gu BJ, Zhang WY, Bendall LJ, Chessell IP, Buell GN, Wiley JS (2000) Expression of $P2X_7$ purinoceptors on human lymphocytes and monocytes: evidence for non-functional $P2X_7$ receptors. Am J Physiol Cell Physiol 279:C1189-C1197

Imai M, Goepfert C, Kaczmarek E, Robson SC (2000). CD39 modulates IL-1 release from activated endothelial cells. Biochem Biophys Res Commun 270:272–278

Jiang Q, Mak D, Devidas S, Schwiebert EM, Bragin A, Zhang Y, Skach WR, Guggino WB, Foskett JK, Engelhardt JF (1998) Cystic fibrosis transmembrane conductance regulator-associated ATP release is controlled by a chloride sensor. J Cell Biol 143:645–657

Johnston JB, Silva C, Gonzales G, Holden J, Warren KG, Metz LM, Power C (2001) Diminished adenosine A1 receptor expression on macrophages in brain and blood of patients with multiple sclerosis. Ann Neurol 49:650–658

Jorgensen NR, Henriksen Z, Sorensen OH, Eriksen EF, Civitelli R, Steinberg TH (2002) Intercellular calcium signalling occurs between human osteoblasts and osteoclasts and requires activation of $P2X_7$. J Biol Chem 277:7574–7580

Khakh B, Bao X, Labarca C, Lester H (1999) Neuronal P2X receptor-transmitter-gated cation channels change their ion selectivity in seconds. Nature Neurosci 2:322–330

Kim M, Jiang LH, Wilson HL, North RA, Surprenant A (2001a) Proteomic and functional evidence for a $P2X_7$ receptor signalling complex. EMBO J 20:6347–6358

Kim M, Spelta V, Sim J, North RA, Surprenant A (2001b) Differential assembly of rat purinergic $P2X_7$ receptor in immune cells of the brain and periphery. J Biol Chem 276:23262–23267

Kim M, Yoo OJ, Choe S (1997) Molecular assembly of the extracellular domain of $P2X_2$, an ATP-gated ion channel. Biochem Biophys Res Commun 240:618–622

Lages B, Weiss HJ (1999) Secreted dense granule adenine nucleotides promote calcium influx and the maintenance of elevated cytosolic calcium levels in stimulated human platelets. Thromb Haemost 81:286–292

Laliberte RE, Eggler J, Gabel CA (1999) ATP treatment of human monocytes promotes caspase-1 maturation and externalisation. J Biol Chem 274:36944–36951

Lammas DA, Stober C, Harvey CJ, Kendrick N, Panchalingam S, Kumararatne DS (1997) ATP-induced killing of mycobacteria by human macrophages is mediated by purinergic P2Z ($P2X_7$) receptors. Immunity 7:433–444

Lazarowski ER, Boucher RC (2001) UTP as an extracellular signalling molecule. News Physiol. Sci. 16:1–5

Lewis C, Neidhart S, Holy C, North RA, Buell G, Surprenant A (1995) Coexpression of $P2X_2$ and $P2X_3$ receptor subunits can account for ATP-gated currents in sensory neurons. Nature 377:423–434

Linden J (2001) Molecular approach to adenosine receptors: receptor-mediated mechanisms of tissue protection. Annu Rev Pharmacol Toxicol 41:775–787

Mancino G, Placido R, Di Virgilio F (2001) $P2X_7$ receptors and apoptosis in tuberculosis infection. J Biol Regul Homeost Agents 15:286–293

Maroto R and Hamill OP (2001) Brefeldin A block of integrin-dependent mechanosensitive ATP release from *Xenopous* oocytes reveals a novel mechanism of mechanotransduction. J Biol Chem 276:23867–23872

McCloskey MA, Fan Y, Luther S (1999) Chemotaxis of mast cells toward adenine nucleotides. J Immunol 163:970–977

McWhinney CD, Dudley MW, Bowlin TL, Peet NP, Schook L, Bradshaw M, De M, Borcherding DR, Edwards CK (1996) Activation of A3 receptors on macrophages inhibits tumor necrosis factor-α. Eur J Pharmacol 310:209–216

Meyers KM, Holmsen H, Seachord CL (1982) Comparative study of platelet dense granule constituents. Am J Physiol 243:R454-R461
Mizuno K, Okamoto H, Horio T (2001) Heigthened ability of monocytes from sarcoidosis patients to form multinucleated giant cells in vitro by supernatants of concanavalin A-stimulated mononuclear cells. Clin Exp Immunol 126:151–156
Molloy A, Laochumroonvorapong P, Kaplan G (1994) Apoptosis, but not necrosis, of infected monocytes is coupled with killing of intracellular Bacillus Calmette-Guerin. J Exp Med 180:1499–1509
Montesinos MC, Cronstein BN (2001) Role of P1 receptors in inflammation. Handbook of Experimental Pharmacology 151/II:303–321
Morabito L, Montesinos MC, Schreibman DM, Balter L, Thompson LF, Resta R, Carlin G, Huie MA, Cronstein BN (1998) Methotrexate and sulfasalazine promote adenosine release by a mechanism that requires ecto-5'-nucleotidase-mediated conversion of adenine nucleotides. J Clin Invest 101:295–300
Morrison MS, Turin L, King BF, Burnstock G, Arnett TR (1998) ATP is a potent stimulator of the activation and formation of rodent osteoclasts. J Physiol 511:495–500
Murgia M, Hanau S, Pizzo P, Rippa M, Di Virgilio F (1993) Oxidized ATP. An irreversibile inhibitor of the macrophage purinergic P2Z receptor. J Biol Chem 268:8199–8203
Murgia M, Pizzo P, Steinberg TH, Di Virgilio F (1992) Characterization of the cytotoxic effect of extracellular ATP in J774 mouse macrophages. Biochem J 288:897–901
Nicke A, Baumert HG, Rettinger J, Eichele A, Lambrecht G, Mutschler E, Schmalzing G (1998) P2X1 and P2X3 receptors for stable trimers: a novel structural motif of ligand-gated ion channels. EMBO J 17:3016–3028
North RA (2002) The molecular physiology of P2X receptors. Physiol Rev 82:1013–1067
North RA, Barnard EA (1997) Nucleotide receptors. Curr Opin Neurobiol 7:346–357
Ohta A, Sitkovsky M (2001) Critical role of G protein-coupled adenosine receptors in down regulation of inflammation and prevention of tissue damage in vivo. Nature 414:916–920
Oike M, Kimura C, Koyama T, Yoshikawa M, Ito Y (2000) Hypotonic stress-induced dual Ca^{2+} responses in bovine aortic endothelial cells. Am J Physiol Heart Circ Physiol 279:H630–H638
Oshimi Y, Miyazaki S, Oda S (1999) ATP-induced Ca^{2+} response mediated by P2U and P2Y purinoceptors in human macrophages: signalling from dying cells. Immunology 98:220–227
Ostrom RS, Gregorian C, Insel PA (2000) Cellular release of and response to ATP as key determinants of the set-point of signal transduction pathways. J Biol Chem 275:11735–11739
Perregaux D, Gabel CA (1994) Interleukin-1β maturation and release in response to ATP and nigericin. Evidence that potassium depletion mediated by these agents is a necessary and common feature of their activity. J Biol Chem 269:15195–15203
Perregaux DG and Gabel CA (1998) Post-translational processing of murine IL-1: evidence that ATP-induced release of IL-1α and IL-1β occurs via a similar mechanism. J Immunol 160:2469–2477
Perregaux DG, McNiff P, Laliberte R, Conklyn M, Gabel CA (2000) ATP acts as an agonist to promote stimulus-induced secretion of IL-1β and IL-18 in human blood. J Immunol 165:4615–4623
Purkiss JR, Boarder MR (1992) Stimulation of phosphatidate synthesis in endothelial cells in response to P2-receptor activation. Evidence for phospholipase C and phospholipase D involvement, phosphatidate and diacylglycerol interconversion and the role of protein kinase C. Biochem J 287:31–36
Ralevic V, Burnstock G (1998) Receptors for purines and pyrimidines. Pharmacol Rev 50:413–492

Ravi RG, Kertesy SB, Dubyal GR, Jacobson KA (2001) Potent $P2X_7$ receptor antagonists: Tyrosil derivatives synthesized using a sequential parallel synthetic approach. Drug Dev Res 54:75–87

Romanello M, Pani B, Bicego M, D' Andrea P (2001) Mechanically induced ATP release from human osteoblastic cells. Biochem Biophys Res Commun 289:1275–1281

Sala A la, Ferrari D, Corinti S, Cavani A, Di Virgilio F, Girolomoni G (2001) Extracellular ATP induces a distorted maturation of dendritic cells and inhibits their capacity to initiate Th1 responses. J Immunol 166:1611–1617

Schmid-Antomarchi H, Schmid-Alliana A, Romey G, Ventura MA, Breittmayer V, Millet MA, Husson H, Moghrabi B, Lazdunski M, Rossi B (1997) Extracellular ATP and UTP control the generation of reactive oxygen intermediates in human macrophages through the opening of a charybdotoxin-sensitive Ca^{2+}-dependent K^+ channel. J Immunol 159:6209–6215

Schwiebert LM, Rice WC, Kudlow BA, Taylor AL, Schwiebert EM (2002) Extracellular ATP signalling and P2X nucleotide receptors in monolayers of primary human vascular endothelial cells. Am J Physiol Cell Physiol 282:C289-C301

Sikora A, Liu J, Brosnan C, Buell G, Chessell I, Bloom BR (1999) Cutting edge: purinergic signaling regulates radical-mediated bacterial killing mechanisms in macrophages through a $P2X_7$-independent mechanism. J Immunol 163:558–561

Soto F, Garcia-Guzman M, Gomez-Hernandez JM, Hollmann M, Karschin C, Stuhmer W (1996) P2X4: an ATP-activated ionotropic receptor cloned from rat brain. Proc Natl Acad Sci USA 93:3684–3688

Sperlagh B, Hasko G, Nemeth Z, Vizi ES (1998) ATP released by LPS increases nitric oxide production in Raw 264.7 macrophage cell line via $P2Z/P2X_7$ receptors. Neurochem Int 33:209–215

Steinberg TH and Silverstein SC (1987) Extracellular ATP^{4-} promotes cation fluxes in the J774 mouse macrophage cell line. J Biol Chem 262:3118–3122

Storey F (2001) The $P2Y_{12}$ receptor as a therapeutic target in cardiovascular disease. Platelets 12:197–209

Surprenant A, Rassendren F, Kawashima E, North RA, Buell, G (1996) The cytolytic P2Z receptor for extracellular ATP identified as a P2X receptor ($P2X_7$). Science 272:735–738

Thompson K (2002) Structure/function studies on $P2X_7$ receptor. Doctoral Thesis Cambridge University

Tonetti M, Sturla L, Giovine M, Benatti U, De Flora A (1995) Extracellular ATP enhances mRNA levels of nitric oxide synthase and TNFα in lipopolysaccharide-treated RAW 264.7 murine macrophages. Biochem Biophys Res Commun 214:125–130

Torres GE, Egan TM, Voigt MM (1999) Hetero-oligomeric assembly of P2X receptor subunits. Specificities exist with regard to possible partners. J Biol Chem 274:6653–6659

Virginio C, MacKenzie A, Rassendren F, North RA, Surprenant A (1999) Pore dilation of neuronal P2X receptor channels. Nature Neurosci 2:315–321

Von Kugelgen I, Wetter A (2000) Molecular pharmacology of P2Y-receptors. Naunyn-Schmiedeberg's Arch Pharmacol 362:310–323

Warny M, Aboudola S, Robson SC, Sevigny J, Communi D, Soltoff SP, Kelly CP (2001) $P2Y_6$ nucleotide receptor mediates monocyte IL-8 production in response to UDP or lipopolysaccharide. J Biol Chem 276:26051–26056

Wilkin F, Duhant X, Bruyns C, Suarez-Huerta N, Boeynaems JM, Robaye B (2001) The $P2Y_{11}$ receptor mediates the ATP-induced maturation of human monocyte-derived dendritic cells. J Immunol 166:7172–7177

Xaus J, Mirabet M, Lloberas J, Soler C, Lluis C, Franco R, Celada A (1999) IFN-γ up-regulates the A_{2B} adenosine receptor expression in macrophages: a mechanism of macrophage deactivation. J Immunol 162:3607–3614

Yang S, Cheek DJ, Westfall DP, Buxton IL (1994) Purinergic axis in cardiac blood vessels. Agonist-mediated release of ATP from cardiac endothelial cells. Circ Res 74:401–407

Yoshioka K, Saitoh O, Nakata H (2001) Heteromeric association creates a P2Y-like adenosine receptor. Proc Natl Acad Sci USA 98:7617–7622
Zimmermann H (2000) Extracellular metabolism of ATP and other nucleotides. Naunyn-Schmiedeberg's Arch Pharmacol 362:299–309
Zimmermann H (2001) Ectonucleotidases: some recent developments and a note on nomenclature. Drug Dev Res 52:44–56

Macrophage Lipid Uptake and Foam Cell Formation

W. J. S. de Villiers · D. R. van der Westhuyzen

Division of Gastroenterology, Department of Internal Medicine,
University of Kentucky Medical Center,
800 Rose Street, Rm MN649, Lexington, KY 40536, USA
e-mail: wdevil0@uky.edu

Abstract The attraction and sub-endothelial migration of monocytes or macrophages (Mϕ) to the early atherosclerotic lesion or fatty streak are central events in atherogenesis. During migration through the endothelium these cells differentiate to Mϕ, which subsequently become activated in the milieu of cytokines and reactive products secreted by the dysfunctional endothelial layer, underlying smooth muscle cells and resident tissue leukocytes. As the lesion advances, the products secreted by activated Mϕ further escalate the inflammatory process, increasing the recruitment of more activated inflammatory cells. Mϕ within the lesion also actively accumulate lipid to become foam cells. Therefore, interventions that target the escalation of the inflammatory cascade, foam cell differentiation, and foam cell lipid metabolism should have distinct therapeutic benefits in altering the progression of CAD. In this chapter, various aspects of Mϕ lipid metabolism and foam cell formation are explored. These include the sources of cholesterol for foam cell formation, receptor-dependent and receptor-independent uptake pathways by which foam cells may accumulate cholesterol, cholesterol trafficking within Mϕ, cholesterol efflux from Mϕ, and foam cell death. All of these processes represent possible therapeutic targets in CAD.

Keywords CD36, Cholesterol efflux, Foam cells, Macrophages, Macrosialin, Scavenger receptors, SR-A, SR-BI

Abbreviations

ABCA1	ATP-binding cassette, subfamily A, member 1
ACAT	Acyl-CoA:cholesterol acyltransferase
AcLDL	Acetylated low-density lipoprotein
aP2	Fatty acid-binding protein
β-VLDL	β-Very low-density lipoprotein
CAD	Coronary artery disease
CE	Cholesteryl ester
FC	Free cholesterol
HDL	High-density lipoprotein
LXR	Liver X receptor
Mϕ	Macrophage(s)
oxLDL	Oxidized low-density lipoprotein
PPAR	Peroxisome proliferator-activated receptor
SR-A	Scavenger receptor class A
SR-BI	Scavenger receptor class B, type I
RXR	Retinoid X receptor
VLDL	Very low-density lipoprotein

1 Introduction

The attraction and sub-endothelial migration of monocytes or macrophages (Mϕ) to the early atherosclerotic lesion or fatty streak are central events in atherogenesis (Ross 1999; Glass and Witztum 2001). One of the earliest observed endothelial changes is the surface arrest of monocytes. During migration through the endothelium these cells differentiate to Mϕ, which subsequently become activated in the milieu of cytokines and reactive products secreted by the dysfunctional endothelial layer, underlying smooth muscle cells and resident tissue leukocytes. As the lesion advances, the products secreted by activated Mϕ further escalate the inflammatory process, increasing the recruitment of more activated inflammatory cells. Mϕ within the lesion also actively accumulate lipid to become foam cells. Therefore, interventions that target the escalation of the inflammatory cascade, foam cell differentiation and foam cell lipid metabolism should have distinct therapeutic benefits in altering the progression of CAD. In this chapter the various aspects of Mϕ lipid metabolism and foam cell formation that may represent therapeutic targets are explored.

2 Cholesterol Sources and Foam Cell Development

Two types of lipid oils, cholesteryl ester and triglyceride, can produce cellular lipid droplet accumulation and cause a foamy appearance in tissue sections prepared with organic solvents. Lipid droplets in tissues of cells prepared without extracting solvents can be stained with lipid-soluble dyes such as oil red O. Chemical studies of human atherosclerotic lesions show that the major lipid oil accumulating in lesions is cholesteryl ester (CE) and not triglyceride (Garner et al. 1997).

Exogenous cholesterol, not endogenously synthesized cholesterol, is the major source of cholesterol for foam cell formation. De novo synthesis of cholesterol functions to maintain cellular cholesterol levels in the absence of an exogenous source, and key enzymes in the cholesterol biosynthetic pathway are downregulated when an exogenous source of cholesterol is present (Goldstein and Brown 1990). Plasma lipoproteins are the major source of the exogenous cholesterol of foam cells, although other sources, such as cholesterol-rich platelet fragments (Curtiss et al. 1987), cause cholesterol loading in cultured Mϕ. Lipoproteins form emulsions of particles of varying densities that consist of a hydrophobic core containing neutral lipids, CE and triglycerides surrounded by an outer phospholipid monolayer containing a small amount of free cholesterol (FC) and protein (apolipoprotein). The bulk of the cholesterol (e.g., 80% in LDL) is carried as CE in the core of the particle.

2.1 Modified Forms of LDL

Although fasting plasma levels of LDL are strongly correlated with the extent of atherosclerosis, native plasma LDL is a poor inducer of foam cell formation in cultured Mϕ (Brown and Goldstein 1983; Steinberg et al. 1989). This finding can be explained by the combination of downregulation of LDL receptors (Brown and Goldstein 1983) and, as described below, specific characteristics of LDL-cholesterol metabolism in Mϕ (Tabas et al. 1990). A possible resolution to the apparent paradox of epidemiological data linking LDL to atherogenesis and foam cell formation is that LDL is modified to an "atherogenic" form in the subendothelium (Steinberg 1995). In an in vitro model to explore this concept, LDL modified by oxidation or acetylation has been widely studied. Native LDL is recognized by the LDL receptor by virtue of specific receptor-binding domains of the apolipoprotein B-100 moiety of LDL (Brown and Goldstein 1986). In contrast, AcLDL, due to acetylation of key lysine residues on apolipoprotein B-100, is not recognized by the LDL receptor but rather by another class of receptors, the so-called scavenger receptors (Goldstein et al. 1979; Krieger and Herz 1994). Scavenger receptor class A (SR-A) is widely expressed on differentiated Mϕ and mediates the constitutive uptake of AcLDL, leading to massive CE accumulation in most cultured Mϕ models (Brown and Goldstein 1983; Krieger and Herz 1994). Although acetylation of LDL does not occur in vivo, oxidation of LDL is a physiologically plausible process that also renders LDL a ligand for scavenger receptors and other cell-surface scavenger receptors that are not downregulated by cellular cholesterol loading (Steinberg 1997). Immunohistochemical studies have documented the presence of oxLDL in atherosclerotic lesions, and a wide variety of cell-culture studies have demonstrated potentially atherogenic effects of oxLDL, such as induction of atherogenic endothelial cell responses and smooth muscle cell proliferation (Steinberg 1997). Further support for the oxidation hypothesis includes recent data showing that deletion of the Mϕ-specific enzyme, 12/15-lipoxygenase, which is involved in the synthesis of oxygenated fatty acids, reduces levels of atherosclerosis in *ApoE* $^{-/-}$ mice (Cyrus et al. 1999).

A widely held misconception, however, is that LDL oxidized by most standard in vitro methods induces foam cell formation (i.e., marked CE accumulation) in cultured Mϕ. In fact, quite a few studies have shown that while Mϕ incubated with oxLDL internalize lipoprotein extensively, the cells accumulate mostly FC (Roma et al. 1990; Ryu et al. 1995; Klinkner et al. 1995; Musanti and Ghiselli 1993). The mechanism may be related to reduced CE content of the lipoprotein as a result of oxidation (Ryu et al. 1995) or to incomplete lysosomal degradation of oxLDL components, which in turn may lead to decreased export of FC from lysosomes (Lougheed et al. 1991; Hoppe et al. 1994; Maor et al. 1995). Additional in vitro modifications of oxLDL, such as by FC enrichment (Greenspan et al. 1997), are able to convert this lipoprotein into one that can cause substantial CE accumulation in cultured Mϕ, but it is still not certain whether these additional modifications of oxLDL occur in vivo. Recently, investigators have shown that SR-A-deficient and CD36-deficient mice have reduced, but by no means absent, foam cell lesions (Suzuki et al. 1997; Febbraio et al. 2000). Whereas proponents of the oxidation hypothesis might argue that the presence of foam cells in these knockout mice implicates a role for other oxLDL receptors in foam cell formation in vivo, an equally plausible interpretation is that a significant portion of foam cell formation involves other types of atherogenic lipoproteins altogether.

2.2 Aggregated Lipoproteins

Aggregated lipoproteins, which are not necessarily recognized by scavenger receptors, have been demonstrated in both early and late atherosclerotic lesions and are very potent inducers of foam cell formation in cultured Mϕ (Khoo et al.

1988; Suits et al. 1989; Hoff et al. 1990; Xu and Tabas 1991). Freeze-fracture images of the subendothelium (Nievelstein et al. 1991) as well as analyses of LDL isolated from atherosclerotic lesions (Hoff and Morton 1985; Guyton and Klemp 1996) indicate that a large proportion of lesional LDL is in the form of aggregate and fused particles averaging approximately 100 nm in diameter. Although the mechanism of LDL aggregation is not known, physiological plausible hypotheses include extensive oxidation (Hoff et al. 1989), hydrolysis by mast cell-derived proteases (Kovanen 1991), and hydrolysis by a Mϕ-derived and endothelial-derived secretory sphingomyelinase (Schissel et al. 1998). Thus, in contrast to the situation with oxLDL, cultured Mϕ incubated with aggregated LDL formed by a variety of methods in vitro accumulate massive amounts of CE.

2.3 Chylomicron Remnants

Post-prandial chylomicron remnants are a class of lipoproteins other than LDL that are epidemiologically associated with atherosclerosis, and are potent inducers of foam cell formation in cultured Mϕ (Havel 1995). Chylomicrons are very buoyant, lipid-rich lipoproteins formed in intestinal epithelial cells during the absorption of dietary fat. Following entry into the circulation, the triglyceride component of the core of chylomicrons is rapidly hydrolyzed by the enzyme lipoprotein lipase bound to the lumenal surface of the endothelium. The resulting chylomicron "remnants" may either be rapidly cleared by the liver or remain in the circulation for an extended period, depending upon specific metabolic and genetic factors that differ among individuals (Havel 1995). Chylomicron remnants remaining in the circulation can become enriched in CE, enter the arterial wall, and enhance lesion development. A widely studied model of CE-rich chylomicron remnants is β-VLDL, a lipoprotein found at very high levels in the plasma of cholesterol-fed rabbits. Chylomicron remnants and β-VLDL lead to marked accumulation of CE when incubated with cultured Mϕ (Goldstein et al. 1980; Mahley et al. 1980).

3 Mechanisms by Which Foam Cells May Accumulate Cholesterol

3.1 Receptor-Mediated Uptake Pathways

Lipoprotein binding to the plasma membrane has been studied with different ligands (e.g., AcLDL, β-VLDL, LDL, oxLDL) in different Mϕ model systems. Generally, Mϕ lipoprotein binding occurs in coated (clathrin-associated) pits on microvillous extensions, and on uncoated regions of plasma membrane ruffles. Both uncoated and coated pits function in the uptake and delivery of bound lipoproteins to lysosomes (van der Schroeff et al. 1983; Traber et al. 1983; Mommaas-Kienhuis et al. 1985). Scavenger receptor-mediated endocytic pathways have received the most interest as a mechanism for cholesterol accumulation and Mϕ foam cell formation. Although several proteins may contribute to this overall process, SR-A and CD36 have been demonstrated to play quantitatively significant roles. Recent evidence suggests that SR-A and CD36 represent two scavenging systems that are physiologically important in foam cell formation and atherogenesis.

3.1.1 SR-A

The overexpression or upregulation of SR-A results in the transformation of Chinese hamster ovary cells or peritoneal murine Mϕ into foam cells in the presence of AcLDL (Freeman et al. 1991; de Villiers et al. 1994; de Winther et al. 1999). However, the natural ligand that promotes Mϕ foam cell formation by SR-A in vivo has not been clearly identified, and ligands other than oxidatively modified lipoproteins may lead to foam cell formation by SR-A in vivo (Tabas 1999). SR-A recognizes only extensively modified LDL, and in vitro studies have demonstrated that peritoneal Mϕ lacking SR-A showed 80% less uptake of AcLDL, whereas the uptake of oxLDL was reduced by only 30% (Lougheed et al. 1997; Terpstra et al. 1997). The pro-atherogenic role of SR-A was confirmed in studies with mice deficient in apoE and SR-A (Suzuki et al. 1997). These animals develop 58% less atherosclerotic lesions than apoE-deficient control mice. LDL-receptor/SR-A double-knockout mice similarly showed less atherosclerotic plaque formation (20%) compared with LDL-receptor knockout animals (Sakaguchi et al. 1998).

Mϕ are the primary, but not the only cell type expressing SR-A (Bickel and Freeman 1992; Hughes et al. 1995). Therefore, to dissect the impact of Mϕ SR-A on atherogenesis, mice chimeric for Mϕ SR-A were generated by transplantation with SR-A$^{-/-}$ fetal liver cells. Lethally irradiated C57BL/6 mice reconstituted with SR-A$^{-/-}$ Mϕ and challenged with a high-fat diet for 16 weeks had a 60% reduction in lesion area compared with SR-A$^{+/+}$→C57BL/6 mice (Babaev et al. 2000). LDLR$^{-/-}$ mice reconstituted with SR-A$^{-/-}$ Mϕ and fed a Western diet for 10 weeks had a similar 60% reduction in atherosclerotic lesions compared with SR-A$^{+/+}$→LDLR$^{-/-}$ mice (Babaev et al. 2000). There were no significant differences in serum lipid levels in these two experimental models, suggesting that SR-A does not affect serum cholesterol and triglyceride levels, but plays a central role in atherogenesis by loading vascular wall Mϕ with oxidized and otherwise modified lipoprotein material.

3.1.2
Class B Scavenger Receptors

CD36

CD36, a member of the class B scavenger receptor family, is structurally unrelated to SR-A and more widely expressed, and has a broader ligand specificity (Febbraio et al. 1999). CD36 was originally identified as an oxLDL receptor by an expression cloning strategy used to isolate murine Mϕ receptors that recognize oxLDL, but not AcLDL (Endemann et al. 1993). A genetic polymorphism in the CD36 gene has been identified in Japanese subjects (Kashiwagi et al. 1995) and shown to result in deficient expression of CD36 (NAK^{a-} phenotype). Monocyte-derived Mϕ isolated from these patients bind $\approx$40% less oxLDL and accumulate $\approx$40% less cholesterol ester from oxLDL than cells derived from normal controls (Nozaki et al. 1995), further implicating CD36 as a physiological oxLDL receptor. In addition, peritoneal Mϕ from CD36 knockout mice showed a 60% reduction in the uptake of oxLDL and a 52% reduction in the uptake of AcLDL (Febbraio et al. 1999, 2000). ApoE-deficient mice lacking CD36 develop significantly less atherosclerosis with a 77% decrease in aortic tree lesion area (Western diet) and a 45% decrease in aortic sinus lesion area (normal chow) when compared with control $apoE^{-/-}$ mice (Febbraio et al. 2000). These data support an important role for CD36 in foam cell formation and atherogenesis in vivo. However, CD36 also has a crucial role in fatty acid transport in heart and other tissues, and this predominant physiological role of CD36 is unlikely to be attenuated by pharmacological interventions without a marked negative effect on cardiac metabolism.

SR-BI

The other member of the class B scavenger receptor family, SR-BI, also binds oxLDL and LDL, but has been better characterized as a physiologically relevant lipoprotein receptor for HDL (Krieger 2001). SR-BI mediates the selective uptake of HDL CE (Acton et al. 1996; Krieger 1999) by a process in which HDL delivers CE to the cell without lysosomal degradation of the whole HDL particle (Pittman et al. 1987). Recent reports indicate that HDL is internalized by SR-BI and that CE is selectively removed from HDL as HDL particles are recycled back to the cell surface (Silver et al. 2000, 2001). Although CD36 can also mediate the selective uptake of lipids from HDL, it is much less efficient than SR-BI, and its predominant physiological role seems to be in oxLDL, rather than LDL or HDL metabolism (de Villiers et al. 2001). SR-BI is predominantly expressed in liver and steroidogenic tissues, precisely those tissues that show the highest levels of selective uptake of HDL cholesteryl esters (Acton et al. 1996; Krieger 1999). Hepatic overexpression of SR-BI by adenoviral-mediated transfer (Kozarsky et al. 1997; Webb et al. 1998) results in decreased plasma levels of HDL cholesterol and increased biliary cholesterol content.

SR-BI expression can be regulated by estrogen, PPAR agonists, vitamin E, polyunsaturated fatty acids, and cholesterol (Fluiter et al. 1998; Krieger 1999; Spady et al. 1999; Trigatti et al. 2000; Witt et al. 2000). The antiatherogenic role of SR-BI in mouse models of atherosclerosis is unequivocal. The absence of SR-BI in knockout mice dramatically accelerates the onset of atherosclerosis (Trigatti et al. 1999; Huszar et al. 2000), whereas atherosclerosis is suppressed by hepatic overexpression of SR-BI (Arai et al. 1999; Kozarsky et al. 2000; Ueda et al. 2000). Because of the antiatherogenic effects of increased hepatic SR-BI expression, SR-BI represents a possible target for therapeutic intervention in CAD. However, the contribution of Mϕ SR-BI to atherogenesis is unclear and may be a confounding factor.

Selective lipid uptake of CE from HDL has been shown for many cell types, including Mϕ (Stein et al. 1987; Panzenboeck et al. 1997; Hirano et al. 1999; Chinetti et al. 2000). SR-BI expression levels correlate with selective CE uptake from HDL in differentiated THP-1 Mϕ, and caveolin-1, an important constituent of caveolae, acts as a negative regulator (Matveev et al. 1999, 2001). In addition to uptake from HDL, Mϕ also take up CE selectively from LDL (Rinninger et al. 1995) and oxLDL (manuscript submitted). In the latter instance, THP-1 cells selectively internalize CE from oxLDL by a process that appears to be independent of SR-BI. Selective uptake of CE in mouse peritoneal Mϕ has also been suggested in the case of aggregated LDL (Rinninger et al. 1995; Rhainds and Brissette 1999), but the receptors involved have not been identified. It remains to be shown, however, that selective uptake from any of the ligands mentioned can induce Mϕ cholesterol accumulation.

3.1.3
Macrosialin

Murine macrosialin and its human homologue CD68 are extensively glycosylated transmembrane proteins expressed in Mϕ and Mϕ-related cells, including liver Kupffer cells (Rabinowitz and Gordon 1991). Macrosialin is predominantly a late endosomal protein but is also found on the cell surface (Rabinowitz and Gordon 1991; Kurushima et al. 2000). Interest in macrosialin as a scavenger and oxLDL receptor arose when, on the basis of ligand blotting, macrosialin was suggested to bind oxLDL (Ramprasad et al. 1995; Ramprasad et al. 1996). Further examples of evidence supporting a role for macrosialin in modified LDL catabolism include its identification in liver Kupffer cells as the major oxLDL binding protein (van Velzen et al. 1997) and its prominent expression in Mϕ in atherosclerotic plaques from apoE knockout mice (de Villiers et al. 1998). However, we found no binding of oxLDL to macrosialin in intact transfected COS-7 and CHO cells, despite significant cell-surface expression of macrosialin (manuscript submitted). Although macrosialin expression in Mϕ and Kupffer cells is responsive to a pro-atherogenic inflammatory diet and to oxLDL, possibly indicating a compensatory protective role, there is no evidence that macrosialin functions as an oxLDL receptor on the cell surface or participates in foam cell formation.

3.1.4 LOX-1 and SR-PSOX

More recently, two additional oxLDL receptors were described that are also expressed on Mϕ. Their precise role in foam cell formation and atherogenesis is unclear and is currently the focus of intense investigation. Lectin-like oxLDL receptor-1 (LOX-1) was identified by expression cloning from a cDNA library of cultured bovine aortic endothelial cells (BAECs) and is a type II membrane protein that belongs to the C-type lectin family of molecules (Sawamura et al. 1997). LOX-1 acts as a cell-surface endocytic receptor by mediating the binding, internalization, and proteolytic degradation of oxLDL, but not AcLDL (Moriwaki et al. 1998). Cell-surface LOX-1 can be cleaved through some protease activities that are associated with the plasma membrane, and released into the culture media (Murase et al. 2000). LOX-1 is present in cultured human and murine Mϕ and activated smooth muscle cells (Yoshida et al. 1998); and its expression is inducible by proinflammatory stimuli, such as tumor necrosis factor (TNF)-α, transforming growth factor (TGF)-β, lipopolysaccharide (LPS), angiotensin II and oxLDL itself (Kume and Kita 2001). In early atherosclerotic lesions in human carotid arteries, LOX-1 is highly expressed in endothelial cells. In more advanced lesions with large atheromatous plaques, LOX-1 expression is more prominent in Mϕ and smooth muscle cells suggesting roles for LOX-1 in foam cell formation and vascular cell dysfunction (Kataoka et al. 2000).

The same group of investigators recently identified another novel cell-surface receptor for oxLDL by expression cloning from phorbol myristate acetate (PMA)-stimulated THP-1 cells, designated SR-PSOX (*s*cavenger *r*eceptor for *p*hosphatidyl*s*erine and *ox*idized lipoprotein). SR-PSOX can specifically bind oxLDL with high affinity, followed by internalization and degradation (Shimaoka et al. 2000). Human and murine SR-PSOX are 30-kDa type I membrane glycoproteins which do not share any homology with other oxLDL receptors, and seem to be identical to the membrane-anchored chemokine CXCL16, which may play a dual role in inflammation and homeostasis (Matloubian et al. 2000). Immunohistochemistry showed that SR-PSOX was predominantly expressed by lipid-laden Mϕ in the intima of atherosclerotic plaques (Minami et al. 2001). Taken together, SR-PSOX may be involved in oxLDL uptake and subsequent foam cell transformation in Mϕ in vivo.

3.1.5 Fatty Acid Binding Protein (aP2)

Another lipid-binding protein that is expressed in both adipocytes and Mϕ also has a role in the development and metabolism of foam cells. Genetic deletion of fatty acid binding protein (aP2), a protein known to have a physiologically important role in regulating lipid metabolism and insulin sensitivity, has a proatherogenic effect in *ApoE* $^{-/-}$ mice. Isolated Mϕ from *aP2* $^{-/-}$ mice secrete lower levels of inflammatory cytokines and accumulate lower levels of intracellular cholesterol esters in response to modified lipoproteins (Layne et al. 2001; Makowski et al. 2001). The introduction of *aP2* $^{-/-}$ Mϕ to *ApoE* $^{-/-}$x*aP2* $^{+/+}$ mice by bone marrow transplantation results in a reduction of atherosclerotic lesions that is comparable to that observed in *ApoE* $^{-/-}$x*aP2* $^{-/-}$ mice, indicating that it is the Mϕ expression of aP2 that is pro-atherogenic.

3.2 Receptor-Independent Uptake Pathways

In contrast to most other cells, Mϕ also have the capacity to take up materials by phagocytosis. Mϕ phagocytose some types of aggregated LDL that then leads to rapid lysosomal degradation of the accumulated aggregated LDL (Hoff et al. 1989; Hoff and Cole 1991). Two lines of evidence support the phagocytosis of chylomicron remnants and aggregated LDL by Mϕ. First, cytochalasin D inhibited the uptake of lipoprotein particles (Khoo et al. 1988; Suits et al. 1989) and second, electron microscopy studies suggested the accumulation of lipoproteins in phagocytic vacuoles (Salisbury et al. 1985).

Other endocytic pathways have been described that may be unique to Mϕ. Large β-VLDL enter peripheral surface-connected wide invaginations, so-called STEMs (*s*urface *t*ubules for *e*ntry into *M*ϕ), prior to undergoing lysosomal degradation (Tabas et al. 1990; Tabas et al. 1991; Myers et al. 1993). Importantly, there are different fates for β-VLDL cholesterol that enters Mϕ through STEMs and LDL cholesterol that enters Mϕ through a coated pit-mediated endocytic pathway. β-VLDL cholesterol delivered through STEMs leads to more efficient cholesterol esterification compared with LDL cholesterol delivered through coated pits.

Patocytosis is a recently described pathway for human monocyte–Mϕ uptake of aggregated lipoproteins, microcrystalline cholesterol, cholesterol-phospholipid liposomes, and other hydrophobic materials. In this pathway, aggregated LDL induces surface invaginations that connect with a labyrinth of interconnected vacuolar compartments within the Mϕ cytoplasm (Kruth et al. 1999a,b; Zhang et al. 2000). The characteristic hallmark of phagocytosis, namely the pinching off of Mϕ plasma membrane to form a phagocytic vacuole, does not occur in patocytosis. During patocytosis the aggregated LDL accumulates within a cytoplasmic labyrinth that remains connected to the Mϕ surface. While some accumulated aggregated LDL subsequently undergoes lysosomal degradation, most aggregated LDL remains in the surface-connected compartments of the labyrinth. The poor degradation of aggregated LDL taken up by patocytosis differs from the rapid degradation of aggregated LDL taken up by phagocytosis. Actin microfilaments function in lipoprotein uptake during both patocytosis and phagocytosis but not during uptake of β-VLDL into STEMs (Tabas et al. 1990). As a result, cytochalasin D, an agent that interferes with actin polymerization, inhibits lipoprotein uptake mediated by patocytosis and phagocytosis.

Pinocytosis (uptake of fluid in small vesicles) and macropinocytosis (uptake of fluid in large vacuoles) also potentially function in lipoprotein uptake by Mϕ. In fibroblasts, pinocytotic uptake of LDL in bulk fluid is linearly related to LDL concentration, and the LDL taken up undergoes lysosomal degradation (Goldstein and Brown 1977). A similar pathway appears to function in Mϕ but has not been extensively studied (Traber et al. 1983). Uptake of lipoproteins bound to plasma membrane areas that then form macropinosomes is another mechanism by which lipoproteins may enter Mϕ (Jones et al. 1999; Jones and Willingham 1999). Indeed, AcLDL and aggregated LDL, both potent inducers of cholesterol esterification, engage in prolonged association with surface invaginations in Mϕ (Zha et al. 1997). The underlying mechanism is unknown, but may be quite relevant in foam cell formation in the subendothelium of developing lesions in vivo, where most of the lipoproteins are not free in solution, bur rather are very tightly bound to extracellular matrix, often in an aggregated form.

4
Cholesterol Trafficking Within Macrophages

Multiple pathways exist for lipoprotein and non-lipoprotein sources of cholesterol to enter Mϕ. Following endocytosis of lipoproteins by Mϕ and foam cells, CE is hydrolyzed in lysosomes by a CE hydrolase (Anderson and Sando 1991). The newly liberated FC may either be retained or released from Mϕ, perhaps depending on the pathway by which it was delivered. In most peripheral cells, intracellular levels of FC are controlled in part by an enzymatically regulated cycle of esterification and hydrolysis (Brown et al. 1980). Excess plasma membrane cholesterol enters the cytoplasm and is delivered to the intracellular cholesterol esterifying enzyme, acyl-CoA:cholesterol acyltransferase (ACAT) located predominantly in the endoplasmic reticulum. Esterified cholesterol is stored in membrane-bound inclusions in the cytoplasm. These CE stores can be rehydrolyzed to FC by cytoplasmic neutral cholesterol-ester hydrolases (Khoo et al. 1993) to complete the cycle. Following hydrolysis, FC traffics back to the plasma membrane where it may undergo efflux from the cell.

Two ACAT isoforms have been described: ACAT-1 is expressed largely in peripheral tissues, including Mϕ; and ACAT-2 is expressed in the intestine and liver (Brewer 2000). ACAT inhibitors have been shown to reduce diet-induced atherosclerosis in rabbits and hamsters with minimal alterations to circulating cholesterol levels (Bocan et al. 2001). The mechanism for this effect is presumed to be a reduction in the differentiation of Mϕ into lipid-laden foam cells, which results from the interruption of the cholesterol esterification-hydrolysis cycle (Rodriguez and Usher 2002). An ACAT inhibitor, avasimibe (Pfizer), is now in phase III clinical trials for the treatment of CAD (Insull et al. 2001).

This process of cholesterol trafficking through cells, including Mϕ, is under intense investigation to learn whether it occurs in association with carrier proteins, such as the NPC1 protein (Carstea et al. 1997; Loftus et al. 1997), and whether cholesterol is transported by membrane vesicles or smaller macromolecular complexes which may contain caveolin-1 and other proteins (Smart et al. 1996; Uittenbogaard et al. 2002). Cholesterol and other molecules that become deposited in ceroid inclusions in Mϕ may not be available for eventual cholesterol efflux. These inclusions contain insoluble oxidized and polymerized proteins and lipid presumably including CE. Ceroid inclusions can be produced in vitro when Mϕ take up oxLDL or lipid particles containing lipids that are especially susceptible to oxidation (Ball et al. 1986, 1987, 1988). Sphingomyelin is a cholesterol-binding lipid that exists with cholesterol in endosomes, lysosomes, and in cholesterol-enriched detergent-insoluble membrane microdomains (DIGS) such as caveolae. Sphingomyelin is rapidly hydrolyzed by lysosomal sphingomyelinase, a product of the acid sphingomyelinase gene. Mϕ deficient in acid sphingomyelinase exhibit defective cholesterol trafficking and efflux, suggesting sphingomyelin plays an important role in cholesterol trafficking from intracellular sites to the plasma membrane (Leventhal et al. 2001).

5 Cholesterol Efflux From Macrophages

Mechanisms mediating cholesterol efflux are of critical importance in foam cell development. As with the control of the cholesterol esterification–hydrolysis cycle, many of the mechanisms that are important for the regulation of cholesterol metabolism and efflux in Mϕ and foam cells are shared by the liver and other peripheral tissues. Mϕ have two potential mechanisms for disposing of excess cholesterol: enzymatic modification to more soluble forms, and efflux via membrane transporters.

5.1 Macrophage Excretion of 27-Oxygenated Cholesterol Metabolites

The enzyme cholesterol 27-hydroxylase is expressed in Mϕ at relatively high levels and could potentially play a role in cholesterol excretion by converting it to the more polar and soluble 27-OH-cholesterol and 3β-OH-cholesterolenoic acid (Babiker et al. 1997). In the absence of cholesterol acceptors such as HDL, these compounds are then excreted from Mϕ (Westman et al. 1998; Brown et al. 2000).

5.2 Plasma-Derived HDL

The major mechanism for cholesterol efflux is likely to be via membrane transporters, with HDL serving as a primary extracellular acceptor. This role of HDL is central to the "reverse cholesterol transport" process and may explain the inverse correlation of HDL levels with the risk of atherosclerosis (Tall et al. 2000). HDLs induce cholesterol efflux when incubated with Mϕ by stimulating translocation of cholesterol from intracellular membranes to the plasma membrane (Aviram et al. 1989; Bierman et al. 1991). The HDL then acquires excess plasma membrane cholesterol. Alternatively, some studies show that HDL enters Mϕ and acquires cholesterol through interactions with lipid droplets. This cholesterol-enriched HDL is then re-secreted by the Mϕ (Schmitz et al. 1985; Takahashi et al. 1989; Takahashi and Smith 1999).

SR-BI, in addition to mediating selective CE uptake from HDL, has also been shown to promote bi-directional flux of FC and phospholipids between cells and HDL (Ji et al. 1997; Jian et al. 1998; Krieger 1999; Rothblat et al. 1999). The physiologic importance of SR-BI for reverse cholesterol transport is suggested by studies that show mice over-expressing hepatic SR-BI have reduced atherosclerosis (Kozarsky et al. 2000). However, the significance of SR-BI-dependent cellular cholesterol efflux from Mϕ has not been established.

5.3 The Cassette Protein ABCA1

A key insight into the molecular mechanisms responsible for cholesterol efflux resulted from studies of patients with Tangier disease, which is characterized by extremely low levels of HDL and cholesterol accumulation in Mϕ. The cause of Tangier disease was found when several different approaches led to the identification of null mutations in the *ABCA1* (ATP-binding cassette, subfamily A, member 1) gene, encoding a member of the ATP binding cassette family of transporters (Bodzioch et al. 1999; Brooks-Wilson et al. 1999; Lawn et al. 1999; Rust et al. 1999). In vitro studies indicate that ABCA1 mediates transport of cholesterol and phospholipids from cells to apoA-I and other apolipoproteins or to lipid-poor pre-β HDL (Oram et al. 2000; Oram and Lawn 2001). In the absence of sufficient lipidation, nascent HDL particles are rapidly cleared, suggesting a probable explanation for the extremely low HDL cholesterol levels in Tangier patients.

Cholesterol efflux occurs with lipid-free amphipathic apolipoproteins of HDL such as apoA-I that associate with Mϕ phospholipid and form nascent HDL particles. Deletion of apoA-I, the major protein component of HDL, is not sufficient to cause atherosclerosis in mice fed a normal chow diet. However, apoA-I deficiency markedly exacerbates atherosclerosis in hypercholesterolemic mouse models (Voyiaziakis et al. 1998). Adenovirus-mediated overexpression of apoA-I protects against the development of atherosclerosis (Benoit et al. 1999; Tangirala et al. 1999).

In addition to ABCA1, Mϕ and adipocytes also express and secrete the amphipathic apolipoprotein, apoE, in response to increases in intracellular cholesterol. Importantly, apoE expression in arterial wall Mϕ is believed to promote cholesterol efflux from lipid-laden foam cells, and to protect against atherogenesis by a mechanism that is distinct from its well-known role in hepatic lipoprotein uptake. Mϕ produce their own HDL particles that may mediate Mϕ cholesterol efflux through an autocrine/paracrine mechanism involving both apoE and ABCA1. This occurs when Mϕ secrete apoE that associates with Mϕ phospholipid to form apoE-phospholipid discoidal complexes (Basu et al. 1982). These apoE–phospholipid particles acquire cholesterol from Mϕ sufficiently to cause a decrease in cholesterol content of human monocyte-derived Mϕ and cyclic adenosine monophosphate (cAMP)-treated RAW mouse Mϕ, but not untreated mouse peritoneal Mϕ (Smith et al. 1996; Zhang et al. 1996). ABCA1 probably plays a major role in this cellular efflux of cholesterol and phospholipid to apoE. Mϕ-specific expression of human apoE reduces atherosclerosis in hypercholesterolemic apoE-null mice, supporting a possible function of Mϕ-produced apoE within lesions in promoting cholesterol efflux (Bellosta et al. 1995).

Recent studies link ABCA1, PPARs, apoE and the nuclear receptors LXRα and LXRβ in cholesterol efflux. These nuclear receptors are important in regulating the expression of various proteins involved in the control of hepatic lipid metabolism, and have significant biological effects on the regulation of systemic lipid and cholesterol levels, as well as on the regulation of cellular cholesterol efflux. Among the main physiological activators of the LXR receptors are oxysterols, and genetic deletion of LXRα in mice attenuates the ability of the liver to regulate the expression of lipid-metabolizing gene products in response to dietary cholesterol (Peet et al. 1998). In contrast, mice in which LXRβ is deleted respond normally to changes in dietary cholesterol (Alberti et al. 2001). These receptors might also have an important role in cholesterol efflux, independent of their role in lipid metabolism. Overexpression of LXRα in fibroblast or Mϕ cell lines, or treatment of these cells with oxysterols, induces expression of the mRNA for ABCA1, indicating that its expression might be under the control of oxysterols through their interaction with LXR (Venkateswaran et al. 2000). Cholesterol regulation of Mϕ and adipocyte expression of apoE is also under the transcriptional control of LXR receptors. It has recently been shown that adipocytes and Mϕ from $Lxr\alpha^{-/-}$, $Lxr\beta^{-/-}$ mice and double-knockout mice have reduced or absent activation of apoE expression by oxysterols. These data indicate that Mϕ LXR receptors are physiologically important in the regulation of cholesterol efflux (Laffitte et al. 2001).

The PPAR class of transcription factors belongs to the broad nuclear hormone receptor superfamily, which includes the steroid, retinoid, and thyroid hormone receptors. The three PPAR isoforms (α, δ, and γ) form a heterodimer with RXR and regulate the transcription of genes that are involved in lipid and glucose metabolism. Binding of lipid ligands to the PPAR–RXR heterodimer activates the transcription factor complex (Kersten et al. 2000). PPAR-γ is most abundantly expressed in adipocytes, but is also present in Mϕ (Ricote et al. 1998). PPAR-γ is the specific target of the insulin-sensitizing thiazolidinediones that are widely used in the treatment of hyperlipidemia and type 2 diabetes mellitus. A role for PPAR-γ in foam cell formation became relevant when it was shown that lipid ligands present in oxLDL, as well as the thiazolidinediones, could activate PPAR-γ in Mϕ, upregulate expression of the oxLDL receptor, CD36, and increase their own uptake (Nagy et al. 1998; Tontonoz et al. 1998). These findings, describing a forward feeding loop resulting in increased oxLDL uptake and foam cell formation, raised serious concerns about the potential long-term side effects of thiazolidinedione treatment of type 2 diabetic patients, who are already at increased risk for developing atherosclerosis (Spiegelman 1998). However, recent studies show a more complex regulation of Mϕ lipid metabolism by PPAR-γ resulting in protection from atherosclerosis. In addition to upregulating CD36 expression, troglitazone treatment of peritoneal Mϕ also downregulates SR-A expression (Moore et al. 2001). This opposing regulation of modified LDL receptors results in the largely unchanged uptake of oxLDL by Mϕ. Studies in apoE and LDL-receptor knockout mice showed that treatment with PPAR-γ agonists (rosiglitazone and troglitazone) had a favorable influence on the development of atherosclerosis (Li et al. 2000; Chen et al. 2001; Collins et al. 2001). PPAR-γ was shown to limit cholesterol accumulation in Mϕ by increasing ABCA1 expression and cholesterol efflux in response to apoA-I. Interestingly, the PPAR-γ induction of ABCA1 appears to be driven through the PPAR-γ induction of the nuclear receptor LXR (Chawla et al. 2001; Chinetti et al. 2001). A major physiologic role of PPAR-γ/LXR signaling may, therefore, be modulation of the reverse cholesterol transport process in Mϕ and the atheroprotective actions of PPAR-γ agonists may be partly related to the induction of cholesterol efflux. Recent data also indicate that selective activation of another PPAR isoform, PPAR-δ, resulted in lipid accumulation in primary Mϕ and THP-1 cells (Vosper et al. 2001). A selective PPAR-δ agonist increased SR-A and CD36 expression, and downregulated gene expression of cholesterol 27-hydroxylase and apoE. The exact significance of PPAR-δ activation as a promoter of Mϕ foam cell formation remains unclear.

6
Foam Cell Death

Foam cell death may interfere with removal of cholesterol from atherosclerotic lesions and contribute to the accumulation of extracellular cholesterol in lesions. There are conflicting reports whether massive cholesterol accumulation within Mϕ actually causes foam cell death. Cholesterol accumulation in Mϕ treated with an ACAT inhibitor is associated with the buildup of unesterified cholesterol in cellular membranes and Mϕ cell death (Warner et al. 1995; Kellner-Weibel et al. 1998). However, Mϕ accumulation of excessive unesterified cholesterol does not always result in foam cell death. Mϕ are able to increase phospholipid content, a natural buffer, in response to excess unesterified cholesterol (Tabas 1997). Even when maintaining normal phospholipid content, human monocyte-derived Mϕ are able to accumulate large amounts of cholesterol crystals without displaying cell death. This may be because these Mϕ sequester excessive cholesterol in protective locations such as surface-connected compartments or lysosomes. The conversion of cholesterol to oxysterols could be another factor determining the toxicity of excess cholesterol, because oxysterols, but not cholesterol, are reported to be toxic to human monocyte-derived Mϕ (Clare et al. 1995).

7 Therapeutic Implications

The data describing the importance of Mϕ expression of various types of scavenger receptors, ACAT, PPARs, LXR, ABCA1, apoE and aP2 for the progression of atherosclerosis in mouse models confirm the central contribution of lipid metabolism in Mϕ and foam cells to the pathology of CAD. Therapeutic interventions aimed at inhibiting the pathways that are involved in the accumulation of lipid by Mϕ as they differentiate into foam cells could complement both existing and developing therapies for the management of lipid and lipoprotein levels.

Important caveats for the benefit of these approaches may be the impact of a particular therapeutic intervention on the normal function of the immune system and its response to invading pathogens, as is evident from the described multi-functional role of scavenger receptors, particularly SR-A in innate immunity (Platt and Gordon 2001). As existing lipid-lowering therapies have an excellent risk/benefit ratio, new therapies for the treatment of CAD will require a similarly excellent safety profile in order to be maximally accepted and used. Significant impairment of the normal immune system, leading to an increase in the rate of infections, will probably be unacceptable for therapies used for the treatment of CAD.

8 Conclusion

The Mϕ foam cell is a prominent and important component of the atherosclerotic lesion, playing roles in both lesion initiation and lesion progression. Foam cell biology as it pertains to both of these processes can be understood only through analysis of how Mϕ interact with and internalize atherogenic lipoproteins and how they metabolize lipoprotein-derived cholesterol. Much of our knowledge of foam cell formation comes from studying cultured Mϕ, often permanent cell lines, interacting with monomeric lipoproteins dissolved in tissue culture medium. Mϕ subtypes are known to differ in important ways, and lesional Mϕ may in fact possess important differences to those studied in the laboratory. Likewise, the form of lipoprotein that interact with Mϕ in lesions is almost certainly different from those investigated in most cell-culture studies. Therefore, studies examining the interaction of Mϕ with aggregated lipoproteins should be pursued in future research.

The conclusion that foam cells promote atherogenesis may require careful examination. Clearly, Mϕ foam cells can secrete molecules, such as oxidants, growth factors, inflammatory cytokines, and metalloproteinases, which may promote lesion development and plaque breakdown. But the ability of Mϕ to scavenge potentially harmful molecules, including oxidized lipids, may be beneficial, such as is often the case in other types of inflammatory and infectious lesions. Thus, as we identify specific molecules related to foam cell biology, and as we increasingly use in vivo systems, such as transgenic and knockout mice, to study these molecules, experimental strategies must specifically address this critical issue. Only through such studies will we be able to use our knowledge of foam cell biology to rationally design anti-atherogenic therapeutic interventions.

9 References

Acton S, Rigotti A, Landschulz KT, Xu S, Hobbs HH, Krieger M (1996) Identification of scavenger receptor SR-BI as a high density lipoprotein receptor, Science 271:518–20

Alberti S, Schuster G, Parini P, Feltkamp D, Diczfalusy U, Rudling M, Angelin B, Bjorkhem I, Pettersson S, Gustafsson JA (2001) Hepatic cholesterol metabolism and resistance to dietary cholesterol in LXRbeta-deficient mice, J Clin Invest 107:565–73

Anderson RA, Sando GN (1991) Cloning and expression of cDNA encoding human lysosomal acid lipase/cholesteryl ester hydrolase. Similarities to gastric and lingual lipases, J Biol Chem 266:22479–84

Arai T, Wang N, Bezouevski M, Welch C, Tall AR (1999) Decreased Atherosclerosis in Heterozygous Low Density Lipoprotein Receptor-deficient Mice Expressing the Scavenger Receptor BI Transgene, J Biol Chem 274:2366–2371

Aviram M, Bierman EL, Oram JF (1989) High density lipoprotein stimulates sterol translocation between intracellular and plasma membrane pools in human monocyte-derived macrophages, J Lipid Res 30:65–76

Babaev VR, Gleaves LA, Carter KJ, Suzuki H, Kodama T, Fazio S, Linton MF (2000) Reduced atherosclerotic lesions in mice deficient for total or macrophage-specific expression of scavenger receptor-A, Arterioscler Thromb Vasc Biol 20:2593–9.

Babiker A, Andersson O, Lund E, Xiu RJ, Deeb S, Reshef A, Leitersdorf E, Diczfalusy U, Bjorkhem I (1997) Elimination of cholesterol in macrophages and endothelial cells by the sterol 27-hydroxylase mechanism. Comparison with high density lipoprotein-mediated reverse cholesterol transport, J Biol Chem 272:26253–61

Ball RY, Bindman JP, Carpenter KL, Mitchinson MJ (1986) Oxidized low density lipoprotein induces ceroid accumulation by murine peritoneal macrophages in vitro, Atherosclerosis 60:173–81

Ball RY, Carpenter KL, Enright JH, Hartley SL, Mitchinson MJ (1987) Ceroid accumulation by murine peritoneal macrophages exposed to artificial lipoproteins, Br J Exp Pathol 68:427–38

Ball RY, Carpenter KL, Mitchinson MJ (1988) Ceroid accumulation by murine peritoneal macrophages exposed to artificial lipoproteins: ultrastructural observations, Br J Exp Pathol 69:43–56

Basu SK, Ho YK, Brown MS, Bilheimer DW, Anderson RGW, Goldstein JL (1982) Biochemical and genetic studies of the apoprotein E secreted by mouse macrophages and human monocytes., J Biol Chem 257:9799–9795

Bellosta S, Mahley RW, Sanan D, Murata J, Newland DL, Taylor JM, Pitas RE (1995) Macrophage-specific expression of human apolipoprotein E reduces atherosclerosis in hypercholesterolemic apolipoprotein E-null mice, J Clin Invest 96:2170–2179

Benoit P, Emmanuel F, Caillaud JM, Bassinet L, Castro G, Gallix P, Fruchart JC, Branellec D, Denefle P, Duverger N (1999) Somatic gene transfer of human ApoA-I inhibits atherosclerosis progression in mouse models, Circulation 99:105–10

Bickel PE, Freeman MW (1992) Rabbit aortic smooth muscle cells express inducible macrophage scavenger receptor messenger RNA that is absent from endothelial cells, J Clin Invest 90:1450–1457

Bierman EL, Oram J, Mendez A (1991) HDL receptor-mediated cholesterol efflux from cells and its regulation, Adv Exp Med Biol 285:81–3

Bocan TM, Krause BR, Rosebury WS, Lu X, Dagle C, Bak Mueller S, Auerbach B, Sliskovic DR (2001) The combined effect of inhibiting both ACAT and HMG-CoA reductase may directly induce atherosclerotic lesion regression, Atherosclerosis 157:97–105

Bodzioch M, Orso E, Klucken J, Langmann T, Bottcher A, Diederich W, Drobnik W, Barlage S, Buchler C, Porsch-Ozcurumez M, Kaminski WE, Hahmann HW, Oette K, Rothe G, Aslanidis C, Lackner KJ, Schmitz G (1999) The gene encoding ATP-binding cassette transporter 1 is mutated in Tangier disease, Nat Genet 22:347–51

Brewer HB, Jr (2000) The lipid-laden foam cell: an elusive target for therapeutic intervention, J Clin Invest 105:703–5

Brooks-Wilson A, Marcil M, Clee SM, Zhang LH, Roomp K, van Dam M, Yu L, Brewer C, Collins JA, Molhuizen HO, Loubser O, Ouelette BF, Fichter K, Ashbourne-Excoffon KJ, Sensen CW, Scherer S, Mott S, Denis M, Martindale D, Frohlich J, Morgan K, Koop B, Pimstone S, Kastelein JJ, Hayden MR, et al (1999) Mutations in ABC1 in Tangier disease and familial high-density lipoprotein deficiency, Nat Genet 22:336–45

Brown AJ, Watts GF, Burnett JR, Dean RT, Jessup W (2000) Sterol 27-hydroxylase acts on 7-ketocholesterol in human atherosclerotic lesions and macrophages in culture, J Biol Chem 275:27627–33

Brown MS, Ho YK, Goldstein JL (1980) The cholesteryl ester cycle in macrophage foam cells. Continual hydrolysis and re-esterification of cytoplasmic cholesteryl esters, J Biol Chem 255:9344–52

Brown MS, Goldstein JL (1983) Lipoprotein metabolism in the macrophage: implications for cholesterol deposition in atherosclerosis, Ann Rev Biochem 52:223–261

Brown MS, Goldstein JL (1986) A receptor-mediated pathway for cholesterol homeostasis, Science 232:34–47

Carstea ED, Morris JA, Coleman KG, Loftus SK, Zhang D, Cummings C, Gu J, Rosenfeld MA, Pavan WJ, Krizman DB, Nagle J, Polymeropoulos MH, Sturley SL, Ioannou YA, Higgins ME, Comly M, Cooney A, Brown A, Kaneski CR, Blanchette-Mackie EJ, Dwyer NK, Neufeld EB, Chang TY, Liscum L, Tagle DA, et al (1997) Niemann-Pick C1 disease gene: homology to mediators of cholesterol homeostasis, Science 277:228–31

Chawla A, Boisvert WA, Lee CH, Laffitte BA, Barak Y, Joseph SB, Liao D, Nagy L, Edwards PA, Curtiss LK, Evans RM, Tontonoz P (2001) A PPAR gamma-LXR-ABCA1 pathway in macrophages is involved in cholesterol efflux and atherogenesis, Mol Cell 7:161–71

Chen Z, Ishibashi S, Perrey S, Osuga J, Gotoda T, Kitamine T, Tamura Y, Okazaki H, Yahagi N, Iizuka Y, Shionoiri F, Ohashi K, Harada K, Shimano H, Nagai R, Yamada N (2001) Troglitazone inhibits atherosclerosis in apolipoprotein E-knockout mice: pleiotropic effects on CD36 expression and HDL, Arterioscler Thromb Vasc Biol 21:372–7

Chinetti G, Gbaguidi FG, Griglio S, Mallat Z, Antonucci M, Poulain P, Chapman J, Fruchart JC, Tedgui A, Najib-Fruchart J, Staels B (2000) CLA-1/SR-BI is expressed in atherosclerotic lesion macrophages and regulated by activators of peroxisome proliferator-activated receptors, Circulation 101:2411–7

Chinetti G, Lestavel S, Bocher V, Remaley AT, Neve B, Torra IP, Teissier E, Minnich A, Jaye M, Duverger N, Brewer HB, Fruchart JC, Clavey V, Staels B (2001) PPAR-alpha and PPAR-gamma activators induce cholesterol removal from human macrophage foam cells through stimulation of the ABCA1 pathway, Nat Med 7:53–8

Clare K, Hardwick SJ, Carpenter KL, Weeratunge N, Mitchinson MJ (1995) Toxicity of oxysterols to human monocyte-macrophages, Atherosclerosis 118:67–75

Collins AR, Meehan WP, Kintscher U, Jackson S, Wakino S, Noh G, Palinski W, Hsueh WA, Law RE (2001) Troglitazone inhibits formation of early atherosclerotic lesions in

diabetic and nondiabetic low density lipoprotein receptor-deficient mice, Arterioscler Thromb Vasc Biol 21:365–71
Curtiss LK, Black AS, Takagi Y, Plow EF (1987) New mechanism for foam cell generation in atherosclerotic lesions, J Clin Invest 80:367–73
Cyrus T, Witztum JL, Rader DJ, Tangirala R, Fazio S, Linton MF, Funk CD (1999) Disruption of the 12/15-lipoxygenase gene diminishes atherosclerosis in apo E-deficient mice, J Clin Invest 103:1597–604
de Villiers WJS, Smith JD, Miyata M, Dansky HM, Darley E, Gordon S (1998) Macrophage phenotype in mice deficient in both macrophage-colony- stimulating factor (op) and apolipoprotein E, Arterioscler Thromb Vasc Biol 18:631–40
de Villiers WJS, Cai L, Webb NR, de Beer MC, van der Westhuyzen DR, de Beer FC (2001) CD36 does not play a direct role in HDL or LDL metabolism, J Lipid Res 42:1231–8.
de Villiers WJS, Fraser IP, Hughes DA, Doyle AG, Gordon S (1994) Macrophage-colony-stimulating factor selectively enhances macrophage scavenger receptor expression and function., J Exp Med 180:705–709
de Winther MP, van Dijk KW, van Vlijmen BJ, Gijbels MJ, Heus JJ, Wijers ER, van den Bos AC, Breuer M, Frants RR, Havekes LM, Hofker MH (1999) Macrophage specific overexpression of the human macrophage scavenger receptor in transgenic mice, using a 180-kb yeast artificial chromosome, leads to enhanced foam cell formation of isolated peritoneal macrophages, Atherosclerosis 147:339–47
Endemann G, Stanton LW, Madden KS, Bryant CM, White RT, Protter AA (1993) CD36 is a receptor for oxidized low density lipoprotein, J Biol Chem 268:11811–11816
Febbraio M, Abumrad NA, Hajjar DP, Sharma K, Cheng W, Pearce SF, Silverstein RL (1999) A null mutation in murine CD36 reveals an important role in fatty acid and lipoprotein metabolism, J Biol Chem 274:19055–62
Febbraio M, Podrez EA, Smith JD, Hajjar DP, Hazen SL, Hoff HF, Sharma K, Silverstein RL (2000) Targeted disruption of the class B scavenger receptor CD36 protects against atherosclerotic lesion development in mice, J Clin Invest 105:1049–56
Fluiter K, van der Westhuijzen DR, van Berkel TJ (1998) In vivo regulation of scavenger receptor BI and the selective uptake of high density lipoprotein cholesteryl esters in rat liver parenchymal and Kupffer cells, J Biol Chem 273:8434–8
Freeman M, Ekkel Y, Rohrer L, Penman M, Freedman NJ, Chisolm GM, Krieger M (1991) Expression of type I and type II bovine scavenger receptors in Chinese hamster ovary cells: lipid droplet accumulation and nonreciprocal cross competition by acetylated and oxidized low density lipoprotein, Proc Natl Acad Sci USA 88:4931–4935
Garner B, Baoutina A, Dean RT, Jessup W (1997) Regulation of serum-induced lipid accumulation in human monocyte-derived macrophages by interferon-gamma. Correlations with apolipoprotein E production, lipoprotein lipase activity and LDL receptor-related protein expression, Atherosclerosis 128:47–58
Glass CK, Witztum JL (2001) Atherosclerosis. the road ahead, Cell 104:503–16
Goldstein JL, Brown MS (1977) The low-density lipoprotein pathway and its relation to atherosclerosis, Ann Rev Biochem 46:897–930
Goldstein JL, Ho YK, Basu SK, Brown MS (1979) Binding site on macrophages that mediates uptake and degradation of acetylated low density lipoprotein producing massive cholesterol deposition, Proc Natl Acad Sci USA 76:333–337
Goldstein JL, Ho YK, Brown MS, Innerarity TL, Mahley RW (1980) Cholesteryl ester accumulation in macrophages resulting from receptor-mediated uptake and degradation of hypercholesterolemic canine beta-very low density lipoproteins, J Biol Chem 255:1839–48
Goldstein JL, Brown MS (1990) Regulation of the mevalonate pathway, Nature 343:425–30
Greenspan P, Yu H, Mao F, Gutman RL (1997) Cholesterol deposition in macrophages: foam cell formation mediated by cholesterol-enriched oxidized low density lipoprotein, J Lipid Res 38:101–9

Guyton JR, Klemp KF (1996) Development of the lipid-rich core in human atherosclerosis, Arterioscler Thromb Vasc Biol 16:4–11

Havel RJ (1995) Chylomicron remnants: hepatic receptors and metabolism, Curr Opin Lipidol 6:312–6

Hirano K, Yamashita S, Nakagawa Y, Ohya T, Matsuura F, Tsukamoto K, Okamoto Y, Matsuyama A, Matsumoto K, Miyagawa J, Matsuzawa Y (1999) Expression of human scavenger receptor class B type I in cultured human monocyte-derived macrophages and atherosclerotic lesions, Circ Res 85:108–16

Hoff HF, Morton RE (1985) Lipoproteins containing apo B extracted from human aortas. Structure and function, Ann N Y Acad Sci 454:183–94

Hoff HF, O'Neil J, Chisolm GM, 3rd, Cole TB, Quehenberger O, Esterbauer H, Jurgens G (1989) Modification of low density lipoprotein with 4-hydroxynonenal induces uptake by macrophages, Arteriosclerosis 9:538–49

Hoff HF, O'Neil J, Pepin JM, Cole TB (1990) Macrophage uptake of cholesterol-containing particles derived from LDL and isolated from atherosclerotic lesions, Eur Heart J 11 Suppl E:105–15

Hoff HF, Cole TB (1991) Macrophage uptake of low-density lipoprotein modified by 4-hydroxynonenal. An ultrastructural study, Lab Invest 64:254–64

Hoppe G, O'Neil J, Hoff HF (1994) Inactivation of lysosomal proteases by oxidized low density lipoprotein is partially responsible for its poor degradation by mouse peritoneal macrophages, J Clin Invest 94:1506–12

Hughes DA, Fraser IP, Gordon S (1995) Murine macrophage scavenger receptor: in vivo expression and function as receptor for macrophage adhesion in lymphoid and non-lymphoid organs., Eur J Immunol 25:466–473

Huszar D, Varban ML, Rinninger F, Feeley R, Arai T, Fairchild-Huntress V, Donovan MJ, Tall AR (2000) Increased LDL cholesterol and atherosclerosis in LDL receptor-deficient mice with attenuated expression of scavenger receptor B1, Arterioscler Thromb Vasc Biol 20:1068–73

Insull W, Jr., Koren M, Davignon J, Sprecher D, Schrott H, Keilson LM, Brown AS, Dujovne CA, Davidson MH, McLain R, Heinonen T (2001) Efficacy and short-term safety of a new ACAT inhibitor, avasimibe, on lipids, lipoproteins, and apolipoproteins, in patients with combined hyperlipidemia, Atherosclerosis 157:137–44

Ji Y, Jian B, Wang N, Sun Y, de la Llera Moya M, Phillips MC, Rothblat GH, Swaney JB, Tall AR (1997) Scavenger receptor BI promotes high density lipoprotein-mediated cellular cholesterol efflux., J Biol Chem 272:20982–20985

Jian B, de la Llera-Moya M, Ji Y, Wang N, Phillips MC, Swaney JB, Tall AR, Rothblat GH (1998) Scavenger receptor class B type I as a mediator of cellular cholesterol efflux to lipoproteins and phospholipid acceptors, J Biol Chem 273:5599–606

Jones NL, Allen NS, Willingham MC, Lewis JC (1999) Modified LDLs induce and bind to membrane ruffles on macrophages, Anat Rec 255:44–56

Jones NL, Willingham MC (1999) Modified LDLs are internalized by macrophages in part via macropinocytosis, Anat Rec 255:57–68

Kashiwagi H, Tomiyama Y, Honda S, Kosugi S, Shiraga M, Nagao N, Sekiguchi S, Kanayama Y, Kurata Y, Matsuzawa Y (1995) Molecular basis of CD36 deficiency. Evidence that a 478C→T substitution (proline90→serine) in CD36 cDNA accounts for CD36 deficiency, J Clin Invest 95:1040–6

Kataoka H, Kume N, Minami M, Moriwaki H, Sawamura T, Masaki T, Kita T (2000) Expression of lectin-like oxidized LDL receptor-1 in human atherosclerotic lesions, Ann N Y Acad Sci 902:328–35

Kellner-Weibel G, Jerome WG, Small DM, Warner GJ, Stoltenborg JK, Kearney MA, Corjay MH, Phillips MC, Rothblat GH (1998) Effects of intracellular free cholesterol accumulation on macrophage viability. A model for foam cell death., Arterioscler Thromb Vasc Biol 18:423–431

Kersten S, Desvergne B, Wahli W (2000) Roles of PPARs in health and disease, Nature 405:421–4

Khoo JC, Miller E, McLoughlin P, Steinberg D (1988) Enhanced macrophage uptake of low density lipoprotein after self-aggregation, Arteriosclerosis 8:348–58

Khoo JC, Reue K, Steinberg D, Schotz MC (1993) Expression of hormone-sensitive lipase mRNA in macrophages, J Lipid Res 34:1969–74

Klinkner AM, Waites CR, Kerns WD, Bugelski PJ (1995) Evidence of foam cell and cholesterol crystal formation in macrophages incubated with oxidized LDL by fluorescence and electron microscopy, J Histochem Cytochem 43:1071–8

Kovanen PT (1991) Mast cell granule-mediated uptake of low density lipoproteins by macrophages: a novel carrier mechanism leading to the formation of foam cells, Ann Med 23:551–9

Kozarsky KF, Donahee MH, Rigotti A, Iqbal SN, Edelman ER, Krieger M (1997) Overexpression of the HDL receptor SR-BI alters plasma HDL and bile cholesterol levels., Nature 387:414–417

Kozarsky KF, Donahee MH, Glick JM, Krieger M, Rader DJ (2000) Gene Transfer and Hepatic Overexpression of the HDL Receptor SR-BI Reduces Atherosclerosis in the Cholesterol-Fed LDL Receptor-Deficient Mouse, Arterioscler Thromb Vasc Biol 20:721–727

Krieger M, Herz J (1994) Structures and functions of multiligand lipoprotein receptors: macrophage scavenger receptors and LDL receptor-related protein (LRP). Annu Rev Biochem 63:601–637

Krieger M (1999) Charting the fate of the "good cholesterol": Identification and characterization of the high-density lipoprotein receptor SR-BI., Annu Rev Biochem 68:523–558

Krieger M (2001) Scavenger receptor class B type I is a multiligand HDL receptor that influences diverse physiologic systems, J Clin Invest 108:793–7

Kruth HS, Chang J, Ifrim I, Zhang WY (1999a) Characterization of patocytosis: endocytosis into macrophage surface-connected compartments, Eur J Cell Biol 78:91–9

Kruth HS, Zhang WY, Skarlatos SI, Chao FF (1999b) Apolipoprotein B stimulates formation of monocyte-macrophage surface-connected compartments and mediates uptake of low density lipoprotein-derived liposomes into these compartments, J Biol Chem 274:7495–500

Kume N, Kita T (2001) Roles of lectin-like oxidized LDL receptor-1 and its soluble forms in atherogenesis, Curr Opin Lipidol 12:419–23

Kurushima H, Ramprasad M, Kondratenko N, Foster DM, Quehenberger O, Steinberg D (2000) Surface expression and rapid internalization of macrosialin (mouse CD68) on elicited mouse peritoneal macrophages, J Leukoc Biol 67:104–8.

Laffitte BA, Repa JJ, Joseph SB, Wilpitz DC, Kast HR, Mangelsdorf DJ, Tontonoz P (2001) LXRs control lipid-inducible expression of the apolipoprotein E gene in macrophages and adipocytes, Proc Natl Acad Sci U S A 98:507–12

Lawn RM, Wade DP, Garvin MR, Wang X, Schwartz K, Porter JG, Seilhamer JJ, Vaughan AM, Oram JF (1999) The Tangier disease gene product ABC1 controls the cellular apolipoprotein-mediated lipid removal pathway, J Clin Invest 104:R25-R31

Layne MD, Patel A, Chen YH, Rebel VI, Carvajal IM, Pellacani A, Ith B, Zhao D, Schreiber BM, Yet SF, Lee ME, Storch J, Perrella MA (2001) Role of macrophage-expressed adipocyte fatty acid binding protein in the development of accelerated atherosclerosis in hypercholesterolemic mice, FASEB J 15:2733–5

Leventhal AR, Chen W, Tall AR, Tabas I (2001) Acid sphingomyelinase-deficient macrophages have defective cholesterol trafficking and efflux, J Biol Chem 276:44976–83

Li AC, Brown KK, Silvestre MJ, Willson TM, Palinski W, Glass CK (2000) Peroxisome proliferator-activated receptor gamma ligands inhibit development of atherosclerosis in LDL receptor-deficient mice, J Clin Invest 106:523–31

Loftus SK, Morris JA, Carstea ED, Gu JZ, Cummings C, Brown A, Ellison J, Ohno K, Rosenfeld MA, Tagle DA, Pentchev PG, Pavan WJ (1997) Murine model of Niemann-Pick C disease: mutation in a cholesterol homeostasis gene, Science 277:232–5

Lougheed M, Zhang HF, Steinbrecher UP (1991) Oxidized low density lipoprotein is resistant to cathepsins and accumulates within macrophages, J Biol Chem 266:14519–25

Lougheed M, Lum CM, Ling W, Suzuki H, Kodama T, Steinbrecher U (1997) High affinity saturable uptake of oxidized low density lipoprotein by macrophages from mice lacking the scavenger receptor class A type I/II, J Biol Chem 272:12938–44

Mahley RW, Innerarity TL, Brown MS, Ho YK, Goldstein JL (1980) Cholesteryl ester synthesis in macrophages: stimulation by beta-very low density lipoproteins from cholesterol-fed animals of several species, J Lipid Res 21:970–80

Makowski L, Boord JB, Maeda K, Babaev VR, Uysal KT, Morgan MA, Parker RA, Suttles J, Fazio S, Hotamisligil GS, Linton MF (2001) Lack of macrophage fatty-acid-binding protein aP2 protects mice deficient in apolipoprotein E against atherosclerosis, Nat Med 7:699–705

Maor I, Mandel H, Aviram M (1995) Macrophage uptake of oxidized LDL inhibits lysosomal sphingomyelinase, thus causing the accumulation of unesterified cholesterol-sphingomyelin-rich particles in the lysosomes. A possible role for 7-Ketocholesterol, Arterioscler Thromb Vasc Biol 15:1378–87

Matloubian M, David A, Engel S, Ryan JE, Cyster JG (2000) A transmembrane CXC chemokine is a ligand for HIV-coreceptor Bonzo, Nat Immunol 1:298–304

Matveev S, van der Westhuyzen DR, Smart EJ (1999) Co-expression of scavenger receptor-BI and caveolin-1 is associated with enhanced selective cholesteryl ester uptake in THP-1 macrophages, J Lipid Res 40:1647–54

Matveev S, Uittenbogaard A, van Der Westhuyzen D, Smart EJ (2001) Caveolin-1 negatively regulates SR-BI mediated selective uptake of high-density lipoprotein-derived cholesteryl ester, Eur J Biochem 268:5609–16

Minami M, Kume N, Shimaoka T, Kataoka H, Hayashida K, Akiyama Y, Nagata I, Ando K, Nobuyoshi M, Hanyuu M, Komeda M, Yonehara S, Kita T (2001) Expression of SR-PSOX, a novel cell-surface scavenger receptor for phosphatidylserine and oxidized LDL in human atherosclerotic lesions, Arterioscler Thromb Vasc Biol 21:1796–800

Mommaas-Kienhuis AM, van der Schroeff JG, Wijsman MC, Daems WT, Vermeer BJ (1985) Conjugates of colloidal gold with native and acetylated low density lipoproteins for ultrastructural investigations on receptor-mediated endocytosis by cultured human monocyte-derived macrophages, Histochemistry 83:29–35

Moore KJ, Rosen ED, Fitzgerald ML, Randow F, Andersson LP, Altshuler D, Milstone DS, Mortensen RM, Spiegelman BM, Freeman MW (2001) The role of PPAR-gamma in macrophage differentiation and cholesterol uptake, Nat Med 7:41–7

Moriwaki H, Kume N, Sawamura T, Aoyama T, Hoshikawa H, Ochi H, Nishi E, Masaki T, Kita T (1998) Ligand specificity of LOX-1, a novel endothelial receptor for oxidized low density lipoprotein, Arterioscler Thromb Vasc Biol 18:1541–7

Murase T, Kume N, Kataoka H, Minami M, Sawamura T, Masaki T, Kita T (2000) Identification of soluble forms of lectin-like oxidized LDL receptor-1, Arterioscler Thromb Vasc Biol 20:715–20

Musanti R, Ghiselli G (1993) Interaction of oxidized HDLs with J774-A1 macrophages causes intracellular accumulation of unesterified cholesterol, Arterioscler Thromb 13:1334–1345

Myers JN, Tabas I, Jones NL, Maxfield FR (1993) Beta-very low density lipoprotein is sequestered in surface-connected tubules in mouse peritoneal macrophages, J Cell Biol 123:1389–402

Nagy L, Tontonoz P, Alvarez JGA, Chen H, Evans RM (1998) Oxidized LDL regulates macrophage gene expression through ligand activation of PPARγ, Cell 93:229–240

Nievelstein PF, Fogelman AM, Mottino G, Frank JS (1991) Lipid accumulation in rabbit aortic intima 2 hours after bolus infusion of low density lipoprotein. A deep-etch and immunolocalization study of ultrarapidly frozen tissue, Arterioscler Thromb 11:1795–805

Nozaki S, Kashiwagi H, Yamashita S, Nakagawa T, Kostner B, Tomiyama Y, Nakata A, Ishigami M, Miyagawa J, Kameda-Takemura K, Kurata Y, Matsuzawa Y (1995) Reduced uptake of oxidized low density lipoproteins in monocyte-derived macrophages from CD36-deficient subjects, J Clin Invest 96:1859–1865

Oram JF, Lawn RM, Garvin MR, Wade DP (2000) ABCA1 is the cAMP-inducible apolipoprotein receptor that mediates cholesterol secretion from macrophages, J Biol Chem 275:34508–11

Oram JF, Lawn RM (2001) ABCA1. The gatekeeper for eliminating excess tissue cholesterol, J Lipid Res 42:1173–9

Panzenboeck U, Wintersberger A, Levak-Frank S, Zimmermann R, Zechner R, Kostner GM, Malle E, Sattler W (1997) Implications of endogenous and exogenous lipoprotein lipase for the selective uptake of HDL3-associated cholesteryl esters by mouse peritoneal macrophages, J Lipid Res 38:239–53

Peet DJ, Turley SD, Ma W, Janowski BA, Lobaccaro JM, Hammer RE, Mangelsdorf DJ (1998) Cholesterol and bile acid metabolism are impaired in mice lacking the nuclear oxysterol receptor LXR alpha, Cell 93:693–704

Pittman RC, Knecht TP, Rosenbaum MS, Taylor CA (1987) A non-endocytotic mechanism for the selective uptake of high density lipoprotein-associated cholesterol esters, J Biol Chem 262:2443–2450

Platt N, Gordon S (2001) Is the class A macrophage scavenger receptor (SR-A) multifunctional? - The mouse's tale, J Clin Invest 108:649–54

Rabinowitz S, Gordon S (1991) Macrosialin, a macrophage-restricted membrane sialoprotein differentially glycosylated in response to inflammatory stimuli, J Exp Med 174:827–836

Ramprasad MP, Fischer W, Witztum JL, Sambrano GR, Quehenberger O, Steinberg D (1995) The 94- to 97-kDa mouse macrophage membrane protein that recognizes oxidized low density lipoprotein and phosphatidylserine-rich liposomes is identical to macrosialin, the mouse homologue of human CD68, Proc Natl Acad Sci USA 92:9580–9584

Ramprasad MP, Terpstra V, Kondratenko N, Quehenberger O, Steinberg D (1996) Cell-surface expression of mouse macrosialin and human CD68 and their role as macrophage receptors for oxidized low density lipoprotein, Proc Natl Acad Sci USA 93:14833–14838

Rhainds D, Brissette L (1999) Low density lipoprotein uptake: holoparticle and cholesteryl ester selective uptake, Int J Biochem Cell Biol 31:915–31

Ricote M, Huang J, Fajas L, Li A, Welch J, Najib J, Witztum JL, Auwerx J, Palinski W, Glass CK (1998) Expression of the peroxisome proliferator-activated receptor γ (PPARγ) in human atherosclerosis and regulation in macrophages by colony stimulating factors and oxidized low density lipoprotein, Proc Natl Acad Sci USA 95:7614–7619

Rinninger F, Brundert M, Jackle S, Kaiser T, Greten H (1995) Selective uptake of low density associated cholesteryl esters by human fibroblasts, human HepG2 hepatoma cells and J774 macrophages in culture, Biochim Biophys Acta 1255:141–153

Rodriguez A, Usher DC (2002) Anti-atherogenic effects of the acyl-CoA:cholesterol acyltransferase inhibitor, avasimibe (CI-1011), in cultured primary human macrophages, Atherosclerosis 161:45–54

Roma P, Catapano AL, Bertulli SM, Varesi L, Fumagalli R, Bernini F (1990) Oxidized LDL increase free cholesterol and fail to stimulate cholesterol esterification in murine macrophages, Biochem Biophys Res Commun 171:123–31

Ross R (1999) Atherosclerosis—an inflammatory disease, N Engl J Med 340:115–126

Rothblat GH, de la Llera-Moya M, Atger V, Kellner-Weibel G, Williams DL, Phillips MC (1999) Cell cholesterol efflux: integration of old and new observations provides new insights, J Lipid Res 40:781–96

Rust S, Rosier M, Funke H, Real J, Amoura Z, Piette JC, Deleuze JF, Brewer HB, Duverger N, Denefle P, Assmann G (1999) Tangier disease is caused by mutations in the gene encoding ATP-binding cassette transporter 1, Nat Genet 22:352–5

Ryu BH, Mao FW, Lou P, Gutman RL, Greenspan P (1995) Cholesteryl ester accumulation in macrophages treated with oxidized low density lipoprotein, Biosci Biotechnol Biochem 59:1619–22

Sakaguchi H, Takeya M, Suzuki H, Hakamata H, Kodama T, Horiuchi S, Gordon S, van der Laan LJ, Kraal G, Ishibashi S, Kitamura N, Takahashi K (1998) Role of macrophage scavenger receptors in diet-induced atherosclerosis in mice, Lab Invest 78:423–34

Salisbury BG, Falcone DJ, Minick CR (1985) Insoluble low-density lipoprotein-proteoglycan complexes enhance cholesteryl ester accumulation in macrophages, Am J Pathol 120:6–11

Sawamura T, Kume N, Aoyama T, Moriwaki H, Hoshikawa H, Aiba Y, Tanaka T, Miwa S, Katsura Y, Kita T, Masaki T (1997) An endothelial receptor for oxidized low-density lipoprotein, Nature 386:73–77

Schissel SL, Jiang X, Tweedie-Hardman J, Jeong T, Camejo EH, Najib J, Rapp JH, Williams KJ, Tabas I (1998) Secretory sphingomyelinase, a product of the acid sphingomyelinase gene, can hydrolyze atherogenic lipoproteins at neutral pH. Implications for atherosclerotic lesion development, J Biol Chem 273:2738–46

Schmitz G, Robenek H, Lohmann U, Assmann G (1985) Interaction of high density lipoproteins with cholesteryl ester-laden macrophages: biochemical and morphological characterization of cell-surface receptor binding, endocytosis and resecretion of high density lipoproteins by macrophages, EMBO J 4:613–22

Shimaoka T, Kume N, Minami M, Hayashida K, Kataoka H, Kita T, Yonehara S (2000) Molecular cloning of a novel scavenger receptor for oxidized low density lipoprotein, SR-PSOX, on macrophages, J Biol Chem 275:40663–6

Silver DL, Wang N, Tall AR (2000) Defective HDL particle uptake in ob/ob hepatocytes causes decreased recycling, degradation, and selective lipid uptake, J Clin Invest 105:151–9

Silver DL, Wang N, Xiao X, Tall AR (2001) High density lipoprotein (HDL) particle uptake mediated by scavenger receptor class B type 1 results in selective sorting of HDL cholesterol from protein and polarized cholesterol secretion, J Biol Chem 276:25287–93

Smart EJ, Ying Y, Donzell WC, Anderson RG (1996) A role for caveolin in transport of cholesterol from endoplasmic reticulum to plasma membrane, J Biol Chem 271:29427–35

Smith JD, Miyata M, Ginsberg M, Grigaux C, Shmookler E, Plump AS (1996) Cyclic AMP induces apolipoprotein E binding activity and promotes cholesterol efflux from a macrophage cell line to apolipoprotein acceptors, J Biol Chem 271:30647–55

Spady DK, Kearney DM, Hobbs HH (1999) Polyunsaturated fatty acids up-regulate hepatic scavenger receptor B1 (SR-BI) expression and HDL cholesteryl ester uptake in the hamster, J Lipid Res 40:1384–1394

Spiegelman BM (1998) PPARgamma in monocytes: less pain, any gain?, Cell 93:153–5

Stein O, Israeli A, Leitersdorf E, Halperin G, Stein Y (1987) Preferential uptake of cholesteryl ester-HDL by cultured macrophages, Atherosclerosis 65:151–8

Steinberg D, Parthasarathy S, Carew TE, Khoo JC, Witztum JL (1989) Beyond cholesterol—modifications of low-density lipoprotein that increase its atherogenicity, New Engl J Med 320:915–924

Steinberg D (1997) Low density lipoprotein oxidation and its pathobiological significance, J Biol Chem 272:20963–20966

Suits AG, Chait A, Aviram M, Heinecke JW (1989) Phagocytosis of aggregated lipoprotein by macrophages: low density lipoprotein receptor-dependent foam-cell formation, Proc Natl Acad Sci U S A 86:2713–7

Suzuki H, Kurihara Y, Takeya M, Kamada N, Kataoka M, Jishage K, Ueda O, Sakaguchi H, Higashi T, Suzuki T, Takashima Y, Kawabe Y, Cynshi O, Wada Y, Honda M, Kurihara H, Aburatani H, Doi T, Matsumoto A, Azuma S, Noda T, Toyoda Y, Itakura H, Yazaki Y, Horiuchi S, Takahashi K, Kruijt JK, van Berkel TJC, Steinbrecher UP, Ishibashi S, Maeda N, Gordon S, Kodama T (1997) A role for macrophage scavenger receptors in atherosclerosis and susceptibility to infection, Nature 386:292–6

Tabas I, Lim S, Xu XX, Maxfield FR (1990) Endocytosed beta-VLDL and LDL are delivered to different intracellular vesicles in mouse peritoneal macrophages, J Cell Biol 111:929–40

Tabas I, Myers JN, Innerarity TL, Xu XX, Arnold K, Boyles J, Maxfield FR (1991) The influence of particle size and multiple apoprotein E-receptor interactions on the endocytic targeting of beta-VLDL in mouse peritoneal macrophages, J Cell Biol 115:1547–60

Tabas I (1997) Phospholipid metabolism in cholesterol-loaded macrophages, Curr Opin Lipidol 8:263–7

Tabas I (1999) Nonoxidative modifications of lipoproteins in atherogenesis, Annu Rev Nutr 19:123–39

Takahashi K, Fukuda S, Naito M, Horiuchi S, Takata K, Morino Y (1989) Endocytic pathway of high density lipoprotein via trans-Golgi system in rat resident peritoneal macrophages, Lab Invest 61:270–7

Takahashi Y, Smith JD (1999) Cholesterol efflux to apolipoprotein AI involves endocytosis and resecretion in a calcium-dependent pathway, Proc Natl Acad Sci USA 96:11358–63

Tall AR, Jiang X, Luo Y, Silver D (2000) 1999 George Lyman Duff memorial lecture: lipid transfer proteins, HDL metabolism, and atherogenesis, Arterioscler Thromb Vasc Biol 20:1185–8

Tangirala RK, Tsukamoto K, Chun SH, Usher D, Pure E, Rader DJ (1999) Regression of atherosclerosis induced by liver-directed gene transfer of apolipoprotein A-I in mice, Circulation 100:1816–22

Terpstra V, Kondratenko N, Steinberg D (1997) Macrophages lacking scavenger receptor A show a decrease in binding and uptake of acetylated low-density lipoprotein and of apoptotic thymocytes, but not of oxidatively damaged red blood cells, Proc Natl Acad Sci U S A 94:8127–31

Tontonoz P, Nagy L, Alvarez JGA, Thomazy VA, Evans RM (1998) PPARγ promotes monocyte/macrophage differentiation and uptake of oxidized LDL., Cell 93:241–252

Traber MG, Kallman B, Kayden HJ (1983) Localization of the binding sites of native and acetylated low-density lipoprotein (LDL) in human monocyte-derived macrophages, Exp Cell Res 148:281–92

Trigatti B, Rayburn H, Vinals M, Braun A, Miettinen H, Penman M, Hertz M, Schrenzel M, Amigo L, Rigotti A, Krieger M (1999) Influence of the high density lipoprotein receptor SR-BI on reproductive and cardiovascular pathophysiology, Proc Natl Acad Sci USA 96:9322–9327

Trigatti B, Rigotti A, Krieger M (2000) The role of the high-density lipoprotein receptor SR-BI in cholesterol metabolism, Curr Opin Lipidol 11:123–31

Ueda Y, Gong E, Royer L, Cooper PN, Francone OL, Rubin EM (2000) Relationship between expression levels and atherogenesis in scavenger receptor class B, type I transgenics, J Biol Chem 275:20368–73

Uittenbogaard A, Everson WV, Matveev SV, Smart EJ (2002) Cholesteryl ester is transported from caveolae to internal membranes as part of a caveolin-annexin II lipid-protein complex, J Biol Chem 277:4925–31

Van der Schroeff JG, Havekes L, Emeis JJ, Wijsman M, van Der Meer H, Vermeer BJ (1983) Morphological studies on the binding of low-density lipoproteins and acetylated low-density lipoproteins to the plasma membrane of cultured monocytes, Exp Cell Res 145:95–103

Van Velzen AG, Da Silva RP, Gordon S, Van Berkel TJ (1997) Characterization of a receptor for oxidized low-density lipoproteins on rat Kupffer cells: similarity to macrosialin, Biochem J 322:411–5.

Venkateswaran A, Laffitte BA, Joseph SB, Mak PA, Wilpitz DC, Edwards PA, Tontonoz P (2000) Control of cellular cholesterol efflux by the nuclear oxysterol receptor LXR alpha, Proc Natl Acad Sci U S A 97:12097–102

Vosper H, Patel L, Graham TL, Khoudoli GA, Hill A, Macphee CH, Pinto I, Smith SA, Suckling KE, Wolf CR, Palmer CN (2001) The peroxisome proliferator-activated receptor delta promotes lipid accumulation in human macrophages, J Biol Chem 276:44258–65

Voyiaziakis E, Goldberg IJ, Plump AS, Rubin EM, Breslow JL, Huang LS (1998) ApoA-I deficiency causes both hypertriglyceridemia and increased atherosclerosis in human apoB transgenic mice, J Lipid Res 39:313–21

Warner GJ, Stoudt G, Bamberger M, Johnson WJ, Rothblat GH (1995) Cell toxicity induced by inhibition of acyl coenzyme A:cholesterol acyltransferase and accumulation of unesterified cholesterol, J Biol Chem 270:5772–8

Webb NR, Connell PM, Graf GA, Smart EJ, de Villiers WJS, de Beer FC, van der Westhuyzen DR (1998) SR-BII, an isoform of the scavenger receptor BI containing an alternate cytoplasmic tail, mediates lipid transfer between high density lipoprotein and cells, J Biol Chem 273:15241–8

Westman J, Kallin B, Bjorkhem I, Nilsson J, Diczfalusy U (1998) Sterol 27-hydroxylase- and apoAI/phospholipid-mediated efflux of cholesterol from cholesterol-laden macrophages: evidence for an inverse relation between the two mechanisms, Arterioscler Thromb Vasc Biol 18:554–61

Witt W, Kolleck I, Fechner H, Sinha P, Rustow B (2000) Regulation by vitamin E of the scavenger receptor BI in rat liver and HepG2 cells, J Lipid Res 41:2009–16

Xu XX, Tabas I (1991) Sphingomyelinase enhances low density lipoprotein uptake and ability to induce cholesteryl ester accumulation in macrophages, J Biol Chem 266:24849–58

Yoshida H, Kondratenko N, Green S, Steinberg D, Quehenberger O (1998) Identification of the lectin-like receptor for oxidized low-density lipoprotein in human macrophages and its potential role as a scavenger receptor, Biochem J 334:9–13

Zha X, Tabas I, Leopold PL, Jones NL, Maxfield FR (1997) Evidence for prolonged cell-surface contact of acetyl-LDL before entry into macrophages, Arterioscler Thromb Vasc Biol 17:1421–31

Zhang WY, Gaynor PM, Kruth HS (1996) Apolipoprotein E produced by human monocyte-derived macrophages mediates cholesterol efflux that occurs in the absence of added cholesterol acceptors, J Biol Chem 271:28641–6

Zhang WY, Ishii I, Kruth HS (2000) Plasmin-mediated macrophage reversal of low density lipoprotein aggregation, J Biol Chem 275:33176–83

Dietary Fatty Acids and Macrophages

P. C. Calder[1] · P. Yaqoob[2]

[1] Institute of Human Nutrition, School of Medicine, University of Southampton, Bassett Crescent East, Southampton, SO16 7PX, UK
e-mail: pcc@soton.ac.uk

[2] Hugh Sinclair Unit of Human Nutrition, School of Food Biosciences, University of Reading, Reading, RG6 6AP, UK

Abstract Fatty acids constitute key components of cell membranes. The fatty acid composition of cell membranes influences membrane fluidity. Membrane phospholipids are substrates for the generation of intracellular and extracellular signalling molecules. The supply of fatty acids to monocytes and macrophages influences their fatty acid composition. Thus, dietary fatty acid composition influences that of monocyte and macrophage membranes. This can have functional consequences. The *n*-6 polyunsaturated fatty acid (PUFA) arachidonic acid is the principal substrate for generation of eicosanoids via cyclooxygenase and lipoxygenase enzymes. Increased availability of *n*-3 PUFAs (found in oily fish and fish oil) can affect chemotaxis, phagocytosis, respiratory burst, eicosanoid production, cytokine production and other monocyte/macrophage functions. Although some of the effects of *n*-3 PUFAs may be brought about by modulation of the amount and types of eicosanoids made, it appears that these fatty acids might elicit some of their effects by eicosanoid-independent mechanisms, including actions upon intracellular signalling pathways and transcription factor activity. The functional effects of *n*-3 PUFAs are generally termed as "anti-inflammatory" and are considered beneficial to health. The effects of dietary fatty acids on monocyte/macrophage function may also be relevant to atherosclerosis, which is now recognised to include an inflammatory component. Fatty acids could potentially affect the degree of oxidation of low-density lipoprotein (LDL), its uptake by vascular cells, aspects of foam cell formation and inflammatory activity within atherosclerotic lesions. These effects might account for the reported protective effects of *n*-3 PUFAs towards cardiovascular mortality.

Keywords Cardiovascular disease, Cholesterol, Cytokine, Eicosanoid, Fish oil, Foam cell, LDL oxidation, Phagocytosis, Polyunsaturated fatty acid, Respiratory burst

Abbreviations

AA	Arachidonic acid
COX	Cyclooxygenase
DGLA	Dihomo-γ-linolenic acid
DHA	Docosahexaenoic acid
EPA	Eicosapentaenoic acid
GLA	γ-Linolenic acid
IFN	Interferon
IL	Interleukin
LOX	Lipoxygenase
LPS	Lipopolysaccharide
LT	Leukotriene
MUFA	Monounsaturated fatty acid
PG	Prostaglandin
PUFA	Polyunsaturated fatty acid
TNF	Tumour necrosis factor

1 Dietary Fatty Acids

1.1 Types and Sources of Fatty Acids in the Human Diet

In Western countries an adult eats on average 75–150 g of fat each day and fat contributes 30%–45% of dietary energy. By far the most important component of dietary fat in quantitative terms is triacylglycerol, which in most diets constitutes more than 95% of dietary fat. Each triacylglycerol molecule is composed of three fatty acids esterified to a glycerol backbone. Thus, fatty acids are major constituents of dietary fat. Because of the wide range of foods consumed, the human diet contains a great variety of fatty acids. It is the nature of the constituent fatty acids (their chain length and degree of unsaturation) that gives a fat its physical properties.

Fatty acids have systematic names, but most also have common names and are described by a shorthand nomenclature, e.g. 18:2*n*-6. This nomenclature indicates the number of carbon atoms in the hydrocarbon chain, the number of double bonds in the hydrocarbon chain, and the position of the first double bond from the methyl terminus of the chain (see Fig. 1). It is the *n*-7, *n*-9, *n*-6 or *n*-3 notation that indicates the position of the first double bond in the hydrocarbon chain for a fatty acid. Thus, an *n*-6 fatty acid has the first double bond on carbon number 6 from the methyl terminus and an *n*-3 fatty acid has the first double bond on carbon number 3 from the methyl terminus. The *n*- notation is sometimes referred to as ω or omega.

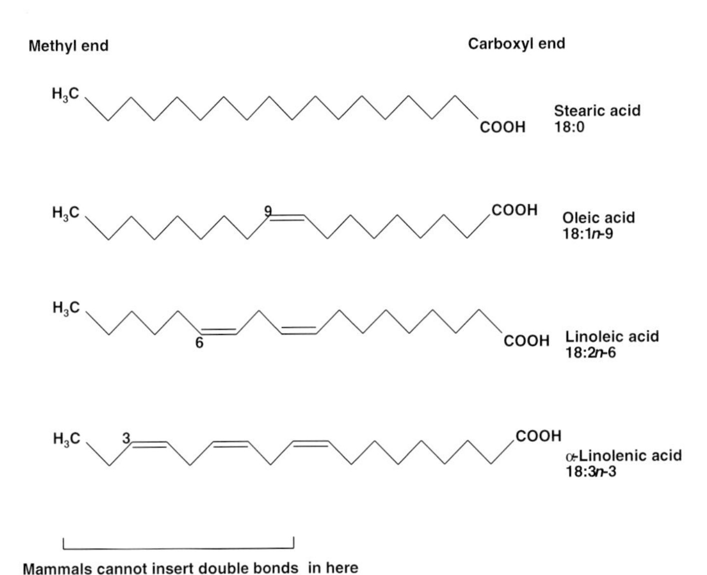

Fig. 1 The structure of fatty acids

Saturated fatty acids and most monounsaturated fatty acids (MUFAs) can be made in mammalian tissues from non-fat precursors, but this does not usually occur in humans eating a Western diet, since the consumption of fat in general, and of saturated and monounsaturated fatty acids in particular, is high. However, mammals cannot insert double bonds between the methyl terminus and carbon number 9 in oleic acid (18:1*n*-9) (Fig. 1). Thus, mammals cannot convert oleic acid into linoleic acid (18:2*n*-6). The Δ12-desaturase enzyme which does this is found only in plants. Likewise, mammals cannot convert linoleic acid into *α*-linolenic acid (18:3*n*-3). The Δ15-desaturase enzyme which does this is again found only in plants. Since these two fatty acids cannot be made by mammals they are termed essential fatty acids. Also, because mammalian tissues do not contain the Δ15-desaturase, they cannot interconvert *n*-6 and *n*-3 fatty acids. Plant tissues and plant oils tend to be rich sources of linoleic and *α*-linolenic acids. These fatty acids are the main polyunsaturated fatty acids (PUFAs) in most human diets: the average intake of linoleic acid among adult males in the United Kingdom is approximately 14 g/day, while that of *α*-linolenic acid is approximately 2 g/day. These intakes are lower than those of saturated and monounsaturated fatty acids (approximately 42 and 32 g/day, respectively).

1.2 Synthesis of Longer Chain Polyunsaturated Fatty Acids

Once consumed in the diet, linoleic acid can be converted via *γ*-linolenic acid (GLA; 18:3*n*-6) and dihomo-*γ*-linolenic acid (DGLA; 20:3*n*-6) to arachidonic acid (AA; 20:4*n*-6) by the pathway outlined in Fig. 2. Using the same pathway (Fig. 2), dietary *α*-linolenic acid can be converted into eicosapentaenoic acid (EPA; 20:5*n*-3), docosapentaenoic acid (22:5*n*-3) and docosahexaenoic acid (DHA; 22:6*n*-3). Thus, there is competition between the *n*-6 and *n*-3 fatty acids for the enzymes which metabolise them. The long chain n-3 PUFAs, EPA and DHA can be obtained directly from the diet since they are found in relatively high proportions in the tissues of so-called "oily fish" (e.g. herring, mackerel, tuna, sardines) and in the commercial products called "fish oils" which are a preparation of the body oils of oily fish; EPA and DHA are also found in high proportions in the oils extracted from the livers of some other species of fish (e.g. cod). EPA and DHA comprise 20%–30% of the fatty acids in a typical preparation of fish oil. The intake of longer chain *n*-3 PUFAs is not clearly known, but it appears that the average adult in the United Kingdom consumes about 250 mg EPA plus DHA per day. In the absence of significant consumption of oily fish, *α*-linolenic acid is the major dietary *n*-3 fatty acid.

1.3 Polyunsaturated Fatty Acid Synthesis by Macrophages

Experiments with murine peritoneal macrophages in culture demonstrated that these cells have limited capacity to carry out the key metabolic transformations

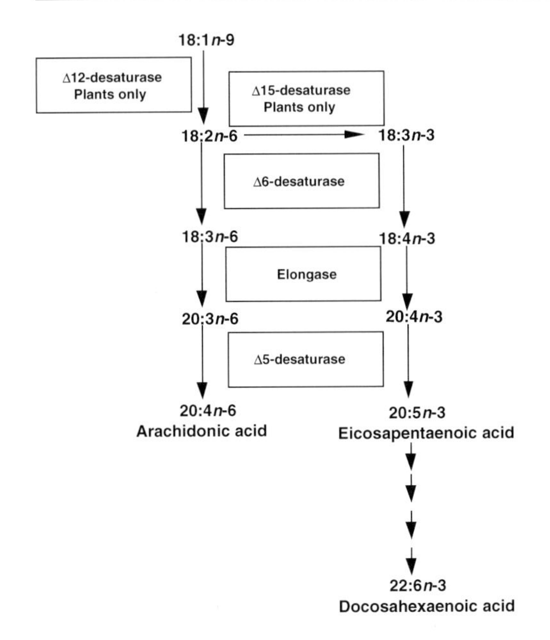

Fig. 2 Outline of the pathway of biosynthesis and metabolism of polyunsaturated fatty acids

shown in Fig. 2. These studies indicate that (murine) macrophages have an efficient fatty acid elongase (capable of converting linoleic acid to 20:2n-6, GLA to DGLA and EPA to 22:5*n*-3), lack Δ6-desaturase activity and have a low activity of Δ5-desaturase (Chapkin et al. 1988a,b; Chapkin and Coble 1991). Thus, the bulk, if not all, of the AA, EPA and DHA in macrophage phospholipids is likely to originate from exogenous sources. PUFAs circulate in the bloodstream as the fatty acid components of phospholipids and triacylglycerols in lipoproteins and as non-esterified fatty acids. These circulating lipid pools provide PUFAs to immune cells such as monocytes and macrophages.

2
Roles of Fatty Acids in Monocytes and Macrophages

The majority of the fatty acids taken up by cultured macrophages are incorporated into phospholipids of the plasma and organelle membranes. There is selective enrichment of particular phospholipid classes with particular fatty acids (e.g. AA is particularly associated with phosphatidylcholine). The fatty acid components of membrane phospholipids are partly responsible for regulating the fluidity of the membrane (Stubbs and Smith 1984). Fluidity ensures the appropriate environment for the function and movement of membrane proteins, and changing fluidity can affect the activities of such proteins (Stubbs and Smith 1984). Membrane phospholipids are also the source of a range of signalling molecules including inositol-1,3,5-trisphosphate, diacylglycerol, phosphatidic acid, lysophosphatidylcholine, choline, ceramide, platelet activating factor and AA. It is now recognised that there are particular regions of the cell

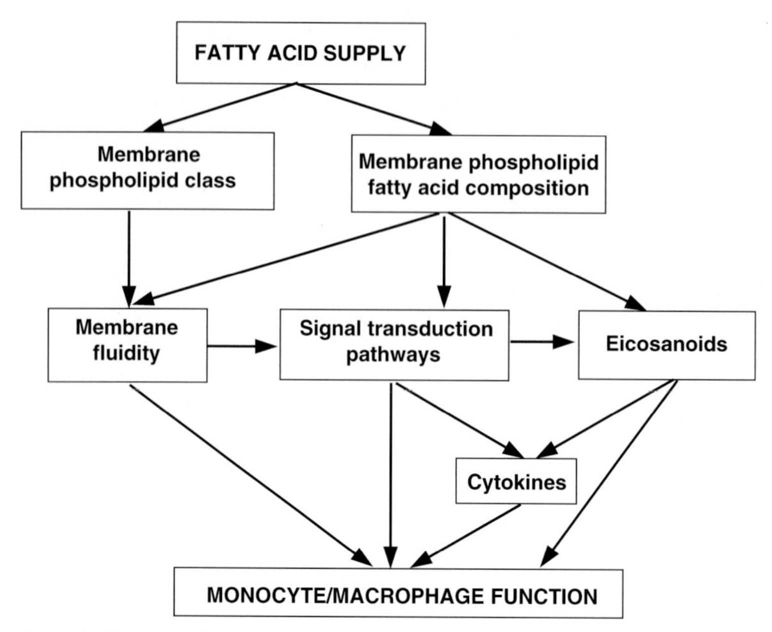

Fig. 3 Mechanisms whereby polyunsaturated fatty acids might exert effects on monocyte/macrophage function

membrane specialised for these roles in cell signalling; these regions are termed rafts, and they are sites where certain receptors and signalling enzymes are clustered (Simons and Toomre 2000). Rafts are also characterised by the presence of a different range of phospholipids from the bulk of the plasma membrane, and this most likely relates to providing the correct environment for the receptors and signalling enzymes and providing substrates for the generation of signalling molecules. The arrangement of the phospholipids in membranes and their fatty acid compositions might have important functional effects for macrophages (Fig. 3).

3
Influence of Altered Supply of Fatty Acids on Monocyte and Macrophage Fatty Acid Composition

Monocyte and macrophage membrane phospholipids are rich in AA, which typically comprises 15%–25% of the fatty acids. Since it appears that these cells have a limited capacity to synthesise AA from linoleic acid (see "Polyunsaturated Fatty Acid Synthesis by Macrophages" above), the AA most likely originates from metabolism of linoleic acid in the liver. The proportions of EPA and DHA in macrophages and monocytes are low, since most diets contain little α-linolenic acid and very little EPA and DHA.

Culture of macrophages with a given fatty acid results in enrichment of that fatty acid in the cells (Calder et al. 1990). The incorporation of exogenously supplied *n*-3 PUFAs (EPA and DHA) is frequently at the partial expense of AA, i.e. increased proportions of *n*-3 PUFAs result in a decreased proportion of AA. This is believed to be of functional significance because of the role of AA as an eicosanoid precursor (see "Arachidonic Acid as an Eicosanoid Precursor" below). Furthermore, changes in macrophage membrane fluidity occur as a result of selective enrichment with certain fatty acids (Mahoney et al. 1980), and this might affect functions such as phagocytosis (see "Phagocytosis" below).

Changing the fatty acid composition of the diet has also been demonstrated to result in a change in the fatty acid composition of macrophages. Feeding mice a diet in which borage oil (contains GLA) was the fat source resulted in a marked increase in the proportion of DGLA in peritoneal macrophage phospholipids (Chapkin et al. 1988c). Feeding rats a diet in which linseed oil was the fat source resulted in a marked increase in the proportions of α-linolenic acid, EPA, docosapentaenoic acid and DHA in peritoneal macrophage phospholipids (Magrum and Johnston 1983). Feeding laboratory animals a diet rich in fish oil results in a marked increase in the proportions of EPA and DHA in macrophage phospholipids (see Calder 1998 for references). The incorporation of *n*-3 PUFAs following feeding of either linseed or fish oil results in a decline (up to 50%) in the proportion of AA in macrophage phospholipids.

Those studies that have examined the effects of providing fish oil to humans on the fatty acid composition of peripheral blood monocytes show a marked increase in the proportion of EPA (e.g. from <0.1 to >1.5% [Lee et al. 1985; Fisher et al.1990]) and DHA. The incorporation is paralleled by a decline in the proportion of AA (e.g. from 22.6% to 15% [Fisher et al. 1990]). The incorporation of dietary EPA and DHA from the diet into human monocytes reaches a plateau within 3 or 4 weeks (see Gibney and Hunter 1993; Yaqoob et al. 2000). However, both the time course of incorporation and the extent of the compositional change depend upon the dose of fish oil provided.

4 Regulation of Monocyte and Macrophage Functions by Dietary Fatty Acids

4.1 Influence of Fatty Acids on Membrane-Mediated Functions of Monocytes and Macrophages

4.1.1 Chemotaxis

Chemotaxis of monocytes and macrophages could be affected by changes in the fatty acid composition of membrane phospholipids which might influence the binding of chemotactic agents to their receptors, the subsequent signalling pathways, or the cytoskeletal rearrangements which occur. Chemotaxis of blood monocytes towards the chemoattractants leukotriene (LT) B_4 and formyl-methionyl-leucyl-phenylalanine was found to be suppressed following supplementation of the human diet with approximately 5.5 g EPA plus DHA per day for 6 weeks (Lee et al. 1985; Endres et al. 1989; Schmidt et al. 1992). There was no effect of a much lower dose of *n*-3 PUFAs (0.65 g/day for 12 weeks) upon monocyte chemotaxis towards pooled human serum (Schmidt et al. 1996).

4.1.2 Phagocytosis

The ability of a cell to perform phagocytosis may be influenced by membrane structure, in particular by the fluidity of the membrane, which may in turn be modulated by the fatty acid composition of membrane phospholipids (Mahoney et al. 1980). Indeed, several studies show that phagocytosis by murine macrophages is influenced by manipulation of their fatty acid composition in culture (see Calder 1998 for references). In general, increasing the macrophage content of saturated fatty acids decreases the ability to perform phagocytosis, while increasing the macrophage content of PUFAs increases the ability to perform phagocytosis. These studies showed phagocytosis to be highly correlated with the degree of phospholipid fatty acid unsaturation, suggesting that membrane fluidity is an important determinant of phagocytosis.

Despite the consistency of the effects of fatty acids on phagocytosis by macrophages following their enrichment with fatty acids in culture, most studies show little or no effect of dietary fatty acid manipulation on phagocytosis by rodent and pig macrophages (see Calder 1998 for references) or human monocytes (Halvorsen et al. 1997; Thies et al. 2001). The differences between the effects of fatty acids delivered in vitro or through the diet most likely relate to the much smaller changes in fatty acid composition observed in the latter case.

4.1.3 Respiratory Burst

There is a large but inconsistent literature on the effects of dietary fish oil on respiratory burst by macrophages. This might in part relate to the different experimental models used, particularly the stimulus used to generate respiratory burst, and to the different ways of expressing the results from such measurements. Some studies demonstrate that feeding fish or linseed oils results in a reduction in the absolute amount of superoxide (and hydrogen peroxide) generated at a given time after stimulation of macrophages with some agents (see Calder 1998 for references). However, Eicher and McVey (1995) reported no effect of dietary fat on the number of murine Kupffer cells engaging in respiratory burst. A detailed study of the hydrogen peroxide generation by peritoneal macrophages from mice fed safflower or fish oil was conducted by Hubbard et al. (1991). These authors found that macrophages from fish oil-fed mice showed lower hydrogen peroxide production in response to unopsonised zymosan, but production in response to phorbol ester was not different from cells of mice fed the different fatty acids. However, macrophages from fish oil-fed mice produced more hydrogen peroxide than those from safflower oil-fed mice following priming with a high concentration of interferon (IFN)-γ (Hubbard et al. 1991). Thus the true impact of dietary *n*-3 PUFAs on respiratory burst remains unclear. There have been few investigations of the influence of dietary fatty acids on respiratory burst by human monocytes. Fisher et al. (1990) reported that giving healthy volunteers 6 g EPA plus DHA per day for 6 weeks resulted in a marked decrease in the production of superoxide by zymosan-stimulated monocytes. In contrast, Halvorsen et al. (1997) reported no effect of 3.8 g EPA or DHA per day for 7 weeks on superoxide production by monocytes in response to *Escherichia coli*. Furthermore, superoxide production by monocytes in response to *E. coli* was not affected by consumption of 2 g α-linolenic acid, 0.75 g DHA or 1.2 g EPA plus DHA per day for 12 weeks by healthy elderly humans (Thies et al. 2001). These data suggest that there is little impact of modestly increased consumption of *n*-3 PUFAs on respiratory burst by human monocytes.

4.2 Influence of Fatty Acids on Eicosanoid Generation by Monocytes and Macrophages

4.2.1 Arachidonic Acid as an Eicosanoid Precursor

Eicosanoids are a family of oxygenated derivatives of DGLA, AA and EPA. Eicosanoids include prostaglandins (PGs), thromboxanes, LTs, lipoxins, hydroperoxyeicosatetraenoic acids and hydroxyeicosatetraenoic acids. Monocytes and macrophages are important sources of eicosanoids. Because the membranes of monocytes and macrophages typically contain large amounts of AA, compared with DGLA and EPA, AA is usually the principal precursor for eicosanoid synthesis. AA in the monocyte/macrophage can be mobilised by various phospholipase enzymes, most notably phospholipase A_2, and the free AA can subsequently act as a substrate for cyclooxygenase (COX), forming 2-series PGs and related compounds, or for one of the lipoxygenase (LOX) enzymes, forming 4-series LTs and related compounds (Fig. 4). There are two forms of COX: COX-1 is a constitutive enzyme and COX-2 is induced in response to stimulation, for example with bacterial lipopolysaccharide (LPS) or tumour necrosis factor (TNF), and is responsible for the marked elevation in production of PG which accompanies such cellular activation. Monocytes and macrophages produce large amounts of PGE_2 and $PGF_{2\alpha}$. The LOX enzymes have different tissue distributions, with 5-LOX being the most important in immune cells including monocytes and macrophages.

4.2.2
Effects of Eicosanoids on Inflammation and Immunity

Eicosanoids are involved in modulating the intensity and duration of inflammatory and immune responses. The effects of PGE_2 and LTB_4 have been studied most widely. PGE_2 has a number of pro-inflammatory effects including inducing fever, increasing vascular permeability and vasodilation and enhancing pain and oedema caused by other agents such as histamine. PGE_2 acts on T cells to suppress proliferation and interleukin (IL)-2 and IFN-γ production. PGE_2 also inhibits natural killer cell activity. Thus, in these respects PGE_2 is immunosuppressive. With respect to cytokine production by monocytes and macrophages, PGE_2 inhibits production of TNF-α, IL-1 and IL-6. Since TNF-α induces COX-2 and so promotes PGE_2 production, the inhibition of synthesis of the classic pro-inflammatory cytokines by PGE_2 forms an important regulatory loop. LTB_4 increases vascular permeability, enhances local blood flow, is a potent chemotactic agent for leukocytes (including monocytes), induces release of lysosomal enzymes by neutrophils, enhances generation of reactive oxygen species, inhibits lymphocyte proliferation and promotes natural killer cell activity. The 4-series LTs also regulate production of pro-inflammatory cytokines; for example LTB_4 enhances production of TNF-α, IL-1 and IL-6. In this latter respect, PGE_2 and LTB_4 are antagonistic. Thus, AA gives rise to mediators which can have opposing effects to one another, so the overall physiological effect will be governed by the concentration of those mediators, the timing of their production and the sensitivities of target cells to their effects.

4.2.3
EPA as an Alternative Eicosanoid Precursor

Since increased consumption of fish oil results in a decrease in the amount of AA in the membranes of monocytes and macrophages (see "Influence of Altered Supply of Fatty Acids on Monocyte and Macrophage Fatty Acid Composition" above), there will be less substrate available for synthesis of eicosanoids from AA. Furthermore, *n*-3 PUFAs inhibit phospholipase A_2 activity in macrophages, competitively inhibit the oxygenation of AA by COX, and inhibit the cytokine-induced upregulation of COX-2 gene expression (Curtis et al. 2000). Thus, fish oil feeding results in a decreased capacity of monocytes and macrophages to synthesise eicosanoids from AA. This has been demonstrated in a variety of animal models and following high-dose fish oil feeding in humans (see Calder 1998 for references).

In addition to effects on generation of eicosanoids from AA, EPA is able to act as a substrate for both COX and 5-LOX (Fig. 4), giving rise to derivatives which have a different structure from those produced from AA (i.e. 3-series PGs and 5-series LTs). Thus, the EPA-induced suppression in the production of AA-derived eicosanoids can potentially be accompanied by an elevation in the production of EPA-derived eicosanoids. Studies in experimental animals have dem-

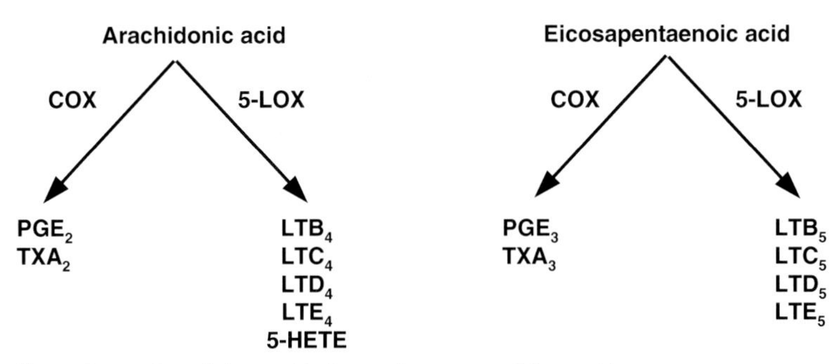

Fig. 4 Generation of eicosanoids from polyunsaturated fatty acids

onstrated that feeding fish oil results in markedly enhanced production of 5-series LT (Chapkin et al. 1990; Whelan et al. 1991). Similarly, dietary fish oil (at a high dose) was demonstrated to significantly increase generation of LTB_5, 6-trans LTB_5 and 5-hydroxyeicosapentaenoic acid by stimulated human monocytes (Lee et al. 1985; Sperling et al. 1993). The generation of EPA-derived COX metabolites following fish oil feeding has not been demonstrated suggesting that, at the concentrations incorporated into membrane phospholipids, EPA is a relatively poor substrate for COX. It is considered that the eicosanoids produced from EPA are less biologically potent than the analogues synthesised from AA, although the full range of biological activities of these compounds has not been investigated.

4.3 Influence of Fatty Acids on Cytokine Generation by Monocytes and Macrophages

Since n-3 PUFAs decrease PGE_2 production and since PGE_2 is an inhibitor of inflammatory cytokine production, it would be predicted that n-3 PUFAs would enhance production of TNF, IL-1 and IL-6. Some animal studies support this prediction as far as TNF and IL-1 are concerned (see Calder 1998 for references). However, other animal studies demonstrate the reverse effect of fish oil feeding: significantly decreased production of TNF, IL-1 and IL-6 following LPS stimulation of macrophages from fish oil-fed rodents has been reported (see Calder 1998 for references). The studies showing reduced cytokine production after fish oil feeding are supported by cell culture studies which demonstrate that EPA or DHA decrease the production of TNF-α, IL-β and tissue factor by monocytic cell lines (Baldie et al. 1993; Chu et al. 1999), and that n-3 PUFAs inhibit the cytokine-induced up-regulation of the TNF-α and IL-1α genes in bovine chondrocytes (Curtis et al. 2000). Whatever the reasons for the differences between study outcomes, it is evident that n-3 PUFAs might affect cytokine production by mechanisms other than a decrease in production of PGE_2.

A large number of studies have now investigated the effect of dietary n-3 PUFAs on ex vivo cytokine production by LPS-stimulated blood mononuclear cells or purified monocytes from healthy human subjects (reviewed by Calder

2001a). A high dose of α-linolenic acid (approximately 15 g/day for 4 weeks) decreased by about 30% the production of IL-1β and TNF-α production by LPS-stimulated human mononuclear cells (Caughey et al. 1996). A lower dose of α-linolenic acid (2 g/day for 12 weeks) did not affect production of TNF-α, IL-1β or IL-6 by such cells (Thies et al. 2001). A number of studies have shown that supplementation of the diet with between 2.4 and 5 g EPA plus DHA per day for a number of weeks leads to a significant reduction in ex vivo production of TNF, IL-1 and IL-6 (e.g. Endres et al. 1989; Meydani et al. 1991; Caughey et al. 1996; see Calder 2001a,b for further references). Similarly high dose fish oil (providing 4.6 g EPA plus DHA/day) resulted in decreased expression of mRNA for platelet-derived growth factors A and B and monocyte chemoattractant protein-1 in unstimulated and adherence-stimulated human monocytes (Baumann et al. 1999). Lower doses of EPA plus DHA appear to be without effect on cytokine production (Schmidt et al. 1996; Blok et al. 1997; Thies et al. 2001).

5 Dietary Fatty Acids and Atherosclerosis

5.1 Dietary Fatty Acids, Blood Cholesterol and Cardiovascular Disease

Atherosclerosis is the leading cause of death in Western populations. One of the few statements concerning the condition that can be made with certainty is that blood cholesterol concentrations play an important role. Blood cholesterol represents one of the "modifiable" risk factors for cardiovascular disease, which means that it can be altered through pharmacological or dietary therapy. The degree to which treatment is able to alter blood cholesterol levels depends to some extent on an individual's "non-modifiable" risk factors, which include family history, race, age and gender. Epidemiological studies (such as the Multiple Risk Factor Intervention Trial 1982) and drug trials suggest that a lowering of blood cholesterol concentration by 1% decreases the risk of a heart attack by 2% (Vines 1989). In most individuals, low-density lipoprotein (LDL) contributes approximately 75% of total circulating cholesterol. With the elucidation of the LDL and scavenger receptor pathways and the discovery that oxidation of LDL is a pre-requisite for uptake by macrophages, a widely accepted hypothesis of the mechanisms underlying atherosclerosis has evolved, whereby an elevated LDL-cholesterol concentration predisposes towards oxidation and subsequent uptake of modified LDL by macrophages (Kruth 2001).

Although dietary cholesterol is involved in blood cholesterol homeostasis, it generally has little impact on total cholesterol concentration, since homeostasis primarily involves regulation of hepatic synthesis and degradation of cholesterol. Dietary fat, especially saturated fat, intake, on the other hand, is correlated with total blood cholesterol concentration in cross-sectional studies of populations with markedly different total fat intakes (Keys 1970). However, while cross-cultural studies tend to find significant associations between dietary fat consumption and blood cholesterol concentration, within-population studies often fail to find this correlation, probably because the variation in fat intake is much smaller (Caggiula and Mustad 1997). Nevertheless, the relationships between intake of each class of dietary fatty acid, blood cholesterol and cardiovascular disease continue to be a subject for debate. The general consensus is that the relationship between dietary fatty acids and coronary heart disease is mediated in part by effects on blood cholesterol concentrations, and, while saturated fatty acids tend to be positively associated with both blood cholesterol and coronary heart disease, relationships between other fatty acids (MUFAS and PUFAs) are less consistent, but tend to be negative (Caggiula and Mustad 1997).

Cholesterol-lowering drug intervention trials, such as those testing the effects of statins, prove beyond doubt that decreasing blood cholesterol concentration in individuals with existing coronary heart disease or with a raised cholesterol concentration reduces the risk of a coronary event. However, paradoxically, while blood cholesterol concentration is a primary risk factor, it does not serve as an adequate *predictor* of coronary risk within populations. This is because a high proportion of individuals with high blood cholesterol concentrations have cardiovascular disease, yet a large proportion of individuals diagnosed with cardiovascular disease have concentrations within the normal range (Griffin 1999). Thus, while the interrelationships between diet and blood cholesterol may be of interest, it is impossible to conclude with absolute certainty the importance of the effects of dietary fatty acids on blood cholesterol with respect to cardiovascular disease in a normal population. This is not to say that dietary fat cannot influence atherosclerosis through other mechanisms. Some fatty acid classes (in particular the *n*-3 PUFAs) have profound effects on blood clotting and, potentially, thrombosis (British Nutrition Foundation 1992, 1999). Furthermore, fatty acids may directly affect the atherogenic process by modifying the susceptibility of LDL to oxidation, by modulating inflammatory functions of macrophages and by altering scavenger receptor expression and foam cell formation.

5.2 Dietary Fatty Acids and LDL Oxidation

Oxidative modification of LDL, which progressively degrades PUFA within the particle, increases its atherogenicity (Steinberg et al. 1989). The susceptibility of LDL to oxidation is influenced by its PUFA content (amount of substrate available for oxidation) and its antioxidant content (confer resistance to oxidation). Thus, at a given antioxidant content, decreasing the PUFA content of LDL should decrease its susceptibility to oxidation. Consumption of a diet rich in oleic acid, a MUFA, has been reported to decrease the susceptibility of LDL to oxidation ex vivo (Parthasarathy et al. 1990; Reaven et al. 1991; Berry et al. 1992), presumably because MUFAs replace PUFAs in LDL and are less susceptible to oxidation. Diets high in linoleic acid increase the linoleic acid content of LDL and in some studies this is associated with increased susceptibility to oxidation ex vivo (Parthasarathy et al. 1990; Abbey et al. 1993; Reaven et al. 1993;

Louheranta et al. 1996). However, some studies report no effect of increased linoleic acid consumption on susceptibility of LDL to oxidation (Nenseter et al. 1992; Suzukawa et al. 1995), although this may be due to the use of lower doses of linoleic acid than in some other studies. The effects of dietary *n*-3 PUFAs on LDL oxidation are unclear. Some studies demonstrate that fish oil supplementation of the diet increases the susceptibility of LDL to oxidation ex vivo (Suzukawa et al. 1995), presumably by increasing the PUFA content. However, other studies, using similar doses and similar measurements of lipid peroxidation, showed no effect of fish oil (Nenseter et al. 1992; Frankel et al. 1994). The discrepancies in the results may be related to the antioxidant content of the LDL, which was not reported in any of the studies and may differ. Thus, despite reported protective effects of fish oil against mortality from cardiovascular disease (Burr et al. 1989; Singh et al. 1997; Gissi Prevenzione 1999), it remains possible that *n*-3 PUFAs may, paradoxically, increase the susceptibility of LDL to oxidation.

5.3
Fatty Acids and the Atherogenicity of Oxidised LDL

The foam cell hypothesis, describing uptake of oxidatively modified LDL by macrophages, which subsequently become foam cells, is widely accepted (Kruth 2001). The foam cell hypothesis goes on to explain that the lipid core of an atherosclerotic plaque is derived from the release of cholesterol by dying foam cells. However, it is interesting to note that the fatty acid composition of cholesteryl esters from plaque regions containing foam cells is different from that in the lipid core region. Foam cells contain cholesteryl esters which are rich in oleic acid, whereas extracellular lipid particles in the lipid core have a high proportion of linoleic acid, similar in fact to the fatty acid profile of LDL cholesteryl esters (Smith and Slater 1972; Chao et al. 1990). Furthermore, a number of experimental studies suggest that subendothelial lipid accumulation precedes entry of monocytes into the subendothelial space and subsequent foam cell formation (Guyton and Kemp 1992). These observations suggest that there may be flaws in some aspects of the rationale for the foam cell hypothesis. It has been suggested as an alternative that the primary function of macrophages in atherosclerotic lesions may in fact be to *remove* cholesterol, since they can excrete the cholesterol they accumulate through many processes (Kruth 2001). Accordingly, pharmacological agents used to modulate foam cell formation can function to limit cholesterol uptake, to alter cholesterol esterification or trafficking within the macrophage (e.g. by modulating lysosomal degradation) or to enhance cholesterol efflux (Kruth 2001). Physiological agents, such as cytokines and hormones, may also modulate lipoprotein uptake and metabolism by macrophages. This suggests that there may be potential for the modulation of foam cell formation by dietary fatty acids, either through their actions on cytokines or through modifications in the fatty acid composition of LDL and/or macrophage lipids, resulting in altered uptake of oxidised LDL. However, this area has been little studied to date. A few studies have examined the effects of *n*-3 PUFAs on scavenger receptor expression by monocytes or macrophages. An animal study demonstrated that feeding a fish-oil-rich diet to mice resulted in down-regulation of macrophage scavenger receptors AI and AII, while coconut oil and sunflower oil had no effect compared with the standard diet fed to the animals (Miles et al. 2000). Pietsch et al. (1995) reported a down-regulation of the expression of CD36 by the human monocytic U937 cell line after incubation with 5 μM EPA or DHA, but not with linoleic acid or AA. In another study, EPA (30–240 μM) was shown to inhibit the proliferation of the same cell line in a dose-dependent manner and, at the highest concentrations, induced apoptosis (Finstad et al. 1998). Expression of CD36 was lower in cells treated with 60 μM EPA or oleic acid compared with untreated cells (Finstad et al. 1998). However, EPA unexpectedly caused greater accumulation of lipid droplets in the cells than oleic acid, although the effects were reversed when cells were re-incubated in EPA-free medium. This leaves the question of the precise nature of the effects of fatty acids on foam cell formation unresolved. It is interesting to note, however, that the transcription factor, peroxisome proliferator activated receptor (PPAR)γ, has been reported to be involved in foam cell formation by virtue of its induction of CD36 (Nagy et al. 1998; Tontonoz et al. 1998). Since oxidised fatty acids are likely candidates as physiological ligands of PPARγ, it is possible that dietary modulation of the fatty acid composition of LDL could generate different patterns of oxidised derivatives, which could have differential effects on PPARγ.

Oxidation of LDL has many physiological effects that may influence atherosclerotic lesion development. In cell culture systems, foam cell formation by monocytic cells is rarely observed (Kruth 2001), reflecting a limitation in experimental techniques. These systems have been used, however, to investigate the pro-atherogenic and pro-inflammatory properties of oxidised LDL. Oxidised LDL is pro-atherogenic by virtue of its upregulated uptake by scavenger receptors. There has often been an assumption that oxidised LDL is pro-inflammatory in nature and that it is primarily the reaction of macrophages to oxidised LDL which confers the chronic inflammation characteristic of atherosclerosis. However, the evidence for this is rather limited. Oxidised LDL is reported to stimulate the expression of monocyte chemotactic protein, adhesion molecules and some cytokines (Berliner and Heinecke 1996). However, a number of studies report anti-inflammatory effects of oxidised LDL, including downregulation of CCR2 (Han et al. 2000) and downregulation of the platelet-activating factor receptor on monocytes (Hourton et al. 2001). There are even reports that oxidised LDL at low concentrations improves the viability of monocytes and that the purpose of this may be to maintain long-term survival of macrophages in lesions (Hamilton et al. 1999). Other studies demonstrate that oxidised LDL at high concentrations leads to apoptosis of vascular cells, which could contribute to plaque instability (Siow et al. 1999). Thus, the nature of the atherogenicity of oxidised LDL remains to be clarified. Since the fatty acid composition of LDL is responsive to diet, it is possible that dietary modification could alter cellular responses to oxidised LDL, but this has not been studied to date.

5.4 Fatty Acids and Plaque Stability

The propensity of atherosclerotic plaques to rupture is influenced by their lipid content and the distribution of lipid within the plaque (Felton et al. 1997). There appears to be a reduction in the proportion of *n*-6 PUFAs and total PUFAs at the edges of disrupted plaques compared to the centres, which may reflect oxidative damage (Felton et al. 1997). It is therefore postulated that oxidised derivatives of PUFA may alter inflammatory activity and connective tissue degradation at the edges of lesions, enhancing the likelihood of disruption at this site (Felton et al. 1997). The effects of individual fatty acids and their oxidised derivatives have not been elucidated. However, given the evidence for the anti-coagulatory, anti-thrombotic and anti-inflammatory properties of *n*-3 PUFAs (British Nutrition Foundation 1992, 1999), it is possible that alteration of the PUFA composition of the diet could affect plaque progression, stability and thrombus formation. This has yet to be demonstrated in humans, but would strengthen the case for the reported protective effects of *n*-3 PUFAs in mortality from cardiovascular disease (Burr et al. 1989; Singh et al. 1997; Gissi Prevenzione 1999).

6 Summary of the Effects of Fatty Acids on Monocyte and Macrophage Functions

Dietary fatty acids, especially *n*-3 PUFAs, can modulate monocyte/macrophage activities. At high intakes *n*-3 PUFAs can affect chemotaxis, phagocytosis, respiratory burst, eicosanoid production, cytokine production and other monocyte/macrophage functions. Although some of the effects of *n*-3 PUFAs may be brought about by modulation of the amount and types of eicosanoids made, it appears that these fatty acids might elicit some of their effects by eicosanoid-independent mechanisms, including actions upon intracellular signalling pathways and transcription factor activity (see Miles and Calder 1998; Calder 2002). These effects of *n*-3 PUFAs are generally termed as "anti-inflammatory" and are considered to be beneficial to health (see Calder 2001a,b). The effects of dietary fatty acids on monocyte/macrophage function may also be relevant to atherosclerosis, which is now recognised to include an inflammatory component. Fatty acids could potentially affect the degree of oxidation of LDL, its uptake by vascular cells, aspects of foam cell formation and inflammatory activity within lesions.

7
References

Abbey M, Belling GB, Noakes M, Hirata F, Nestel PJ (1993) Oxidation of low density lipoproteins: intraindividual variability and the effect of dietary linoleate supplementation, Am J Clin Nutr 57:391–398

Baldie G, Kaimakamis D, Rotondo D (1993) Fatty acid modulation of cytokine release from human monocytic cells, Biochim Biophys Acta 1179:125–133

Baumann KH, Hessel F, Larass I, Muller T, Angerer P, Kiefl R, von Schacky C (1999) Dietary ω-3, ω-6, and ω-9 unsaturated fatty acids and growth factor and cytokine gene expression in unstimulated and stimulated monocytes, Arterioscler Thromb Vasc Biol 19:59–66

Berry EM, Eisenberg S, Friedlander M, Harats D, Kaufmann NA, Norman Y, Stein Y (1992) Effects of diets rich in MUFA on plasma lipoproteins- the Jerusalem Nutrition Study, Am J Clin Nutr 56:394–403

Berliner JA, Heinecke JW (1996) The role of oxidized lipoproteins in atherogenesis, Free Rad Biol Med 20:707–727

Blok WL, Deslypere J-P, Demacker PNM, van der Ven-Jongekrijg J, Hectors MPC, van der Meer JMW, Katan MB (1997) Pro- and anti-inflammatory cytokines in healthy volunteers fed various doses of fish oil for 1 year, Eur J Clin Invest 27:1003–1008

British Nutrition Foundation (1992) Unsaturated Fatty Acids: Nutritional and Physiological Significance, Task Force Report, Chapman and Hall, London

British Nutrition Foundation (1999) Briefing Paper: N-3 Fatty Acids and Health, British Nutrition Foundation, London

Burr ML, Gilbert JF, Holliday RM, Elwood PC, Fehily AM, Rogers S, Sweetman PM, Deadman NM (1989) Effects of changes in fat, fish and fibre intake on death and myocardial reinfarction: diet and reinfarction trial (DART), Lancet ii:757–761

Caggiula AW, Mustad VA (1997) Effects of dietary fat and fatty acids on coronary artery disease risk and total and lipoprotein cholesterol concentrations: epidemiologic studies, Am J Clin Nutr 65:1597–1610

Calder PC (1998) N-3 fatty acids and mononuclear phagocyte function, In: Kremer JM (ed) Medicinal Fatty Acids. Birkhauser, Basel, pp 1–27

Calder PC (2001a) N-3 polyunsaturated fatty acids, inflammation and immunity: pouring oil on troubled waters or another fishy tale? Nutr Res 21:309–341

Calder PC (2001b) Polyunsaturated fatty acids, inflammation and immunity, Lipids 36:1007–1024

Calder PC (2002) Dietary modification of inflammation with lipids, Proc Nutr Soc 61:345–358

Calder PC, Bond JA, Harvey DJ, Gordon S, Newsholme EA (1990) Uptake of saturated and unsaturated fatty acids into macrophage lipids and their effect upon macrophage adhesion and phagocytosis, Biochem J 269:807–814

Caughey GE, Mantzioris E, Gibson RA, Cleland LG, James MJ (1996) The effect on human tumor necrosis factor α and interleukin 1β production of diets enriched in n-3 fatty acids from vegetable oil or fish oil, Am J Clin Nutr 63:116–122

Chao FF, Blanchette-Mackie EJ, Chen YJ, Dickens BF, Berlin E, Amende LM, Skarlatos SI, Gamble W, Resau JH, Mergner WT (1990) Characterization of two unique cholesterol-rich lipid particles isolated from human atherosclerotic lesions, Am J Pathol 136:169–179

Chapkin RS, Coble KJ (1991) Utilization of gammalinolenic acid by mouse peritoneal macrophages, Biochim Biophys Acta 1085:365–370

Chapkin RS, Hubbard NE, Erickson KL (1990) 5-Series peptido-leukotriene synthesis in mouse peritoneal macrophages: modulation by dietary n-3 fatty acids, Biochem Biophys Res Commun 171:764–769

Chapkin RS, Miller CC, Somers SD, Erickson KL (1988a) Utilization of dihomo-γ-linolenic acid (8,11,14-eicosatrienoic acid) by murine peritoneal macrophages, Biochim Biophys Acta 959:322–331

Chapkin RS, Somers SD, Erickson KL (1988b) Inability of murine peritoneal macrophages to convert linoleic acid into arachidonic acid, J Immunol 140:2350–2355

Chapkin RS, Somers SD, Erickson KL (1988c) Dietary manipulation of macrophage phospholipid classes: selective increase of dihomogammalinolenic acid, Lipids 23:766–770

Chu AJ, Walton MA, Prasad JK, Seto A (1999) Blockade by polyunsaturated n-3 fatty acids of endotoxin-induced monocytic tissue factor activation is mediated by the depressed receptor expression in THP-1 cells, J Surg Res 87:217–224

Curtis CL, Hughes CE, Flannery CR, Little CB, Harwood JL, Caterson B (2000) n-3 Fatty acids specifically modulate catabolic factors involved in articular cartilage degradation, J Biol Chem 275:721–724

Eicher SD, McVey DS (1995) Dietary modulation of Kupffer cell and splenocyte function during a Salmonella typhimurium challenge in mice, J Leuk Biol 58:32–39

Endres S, Ghorbani R, Kelley VE, Georgilis K, Lonnemann G, van der Meer JMW, Cannon, JG, Rogers TS, Klempner MS, Weber PC, Schaeffer EJ, Wolff SM, Dinarello CA (1989) The effect of dietary supplementation with n-3 polyunsaturated fatty acids on the synthesis of interleukin-1 and tumor necrosis factor by mononuclear cells, N Eng J Med 320:265–271

Felton CV, Crook D, Davies MJ, Oliver MF (1997) Relation of plaque lipid composition and morphology to the stability of human aortic plaques, Arterioscler Thromb Vasc Biol 17:1337–1345

Finstad HS, Drevon CA, Kulseth MA, Synstad AV, Knudsen E, Kolset SO (1998) Cell proliferation, apoptosis and accumulation of lipid droplets in U937-1 cells incubated with eicosapentaenoic acid, Biochem J 336:451–459

Fisher M, Levine PH, Weiner BH, Johnson MH, Doyle EM, Ellis PA, Hoogasian JJ (1990) Dietary n-3 fatty acid supplementation reduces superoxide production and chemiluminescence in a monocyte-enriched preparation of leukocytes, Am J Clin Nutr 51:804–808

Frankel EN, Parks EJ, Xu R, Schneeman BO, Davies PA, German JB (1994) Effect of n-3 fatty acid rich fish oil supplements on the oxidation of low density lipoproteins, Lipids 29:233–236

Gibney MJ, Hunter B (1993): The effects of short- and long-term supplementation with fish oil on the incorporation of n-3 polyunsaturated fatty acids into cells of the immune system in healthy volunteers, Eur J Clin Nutr 47:255–259

GISSI Prevenzione (1999) Dietary supplementation with n-3 polyunsaturated fatty acids and vitamin E after myocardial infarction: results of the GISSI-Prevenzione trial, Lancet 354:447–455

Griffin BA (1999) Lipoprotein atherogenicity: an overview of current mechanisms, Proc Nutr Soc 58:163–169

Guyton JR, Kemp KF (1992) Early extracellular and cellular lipid deposits in aorta of cholesterol-fed rabbits, Am J Pathol 141:925–936

Halvorsen DA, Hansen J-B, Grimsgaard S, Bonaa KH, Kierulf P, Nordoy A (1997) The effect of highly purified eicosapentaenoic and docosahexaenoic acids on monocyte phagocytosis in man, Lipids 32:935–942

Hamilton JA, Myers D, Jessup W, Cochrane F, Byrne R, Whitty G, Moss S (1999) Oxidized LDL can induce macrophage survival, DNA synthesis and enhanced proliferative response to CSF-1 and GM-CSF, Arterioscler Thromb Vasc Biol 19:98–105

Han KH, Chang MK, Boullier A, Green SR, Li A, Glass CK, Quehenberger O (2000) Oxidized LDL reduces monocyte CCR2 expression through pathways involving peroxisome proliferator-activated receptor-γ, J Clin Invest 106:793–802

Hourton D, Delerive P, Stankova J, Staels B, Chapman MJ, Ninio E (2001) Oxidized low-density lipoprotein and peroxisome proliferator-activated receptor a down-regulate

platelet-activating-factor receptor expression in human macrophages, Biochem J 354:225–232
Hubbard NE, Somers SD, Erickson KL (1991) Effect of dietary fish oil on development and selected functions of murine inflammatory macrophages, J Leuk Biol 49:592–598
Keys A (1970) The Seven Countries Study, Circulation 41:1–211
Kruth HS (2001) Macrophage foam cells and atherosclerosis, Front Biosci 6:429–455
Lee TH, Hoover RL, Williams JD, Sperling RI, Ravalese J, Spur BW, Robinson DR, Corey EJ, Lewis RA, Austen KF (1985) Effects of dietary enrichment with eicosapentaenoic acid and docosahexaenoic acid on in vitro neutrophil and monocyte leukotriene generation and neutrophil function, N Engl J Med 312:1217–1224
Louheranta AM, Porkkala-Sarataho EK, Nyyssonen MK, Salonen RM, Salonen JT (1996) Linoleic acid intake and susceptibility of very low density and low density lipoproteins to oxidation in men, Am J Clin Nutr 63:698–703
Magrum LJ, Johnston PV (1983) Modulation of prostaglandin synthesis in rat peritoneal macrophages with ω-3 fatty acids, Lipids 18:514–521
Mahoney EM, Scott WA, Landsberger FR, Hamill AL, Cohn ZA (1980) Influence of fatty acyl substitution on the composition and function of macrophage membranes, J Biol Chem 255:4910–4917
Meydani SN, Endres S, Woods MM, Goldin BR, Soo C, Morrill-Labrode A, Dinarello C, Gorbach SL (1991) Oral (n-3) fatty acid supplementation suppresses cytokine production and lymphocyte proliferation: comparison between young and older women, J Nutr 121:547–555
Miles EA, Calder PC (1998) Modulation of immune function by dietary fatty acids, Proc Nutr Soc 57:277–292
Miles EA, Wallace FA, Calder PC (2000) Dietary fish oil reduces intercellular adhesion molecule 1 and scavenger receptor expression on murine macrophages, Atherosclerosis 152:43–50
Multiple Risk Factor Intervention Trial Research Group (1982) Multiple risk factor intervention trial, JAMA 248:1465–1477
Nagy L, Tontonoz P, Alvarez JGA, Chen H, Evans RM (1998) Oxidized LDL regulates macrophage gene expression through ligand activation of PPARγ, Cell 93:229–240
Nenseter MS, Rustan AC, Lend-Katz S, Soyland E, Maelandsmo G, Phillips MC, Drevon CA (1992) Effect of dietary supplementation with n-3 polyunsaturated fatty acids on physical properties and metabolism of low desnity lipoproteins in humans, Arterioscler Thromb Vasc Biol 12:369–379
Parthasarathy S, Khoo JC, Miller E, Narnett J, Wiztum JL (1990) Low density lipoprotein rich in oleic acid is protected against oxidative modification: implications for dietary prevention of atherosclerosis, Proc Natl Acad Sci USA 87:3894–3898
Pietsch A, Weber C, Goretzki M, Weber PC, Lorenz RL (1995) N-3 but not n-6 fatty acids reduce the expression of the combined adhesion and scavenger receptor CD36 in human monocytic cells, Cell Biochem Func 13:211–216
Reaven P, Parthasarathy S, Grasse BJ, Miller E, Steinberg D, Wiztum JL (1993) Effects of oleate rich and linoleate rich diets on the susceptibility of low density lipoprotein to oxidative modification in mildy hypercholesterolaemic subjects, J Clin Invest 91:668–676
Schmidt EB, Varming K, Moller JM, Bulow Pederson I, Madsen P, Dyerberg J (1996) No effect of a very low dose of n-3 fatty acids on monocyte function in healthy humans, Scand J Clin Invest 56:87–92
Schmidt EB, Varming K, Pederson JO, Lervang HH, Grunnet N, Jersild C, Dyerberg J (1992) Long term supplementation with n-3 fatty acids. II. Effect on neutrophil and monocyte chemotaxis, Scand J Clin Lab Invest 52:229–236
Singh RB, Niaz MA, Sharma JP, Kumar R, Rastogi V, Moshiri M (1997) Randomised double-blind, placebo-controlled trial of fish oil and mustard oil in patients with suspect-

ed acute myocardial infarction: the Indian experiment of infarct survival, Cardiovasc Drugs Ther 11:485–491

Simons K, Toomre D (2000) Lipid rafts and signal transduction, Nature Rev Mol Cell Biol 1:31–40

Siow RCM, Richards JP, Pedley KC, Leake DS, Mann GE (1999) Vitamin C protects human vascular smooth muscle cells against apoptosis induced by moderately oxidized LDL containing high levels of lipid hydroperoxides, Arterioscler Thromb Vasc Biol 19:2387–2394

Smith EB, Slater RS (1972) The microdissection of large atherosclerotic plaques to give morphologically and topographically defined fractions for analysis. 1. The lipids in the isolated fractions, Atherosclerosis 15:37–56

Sperling RI, Benincaso AI, Knoell CT, Larkin JK, Austen KF, Robinson DR (1993) Dietary ω-3 polyunsaturated fatty acids inhibit phosphoinositide formation and chemotaxis in neutrophils, J Clin Invest 91:651–660

Steinberg D, Parthasarathy S, Carew TE, Khoo JC, Wiztum JL (1989) Beyond cholesterol; modifications of low-density lipoprotein that increases its atherogenicity, N Engl J Med 320:915–924

Stubbs CD, Smith AD (1984) The modification of mammalian membrane polyunsaturated fatty acid composition in relation to membrane fluidity and function, Biochim Biophys Acta 779:89–137

Suzukawa M, Abbey M, Howe PR, Nestel PJ (1995) Effects of fish oil fatty acids on low density lipoprotein size, oxidisability and uptake by macrophages, J Lipid Res 36:473–484

Thies F, Miles EA, Nebe-von-Caron G, Powell JR, Hurst TL, Newsholme EA, Calder PC (2001) Influence of dietary supplementation with long chain n-3 or n-6 polyunsaturated fatty acids on blood inflammatory cell populations and functions and on plasma soluble adhesion molecules in healthy adults, Lipids 36:1183–1193

Tontonoz P, Nagy L, Alvarez JGA, Thomazy VA, Evans RM (1998) PPARγ promotes monocyte/macrophage differentiation and uptake of oxidized LDL, Cell 93:241–252

Turek JJ, Schoenlein IA, Clark LK, van Alstine WG (1994) Dietary polyunsaturated fatty acids effects on immune cells of the porcine lung, J Leuk Biol 56:599–604

Vines G (1989) Diet, drugs and heart disease, New Sci 44–49

Whelan J, Broughton KS, Lokesh B, Kinsella JE (1991) In vivo formation of leukotriene E5 by murine peritoneal cells, Prostaglandins 41:29–42

Yaqoob P, Pala HS, Cortina-Borja M, Newsholme EA, Calder PC (2000) Encapsulated fish oil enriched in α-tocopherol alters plasma phospholipid and mononuclear cell fatty acid compositions but not mononuclear cell functions, Eur J Clin Invest 30:260–274

Macrophages as Therapeutic Targets in Lysosomal Storage Disorders

J. M. Aerts · C. Hollak · R. Boot · A. Groener

Department of Biochemistry and Internal Medicine, Academic Medical Center, University of Amsterdam,Meibergdreef 15, 1105 AZ Amsterdam, The Netherlands
e-mail: j.m.aerts@amc.uva.nl

Abstract Gaucher disease (lysosomal glucocerebrosidase deficiency) is a rare inborn error of metabolism. The type 1 variant is characterised by lysosomal storage of glucosylceramide in tissue macrophages exclusively. The accumulation of storage cells (Gaucher cells) results in pronounced hepatosplenomegaly, haematological abnormalities and deterioration of the skeleton. Type 1 Gaucher disease should be considered as a true macrophage disorder. Specific markers for Gaucher cells, like a hitherto unknown chitinase, have been identified and are commonly used to monitor progression of disease and efficacy of therapies. A spectacular correction in clinical symptoms of type 1 patients can be accomplished by chronic intravenous administration of human glucocerebrosidase containing glycans with terminal mannose-moieties. Currently, about 3,000 patients are treated worldwide with recombinant enzyme (Cerezyme). Enzyme replacement therapy (ERT) is not able to prevent glucosylceramide accumulation in the brain of patients suffering from the severe type 2 variant of Gaucher disease. Recently, oral administration of *N*-butyl-deoxynojirimycin has been registered in the EU for treatment of type 1 Gaucher patients that are unsuitable for ERT. The iminosugar inhibits the synthesis of glucosylceramide and thus prevents massive accumulation of the lipid.

Keywords Chitinase, Deoxynojirimycin, Enzyme replacement, Gaucher disease, Glucocerebrosidase, Glucosylceramide, Lysosome, Substrate reduction

1 Introduction to Lysosomes and Lysosomal Storage Disorders

1.1 Lysosomes

The continuous recycling of their macromolecular constituents is a hallmark of the long-lived eukaryotic cell. For this reason mammalian cells contain single membrane-enclosed compartments in which a variety of biological macromolecules can be safely and efficiently degraded. Based on their lytic function, these acid organelles have been named lysosomes (De Duve et al. 1955). Substrates for lysosomal degradation can enter the organelles via different routes, such as endocytosis, pinocytosis, phagocytosis and autophagocytosis. In addition, direct chaperon-mediated import of specific proteins from the cytoplasm has been reported (Holzmann 1989; Dice et al. 1990). Lysosomes are equipped with a set of about 60 acid hydrolases and a dozen accessory proteins that allow sequential degradation of almost all natural macromolecules, including lipids, glycosaminoglycans, oligosaccharides, proteins and nucleic acids. Mediated by specific carriers in the lysosomal membrane, the products of intralysosomal catabolism are exported to the cytoplasm where they can be re-utilised. The lysosomal membrane is effectively protected against self-digestion by the presence of transmembrane proteins with large, highly glycosylated intralysosomal domains (Peters and von Figura 1994).

The prominent lectin-based mechanism that governs the selective routing of newly formed acid hydrolases to lysosomes was elucidated two decades ago (Kornfeld and Mellman 1989). Upon co-translational translocation of lysosomal enzymes into the lumen of the endoplasmic reticulum, their signal peptide is removed and specific asparagine residues are glycosylated by transfer of a preformed oligosaccharide from a dolichol phosphate lipid carrier. The glycoproteins are folded, assembled in correct multimeric structures, and terminal glucose moieties are removed from their glycans, an important checkpoint in the quality control of protein folding (Helenius 1994). Next, the glycoproteins are exported to the Golgi apparatus where some of their oligosaccharide chains exclusively obtain mannose-6-phosphate moieties by a two-step process. The phosphomannosyl moieties act as a specific recognition signal. Selective binding of a major fraction of most lysosomal enzymes to cation-dependent or cation-independent mannose-6-phosphate receptors (MPRs), allows their segregation from the secretory proteins in the trans-Golgi network. In endosomal compartments, dissociation of mannose-6-phosphate receptors and lysosomal protein ligands occurs due to local acidity. Following uncoupling, the receptor recycles to the Golgi apparatus and the newly formed hydrolases are delivered into lysosomes. The cation-independent MPR is also involved in the delivery to lysosomes of extracellular soluble acid hydrolases containing mannose-6-phosphate residues. In contrast, targeting of integral lysosomal membrane proteins is not mediated by phosphomannosyl moieties but by specific motifs in their cytoplasmic domains. Further alternative targeting mechanism to lysosomes have to exist. Some membrane-associated lysosomal enzymes like glucocerebrosidase do not acquire phosphomannosyl moieties at all in their glycans but are nevertheless efficiently targeted to lysosomes by still unknown mechanisms (Aerts et al. 1988). The lysosomal targeting of lysozyme and chitotriosidase in macrophages is also independent of lectin receptors, since these enzymes completely lack N-linked glycans (Renkema et al. 1997). Moreover, investigations on patients suffering from I-cell disease, in which formation of phosphomannosyl moieties is impaired, have indicated that in hepatocytes and lymphocytes very efficient intracellular sorting of newly formed soluble acid hydrolases can also occur independently of mannose-6-phosphate receptors (Owada and Neufeld 1982). The precise mechanism of the mannose-6-phosphate independent targeting of soluble acid hydrolases is yet unknown, but it has been suggested that it involves a transient membrane association in the Golgi apparatus (Rijnboutt et al. 1991).

1.2 Lysosomal Storage Disorders

The physiological importance of the degradative processes in lysosomes is revealed by the existence of a group of at least 40 distinct inherited diseases, the so-called lysosomal storage disorders (Neufeld 1991; Gieselmann 1995). Most of these diseases are caused by a deficiency in a single lysosomal enzyme or essential cofactor and result in the lysosomal accumulation of one or sometimes several natural compounds. According to the prevailing stored compound, the lysosomal storage diseases are grouped as mucopolysaccharidoses, sphingolipidoses, mucolipidoses, lipidoses, glycoproteinoses, glycogenosis and ceroid lipofuscinoses. Some lysosomal storage disorders are not single enzymopathies but based on defects in transport of hydrolytic products across the lysosomal membrane, deficiencies in non-lysosomal proteins involved in lysosome biogenesis or post-translational modification of lysosomal enzymes or inherited abnormalities in intracellular membrane flow.

All lysosomal storage diseases are relatively rare with an overall incidence for the whole group of 1:5,000–1:10,000. The individual incidence of the more prominent lysosomal diseases is between 1:20,000 and 1:100,000 in most populations (Meikle et al. 1999; Poorthuis et al. 1999). Genetic drift and founder effects have led to unusually high incidences of specific lysosomal storage diseases in some populations. The best examples of this are Gaucher and Tay-Sachs disease among Ashkenazim, and aspartylglucosaminuria, Salla disease and infantile neuronal ceroid lipofuscinosis in Finland (Peltonen 1997). As an example, Table 1 summarises the nature and prevalence of one subgroup of the lysosomal storage disorders: the sphingolipidoses.

Table1 Birth prevalence (per 100,000)of sphingolipidoses in The Netherlands

Disease	Prevalence
Fabry	0.21
Gaucher	1.16
Niemann-Pick type A and B	0.53
Niemann-Pick type C	0.35
Krabbe	1.35
Sandhoff	0.34
Tay-Sachs	0.41
GM1-gangliosidosis	0.41

The clinical manifestation of lysosomal storage disorders is remarkably heterogeneous, contributing to the limited awareness of these diseases. Age of onset and progression of disease vary considerably for almost each individual storage disorder. This remarkable phenotypic variability is usually linked to the extent of the deficiency that is determined by the exact nature of the underlying genetic defect. In the case of some lysosomal enzymopathies, a strict correlation between residual enzyme activity and severity of disease manifestation exists. A common feature of lysosomal storage disorders is that accumulation of storage material is generally restricted to lysosomes of particular cell types. The nature and residual capacity of the defective metabolic pathway in combination with the actual flux through this pathway in various cell types determine the chance that particular cell types are affected. This phenomenon explains why in some lysosomal storage disorders, external genetic or environmental factors that influence the flux through the defective pathway also have a major impact on disease manifestation. The genotype–phenotype relation is therefore not strict in many lysosomal storage disorders.

The lysosomal apparatus of tissue macrophages fulfils many important degradative functions. Macrophages participate in the degradation of invading microbes, the natural turnover of blood cells and tissue modelling. In view of this, it is not surprising that in a considerable number of the lysosomal storage disorders accumulation of storage material also takes place prominently in tissue macrophages. The relatively common type 1 variant of Gaucher disease is unique with respect to the fact that lysosomal storage occurs exclusively in macrophages. The remainder of this review will deal with Gaucher disease and the progress that has been made regarding therapeutic correction of this macrophage disorder.

2
Gaucher Disease: A Macrophage Disorder

Gaucher disease is the most frequently encountered lysosomal storage disorder in man (Barranger and Ginns 1989; Beutler and Grabowski 1995). In 1882 the clinical features of the disease were first described in detail by the French medical student Philippe C.E. Gaucher, reporting the presence of large unusual cells in a 32-year-old female with an enlarged spleen. Already at the beginning of the last century it was suggested that the disease was a familial disorder. In 1934 the primary storage material in Gaucher disease was finally identified as glucocerebroside (glucosylceramide). The glycosphingolipid glucocerebroside is the common intermediate in the synthesis and degradation of gangliosides and globosides. In 1965 Patrick and Brady et al. independently showed that the primary defect in Gaucher disease is a marked deficiency in activity of the lysosomal enzyme glucocerebrosidase (EC. 3.2.1.45) (Brady et al. 1965; Patrick 1965). Inherited deficiencies in glucocerebrosidase result in accumulation of its lipid substrate in the lysosomal compartment of macrophages throughout the body. Three different phenotypes are recognised, which are differentiated on the basis of the presence or absence of neurological symptoms. The most prevalent variant of the disease is the non-neuronopathic form, named type 1 Gaucher disease. The age of onset and clinical manifestations of type 1 Gaucher disease are highly variable. The most common symptoms include splenomegaly with anaemia and thrombocytopaenia, mostly due to hypersplenism, hepatomegaly and bone disease. Anaemia may contribute to chronic fatigue. Thrombocytopaenia and prolonged clotting times may lead to an increase in bleeding tendency. Atypical bone pain, pathological fractures, avascular necrosis and extremely painful bone crises may also have a great impact on the quality of life. Type 1 Gaucher disease is relatively common in all ethnic groups. It is prevalent among Ashkenazim with a carrier frequency as high as about 1 in 10 and an incidence of about 1 in 450. The most common mutation in the glucocerebrosidase gene of Caucasians, including Ashkenazim, encodes the amino acid substitution N370S. The heteroallelic presence of the N370S mutation is always associated with a non-neuronopathic course (Jonsson et al. 1987). It has been demonstrated that the N370S glucocerebrosidase is normally produced and present in lysosomes. Its catalytic activity is only severely impaired at pH values above 5.0, illustrating the subtle nature of the mutation (Van Weely et al. 1993). Most, but not all, homozygotes for the N370S mutation do not develop significant clinical symptoms. Twin studies and the poor predictive power of phenotype–genotype investigations in Gaucher disease have clearly pointed out that epigenetic factors also play a key role in Gaucher disease manifestation (Aerts et al. 1993; Cox and Schofield 1997).

Although glucocerebrosidase is present in lysosomes of all cell types, type 1 Gaucher disease patients develop storage of glucocerebroside solely in cells of the mononuclear phagocyte system. It is believed that the storage material stems from the breakdown of exogenous lipids derived from the turnover of blood cells. The glucocerebroside-loaded cells show a characteristic morphology with a 'wrinkled paper'-like appearance of their cytoplasm, which contains lysosomal inclusion bodies; these cells are referred to as Gaucher cells. In the last decades it has become apparent that Gaucher cells are not inert containers of storage material but viable, chronically activated macrophages that secrete various factors that contribute to the diverse clinical manifestations of Gaucher disease. Increased circulating levels of several pro-inflammatory cytokines [tumour necrosis factor (TNF)-α, interleukin (IL)-1β, IL-6 and IL-8], the anti-inflammatory cytokine IL-10, and macrophage colony-stimulating factor (M-CSF) have been reported (Aerts and Hollak 1997; Cox 2001). It has been hypothesised that cytokine abnormalities may play a crucial role in the development of common clinical abnormalities in Gaucher patients such as osteopaenia, activation of coagulation, hypermetabolism, gammopathies and multiple myeloma and hypolipoproteinaemias. More recently, examination of gene expression profiles by suppressive subtraction hybridisation analysis of Gaucher and control spleens has led to the identification of over-expression by Gaucher cells of transcripts for cathepsins B, K and S (Moran et al. 2000). It is of interest to note that osteoclast-derived cathepsin K is prominently involved in osseous type I collagen destruction. Local release of this cathepsin by Gaucher cells may contribute to the osteolysis in Gaucher disease.

3 Therapy of Type 1 Gaucher Disease

Type 1 Gaucher disease has generally been considered to be the most attractive candidate among the inherited lysosomal storage disorders for developing effective therapeutic interventions. First, the molecular basis of the underlying genetic defect had been already established in detail at the gene and protein level. Second, just a single cell type, the tissue macrophage, is primarily implicated in the pathophysiology of the disorder. The rationale for therapeutic intervention of type 1 Gaucher disease is therefore relatively simple: correction (or prevention of ongoing formation) of Gaucher cells. This could be accomplished by supplementation of macrophages with the enzyme glucocerebrosidase (enzyme replacement therapy), by reduction of glycolipid synthesis with specific inhibitors (substrate deprivation or substrate balancing therapy), or by introduction of glucocerebrosidase cDNA in haematopoietic progenitors of macrophages (gene therapy).

3.1 Enzyme Therapy

Thanks to the pioneering work of Brady, Barranger and co-workers at the National Institutes of Health (Bethesda, USA) as well as valuable contributions by many others, a highly effective treatment of type 1 Gaucher disease is now feasible based on chronic intravenous administration of human glucocerebrosidase (Brady 1997; Barranger and O'Rourke 2001). The first attempts to treat type 1 Gaucher disease by infusions with glucocerebrosidase isolated from human placenta were already started in the early 1970s at the National Institutes of Health. Unfortunately, these did not result in an effective therapy for two compelling reasons. In the first place, too little and insufficiently pure glucocerebrosidase could be generated with the existing technology. In the second place, most of

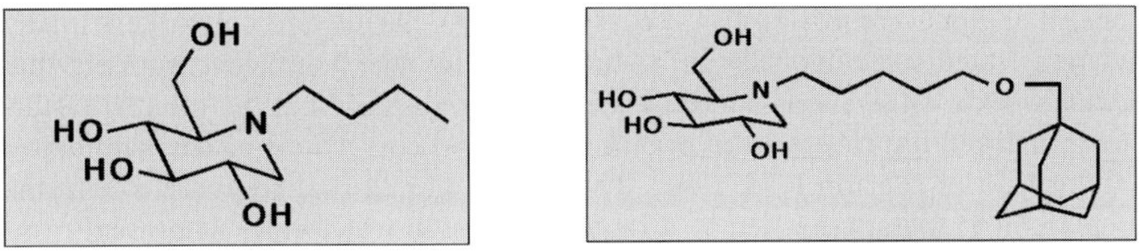

Butyl-DNM
IC_{50} 25000 nM

AMP-DNM
IC_{50} 25 nM

Fig. 1 Structures of deoxynojirimycin-type inhibitors of glycosylceramide synthase. *Butyl-DNM*, butyl-deoxynojirimycin; *AMP-DNM*, adamantane-pentyl-deoxynojirimycin

the administered enzyme was not delivered to macrophages but to other cell types such as hepatocytes. The final development of an effective enzyme replacement therapy for type 1 Gaucher disease relied on a fortunate intersection of scientific disciplines: the discovery of receptors for glycoproteins and the complete purification of glucocerebrosidase. Purification of the protein to homogeneity was achieved in 1977 and subsequently isolation procedures were markedly improved (Murray et al. 1985; Aerts et al. 1986). In 1974, the first mammalian cell lectin, the asialoglycoprotein receptor, was described and next a mannose-specific lectin on Kupffer cells in the liver was identified (Ashwell and Morell 1974). The mannose receptor was shown to interact avidly with mannose-terminal glycoconjugates and mediate their delivery into lysosomes (Stahl et al. 1978). It was realised by Barranger, Brady and co-workers that this receptor-mediated uptake mechanism could be exploited for therapy of Gaucher disease. Analysis of the carbohydrate composition of placental glucocerebrosidase showed the presence of three complex-type glycans and a single high mannose-type glycan per molecule (Takasaki 1984). The presence of terminal galactose moieties in the glycans of placental glucocerebrosidase provided an explanation for its undesired preferential targeting to hepatocytes. To increase the amount of terminal mannose moieties in placental glucocerebrosidase, an in vitro method based on sequential enzymatic removal of *N*-acetylneuraminic acid, galactose and *N*-acetylglucosamine moieties with exoglycosidases was developed (Furbish et al. 1981) (see Fig. 1). The modified 'mannose-terminated' glucocerebrosidase remained fully enzymatically active. It has been further demonstrated that a similar mannose-terminated form of the enzyme is generated by sequential action of lysosomal exoglycosidases during maturation of endogenous glucocerebrosidase in lysosomes of human fibroblasts (Van Weely et al. 1990). Animal studies with the mannose-terminated placental glucocerebrosidase revealed that the enzyme was delivered differentially to Kupffer cells in comparison with hepatocytes (Furbish et al. 1981). Upon treating a 5-year-old Ashkenazi Jewish boy with the modified placental enzyme, Barranger and co-workers noted promising clinical improvements. In subsequent years the involvement of industry (Genzyme Corporation, Boston, USA) was required to produce sufficient enzyme for further clinical studies with mannose-terminated placental glucocerebrosidase (Ceredase). In 1990, Barton and co-workers finally demonstrated unequivocally in a study with 12 type 1 Gaucher patients that twice-weekly intravenous administration of Ceredase (130 IU/kg/month) resulted in marked improvement in organomegaly and corrections of haematological abnormalities (Barton et al. 1991). The spectacular clinical response to enzyme replacement therapy has led to a rapid application worldwide. At present close to 3,000 type 1 Gaucher patients benefit from therapeutic intervention with Cerezyme, the recombinant form of glucocerebrosidase that has superseded the placenta-derived Ceredase (Grabowski et al. 1995).

The introduction of Ceredase was associated with considerable controversy regarding optimal dosing regimens, further stimulated by concerns regarding the safety of the incompletely pure placental enzyme preparation and the extreme costs for treatment of adult patients (US $50,000–$500,000 per patient per year). The availability of pure recombinant glucocerebrosidase and clinical investigations on optimal individualised dosing regimens have resolved most of the debate (Hollak et al. 1995). However, still at this moment little is known about the true efficacy of targeting of mannose-terminated glucocerebrosidase to macrophages or Gaucher cells. Investigations in rats have revealed that a major fraction of Ceredase is not delivered to macrophages but endocytosed by liver endothelial cells (Bijsterbosch et al. 1996). This is not unexpected, since it had been earlier demonstrated that the mannose-receptor is also expressed on these cells (see Linehan et al. 1999). Although elegant studies with radiolabelled enzyme in volunteers have been conducted by Mistry, it remains an unanswered question to which cells precisely mannose-terminated glucocerebrosidase is delivered in Gaucher patients (Mistry et al. 1996).

Systemically administered glucocerebrosidase, a glycoprotein of about 52 kDa, is unable to pass the blood–brain barrier. The outcome of enzyme replacement therapy for acute neuronopathic (type 2) and severe forms of chronic neuronopathic (type 3) Gaucher disease is disappointing (Erikson 2001). Several clinical investigations have revealed that in the severe neuronopathic Gaucher patients, the effects of enzyme replacement therapy on visceral and haematological symptoms are good, but the fatal neurological deterioration continues. Accumulation of glucocerebroside and its metabolite glucosylsphingosine inside the brain underlies the severe neuropathology of these patients. Importantly, milder forms of type 3 Gaucher disease, where the chronic neuronopathic disease is primarily caused by perivascular storage cells, respond well to enzyme replacement therapy; and treatment with a high-dose enzyme regimen is recommended by the European Working Group on Gaucher Disease (Vellodi et al. 2001). Perivascular macrophages in the brain are known to express mannose receptor (Linehan et al. 1999).

3.2 Substrate Deprivation Therapy

An alternative approach for therapeutic intervention of type 1 Gaucher and other glycosphingolipidoses is substrate deprivation (also termed substrate reduction) therapy. Radin and coworkers firstly formulated the challenging concept (see for a review Radin 1996). The approach aims to reduce the rate of glycosphingolipid biosynthesis to levels which match the impaired catabolism. It is conceived that patients who have a significant residual lysosomal enzyme activity could gradually clear lysosomal storage material and therefore should profit most from reduction of substrate biosynthesis.

Two main classes of inhibitors of glycosphingolipid biosynthesis have been described, both of which inhibit the ceramide-specific glucosyltransferase (also termed glucosylceramide synthase; GlcT-1; UDP-glucose: *N*-acylsphingosine d-glucosyl-transferase, EC 2.4.1.80). The enzyme catalyses the transfer of glucose to ceramide, the first step of the biosynthesis of glucosphingolipids. The first class of inhibitors is formed by analogues of ceramide. The prototype inhibitor is PDMP (d, l-threo-1-phenyl-2-decanoylamino-3-morpholino-1-propanol). More specific and potent analogues have been subsequently developed based on substituting the morpholino group for a pyrrolidino function and by substitutions at the phenyl group: 4-hydroxy-1-phenyl-2-palmitoylamino-3-pyrrolidono-1-propanol (p-OH-P4) and ethylenedioxy-1-phenyl-palmitoylamino-3-pyrrolidino-1-propanol (EtDo-P4) (Lee et al. 1999). Studies in a knock-out mouse model for Fabry disease have shown that oral administration of the compounds can result in a marked reduction of the accumulating glycosphingolipid globotriaosylceramide (Abe et al. 2000). The second class of inhibitors of glucosylceramide synthase is formed by *N*-alkylated iminosugars. Such compounds were already in common use as inhibitors of N-glycan processing enzymes, and the potential application of *N*-butyl-deoxynojirimycin (NB-DNJ) as HIV inhibitor had been studied in AIDS patients. Platt and Butters at the Glycobiology Institute in Oxford were the first to recognise the ability of NB-DNJ to inhibit glycosylceramide synthesis at low micromolar concentrations (Platt and Butters 1995). The same researchers demonstrated in knock-out mouse models of Tay-Sachs disease and Sandhoff disease significant reductions in glycosphingolipid storage in the brain (Jeyakumar et al. 1999). Preclinical studies in animals and the previous clinical trial in AIDS patients have indicated (transient) adverse effects in the gastrointestinal tract, probably related to the ability of NB-DNJ to inhibit disaccharidases on the intestinal brush border. Animal studies have shown that the galactose analogue *N*-butyl-deoxygalactonojirimycin (NB-DGJ) may have the same therapeutic efficacy as NB-DNJ but does not cause gastrointestinal side-effects (Andersson et al. 2000). Overkleeft and coworkers in their search for inhibitors of glucosidases have serendipitously developed a more potent inhibitor of glucosylceramide synthase. Adamantane-pentyl-deoxynojirimycin (AMP-DNM) was found to inhibit glycosphingolipid biosynthesis at low nanomolar concentrations (Overkleeft et al. 1998) and able to prevent globotriaosylceramide accumulation in a Fabry knock-out mouse model without overt side-effects (D. Copeland, personal communication).

The first clinical study of the use of NB-DNJ to treat a glycosphingolipid storage disorder was reported recently (Cox et al. 2000). In an open-label phase I/II trial, 28 adult type 1 Gaucher patients received three times daily 100 mg NB-DNJ (OGT918; Oxford GlycoSciences). Improvements in visceromegaly and haematological abnormalities as well as corrections in plasma levels of glucosylceramide and biomarkers of Gaucher disease activity have been described, although the extent of the response is less spectacular than generally observed with high-dose enzyme replacement therapy. Provided that iminosugars or other inhibitors of glucosylceramide synthase prove to be safe in the long term, they will have an important role to play in the management of glycosphingolipid storage disorders, including Gaucher disease.

3.3 Gene Therapy

Since tissue macrophages are derived from bone marrow, it is logical that curative bone marrow transplantations have been reported for some patients with Gaucher disease (Ringden et al. 1995). The risks of allogeneic transplantation, however, do not justify this approach in patients with milder forms of the disease. The observed efficacy of enzyme replacement therapy and bone marrow transplantation has stimulated the pursuit of gene therapy for Gaucher disease. Three independent studies of gene transfer to the haematopoietic cells of Gaucher patients have been conducted but none produced encouraging results (Richter and Karlsson 2001). Low transduction efficiencies of CD34 cells and no sustained expression of glucocerebrosidase in white blood cells have contributed to this. The development of gene therapy strategies to correct haematological and genetic disorders has been hampered by the low levels of gene transfer into human stem cells using vectors derived from oncoretroviruses. Much interest has been recently focused on vectors derived from lentiviruses that have been shown to transduce a variety of nondividing cells, including haematopoietic cells (Richter and Karlsson 2001). The use of such vectors and new developments with respect to macrophage-specific gene targeting (see Chap. 6, this volume) may open novel possibilities for effective gene therapy of Gaucher disease in the future.

4 Monitoring of Therapeutic Correction

Considerable attention has been paid in relation to type 1 Gaucher disease to treatment goals and the monitoring of response to therapeutic interventions (Cox 2001; Hollak et al. 2001). The definition of treatment goals has to depend on clinical endpoints or surrogate endpoints that can predict clinical benefit based on epidemiological, pathophysiological or other scientific evidence. In

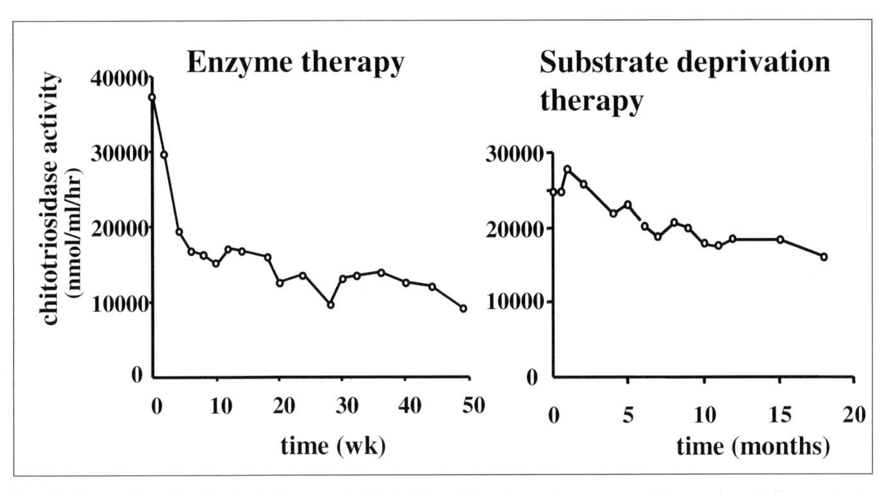

Fig. 2 Corrections in elevated plasma chitotriosidase following therapeutic intervention. *Left panel*: Response of the first type 1 Gaucher patient treated in continental Europe by intravenous administration of Ceredase (48 IU/kg/month). *Right panel*: Response of first type 1 Gaucher disease patient treated in continental Europe by oral administration of butyl-deoxynojirimycin (300 mg/day)

view of the burden imposed by chronic intravenous infusions and the high costs associated with enzyme therapy as well as the uncertainty regarding dose-dependent, long-term adverse effects of iminosugar therapy, it seems wise to establish for the individual Gaucher patient the minimal dose of drug required for effective intervention. In severely affected patients, the initial response to therapy can be accurately assessed by determination of spleen and liver volumes, haemoglobin level and platelet count. During maintenance therapy, however, these clinical parameters are of little value. Monitoring the effect of therapy on bone disease is complicated and has usually been restricted to documentation of the occurrence of bone crises, pathological fractures or the need for surgical intervention. More recently, quantitative chemical shift imaging (QCSI) has been applied to study the triglyceride content of lumbar bone marrow (Hollak et al. 2001). The fat fraction of the bone marrow is variably reduced in Gaucher disease due to displacement of normal triglyceride-rich adipocytes by Gaucher cells. It has been noted that a marked reduction in bone marrow fat fraction is predictive for the occurrence of bone complications. A marked correction in bone marrow fat content following therapy can be therefore defined as a treatment goal.

A search for plasma abnormalities in Gaucher disease has led to the discovery of a marked elevation in chitotriosidase, a hitherto unknown human chitinase (Hollak et al. 1994). In symptomatic Gaucher patients, plasma chitotriosidase levels were found to be about 1,000-fold higher than in normal individuals. It has been subsequently shown that Gaucher cells are the source of this hydrolase in plasma and that the elevated levels are an indicator of the burden of storage cells in a patient. Chitotriosidase is synthesised in the pathological macrophages, and its elevated activity correlates with tissue glucosylceramide storage as well as clinical parameters of disease severity. Enzyme replacement therapy, substrate deprivation therapy or bone marrow transplantation rapidly reduces the plasma chitotriosidase activity (see Fig. 2). To assess the utility of chitotriosidase activity measurements as a biomarker for treatment efficacy, the relationship between and clinical parameters has been studied (Hollak et al. 2001). On the basis of this investigation, it has been proposed that in patients in whom initiation of treatment is questionable, based solely on clinical parameters, a chitotriosidase activity above 15,000 nmol/ml/h may serve as an indicator of a high Gaucher cell burden and an indication for the initiation of treatment. A reduction of less than 15% after 1 year of treatment should be a reason to consider a dose increase. Furthermore, a sustained increase in chitotriosidase at any point during treatment should alert the physician to the possibility of clinical deterioration and the need for dose adjustment. The assay of chitotriosidase activity is complicated by the existence of apparent substrate inhibition due to transglycosidase activity (J.M. Aerts, manuscript in preparation). Another pitfall results from the complete absence of the enzymatic activity in about 6% of all individuals. This results from homozygosity for a null allele of the chitotriosidase gene (Boot et al. 1998). Plasma chitotriosidase levels in heterozygotes for this mutation (about 35% of all individuals) can lead to an underestimation of the actual presence of Gaucher cells in patients. Determination of chitotriosidase genotype in Gaucher patients is therefore recommended.

Chitotriosidase has been characterised in detail at the gene and protein level (Boot et al. 1995, 1998; Renkema et al. 1995, 1997). The enzyme mimics lysozyme in several aspects. It is also selectively expressed in phagocytes, particularly in chronically activated macrophages, and likewise is a compact globular endoglucosaminidase lacking N-linked glycans. The physiological role of chitotriosidase seems to be found also in innate immunity. It has been observed in studies with *Candida albicans* and *Aspergillus fumigatus* that the enzyme exerts a potent fungistatic effect by selective lysis of the growth tip of hyphae. The molecular basis for the massive overexpression of chitotriosidase in Gaucher cells and in related foam cells observed in arteriosclerosis, sarcoidosis, Wolman disease and Niemann-Pick disease is still unknown and the subject of ongoing investigation.

5 Prospects

In the last decade enormous progress has been made in therapy of type 1 Gaucher disease, a severely debilitating disorder characterised by intralysosomal storage of glucocerebroside in tissue macrophages. A highly effective therapy based on chronic intravenous administration of mannose-terminated recombinant human glucocerebrosidase is available. During the past decade this therapy has been applied in several thousand patients without serious adverse effect. Moreover, for the same orphan disease, promising clinical responses have been observed upon oral administration of an iminosugar inhibitor of glucosylceramide synthesis. Provided that long-term treatment with such inhibitors is without adverse effects, substrate deprivation therapy (in conjunction with enzyme replacement therapy) may play an important role in the future clinical management of patients suffering from glycosphingolipid storage disorders. Progress in vector technology and selective expression of the transgene in macrophages seem to be essential requirements before gene therapy can fulfil its promise as cure for type 1 Gaucher disease.

Despite the success of the present enzyme replacement therapy with Cerezyme, the question should be raised whether the enzyme supplementation treatment can be further improved in order to be more economic and widely available. For example, it is unclear which percentage of the mannose-terminated Cerezyme is actually endocytosed by tissue macrophages and storage cells and which percentage is 'wasted' in other cell types such as liver endothelial cells. The occurrence and consequences of binding of the therapeutic enzyme to receptors other than the mannose receptor, to soluble receptor fragments like soluble mannose receptor (sMR), or to serum mannose-binding lectins still warrants further examination. Little attention has so far been paid to the expression of the mannose receptor on macrophages and other cells types of Gaucher patients. Increased knowledge about this matter may give valuable clues for further improvement of the current enzyme replacement therapy. Similar considerations can be made with respect to substrate deprivation therapy. In the case of type 1 Gaucher disease, one would prefer to inhibit selectively the synthesis of glucosylceramide in blood cells. More selective targeting of drugs to blood cells might therefore result in major improvement of efficacy and reduce the risk for side-effects.

It seems likely that Gaucher disease will also serve in the future as an interesting and challenging model for developing new or improved therapy modalities for the correction of lipid-laden tissue macrophages.

6 References

Abe A, Gregory S, Lee L, Killen PD, Brady RO, Kulkarni A, Shayman JA (2000) Reduction of globotriaosylceramide in Fabry disease mice by substrate deprivation. J Clin Invest 105, 1563–1567

Aerts JM, van Weely S, Boot R, Hollak CE, Tager JM (1993) Pathogenesis of lysosomal storage disorders as illustrated by Gaucher disease. J Inherit Met Dis 16, 288–291

Aerts JMFG, Donker-Koopman WE, Murray GJ, Barranger JA, Tager JM, Schram AW (1986) A procedure for the rapid purification in high yield of human glucocerebrosidase using immunoaffinity chromatography with monoclonal antibodies. Anal Biochem 154, 655–663

Aerts JMFG, Schram AW, Strijland A, van Weely S, Jonsson LMV, Tager JM, Sorrell SH, Ginns EI, Barranger JA, Murray GJ (1988) Glucocerebrosidase, a lysosomal enzyme that does not undergo oligosaccharide phosphorylation. Biochem Biophys Acta 964, 303–308

Aerts JMFG, Hollak CEM (1997) Plasma and metabolic abnormalities in Gaucher's disease. Baillieres Clin Haematol 10, 691–709

Andersson U, Butters TD, Dwek RA, Platt FM (2000) N-butyldeoxygalactonojirimycin: a more selective inhibitor of glycosphingolipid biosynthesis than N-butyldeoxynojirimycin, in vitro and in vivo. Biochem Pharmacol 49, 821–829

Ashwell G, Morell AG (1974) The role of surface carbohydrates in the hepatic recognition and transport of circulating glycoproteins In: Wood WA (ed) Advances in Enzymology, vol. 41. Academic Press, New York, p 99–128

Barranger JA, Ginns EI (1989) Glycosylceramide lipidosis: Gaucher disease. In: Scriver CR, Beaudet AL, Sly WS, Vall D (eds) The metabolic basis of inherited disease. McGraw-Hill, New York, p 1677–1698

Barranger JA, O'Rourke E (2001) Lessons learned from the development of enzyme therapy for Gaucher disease J Inherit Met Dis 24 (suppl 2), 89–96

Barton NW, Brady RO, Dambrosia JM, Di Bisceglie AM, Doppelt SH, Hill SC, Mankin HJ, Murray GJ, Parker RI, Argoff CE, Grewal RP, Yu K-T (1991) Replacement therapy for inherited enzyme deficiency: macrophage-targeted glucocerebrosidase for Gaucher's disease. N Eng J Med 324, 1464–1470

Beutler E, Grabowski GA (1995) Gaucher's disease. In: Scriver CR, Beaudet AL, Sly WS, Valle D (eds) The metabolic and molecular bases of inherited disease. McGraw-Hill, New York, p 2641–2670

Boot RG, Renkema GH, Strijland A, van Zonneveld AJ, Muysers AO, Aerts JMFG. (1995) Cloning of cDNA encoding chitotriosidase, a human chitinase produced by macrophages. J Biol Chem 270, 26252–26256

Boot RG, Renkema GH, Verhoek M, Strijland A, Bliek J, de Meulemeester TMAMO, Mannens MMAM, Aerts JMFG (1998) The human chitotriosidae gene. Nature of inherited enzyme deficiency. J Biol Chem 273, 25680–25685

Brady RO, Kanfer JN, Bradley RM, Shapiro D (1966) Demonstration of a deficiency of glucocerebroside-cleaving enzyme in Gaucher's disease. J Clin Invest 45, 1112–1115

Brady RO (1997) Gaucher's disease: past, present and future. Baillieres Clin Haematol 10, 621–634

Bijsterbosch MK, Donker W, van de Bilt H, van Weely S, van Berkel TJ, Aerts JM (1996) Quantitative analysis of the targeting of mannose-terminal glucocerebrosidase. Predominant uptake by liver endothelial cells. Eur J Biochem 237, 344–349

Cox TM, Schofield JP (1997) Gaucher's disease: clinical features and natural history Baillieres Clin.Hematol.10, 657–689

Cox TM (2001) Gaucher disease: understanding the molecular pathogenesis of sphingolipidoses. J Inherit Met Dis 24 (suppl 2), 106–121

De Duve C, Pressman BC, Gianetto R, Wattiaux R, Appelmans F (1955) Tissue fractionation studies. Intracellular distribution patterns of enzymes in rat liver tissue. Biochem J 60, 604–617

Dice JF, Terlecky SR, Chaing HL, Olsen TS, Isenman LD, Short-Russel SR, Freundlieb S, Terlecky LJ (1990) A selective pathway for degradation of cytosolic proteins by lysosomes. Sem Cell Biol 1, 449–455

Erikson A (2001) Remaining problems in the management of patient with Gaucher disease. J Inherit Met Dis 24 (suppl 2),122–126

Furbish FS, Steer CJ, Krett NL, Barranger JA (1981) Uptake and distribution of placental glucocerebrosidase in rat hepatic cells and effects of sequential deglycosylation. Biochim Biophys Acta 673, 425–434

Grabowski GA, Barton NW, Pastores G, Dambrosia JM, Banerjee TK, McKe MA, Parker C, Schifmann R, Hill SC, Brady RO (1995) Enzyme therapy in type 1 Gaucher disease: comparitive efficacy of mannose-terminated glucocerebrosidase from natural and recombinant sources. Ann Intern Med 122, 33–39

Helenius A (1994) How N-linked oligosaccharides affect glycoprotein folding in the endoplasmic reticulum. Mol Biol Cell 5, 253–265

Hollak CEM, van Weely S, van Oers MHJ, Aerts JMFG (1994) Marked elevation of plasma chitotriosidase activity. A novel hallmark of Gaucher disease. J Clin Invest 93, 1288–1292

Hollak CEM, Aerts JMFG, Goudsmit ER, Phoa SS, Ek M. van Weely S, von dem Borne AE, van Oers MH (1995) Individualised low-dose alglucerase therapy for type 1 Gaucher's disease. Lancet 345, 1474–1478

Hollak CEM, Aerts JMFG (2001) Clinically relevant therapeutic endpoints in type 1 Gaucher disease. J Inherit Met Dis 24 (suppl 2), 97–105

Holzmann E (1989) Lysosomes. Plenum Press, New York

Jeyakumar M, Butters TD, Cortina-Borja M, Hunnam V, Proia RL, Perry VH, Dwek RA, Platt FM (1999) Delayed symptom onset and increased life expectancy in Sandhoff mice treated with N-butyl-deoxynojirimycin. Proc Natl Acad Sci USA 96, 6388–6393

Jonsson LMV, Murray GJ, Sorrell SH, Strijland A, Aerts JMFG, Ginns EI, Barranger JA, Tager JM, Schram AW (1987) Biosynthesis and maturation of glucocerebrosidase in Gaucher fibroblasts. Eur J Biochem 164,171–179

Kornfeld S, Mellman I (1989) The biogenesis of lysosomes. Ann Rev Cell Biol. 5, 483–525

Lee L, Abe A, Shayman JA (1999) J Biol Chem 274, 146662–14665 Improved inhibitors of glucosylceramide synthase.

Linehan SA, Martinez-Pomares L, Stahl PD, Gordon S (1999) Mannose receptor and its putative ligands in normal murine lymphoid and nonlymphoid organs. In situ expression of mannose receptor by selected macrophages, endothelial cells, perivascular microglia and mesanglial cells, but not dendritic cells. J Exp Med 189, 1961–1972

Meikle PF, Hopwood JJ, Clague AE, Carety WF (1999) Prevalence of lysosomal storage disorders. JAMA 281, 249–254

Mistry PK, Wraight EP, Cox TM (1996) Therapeutic delivery of proteins to macrophages: implications for treatment of Gaucher's disease. Lancet 348, 1555–1559

Moran MT, Schofield JP, Hayman AR, Shi G-P, Young E, Cox TM (2000) Pathologic gene expression in Gaucher disease: upregulation of cysteine proteinases including osteoclastic catepsin K. Blood 96, 1969–1978

Murray GJ, Youle RJ, Gandy SE, Zirzow GC, Barranger JA (1985) Purification of beta-glucocerebrosidase by preparative-scale high-performance liquid chromatography. The use of ethylene glycol-containing buffers for chromatography of hydrophobic glycoprotein enzymes. Anal Biochem 147, 301–310

Neufeld EF (1991) Lysosomal storage disorders. Annu Rev Biochem. 60, 257–280

Overkleeft HS, Renkema GH, Neele J, Vianello P, Hung IO, Strijland A, van den Burg A, Koomen, GJ, Pandit UK, Aerts J (1998) Generation of specific deoxynijirimycin-type inhibitors of the non-lysosomal glucosylceramidase. J Biol Chem 273, 26522–26527

Owada M, Neufeld EF (1982) Is there a mechanism for introducing acid hydrolases into liver lysosomes that is independent of mannose-6-phosphate recognition? Evidence from I-cell disease. Biochem Biophys Res Commun 105, 814–820

Patrick AD (1965) A deficiency of glucocerebrosidase in Gaucher's disease. Biochem J 97, 17c-18c

Peltonen L (1997) Molecular background of Finnish disease heritage. Ann Medicine 29, 553–556

Peters C, von Figura K (1994) Biogenesis of lysosomal membranes. FEBS Lett 346, 108–114

Platt FM, Neises GR, Dwek RA, Butters TD (1994) N-butyl-deoxynojirimycin is a novel inhibitor of glycolipid biosynthesis. J Biol Chem 269, 8362–8365

Poorthuis BJHM, Wevers RA, Kleijer WJ, Groener JEM, de Jong JGN, van Weely S, Niezen-

Koning KE, van Doggelen OP (1999) The frequency of lysosomal storage diseases in The Netherlands. Human Genet 105, 151–156

Radin NS (1996) Treatment of Gaucher disease with an enzyme inhibitor. Glycoconj J 13, 153–157

Renkema GH, Boot RG, Muysers AO, Donker-Koopman W, Aerts, JMFG. (1995) Purification and characterisation of human chitotriosidase, a novel member of the chitinase family of proteins. J Biol Chem 270, 2198–2202

Renkema GH, Boot RG, Strijland A, Donker-Koopman W, van den Berg M, Muysers AO, Aerts JMFG (1997) Synthesis, sorting and processing into distinct isoforms of human macrophage chitotriosidase. Eur J Biochem 244, 279–285.

Richter J, Karlsson S (2001) Clinical gene therapy in hematology: past and future. Int. J Hematol 73, 162–169

Rijnboutt S, Aerts JMFG, Geuze HJ, Tager JM, Strous GJ (1991) Mannose-6-phosphate independent membrane association and maturation of cathepsin D, glucocerebrosidase and sphingolipid-activating protein in HepG2 cells. J Biol Chem 266, 4862–4868

Ringden O, Groth CG, Erikson A, Granqvist S, Mansson JE, Sparrelid E (1995) Ten years' experience of bone marrow transplanatation for Gaucher disease. Transplantation 56, 864–870.

Stahl PD, Rodman JS, Miller MJ, Schlesinger PH (1978) Evidence for receptor-mediated binding of glycoproteins and lysosomal glycosidases by alveolar macrophages. Proc Natl Acad Sci USA 75, 1399–1430

Takasaki S, Murray GJ, Furbish FS, Brady RO, Barranger JA, Kobata A (1984) Structure of the N-asparagine linked oligosaccharide units of human placental glucocerebrosidase. J Biol Chem 259, 10112–10117

Vellodi A, Bembi B, de Villemeur TB, Collin-Histed T, Erikson A, Mengel E, Rolfs A, Tylki-Szymanska A (2001) Management of neuronopathic Gaucher disease: a European consensus J Inherit Metab Dis 24, 319–327

Van Weely S, Aerts, van Leeuwen MB, Heikoop JC, Donker-Koopman WE, Barranger JA, Tager JM, Schram AW (1990) Function of oligosaccharide modification in glucocerebrosidase, a membrane-associated lysosomal hydrolase. Eur J Biochem 191, 669–677

Van Weely S, van den Berg M, Barranger JA, Sa Miranda MC, Tager JM, Aerts JMFG (1993) Role of pH in determining the cell-type specific residual activity of glucocerebrosidase in type 1 Gaucher disease. J Clin Invest 91, 1167–1175

Nuclear Receptors as Regulators of Macrophage Homeostasis and Function

J. S. Welch · C. K. Glass

Department of Cellular and Molecular Medicine, Medical Scientist Training Program,
University of California, San Diego,
9500 Gilman Drive, La Jolla, CA 92093–0651, USA
e-mail: cglass@ucsd.edu

Abstract Nuclear hormone receptors comprise a superfamily of ligand-dependent transcription factors that regulate diverse aspects of development and homeostasis. Several members of this superfamily play important roles in the regulation of inflammatory responses and lipid homeostasis in macrophages. These include the glucocorticoid receptor, which acts to inhibit inflammatory programs of gene expression in response to natural corticosteroids and synthetic anti-inflammatory agents such as dexamethasone, peroxisome proliferator-activated receptors (PPARs) that regulate fatty acid homeostasis and inflammation in response to endogenous eicosanoids, and liver X receptors (LXRs) that regulate cholesterol efflux in response to endogenous oxysterols. Recent progress in defining the physiological roles of these receptor systems in macrophages and understanding their mechanisms of action suggest that they may be important targets for the development of new classes of pharmaceuticals that will be useful for treating human diseases in which macrophages play critical pathogenic roles, such as atherosclerosis and arthritis.

Keywords Atherosclerosis, Cholesterol, Foam cell, Glucocorticoid receptor, Inflammation, Liver X receptor, NF?B, Nuclear receptor, Peroxisome proliferator-activated receptor, Transcription, Vitamin D receptor

1 Introduction

Nuclear receptors comprise a superfamily of ligand-dependent transcription factors that regulate diverse aspects of development, homeostasis, and immune function (Evans 1988; Kastner et al. 1995; Mangelsdorf and Evans 1995; Chawla et al. 2001). Forty-eight distinct genes encoding nuclear receptors have been identified in humans and mice (Maglich et al. 2001). Several members of the nuclear receptor superfamily have been shown to play important physiologic roles in macrophages. This chapter will focus on the most extensively characterized of these: the glucocorticoid receptor α (GRα), estrogen receptor α (ERα), vitamin D receptor (VDR), retinoid X receptor α (RXRα), peroxisome proliferator-activated receptors α and γ (PPARα and γ), and the liver X receptors α and β (LXRα and LXRβ). Each of these receptors are regulated by small molecular weight ligands and are well-established or emerging targets of drugs that are used to treat human diseases in which macrophages play prominent pathogenic roles. Recent advances in the understanding of nuclear receptor biology and function raise the prospect that new generations of nuclear receptor ligands can be developed with improved therapeutic activities.

1.1 Domain Structure

Nuclear receptors share conserved modular domains that are illustrated in Fig. 1. These include a variable N-terminal activation domain (AF1), a highly conserved DNA binding domain (DBD), and a C-terminal ligand-binding domain (LBD) (Evans 1988). The DBD consists of two interdependent zinc fingers that mediate specific DNA binding of nuclear receptor monomers, dimers, and heterodimers to hormone response elements (HREs) in direct target genes. Crystal structures of several nuclear receptor DBDs bound to their respective HREs indicate that the DBD provides both specific and non-specific DNA interactions that allow for general DNA binding and recognition of receptor-specific sequences (Luisi et al. 1991; Rastinejad et al. 1995). Most hormone response elements contain two or more closely spaced core recognition motifs, each of which is contacted by a single DBD. Steroid hormone receptors recognize palindromic inverted repeats as homodimers. An important subset of nuclear receptors that includes the retinoic acid receptors, PPARs and LXRs bind to DNA as heterodimers with a common partner, RXR (Yu et al. 1991; Kliewer et al. 1992; Chawla et al. 2001). In contrast to steroid hormone receptors, RXR heterodimers generally bind to target genes in the presence or absence of ligand. In

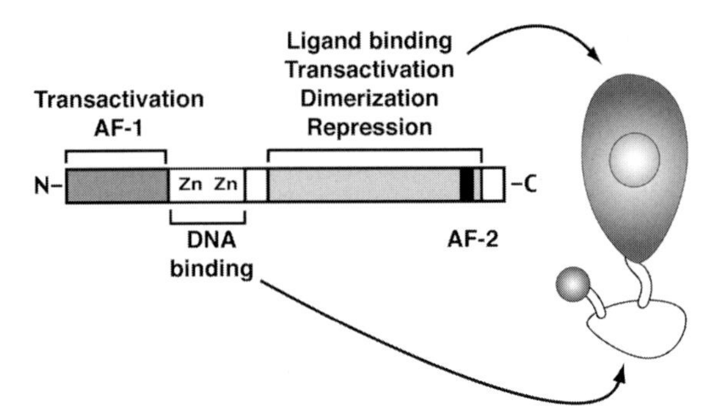

Fig. 1 Domain structure of nuclear receptors. Nuclear receptors exhibit a common modular domain structure. The N-terminus encodes a variable activation domain (*AF-1*) that is particularly important for the function of steroid hormone receptors. The DNA-binding domain is highly conserved and mediates specific interaction with a 6-bp core recognition sequence in hormone response elements. The ligand-binding domain determines the specific ligand-binding properties of each nuclear receptor and undergoes ligand-dependent conformational changes that control coactivator, corepressor, and heat shock protein interactions

addition, RXR heterodimers prefer HREs that consist of direct repeat organizations of the core recognition motifs rather than the palindromic HREs recognized by steroid hormone receptors (Umesono and Evans 1989; Näär et al. 1991; Umesono et al. 1991). Protein–protein interaction surfaces within the DBD determine the optimum spacing and orientation of core recognition motifs for each nuclear receptor dimer or heterodimer, and thereby play important roles in determining specificity of HRE recognition (Rastinejad et al. 1995).

The ligand-binding domain integrates several nuclear receptor functions. In addition to binding hormone, the LBD plays roles in dimerization/heterodimerization, subcellular localization, and ligand-dependent transcriptional activation and repression. In the case of steroid hormone receptors, the LBD also mediates interactions of unliganded receptors with heat shock protein complexes (Fig. 2). Hormone binding results in dissociation of steroid receptors from these complexes and acquisition of transcriptional activity (Evans 1988). Within the nucleus, the major transcriptional role of the LBD is to mediate interactions with coactivator and corepressor proteins in a ligand-dependent manner. This is thought to occur primarily through ligand-dependent conformational changes in an alpha helical region in the C-terminus referred to as activation function 2 (AF2). Crystal structures of a number of nuclear receptor LBDs in the absence and presence of cognate ligands suggest that ligand binding causes the AF2 helix to go from an extended or relatively mobile position to an "active" position, in which it is tightly bound to the LBD and in some cases interacts directly with the ligand (Bourguet et al. 1995; Renaud et al. 1995; Wagner et al. 1995; Moras and Gronemeyer 1998). This structural shift creates a "charge clamp" that interacts with a short helical motif in nuclear receptor coactivators that contains the consensus LxxLL, where L is leucine and x is any amino acid (Heery et al. 1997; Torchia et al. 1997). Because the position of the AF2 domain relative to the LBD

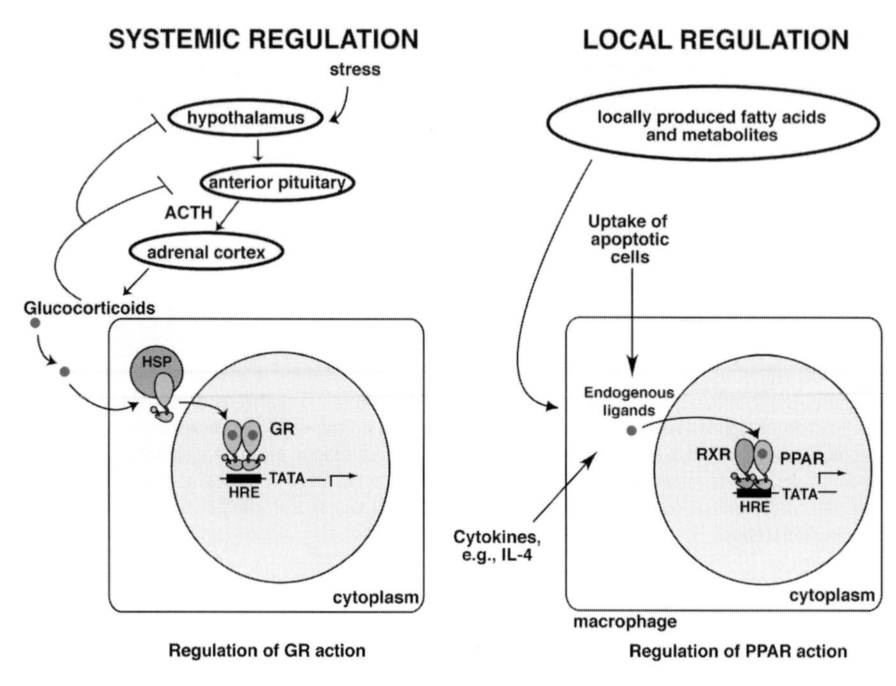

Fig. 2 Systemic and local regulation of macrophage gene expression by nuclear receptors. Classical steroid hormone receptors, such as the glucocorticoid receptor, regulate gene expression in response to circulating hormones that are produced under the control of the hypothalamic/pituitary axis. Adopted orphan receptors, such as PPARs and LXRs (not shown), regulate macrophage gene expression in response to intracellular metabolites of fatty acids and cholesterol, respectively. The production of these ligands may be influenced by local cytokines and other classes of signaling molecules

is essential for coactivator recruitment, ligands that fit in the ligand-binding pocket, but distort the AF2 position, offer the potential to act as antagonists or as selective modulators of nuclear receptor activity (Brzozowski et al. 1997; Shiau et al. 1998). While this principle has been clearly demonstrated for the anti-estrogens raloxifene and raloxifene, it remains to be generalized to the entire nuclear receptor family.

1.2
Functional Classification

The spectrum of nuclear receptors can be subdivided into three sub-families based on their ligand-binding properties and physiological roles (Chawla et al. 2001). The classical steroid/thyroid hormone receptors, exemplified by glucocorticoid and estrogen receptors, define the first and most extensively characterized subfamily (Evans 1988). The so-called orphan receptors define a subfamily at the opposite end of the nuclear receptor spectrum in that they exhibit conserved features of the nuclear receptor superfamily, but have not yet been linked to naturally occurring ligands. A subset of orphan nuclear receptors, such as RORα and Nurr77, have been suggested to play important roles in regulating lymphocyte function and survival (Winoto and Littman 2002), but roles in macrophages remain relatively unexplored. The third sub-family consists of "adopted" orphan receptors. These receptors were initially identified as orphan receptors, with subsequent studies leading to the identification of naturally occurring ligands and physiological roles. Members of this subfamily that have been recently linked to regulation of macrophage gene expression include the PPARs and LXRs (Chawla et al. 2001).

The transcriptional activities of steroid hormone and adopted orphan receptors are regulated by small lipophilic molecules. In the case of the steroid hormone receptors, the regulatory ligands are classic endocrine hormones, produced in glandular tissues in response to systemic physiologic circuits [e.g., the hypothalamic-pituitary-adrenal axis Fig. 2)]. Steroid hormones diffuse into cells and bind to their cognate receptors in target tissues with sub-nanomolar affinities. In contrast, the adopted orphan receptors tend to be activated by metabolites of cholesterol and fatty acids that are produced within the cell and bind with relatively low affinity. Thus, adopted orphan receptors appear to function as effectors of autocrine or paracrine signaling events (Fig. 2). For both steroid hormone and adopted orphan receptors, numerous high-affinity synthetic ligands have been developed, some of which are in widespread clinical use. For example, the synthetic GR agonists dexamethasone and prednisone are extensively utilized as potent anti-inflammatory agents, and the synthetic PPARγ agonist rosiglitazone is used for treatment of type 2 diabetes mellitus.

1.3 Transcriptional Activities of Nuclear Receptors

Nuclear receptors have been shown to regulate transcription by three general mechanisms. The prototypic activity of nuclear receptors is to activate transcription in a ligand-dependent manner following direct binding to DNA response elements in promoter or enhancer regions of target genes (Figs. 2 and 3, *ligand-dependent transactivation*). Ligand-dependent transactivation has been linked to the recruitment of coactivator complexes that modify chromatin structure and facilitate assembly of general transcriptional machinery at the promoter (Glass and Rosenfeld 2000; McKenna and O'Malley 2002). A large number of coactivator complexes have been identified, and it is hypothesized that combinatorial usage of these complexes provides the basis for cell type-specific, gene-specific, and signal-specific transcriptional responses. Nuclear receptors can negatively regulate gene expression by inhibiting the activities of other classes of signal-dependent transcription factors, such as members of the NF-κB and AP-1 families (Jonat et al. 1990; Yang-Yen et al. 1990; Schule et al. 1991; Ray and Prefontaine 1994; Helmberg et al. 1995) (Fig. 3, *ligand-dependent transrepression*). Several mechanisms have been suggested to account for this activity, but unifying principles remain to be elucidated. Third, several of the adopted or-

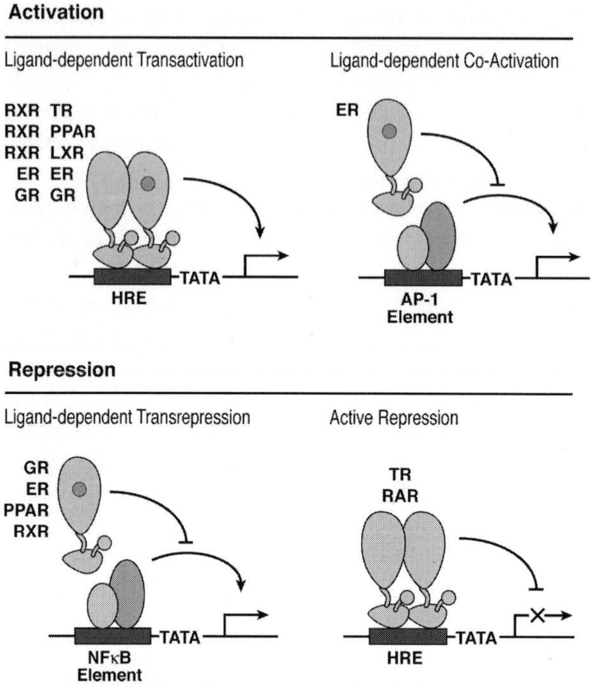

Fig. 3 Transcriptional activities of nuclear receptors. *Ligand-dependent transactivation* is mediated by the binding of nuclear receptor dimers or heterodimers to hormone response elements in target genes. Nuclear receptors can directly repress transcription in the absence of ligand (*ligand-independent repression*) or indirectly inhibit transcription in a ligand-dependent manner by inhibiting the activities of other transcription factors such as nuclear factor (NF)-κB (*ligand-dependent transrepression*). Other mechanisms of action may also exist. For example, the estrogen receptor has been demonstrated to activate transcription from activator protein (AP)-1 elements by serving as a coactivator (*ligand-dependent coactivation*)

phan receptors, exemplified by retinoid acid receptor (RAR) and thyroid receptor (TR), actively repress transcription of direct target genes in the absence of ligands (Fig. 3, *ligand-independent repression*) (Damm et al. 1989; Sap et al. 1989; Datta et al. 1992). This activity has been linked to the recruitment of corepressor complexes that function to antagonize the actions of coactivator complexes (Chen and Evans 1995; Horlein et al. 1995). Not all nuclear receptors exhibit this entire spectrum of transcriptional activities and the possible transcriptional effects of a single nuclear receptor vary in a cell-specific manner. In addition to these general mechanisms of nuclear receptor action, specific nuclear receptors have been demonstrated to regulate transcription by other mechanisms. For example, the estrogen receptor can in some contexts interact with AP-1 proteins and function as a coactivator (Fig. 3, *ligand-dependent coactivation*) (Paech et al. 1997).

2
Nuclear Receptor Functions in Macrophages

Studies of nuclear receptor function in the macrophage suggest three general physiologic roles. One role is the negative regulation of inflammatory responses mediated by AP-1 and NF-κB family members. Emerging evidence suggests that these actions represent important functions of GRα, ERα, and PPARs in the macrophage. A second major role that has emerged involves regulation of lipid homeostasis by LXRs and PPARs. Third, a smaller subset of nuclear receptors, exemplified by the vitamin D receptor, have been found to influence specialized programs of macrophage differentiation, such as osteoclast formation.

While GRα, VDR, RXRα, and LXRβ appear to be constitutively expressed in macrophages, ERα and PPARγ have both been shown to be increased during macrophage differentiation (Ricote et al. 1997; Cutolo et al. 2001). In addition, PPARγ is upregulated by multiple stimuli in the macrophage, including oxidized low-density lipoprotein (LDL), granulocyte macrophage colony-stimulating factor (GM-CSF), macrophage colony-stimulating factor (M-CSF) and interleukin (IL)-4 (Nagy et al. 1998; Ricote et al. 1998; Huang et al. 1999). LXRα expression is positively regulated by its oxysterol ligands, providing a positive feedback mechanism for maintenance of cholesterol homeostasis (Kohro et al. 2000; Laffitte et al. 2001).

2.1
Systemic and Endogenous Ligands

Ligand availability represents one of the most important determinants of nuclear receptor activity. Thus far, the macrophage appears to possess mechanisms for the autocrine or paracrine production of three distinct classes of ligands. First, although estrogen is largely an endocrine hormone regulated by the hypothalamic-gonadal axis, differentiated macrophages express and regulate the enzyme aromatase, capable of converting serum dehydroepiandrosterone (DHEA) into the immunomodulatory steroids estrogen, 3β,17β-androstenediol, and androstenedione (Schmidt et al. 2000). Second, PPARγ can be activated by a variety of fatty acid metabolites and oxidation products. These include the linoleic and arachidonic acid metabolites 13-HODE, 15-HETE and 15-deoxy-$\Delta^{12,14}$ prostaglandin J_2 (Forman et al. 1995; Kliewer et al. 1995; Forman et al. 1997; Kliewer et al. 1997; Nagy et al. 1998). Both 13-HODE and 15-HETE have been identified as components of oxidized LDL, which has itself been shown to upregulate PPARγ expression and induce its transcriptional activity (Nagy et al. 1998; Ricote et al. 1998). In addition, 13-HODE and 15-HETE can be enzymatically generated by the IL-4 inducible 12/15-lipoxygenase, suggesting that IL-4 can coordinately regulate both the expression and activity of PPARγ (Huang et al. 1999). Third, LXRs are activated by cholesterol metabolites (Peet et al. 1998). In the macrophage, the enzyme 27-hydroxylase is able to modify cholesterol into a ligand for both LXRα and LXRβ (Babiker et al. 1997; Fu et al. 2001). Therefore, nuclear receptors provide mechanisms for regulating macrophage gene expression in response to changes in cellular lipid homeostasis and the production of arachidonic acid metabolites that occur during the evolution of inflammatory responses.

2.2 Anti-inflammatory Effects of GRα, ERα, PPARγ, and VDR

The striking ability of GRα agonists to inhibit inflammatory responses is one of the best-documented effects of nuclear receptor ligands on macrophage physiology. Endogenous glucocorticoids are released in response to a variety of stressors (starvation, pain, trauma, infection, etc.) and are essential for maintenance of homeostatic functions (Fig. 2) (Schimmer and Parker 1996). Although in widespread clinical use for the treatment of a variety of inflammatory diseases including rheumatoid arthritis, systemic lupus erythematosus, inflammatory bowel disease, psoriasis, eczema, asthma and transplant rejection, the mechanistic basis of the anti-inflammatory actions of glucocorticoids remains poorly understood (Barnes 1995; McKay and Cidlowski 1999). Accumulating evidence suggests that these effects largely result from inhibition of signal-dependent transcription factors that mediate inflammatory programs of gene activation, particularly NF-κB and AP-1 family members (Herrlich and Ponta 1994; McKay and Cidlowski 1999). NF-κB is a transcription factor that is activated in multiple cell types by inflammatory stimuli such as bacterial lipopolysaccharides (LPS) or interleukins such as IL-1β (Barnes and Karin 1997). Like GRα, classical NF-κB activation involves the removal of a cytoplasmic inhibitor (IκB) and the nuclear translocation of NF-κB dimers or heterodimers to cognate κB elements in the promoter or enhancer regions of target genes. AP-1 transcriptional activity can also be activated by a number of inflammatory stimuli as well as by phorbol esters [e.g., 12-*O*-tetradecanoylphorbol-13-acetate (TPA)]. These general mechanisms allow for the regulated expression of a wide variety of growth factors (GM-CSF), cytokines (IL-1, IL-6, IL-8, TNF-α, MCP-1), and inflammatory mediators [inducible nitric oxide synthase (iNOS), cyclooxygenase (COX)2] in macrophages and other cell types (McKay and Cidlowski 1999). Many of these inflammatory genes can be repressed by GRα activity. Genes that are strongly repressed by GRα agonists include GM-CSF (Adcock and Barnes 1996), TNF-α, (Joyce et al. 1997), IL-1, IL-6, IL-8, IL-12 (Almawi et al. 1996), iNOS (Kleinert et al. 1996; Tanaka and Fujita 1997) and COX2 (Koehler et al. 1990). Despite the requirement for both the GRα, LBD, and DBD for transrepression activity, the majority of the promoters and enhancers for these genes do not contain functional glucocorticoid response elements (GREs) (Caldenhoven et al. 1995; Scheinman et al. 1995; McKay and Cidlowski 1999). These observations suggest that GRα-mediated transrepression of NF-κB and AP-1 involves mechanisms that are distinct from the classical GRα transactivation of target genes (Heck et al. 1997) (Fig. 3). Interestingly, activation of the AP-1 and NF-κB pathways results in reciprocal antagonism of the classical GRα transactivation of HRE-containing promoters (Jonat et al. 1990; Caldenhoven et al. 1995). This has raised the possibility that NF-κB and GRα compete for a limiting supply of co-activator complexes and use similar mechanisms for mutual repression (Kamei et al. 1996; Sheppard et al. 1998; McKay and Cidlowski 2000). Modification of the degree of phosphorylation of the C-terminal repeat of RNA polymerase II at NFκB target genes has also been suggested as the basis for GR-mediated transrepression (Nissen and Yamamoto 2000). Regardless of the mechanistic details, these opposing transcription factors appear to mutually regulate the other's function, maintaining a balance of inflammatory and anti-inflammatory responses in the macrophage and other cells that regulate immune responses, including endothelial cells and lymphocytes.

Like GRα, ERα and PPARγ have also been shown capable of antagonizing the expression of an overlapping set of NF-κB and AP-1 regulated genes in macrophages (Frazier-Jessen and Kovacs 1995; Stein and Yang 1995; Jiang et al. 1998; Ricote et al. 1998). This transrepression function has also been shown to require elements of both the LBD and DBD, but not direct binding to HREs in the enhancer or promoter regions of these genes (Li et al. 2000; Valentine et al. 2000). VDR also inhibits inflammatory gene expression, but has been suggested to exert these effects by inducing the expression of both transforming growth factor (TGF)-β and IL-4 (Deluca and Cantorna 2001). These factors modulate transcriptional programs that evolve during an immune response and act to antagonize the effects of many classical inflammatory signals.

The findings that nuclear receptors possess the ability to modulate inflammatory immune responses have raised the possibility that agonists for ERα, PPARγ, or VDR might have clinical applications as anti-inflammatory drugs. Currently, PPARγ ligands have been shown to ameliorate inflammation in animal models of inflammatory bowel disease (Su et al. 1999; Desreumaux et al. 2001), atherosclerosis (Li et al. 2000; Chen et al. 2001; Claudel et al. 2001; Collins et al. 2001), experimental autoimmune encephalomyelitis (Diab et al. 2002), arthritis (Setoguchi et al. 2001), and psoriasis (Ellis et al. 2000). Likewise, clinical evidence has long been mounting that estrogens play a key role in modulating atherogenesis independent of their effects on lipid metabolism (Reckless et al. 1997; Cushman et al. 2001). In addition, 1,25-dihydroxyvitamin D_3 has been shown to ameliorate experimental autoimmune encephalomyelitis, rheumatoid arthritis, systemic lupus erythematosus, and inflammatory bowel disease (Deluca and Cantorna 2001).

2.3 Regulation of Cholesterol Homeostasis by LXRs

Tight regulation of cellular cholesterol levels is essential for the maintenance of a diverse range of normal cellular functions. In most cells, cholesterol availability is determined by the sum of de novo biosynthesis and uptake from lipoproteins via the LDL receptor (Brown and Goldstein 1986). These processes are regulated by a negative feedback system involving the sterol response element binding proteins (SREBPs), which are transcription factors that stimulate expression of the LDL receptor gene as well as genes involved in cholesterol and fatty acid biosynthesis (Brown and Goldstein 1997). When cellular cholesterol levels become elevated, SREBPs are inactive, leading to decreased cholesterol availability. In addition, elevated cholesterol levels have recently been demonstrated to stimulate cholesterol efflux by inducing the expression of sterol transporters such as ABCA1. Stimulation of efflux pathways by elevated cholesterol levels has been linked to the adopted orphan receptors, LXRα and LXRβ. These receptors bind with RXRs to response elements in target genes such as ABCA1 and are activated by oxysterols that are thought to accumulate in hypercholesterolemic cells (Peet et al. 1998; Chawla et al. 2001). The LXR–ATP-binding cassette (ABC) pathway appears to be particularly important in the macrophage. Mutations in the ABCA1 gene result in Tangier disease, which is characterized by lipid filled macrophages in tissues such as the tonsils, and extremely low levels of circulating high-density lipoproteins (HDL) (Young and Fielding 1999; Tall and Wang 2000). Because one essential function of the macrophage is the phagocytosis and degradation of cholesterol-containing apoptotic and necrotic cells, the macrophage may utilize the LXR-ABC pathway as a critical feed-forward mechanism for disposing of such cellular by-products. This system appears to be overwhelmed or inactivated in foam cells of atherosclerotic lesions, which contain massive amounts of cholesterol derived from modified lipoproteins that are taken up by scavenger receptors. LXR alpha expression is induced in macrophages by PPARγ ligands, which may account for some to the antiatherogenic effects observed in murine models (Chawla et al. 2001).

2.4 Effects of Nuclear Receptors on Macrophage Differentiation and Specialized Functions

One of the most intriguing characteristics of the macrophage lineage is its ability to give rise to a family of related cells that execute specialized roles, such as Kupffer cells, osteoclasts, and microglial cells (Gordon 1995). For example, recent work has found that mature bone resorbing osteoclasts can be induced to differentiate from cells of the monocyte lineage when stimulated with osteoblast products M-CSF and receptor activator of NF-κB ligand (RANKL) (Roodman 1999). This differentiation program can be inhibited by the addition of either an estrogenic ligand (Shevde et al. 2000) or a PPARγ ligand (Bendixen et al. 2001). In contrast, glucocorticoids also decrease bone resorption, but appear to do so by increasing osteoclast apoptosis (Dempster et al. 1997). While mechanisms for these effects are not yet fully elucidated, current models have focused on the inhibition of NF-κB activity, an essential stimulus for osteoclastogenesis.

Like osteoclasts, dendritic cells can be induced to differentiate from macrophages in vitro, but in the presence of different stimuli: GM-CSF and IL-4. This differentiation program can be largely prevented by either the presence of corticosteroids, anti-estrogens (Komi and Lassila 2000) or vitamin D analogs (Griffin et al. 2000; Piemonti et al. 2000). Vitamin D analogs have also been shown to enhance monocyte-to-macrophage differentiation, suggesting that they play a role in determining a balance in the monocyte lineage developmental choices (Nakajima et al. 1996). The physiological significance of these findings remains to be established in vivo.

Recent genetic screens have also identified mutations in the VDR as susceptibility markers for infection with the intracellular pathogens *Mycobacterium tuberculosis* and *M. leprae* (Roy et al. 1999; Bellamy 2000). This relationship could result from VDR biasing a T helper (Th)2 immune response or from as yet uncharacterized mechanisms. However, it does suggest that normal VDR contributions to macrophage function are an important part of regulatory mechanisms necessary to deal with these and perhaps other pathogens.

3
Conclusions and Future Directions

Several nuclear receptors are expressed in macrophages and ligands for these receptors have been documented to influence inflammatory responses, specialized macrophage functions, and lipid homeostasis. While these findings have important biological and pharmacological implications, regulation of macrophage gene expression by members of the extended nuclear receptor family remains relatively unexplored. Even for the best-characterized nuclear receptors, emerging information on the ability of synthetic ligands to alter the specificity of coactivator and corepressor recruitment raises new possibilities for the development of novel pharmaceutical agents. Advances in the understanding of mechanisms responsible for transcriptional activation and repression by nuclear receptors may allow the development of selective nuclear receptor modulators that regulate a defined subset of target genes. For example, it may be possible to develop ligands for the glucocorticoid receptor that retain the ability to inhibit NF-κB, but do not have gluconeogenic activities (Vayssière et al. 1997). Such ligands would be likely to exert anti-inflammatory effects without many of the limiting side effects of currently available steroid hormone analogs. The ability to selectively modulate nuclear receptor function may prove to be of therapeutic benefit in a wide range of human diseases in which macrophages play important roles, including atherosclerosis, osteoporosis, and chronic inflammatory diseases.

4
References

Adcock, I. M., and Barnes, P. J. (1996). Ligand-induced differentiation of glucocorticoid receptor (GR) trans- repression and transactivation. Biochem Soc Trans 24, 267S

Almawi, W. Y., Beyhum, H. N., Rahme, A. A., and Rieder, M. J. (1996). Regulation of cytokine and cytokine receptor expression by glucocorticoids. J Leukoc Biol 60, 563–72

Babiker, A., Andersson, O., Lund, E., Xiu, R. J., Deeb, S., Reshef, A., Leitersdorf, E., Diczfalusy, U., and Bjorkhem, I. (1997). Elimination of cholesterol in macrophages and endothelial cells by the sterol 27-hydroxylase mechanism. Comparison with high density lipoprotein-mediated reverse cholesterol transport. J Biol Chem 272, 26253–61

Barnes, P. J. (1995). Inhaled glucocorticoids for asthma. N Engl J Med 332, 868–75

Barnes, P. J., and Karin, M. (1997). Nuclear factor-kappaB: a pivotal transcription factor in chronic inflammatory diseases. N Engl J Med 336, 1066–1071

Bellamy, R. (2000). Identifying genetic susceptibility factors for tuberculosis in Africans: a combined approach using a candidate gene study and a genome- wide screen. Clin Sci (Lond) 98, 245–50

Bendixen, A. C., Shevde, N. K., Dienger, K. M., Willson, T. M., Funk, C. D., and Pike, J. W. (2001). IL-4 inhibits osteoclast formation through a direct action on osteoclast precursors via peroxisome proliferator-activated receptor gamma 1. Proc Natl Acad Sci U S A 98, 2443–8

Bourguet, W., Ruff, M., Chambon, P., Gronemeyer, H., and Moras, D. (1995). Crystal structure of the ligand-binding domain of the human nuclear receptor RXR-α. Nature 375, 377–382

Brown, M. S., and Goldstein, J. L. (1986). A receptor-mediated pathway for cholesterol homeostasis. Science 232, 34–47

Brown, M. S., and Goldstein, J. L. (1997). The SREBP pathway: regulation of cholesterol metabolism by proteolysis of a membrane-bound transcription factor. Cell 89, 331–340

Brzozowski, A. M., Pike, A. C. W., Dauter, Z., Hubbard, R. E., Bonn, T., Engström, O., Öhman, L., Greene, G. L., Gustafsson, J.-Å., and Carlquist, M. (1997). Molecular basis of agonism and antagonism in the oestrogen receptor. Nature 389, 753–758

Caldenhoven, E., Liden, J., Wissnik, S., Van de Stoipe, A., Raaijmakers, J., Koenderman, L., Okret, S., Gustafsson, J.-A., and Van der Sagg, P. T. (1995). Negative cross-talk between RelA and the glucocorticoid receptor: a possible mechanism for the antiinflammatory action of glucocorticoids. Mol. Endocrinol. 9, 401–412

Chawla, A., Boisvert, W. A., Lee, C. H., Laffitte, B. A., Barak, Y., Joseph, S. B., Liao, D., Nagy, L., Edwards, P. A., Curtiss, L. K., Evans, R. M., and Tontonoz, P. (2001). A PPAR gamma-LXR-ABCA1 pathway in macrophages is involved in cholesterol efflux and atherogenesis. Mol Cell 7, 161–71

Chawla, A., Repa, J., Evans, R., and Mangelsdorf, D. (2001). Nuclear Receptors and Lipid Physiology: Opening the X-Files. Science 294, 1866–1870

Chen, J. D., and Evans, R. M. (1995). A transcriptional co-repressor that interacts with nuclear hormone receptors. Nature 377, 454–457

Chen, Z., Ishibashi, S., Perrey, S., Osuga, J., Gotoda, T., Kitamine, T., Tamura, Y., Okazaki, H., Yahagi, N., Iizuka, Y., Shionoiri, F., Ohashi, K., Harada, K., Shimano, H., Nagai, R., and Yamada, N. (2001). Troglitazone inhibits atherosclerosis in apolipoprotein E-knockout mice: pleiotropic effects on CD36 expression and HDL. Arterioscler Thromb Vasc Biol 21, 372–7

Claudel, T., Leibowitz, M. D., Fievet, C., Tailleux, A., Wagner, B., Repa, J. J., Torpier, G., Lobaccaro, J. M., Paterniti, J. R., Mangelsdorf, D. J., Heyman, R. A., and Auwerx, J. (2001). Reduction of atherosclerosis in apolipoprotein E knockout mice by activation of the retinoid X receptor. Proc Natl Acad Sci U S A 98, 2610–5

Collins, A. R., Meehan, W. P., Kintscher, U., Jackson, S., Wakino, S., Noh, G., Palinski, W., Hsueh, W. A., and Law, R. E. (2001). Troglitazone inhibits formation of early atherosclerotic lesions in diabetic and nondiabetic low density lipoprotein receptor-deficient mice. Arterioscler Thromb Vasc Biol 21, 365–71

Cushman, M., Costantino, J. P., Tracy, R. P., Song, K., Buckley, L., Roberts, J. D., and Krag, D. N. (2001). Tamoxifen and cardiac risk factors in healthy women: Suggestion of an anti-inflammatory effect. Arterioscler Thromb Vasc Biol 21, 255–61

Cutolo, M., Carruba, G., Villaggio, B., Coviello, D. A., Dayer, J. M., Campisi, I., Miele, M., Stefano, R., and Castagnetta, L. A. (2001). Phorbol diester 12-O-tetradecanoylphorbol

13-acetate (TPA) up-regulates the expression of estrogen receptors in human THP-1 leukemia cells. J Cell Biochem 83, 390–400

Damm, K., Thompson, C. C., and Evans, R. M. (1989). Protein encoded by v-erbA functions as a thyroid-hormone antagonist. Nature 339, 593–597

Datta, S., Magge, S. N., Madison, L. D., and Jameson, J. L. (1992). Thyroid hormone receptor mediates transcriptional activation and repression of different promoters in vitro. Mol.Endocrinol. 6, 815–25X

Deluca, H. F., and Cantorna, M. T. (2001). Vitamin D: its role and uses in immunology. Faseb J 15, 2579–85

Dempster, D. W., Moonga, B. S., Stein, L. S., Horbert, W. R., and Antakly, T. (1997). Glucocorticoids inhibit bone resorption by isolated rat osteoclasts by enhancing apoptosis. J Endocrinol 154, 397–406

Desreumaux, P., Dubuquoy, L., Nutten, S., Peuchmaur, M., Englaro, W., Schoonjans, K., Derijard, B., Desvergne, B., Wahli, W., Chambon, P., Leibowitz, M. D., Colombel, J. F., and Auwerx, J. (2001). Attenuation of colon inflammation through activators of the retinoid X receptor (RXR)/peroxisome proliferator-activated receptor gamma (PPARgamma) heterodimer. A basis for new therapeutic strategies. J Exp Med 193, 827–38

Diab, A., Deng, C., Smith, J. D., Hussain, R. Z., Phanavanh, B., Lovett-Racke, A. E., Drew, P. D., and Racke, M. K. (2002). Peroxisome proliferator-activated receptor-gamma agonist 15-deoxy- Delta(12,14)-prostaglandin J(2) ameliorates experimental autoimmune encephalomyelitis. J Immunol 168, 2508–15

Ellis, C. N., Varani, J., Fisher, G. J., Zeigler, M. E., Pershadsingh, H. A., Benson, S. C., Chi, Y., and Kurtz, T. W. (2000). Troglitazone improves psoriasis and normalizes models of proliferative skin disease: ligands for peroxisome proliferator-activated receptor-gamma inhibit keratinocyte proliferation. Arch Dermatol 136, 609–16

Evans, R. M. (1988). The steroid and thyroid hormone receptor superfamily. Science 240, 889–895

Forman, B. M., Chen, J., and Evans, R. M. (1997). Hypolipidemic drugs, polyunsaturated fatty acids, and eicosanoids are ligands for peroxisome proliferator-activated receptors α and δ. Proc. Natl. Acad. Sci. USA 94, 4312–4317

Forman, B. M., Tontonoz, P., Chen, J., Brun, R. P., Spiegelman, B. M., and Evans, R. M. (1995). 15-Deoxy-$\Delta^{12,14}$ - prostaglandin J_2 is a ligand for the adipocyte determination factor PPARγ. Cell 83, 803–812

Frazier-Jessen, M. R., and Kovacs, E. J. (1995). Estrogen modulation of JE/monocyte chemoattractant protein-1 mRNA expression in murine macrophages. J Immunol 154, 1838–45

Fu, X., Menke, J. G., Chen, Y., Zhou, G., MacNaul, K. L., Wright, S. D., Sparrow, C. P., and Lund, E. G. (2001). 27-hydroxycholesterol is an endogenous ligand for liver X receptor in cholesterol-loaded cells. J Biol Chem 276, 38378–87

Glass, C. K., and Rosenfeld, M. G. (2000). The coregulator exchange in transcriptional functions of nuclear receptors. Genes and Dev. 14, 121–141

Gordon, S. (1995). The macrophage. Bioessays 17, 977–986

Griffin, M. D., Lutz, W. H., Phan, V. A., Bachman, L. A., McKean, D. J., and Kumar, R. (2000). Potent inhibition of dendritic cell differentiation and maturation by vitamin D analogs. Biochem Biophys Res Commun 270, 701–8

Heck, S., Bender, K., Kullmann, M., Gottlicher, M., Herrlich, P., and Cato, A. C. (1997). I kappaB alpha-independent downregulation of NF-kappaB activity by glucocorticoid receptor. Embo J 16, 4698–707

Heery, D. M., Kalkhoven, E., Hoare, S., and Parker, M. G. (1997). A signature motif in transcriptional co-activators mediates binding to nuclear receptors. Nature 387, 733–736

Helmberg, A., Auphan, N., Caelles, C., and Karin, M. (1995). Glucocorticoid-induced apoptosis of human leukemic cells is caused by the repressive function of the glucocorticoid receptor. EMBO J. 14, 452–460

Herrlich, P., and Ponta, H. (1994). Mutual cross-modulation of steroid-retinoic acid receptor and AP-1 transcription factor activities: a novel property with practical implications. Trends Endocrinol. Metab. 5, 341–356

Horlein, A. J., Naar, A. M., Heinzel, T., Torchia, J., Gloss, B., Kurokawa, R., Ryan, A., Kamei, Y., Soderstrom, M., Glass, C. K., and et al. (1995). Ligand-independent repression by the thyroid hormone receptor mediated by a nuclear receptor co-repressor [see comments]. Nature 377, 397–404

Huang, J. T., Welch, J. S., Ricote, M., Binder, C. J., Willson, T. M., Kelly, C., Witztum, J. L., Funk, C. D., Conrad, D., and Glass, C. K. (1999). Interleukin-4-dependent production of PPAR-γ ligands in macrophages by 12/15-lipoxygenase. Nature 400, 378–382

Jiang, C., Ting, A. T., and Seed, B. (1998). PPAR-gamma agonists inhibit production of monocyte inflammatory cytokines. Nature 391, 82–86

Jonat, C., Rahmsdorf, H. J., Park, K.-K., Ponta, H., and Herrlich, P. (1990). Anti-tumor promotion and antiinflammation: down-modulation of AP-1 (Fos/Jun) activity by glucocorticoid hormone. Cell 62, 1189–1204

Joyce, D. A., Steer, J. H., and Abraham, L. J. (1997). Glucocorticoid modulation of human monocyte/macrophage function: control of TNF-alpha secretion. Inflamm Res 46, 447–51

Kamei, Y., Xu, L., Heinzel, T., Torchia, J., Kurokawa, R., Gloss, B., Lin, S.-C., Heyman, R., Rose, D., Glass, C., and Rosenfeld, M. (1996). A CBP integrator complex mediates transcriptional activation and AP-1 inhibition by nuclear receptors. Cell 85, 403–414

Kastner, P., Mark, M., and Chambon, P. (1995). Nonsteroid nuclear receptors: what are genetic studies telling us about their role in real life? Cell 83, 859–869

Kleinert, H., Euchenhofer, C., Ihrig-Biedert, I., and Forstermann, U. (1996). Glucocorticoids inhibit the induction of nitric oxide synthase II by down-regulating cytokine-induced activity of transcription factor nuclear factor-kappa B. Mol Pharmacol 49, 15–21

Kliewer, S. A., Lenhard, J. M., Willson, T. M., Patel, I., Morris, D. C., and Lehmann, J. M. (1995). A prostaglandin J_2 metabolite binds peroxisome proliferator-activated receptorγ and promotes adipocyte differentiation. Cell 83, 813–819

Kliewer, S. A., Sundseth, S. S., Jones, S. A., Brown, P. J., Wisely, G. B., Koble, C. S., Devchand, P., Wahli, W., Willson, T. M., Lenhard, J. M., and Lehman, J. M. (1997). Fatty acids and eicosanoids regulate gene expression through direct interactions with peroxisome proliferator-activated receptors α and γ. Proc. Natl. Acad. Sci. U.S.A. 94, 4318–4323

Kliewer, S. A., Umesono, K., Mangelsdorf, D. J., and Evans, R. M. (1992). Retinoid X receptor interacts with nuclear receptors in retinoic acid, thyroid hormone and vitamin D3 signalling. Nature 355, 446–449

Koehler, L., Hass, R., DeWitt, D. L., Resch, K., and Goppelt-Struebe, M. (1990). Glucocorticoid-induced reduction of prostanoid synthesis in TPA- differentiated U937 cells is mainly due to a reduced cyclooxygenase activity. Biochem Pharmacol 40, 1307–16

Kohro, T., Nakajima, T., Wada, Y., Sugiyama, A., Ishii, M., Tsutsumi, S., Aburatani, H., Imoto, I., Inazawa, J., Hamakubo, T., Kodama, T., and Emi, M. (2000). Genomic structure and mapping of human orphan receptor LXR alpha: upregulation of LXRa mRNA during monocyte to macrophage differentiation. J Atheroscler Thromb 7, 145–51

Komi, J., and Lassila, O. (2000). Nonsteroidal anti-estrogens inhibit the functional differentiation of human monocyte-derived dendritic cells. Blood 95, 2875–82

Laffitte, B. A., Joseph, S. B., Walczak, R., Pei, L., Wilpitz, D. C., Collins, J. L., and Tontonoz, P. (2001). Autoregulation of the human liver X receptor alpha promoter. Mol Cell Biol 21, 7558–68

Li, A., Brown, K., Silvestre, M., Willson, T., Palinski, W., and Glass, C. (2000). Peroxisome proliferator-activated receptor γ ligands inhibit development of atherosclerosis in LDL receptor-deficient mice. J. Clin. Invest. 106, 523–531

Li, M., Pascual, G., and Glass, C. (2000). Peroxisome Proliferator-Activated Receptor γ-Dependent Repression of the Inducible Nitric Oxide Synthase Gene. Mol. Cell. Biol. 20, 4699–4707

Luisi, B. F., Xu, W. X., Otwinowski, Z., Freedman, L. P., Yamamoto, K. R., and Sigler, P. B. (1991). Crystallographic analysis of the interaction of the glucocorticoid receptor with DNA. Nature 352, 497–505

Maglich, J. M., Sluder, A., Guan, X., Shi, Y., McKee, D. D., Carrick, K., Kamdar, K., Willson, T. M., and Moore, J. T. (2001). Comparison of complete nuclear receptor sets from the human, Caenorhabditis elegans and Drosophila genomes. Genome Biol 2

Mangelsdorf, D. J., and Evans, R. M. (1995). The RXR heterodimers and orphan receptors. Cell 83, 841–850

McKay, L., and Cidlowski, J. (2000). CBP (CREB Binding Protein) Integrates NF-κB Nuclear Factor-κB) and Glucocorticoid Receptor Physical interactions and Antagonism. Molecular Endocrinology 14, 1222–1234

McKay, L. I., and Cidlowski, J. A. (1999). Molecular control of immune/inflammatory responses: interactions between nuclear factor-kappa B and steroid receptor-signaling pathways. Endocrine Reviews 20, 435–59

McKenna, N. J., and O'Malley, B. W. (2002). Combinatorial control of gene expression by nuclear receptors and coregulators. Cell 108, 465–74

Moras, D., and Gronemeyer, H. (1998). The nuclear receptor ligand-binding domain: structure and function. Curr. Opin. Cell. Biol. 10, 384–391

Näär, A. M., Boutin, J. M., Lipkin, S. M., Yu, V. C., Holloway, J. M., Glass, C. K., and Rosenfeld, M. G. (1991). The orientation and spacing of core DNA-binding motifs dictate selective transcriptional responses to three nuclear receptors. Cell 65, 1267–1279

Nagy, L., Tontonoz, P., Alvarez, J. G. A., Chen, H., and Evans, R. M. (1998). Oxidized LDL Regulates Macrophage Gene Expression through Ligand Activation of PPAR-gamma. Cell 93, 229–240

Nakajima, H., Kizaki, M., Ueno, H., Muto, A., Takayama, N., Matsushita, H., Sonoda, A., and Ikeda, Y. (1996). All-trans and 9-cis retinoic acid enhance 1,25-dihydroxyvitamin D3- induced monocytic differentiation of U937 cells. Leuk Res 20, 665–76

Nissen, R. M., and Yamamoto, K. R. (2000). The glucocorticoid receptor inhibits NFkappaB by interfering with serine-2 phosphorylation of the RNA polymerase II carboxy-terminal domain. Genes Dev 14, 2314–29

Paech, K., Webb, P., Kuiper, G. G., Nilsson, S., Gustafsson, J., Kushner, P. J., and Scanlan, T. S. (1997). Differential ligand activation of estrogen receptors ERalpha and ERbeta at AP1 sites. Science 277, 1508–10

Peet, D. J., Janowski, B. A., and Mangelsdorf, D. J. (1998). The LXRs: a new class of oxysterol receptors. Curr Opin Genet Dev 8, 571–5

Piemonti, L., Monti, P., Sironi, M., Fraticelli, P., Leone, B. E., Dal Cin, E., Allavena, P., and Di Carlo, V. (2000). Vitamin D3 affects differentiation, maturation, and function of human monocyte-derived dendritic cells. J Immunol 164, 4443–51

Rastinejad, F., Perlmann, T., Evans, R. M., and Sigler, P. B. (1995). Structural determinants of nuclear receptor assembly on DNA direct repeats. Nature 375 203–211

Ray, A., and Prefontaine, K. E. (1994). Physical association and function antagonism between the p65 subunit of transcription factor NF-κB and the glucocorticoid receptor. Proc. Natl. Acad. Sci. U.S.A. 91, 752–756

Reckless, J., Metcalfe, J. C., and Grainger, D. J. (1997). Tamoxifen decreases cholesterol sevenfold and abolishes lipid lesion development in apolipoprotein E knockout mice. Circulation 95, 1542–8

Renaud, J.-P., Rochel, N., Ruff, M., Vivat, V., Chambon, P., Gronemeyer, H., and Moras, D. (1995). Crystal structure of the RAR-γ ligand-binding domain bound to all-trans retinoic acid. Nature 378, 681–689

Ricote, M., Geller, P., and Perucho, M. (1997). Frequent alterations in gene expression in colon tumor cells of the microsatellite mutator phenotype. Mutation Res. 374, 153–167

Ricote, M., Huang, J., Fajas, L., Li, A., Welch, J., Najib, J., Witztum, J. L., Auwerx, J., Palinski, W., and Glass, C. K. (1998). Expression of the peroxisome proliferator-activated receptor γ (PPARγ) in human atherosclerosis and regulation in macrophages by colony stimulating factors and oxidized low density lipoprotein. Proc. Natl. Acad. Sci. USA. 95, 7614–7619

Ricote, M., Li, A. C., Willson, T. M., Kelly, C. J., and Glass, C. K. (1998). The peroxisome proliferator-activated receptor-γ is a negative regulator of macrophage activation. Nature 391, 79–82

Roodman, G. D. (1999). Cell biology of the osteoclast. Exp Hematol 27, 1229–41

Roy, S., Frodsham, A., Saha, B., Hazra, S. K., Mascie-Taylor, C. G., and Hill, A. V. (1999). Association of vitamin D receptor genotype with leprosy type. J Infect Dis 179, 187–91

Sap, J., Munoz, A., Schmitt, J., Stunnenberg, H., and Vennstrom, B. (1989). Repression of transcription mediated at a thyroid hormone response element by the v-erb-A oncogene product. Nature 340, 242–244

Scheinman, R. I., Gualberto, A., Jewell, C. M., Cidlowski, J. A., and Baldwin, A. S., Jr. (1995). Characterization of mechanisms involved in transrepression of NF-κB activated glucocorticoid receptors. Mol. Cell. Biol. 15, 242–244

Schimmer, B., and Parker, K. (1996). ACTH, adrenocortical steroids and their synthetic analogs. Goodman and Gilman's Pharmacologiocal Basis of Therapeutics

H. JG and L. LE New York, McGraw Hill. -, -

Schmidt, M., Kreutz, M., Loffler, G., Scholmerich, J., and Straub, R. H. (2000). Conversion of dehydroepiandrosterone to downstream steroid hormones in macrophages. J Endocrinol 164, 161–9

Schule, R., Rangarajan, P., Yang, N., Kliewer, S., Ransone, L. J., Bolado, J., Verma, I. M., and Evans, R. M. (1991). Retinoic acid is a negative regulator of AP-1-responsive genes. Proc.Natl.Acad.Sci.U.S.A. 88, 6092–6096

Setoguchi, K., Misaki, Y., Terauchi, Y., Yamauchi, T., Kawahata, K., Kadowaki, T., and Yamamoto, K. (2001). Peroxisome proliferator-activated receptor-gamma haploinsufficiency enhances B cell proliferative responses and exacerbates experimentally induced arthritis. J Clin Invest 108, 1667–75

Sheppard, K.-A., Phelps, K. M., Williams, A. J., Tbano, D., Rosenfeld, M. G., Glass, C. K., Gerritsen, M. E., and Collins, T. (1998). Nuclear integration of glucocorticoid receptor and nuclear factor-κB signaling by CREB-binding protein and steroid receptor coactivator-1. J. Biol. Chem. 273, 29291–29294

Shevde, N. K., Bendixen, A. C., Dienger, K. M., and Pike, J. W. (2000). Estrogens suppress RANK ligand-induced osteoclast differentiation via a stromal cell independent mechanism involving c-Jun repression. Proc Natl Acad Sci U S A 97, 7829–34

Shiau, A. K., Barstad, D., Loria, P. M., Cheng, L., Kushner, P. J., Agard, D. A., and Greene, G. L. (1998). The structural basis of estrogen receptor/coactivator recognition and the antagonism of this interaction by tamoxifen. Cell 95, 927–937

Stein, B., and Yang, M. X. (1995). Repression of the interleukin-6 promoter by estrogen receptor is mediated by NF-κB and C/EBPβ. Mol. Cell. Biol. 15, 4971–4979

Su, C. G., Wen, X., Bailey, S. T., Jiang, W., Rangwala, S. M., Keilbaugh, S. A., Flanigan, A., Murthy, S., Lazar, M. A., and Wu, G. D. (1999). A novel therapy for colitis utilizing PPAR-γ ligands to inhibit the epithelial inflammatory response. J. Clin. Invest. 104, 383–389

Tall, A. R., and Wang, N. (2000). Tangier disease as a test of the reverse cholesterol transport hypothesis. J Clin Invest 106, 1205–7

Tanaka, J., and Fujita, H. (1997). Glucocorticoid- and mineralocorticoid receptors in microgial cells: the two receptors mediate differential effects of corticosteroids. Glia 20, 23–27

Torchia, J., Rose, D. W., Inostroza, J., Kamei, Y., Westin, S., Glass, C. K., and Rosenfeld, M. G. (1997). The transcriptional co-activator p/CIP binds CBP and mediates nuclear-receptor function. Nature 387, 677–684

Umesono, K., and Evans, R. M. (1989). Determinants of Target Gene Specificity for Steroid/Thyroid Hormone Receptors. Cell 57, 1139–1146

Umesono, K., Murakami, K. K., Thompson, C. C., and Evans, R. M. (1991). Direct repeats as selective response elements for the thyroid hormone, retinoic acid,and vitamin D3 receptors. Cell 65, 1255–1266

Valentine, J. E., Kalkhoven, E., White, R., Hoare, S., and Parker, M. G. (2000). Mutations in the estrogen receptor ligand binding domain discriminate between hormone-dependent transactivation and transrepression. J Biol Chem 275, 25322–9

Vayssière, B. M., Dupont, S., Choquart, A., Petit, F., Garcia, T., Marchandeau, C., Gronemeyer, H., and Resche-Rigon, M. (1997). Synthetic glucocorticoids that dissociate transactivation and AP-1 transrepression exhibit antiinflammatory activity in vivo. Mol. Endo. 11, 1245–1255

Wagner, R. L., Apriletti, J. W., McGrath, M. E., West, B. L., Baxter, J. D., and Fletterick, R. J. (1995). A structural role for hormone in the thyroid hormone receptor. Nature 378, 690–697

Winoto, A., and Littman, D. R. (2002). Nuclear hormone receptors in T lymphocytes. Cell 109 Suppl, S57–66

Yang-Yen, H.-F., Chambard, J.-C., Sun, Y.-L., Smeal, T., Schmidt, T. J., Drouin, J., and Karin, M. (1990). Transcriptional interference between c-Jun and the glucocorticoid receptor: mutual inhibition of DNA binding due to direct protein-protein interaction. Cell 62, 1205–1215

Young, S. G., and Fielding, C. J. (1999). The ABCs of cholesterol efflux. Nat Genet 22, 316–8

Yu, V. C., Delsert, C., Andersen, B., Holloway, J. M., Devary, O. V., Näär, A. M., Kim, S. Y., Boutin, J. M., Glass, C. K., and Rosenfeld, M. G. (1991). RXR beta: a coregulator that enhances binding of retinoic acid, thyroid hormone,and vitamin D receptors to their cognate response elements. Cell 67, 1251–1266

Relationships Between Reactive Oxygen Species and Reactive Nitrogen Oxide Species Produced by Macrophages

M. G. Espey

Radiation Biology Branch,
National Cancer Institute, NIH,
Building 10, Room B3-B69, Bethesda, MD 20892, USA
e-mail: SP@nih.gov

Abstract The armature of macrophages includes systems that catalytically produce a variety of chemical agents that help form the core of the innate inflammatory response. Compounds derived from oxygen are collectively known as reactive oxygen species (ROS), while those generated from nitrogen and oxygen are termed reactive nitrogen oxide species (RNOS). Both ROS and RNOS participate in cytotoxic mechanisms designed to kill pathogens. These systems require tight regulatory control to deter vascular abnormalities and host cell damage. ROS and RNOS also function to modulate a broad array of signaling pathways that shape adaptive immune responses. Effective pharmacological intervention of bystander injury elicited by ROS and RNOS will require an understanding of the specific interrelationships between these agents and how they factor into the various phases of inflammatory responses, which may be unique to different leukocyte subpopulations within each organ system.

Keywords Hydrogen peroxide, Myeloperoxidase, NADPH oxidase, Nitric oxide, Nitric oxide synthase, Nitrogen dioxide, Nitrosation, Nitration, Nitrite, Oxidation, Peroxynitrite, Superoxide

1 Introduction

The reactive oxygen species (ROS) and reactive nitrogen oxide species (RNOS) produced by macrophages play key roles in innate immune responses and the development of specific adaptive immunity. This review will focus on the major enzymatic systems in macrophages that generate ROS and RNOS. The factors that influence the secondary interactions of ROS and RNOS following their formation will be emphasized. An excellent overview of ROS and RNOS interactions has been previously presented in the *Handbook of Experimental Pharmacology* series (Wink et al. 2000).

2 Reactive Oxygen Species

2.1 Superoxide and NADPH Oxidase

The major source of ROS produced by activated leukocytes is the multicomponent enzyme nicotinamide adenine dinucleotide phosphate (NADPH) oxidase, which catalyzes the production of superoxide ($O_2^{\cdot -}$) by the one-electron reduction of oxygen (Sbarra and Karnovsky 1959; Babior 1973, 1999; Root and Cohen 1981; Tauber et al. 1983; Garcia and Segal 1988; Sies and de Groot 1992; Bastian and Hibbs 1994; Robinson and Badwey 1994, 1995; Rosen et al. 1995; Clark 1999). Although the preponderance of data are derived from neutrophils, the pivotal importance of NADPH oxidase in combating pathogens has been amply demonstrated in macrophages. In contrast to vanquishing pathogens, $O_2^{\cdot -}$ production from NAD(P)H oxidase in non-phagocytes participates in signal transduction and the regulation of blood pressure as well as atherosclerotic proliferation. Activation of NADPH oxidase in lymphocytes may largely serve a mitogenic role (Lee et al. 1998; Devadas et al. 2002). Endothelium and vascular smooth muscle cells in kidney (Wilcox 2001), heart (Griendling et al. 2000), and lung (Brar et al. 2002) also possess a relatively lower level of NAD(P)H oxidase activity that is somewhat homologous to the enzyme assembly in leukocytes.

Phagocytes increase oxygen consumption in response to a variety of stimuli; however, the term respiratory burst is a misnomer in that oxygen usage is not strictly related to mitochondrial respiration. Rather, glucose is consumed via the pentose phosphate pathway forming NADPH, which donates electrons for the oxidase reaction. For each NADPH, two electrons are shuttled through the flavin portion of the complex for the univalent reduction of oxygen (Eq. 1).

$$NADPH + 2O_2 \rightarrow NADP^+ + H^+ + 2O_2^- \quad (1)$$

A commonly used inhibitor of the electron transfer reaction is diphenylene iodonium (Cross and Jones 1986); however, this reagent also has inhibitory interaction with mitochondrial complex I (Ragan and Bloxham 1977).

A multitude of stimuli can prompt assembly and activation of the NADPH oxidase complex, which in phagocytes comprises five phox (phagocyte oxidase) components. An initial step involves phosphorylation of cytosolic $p47^{phox}$, which in association with $p40^{phox}$ and $p67^{phox}$ is mobilized to the plasma membrane. These subunits subsequently associate with the $p22^{phox}$and $p91^{phox}$ dimer, which comprise the cytochrome b_{558}. Alternatively, subunits may segment via intracytoplasmic vesicles directly to phagolysosomes. Activity of the NADPH oxidase complex is further modified by interaction with small G proteins Rac (Rho family, cytosolic) and Rap (Ras family, membrane). Therefore, NADPH oxidase activity can be regulated through a multitude of kinase and G protein signaling cascades (Bromberg et al. 1991; Babior 1999; Prada-Delgado et al. 2001). Stimulation of neutrophils with phorbol ester resulted in a rapid onset of O_2^- formation with a rate of 0.78 nmol/min/10^6 cells, while zymosan exposure elicited a slower onset and 10-fold lower rate of production (Roubaud et al. 1998). Cytokines, such as interferon (IFN)-γ or tumor necrosis factor (TNF)-α, can prime expression of signal transduction components in phagocytes resulting in augmented NADPH oxidase activity upon a secondary stimulus (Robinson and Badwey 1994). In addition to soluble factors, cues from the surrounding matrix in conjunction with changes in cell morphology influence trafficking, activation, and catalytic efficiency of NADPH oxidase (Berton and Gordon 1983; Wymann et al. 1989; Nathan et al. 1989; Nauseef et al. 1991; Zhou and Brown 1993; Berton et al. 1996; Hampton et al. 2002).

2.2 Hydrogen Peroxide

2.2.1 Haber-Weiss Cycle

In general, activated NADPH oxidase complex vectorially generates O_2^- on the extracellular face of the plasma membrane. Within minutes of pathogen internalization, fusion events occur to join the NADPH oxidase complex with the phagolysosome (DeLeo et al. 1999). Alternatively, subunits may segment via intracytoplasmic vesicles directly to phagolysosomes. Within this compartment, conditions are favorable for the dismutation of O_2^- to form hydrogen peroxide (H_2O_2; Eq. 2).

$$2O_2^- + 2H^+ \rightarrow H_2O_2 + O_2 \quad (2)$$

Much of the cytotoxicity associated with NADPH oxidase activity is due to secondary generation of H_2O_2 in conjunction with other components of the phagolysosome (Klebanoff 1967; Nanda et al. 1994; Pacelli et al. 1995; Reeves et al. 2002). Microbes have evolved specific defenses against ROS (Prada-Delgado et al. 2001; Vazquez-Torres and Fang 2001).

Dismutation of O_2^- to H_2O_2 is catalytically accelerated to near diffusion control by the superoxide dismutase (SOD) family of isoenzymes (Fridovich 1995, 1999). Cytosolic and mitochondrial SOD in conjunction with peroxisomal catalase, which catalytically forms water from H_2O_2, categorically serves a protective role. Extracellular Cu/Zn SOD (tetrameric glycoprotein isoforms called type A, B, and C) is also present in the tissue matrix and bound to heparin sulfate proteoglycans on the surface of various cell types (Halliwell and Gutteridge 1990; Abrahamsson et al. 1992). Therefore, under inflammatory conditions, generation of O_2^- by NADPH oxidase may lead to formation of H_2O_2 at these intercellular sites, which may influence the redox environment and shape the inflammatory response. Low-dose exposure of macrophages to exogenous H_2O_2 has been shown to elicit specific functional changes (Gamaley et al. 1994, Rhee 1999; Forman and Torres 2001a,b).

In contrast to O_2^- (pK_a=4.3; Fridovich 1995), H_2O_2 may be considered the traveling ROS in that it is freely permeates through cellular membranes. Generation of H_2O_2 through O_2^- formation by NADPH oxidase can cause tissue injury when free-transition metal (M^n; such as Fe^{3+}, Cu^{2+}) catalysts are present (Eqs. 3, 4; Halliwell and Gutteridge 1984, 1999; Hibbs et al. 1984).

$$M^n + O_2^- \rightarrow M^{n-1} + O_2 \tag{3}$$

$$M^{n-1} + H_2O_2 \rightarrow M^n + {}^{\cdot}OH + OH^- + O_2 \text{ (Fenton reaction)} \tag{4}$$

This reaction pathway has become known as the Haber-Weiss cycle (Haber and Weiss 1934), which links O_2^- generation of reduced transition metals (M^{n-1}) to putative formation of the highly reactive hypervalent metal species or hydroxyl radical ($^{\cdot}OH$) via H_2O_2 (Fenton 1894). However, the existence of $^{\cdot}OH$ versus production of alternate oxidizing intermediates remains an area of debate (Koppenol 1985; Wink et al. 1994). For $^{\cdot}OH$ to be an effective bactericide, this ROS would likely need to be generated in the immediate vicinity of the bacteria within the phagosome (Lymar and Hurst 1995). These data emphasize the differences in relative toxicities of ROS and RNOS when determined in the context of cellular and extracellular compartments. In addition to the NADPH oxidase system, lipoxygenase-catalyzed reactions may account for a significant portion of ROS formation by activated macrophages (e.g., 30%; Hume et al. 1983).

2.2.2 Myeloperoxidase

Myeloperoxidase (MPO) is a tetrameric glycosylated heme protoporphyrin-containing enzyme that is stored primary in azurophilic granules (Nauseef and Malech 1986; Heinecke et al. 1993; Kettle and Winterbourn 1997; Podrez et al.

2000). Upon stimulation, MPO can be released into the intercellular space or transport into the phagolysomal pathway and works in tandem with NAD(P)H oxidase pathways. Hydrogen peroxide is required to convert the resting-state, ferric heme of MPO by two electron equivalents to the hypervalent ferryl π cation radical termed compound I. In the presence of halides such as Cl^-, Br^-, and I^-, and the pseudohalide SCN^-, compound I is reduced in a single two-electron step to regenerate resting-state (MPO-Fe^{3+}) and form the corresponding hypohalous acid (HOX). Alternatively, MPO can oxidize nitrite (NO_{2-}) by one electron to give the RNOS nitrogen dioxide (NO_2) and the MPO intermediate compound II. Subsequently, an additional nitrite molecule will reduce compound II by one electron to regenerate the resting state and produce a second NO_2 (Eiserich et al. 1996; Abu-Soud and Hazen 2000; van Dalen et al. 2000). MPO-catalyzed formation of NO_2 can lead to nitration of aromatic compounds such as tyrosine (see sections entitled "NO and O_2^-" and "NO_2^- and H_2O_2"). Numerous other unknown substrates for peroxidation by MPO likely exist. Similar to NADPH oxidase, the presence of transition metals can enhance MPO-catalyzed oxidant production (Ramos et al. 1992).

The roles MPO plays in macrophage biology remain an area of debate. MPO activity in monocytes declines rapidly upon adherence and differentiation in vitro (Nakagawara et al. 1981). However, development of peroxidatic activity in the rough endoplasmic reticulum (ER) and perinuclear cisternae has been observed 2 h after monocyte adherence to serum- or fibrin-coated surfaces (Bodel et al. 1977). MPO immunoreactivity is associated with activated macrophages within human atherosclerotic vascular tissue (Daugherty et al. 1994). Subsequent studies have shown that macrophage MPO mediates modification of LDL cholesterol causing aberrant lipoprotein oxidation, aggregation, and uptake (Hazen et al. 1996, 1999; Chisolm et al. 1999; Hazen 2000; Podrez et al. 2000).

In addition to adherence, MPO activity in monocytes/macrophages is strongly influenced by numerous soluble immune factors. CD14-positive monocytes purified from peripheral blood monocytes developed into dendritic cells with potent antigen-presenting capacity following exposure to cytokines granulocyte-monocyte colony-stimulating factor and interleukin (IL)-4, but also retained significant amounts of MPO (Pickl et al. 1996), while a downregulation of expression has been observed in other preparations (Tsuruta et al. 1996). Likewise, macrophage exposure to exogenous MPO results in alterations in function, such as TNF-α production (Shepherd and Hoidal 1990; Lefkowitz et al. 1992; Lefkowitz and Lefkowitz 2001). Macrophages play an important role in the clearance of neutrophils (Savill et al. 1989) and remove neutrophil-derived MPO via mannose receptor-mediated uptake (Biggar and Sturgess 1976; Shepherd and Hoidal 1990). MPO of both endogenous (Rodrigues et al. 2002) and exogenous origin (Leung and Goren 1989; Lincoln et al. 1995) may incorporate into the macrophage phagosome and augment bacterial killing. A mixture of enzymatically active and effete forms of MPO released during neutrophil degranulation may coordinate a variety of macrophage functions (Bradley et al. 1982; Lefkowitz et al. 2000; Lefkowitz and Lefkowitz 2001).

These data raise an important point for consideration. The role of ROS and RNOS (see "Reactive Nitrogen Oxide Species" below) during an immune response is highly dependent on the tissue and the phase of the response. Activation of NADPH oxidase (either in the absence or presence of SOD) results in formation of H_2O_2, which in turn, provides electrons for MPO-catalyzed oxidation of substrates. However, each of these components does not necessarily have to present within the macrophage. Macrophage interaction with surrounding leukocytes, parenchymal cells, and their products profoundly shapes the pattern of ROS and RNOS that are subsequently formed, thereby impacting the immune response on multiple levels (Klebanoff 1980).

3
Reactive Nitrogen Oxide Species

3.1
Nitric Oxide and Nitric Oxide Synthase

RNOS originate with nitric oxide (NO) biosynthesis catalyzed by the nitric oxide synthase (NOS) family of isozymes, an exception being a potential contribution from dietary nitrite (NO_2^-) and nitrate (NO_3^-). The discovery of NO in biology and pathophysiology was borne out of metabolic studies with macrophages. In 1985, Stuehr and Marletta showed that activation of macrophages with bacterial lipopolysaccharide (LPS) caused a marked increase in NO_{2-} and NO_{3-} formation (Stuehr and Marletta 1985). Hibbs and colleagues subsequently observed that the cytostatic and tumoricidal capacity of macrophages stimulated with cytokines in vitro was dependent on the presence of L-arginine in the culture medium (Hibbs et al. 1987b). Moreover, this capacity was linked to the conversion of L-arginine to L-citrulline with formation of NO_2^- rather than the known urea cycle pathway involving loss of the guanidino-carbon of L-arginine to urea via arginase and subsequent conversion of L-ornithine to L-citrulline by ornithine transcarbamoylase (Hibbs et al. 1987a). These studies paved the way for the identification of NO as the effector molecule (Hibbs et al. 1988; Marletta et al. 1988) and cloning of the inducible isoform of nitric oxide synthase (iNOS, often termed NOS II) from macrophages (Lowenstein et al. 1992; Lyons et al. 1992; Xie et al. 1992). The biosynthesis and function of NO in the immune system has been extensively reviewed in the *Handbook of Experimental Pharmacology* series (Bogdan 2000; Zamora and Billiar 2000). Several excellent sources on the structure and function of NOS are available (Geller and Billiar 1998; Weinberg 1998; Stuehr 1999; Stuehr and Ghosh 2000; Alderton et al. 2001).

The human iNOS gene contains 26 exons and encodes a 131-kDa protein (Geller et al. 1993; Chartrain et al. 1994). Interestingly, an alternative splice variant of iNOS has been detected in human lung epithelia (Eissa et al. 1998). NOS can be divided into three functional domains: (1) an amino-terminal oxygenase domain that binds heme, L-arginine, and tetrahydrobiopterin (BH_4); (2) a calmodulin (CaM)-binding domain; and (3) a carboxy-terminal reductase domain that binds flavin mononucleotide (FMN), flavin adenine dinucleotide (FAD), and NADPH, which facilitates electron transfer to the oxygenase domain. The catalytic cycle of NOS proceeds in two separate successive monooxygenase reactions with the overall stoichiometry given in Eqs. 5 and 6).

$$\begin{aligned} &\mathrm{L-arginine + NADPH + H^+} \\ &\quad \mathrm{+O_2 \rightarrow N^w{-}hydroxy - L - arginine + NADP^+ + H_2O} \end{aligned} \tag{5}$$

$$\begin{aligned} &N^w\mathrm{{-}hydroxy - L - arginine + 0.5NADPH} \\ &\quad \mathrm{+0.5H^+ + O_2 \rightarrow L - citrulline + 0.5NADP^+ + H_2O + NO} \end{aligned} \tag{6}$$

To balance these equations for NO formation, NADPH must contribute 1.5 reducing equivalents. Several groups have proposed that the true stoichiometry is one NADPH consumed per monooxygenase cycle resulting in HNO (nitroxyl) as the product (Hobbs et al. 1994; Schmidt et al. 1996; Adak 2000). HNO has a reactivity pattern uniquely different from NO (Hughes 1999; Bartberger et al. 2001; Miranda et al. 2001, 2002; Espey et al. 2002a). Consequently, catalytic production of HNO by NOS in lieu of NO would influence a different array of cellular targets (Ma et al. 1999; Colton et al. 2001; Paolocci et al. 2001; Espey et al. 2002a).

All NOS isoforms require CaM binding for enzymatic activity. However, iNOS is distinguished from the neuronal and endothelial NOS isoforms (NOS-I, NOS-III) in that CaM is tightly bound at basal calcium levels and therefore does not require higher calcium transients for activation. In addition, iNOS lacks sequences present in the FMN-binding subdomain of constitutive NOS isoforms that may serve to destabilize electron transfer to the heme pocket at lower basal calcium levels (Salerno et al. 1997). Indeed, only an approximately 50% homology exits between iNOS and the neuronal and endothelial NOS isoforms (Ganster and Geller 2000; Zamora et al. 2000). Further distinction of iNOS resides in formation of the homodimer, a requisite step for l-arginine catalysis. In addition to dimer stabilization by heme and l-arginine, iNOS may be particularly sensitive to BH_4 binding, which facilitates alignment of the subunits in a head-to-head manner (Venema et al. 1997; Stuehr 1999; Chen et al. 2002).

While expression and activity of iNOS in rodents can be readily achieved in the laboratory, demonstration has been relatively more difficult in human leukocyte preparations. However, numerous studies have clearly shown iNOS plays a pivotal role in human macrophages (MacMicking et al. 1997). In septic patients, iNOS activity was markedly increased in putrescent areas of the systemic vasculature containing macrophages and expression of TNF-α and IL-1β (Annane et al. 2000). Human macrophages treated ex vivo with LPS and IFN-γ yielded an iNOS-specific amplification product by reverse transcriptase polymerase chain reaction (RT-PCR) (Reiling et al. 1994). IFN-γ, α or IL-4 augmented polyribonucleotide-induced NO_2^- production by human monocyte-derived macrophages (Snell et al. 1997). Alveolar macrophages isolated from lavage fluid of patients with tuberculosis have been shown to produce NO (Steiner et al. 1996). IFN-α treatment of macrophages or administration of IFN-α to hepatitis C patients in vivo increased expression of iNOS mRNA, protein, and activity concomitant with NO production (Sharara et al. 1997; Weinberg 1998). Apolipoprotein-E augmented NO production from activated human microglia, while amyloid-beta peptide blocked this effect (Vitek et al. 1997). Many cells in addition to macrophages are capable of iNOS expression including astrocytes, chondrocytes, hepatocytes, neurons, neutrophils, skeletal muscle, vascular smooth muscle, vascular endothelial cells, and several cancers.

In general, macrophages initiate synthesis of iNOS after immune stimulation (Nathan and Xie 1994; MacMicking et al. 1997; Bogdan 2000; Clancy et al. 1998; Ganster and Geller 2000; Taylor and Geller 2000; Zamora et al. 2000). However, constitutively expressed iNOS has been observed in some cell types (Mannick et al. 1994; Guo et al. 1995) including macrophages (Amin et al. 1995). The signal transduction pathways mediated by IFN are of particular importance. The Jak-STAT cascade is initiated by IFN binding to cell surface receptors with intrinsic or receptor-associated tyrosine kinase activity (Jak), which mediates phosphorylation of cytosolic proteins collectively known as STATs. STAT members activated in this manner form homo- and heterodimeric complexes that translocate to the nucleus, interact with other DNA-binding proteins (e.g., interferon regulatory factors) and bind specific sequences in the iNOS promoter region (e.g., gamma activation sequences, interferon response sequence elements). Exposure to IFN primes macrophages for expression of iNOS upon secondary activation (Espey et al. 2000b). Additional immune modulators such as TNF-α, IL-1β and pathogen products (e.g., LPS) are potent secondary stimulants. These agents predominantly regulate transcription of iNOS through the NF-κB family of DNA-binding proteins; however, numerous routes for gene expression are likely deployed by discrete macrophage subpopulations dependent on their repertoire of receptors and signaling cascades. Engagement of membrane CD23 on human macrophages by IgE complexes stimulates expression of iNOS (Paul-Eugene et al. 1995). Glucocorticoids, transforming growth factor (TGF)-β, IL-4, IL-10 and IL-13 act to negatively regulate iNOS at the levels of transcription and mRNA stability (Doyle et al. 1994; Vodovotz 1997; Diaz-Guerra et al. 1999).

While numerous factors have been found to regulate transcription, deciphering mechanisms for posttranslational control of iNOS remains more elusive. Relative to the constitutive isoforms, iNOS is less susceptible to feedback inhibition from NO reaction with the heme moiety. The factors that control the iNOS subunit dimerization are an important determinant in the biochemistry of the enzyme (Venema et al. 1997; Stuehr 1999; Chen et al. 2002). However, the dynamics of iNOS monomer and dimer pools that likely exist within macrophages have not been determined (Baek et al. 1993). L-Arginine transport and intracellular metabolism are important regulators of NO biosynthesis amenable to pharmacological exploitation (Morris 1999; Closs et al. 2000). Uptake of L-arginine via the cationic amino acid transport systems are a rate-limiting factor for sustained iNOS activity (Hibbs et al. 1987a,b; Bogel et al. 1992). L-Arginine is subsequently mobilized within macrophages differentially dependent upon the stimulant conditions (Morris et al. 1998). Concomitant changes in argininosuccinate synthase, arginase isoforms and iNOS expression upon activation point to the complexity of L-arginine metabolic cycles in macrophages. An intriguing feedback loop is the inhibitory action of N^{w}-hydroxy-L-arginine, the initial oxidation product of L-arginine formed in the iNOS catalytic cycle (Eq. 5), on arginase (Chenais et al. 1993). *Helicobacter pylori* has evolved a survival mechanism involving inhibition of NO production by activated macrophages by usurping L-arginine availability through constitutive bacterial arginase activity (Gobert et al. 2001).

The fate of L-arginine in macrophages is controlled in part by distinct subcellular compartmentalization of enzymes and substrate pools. For instance, arginase I resides in the cytosol, while arginase II is present in the mitochondrial matrix (Morris 1999). The localization of iNOS, particularly in association with cytoskeletal elements (Webb et al. 2001; Zeng and Morrison 2001), may also play a prominent role in the effector functions of NO and related nitrogen oxides derived from macrophages (Vodovotz et al. 1995; Espey et al. 2001b).

3.2 RNOS and ROS Interactions

3.2.1 NO and O_{2-}

Peroxynitrite is a highly reactive RNOS formed by the reaction between NO and O_2^- (Eq. 7; Koppenol et al. 1992; Crow and Beckman 1995; Pryor and Squadrito 1995; Greenacre and Ischiropoulos 2001; Radi et al. 2001).

$$NO + O_2^- \rightarrow ONOO^- \qquad (7)$$

Selective chemical modifications are observed upon exposure of many biological substances to $ONOO^-$ in vitro. Dependent on the constituents of the target, $ONOO^-$ can mediate both one- and two-electron oxidation as well as nitration reactions. Nitration involves the electrophilic addition of a NO_2^+ equivalent (nitronium, an electron acceptor) to a site of electron density, such as an aromatic ring. Transition metal catalysts and physiological levels of carbon dioxide augment nitration and oxidation yields derived from $ONOO^-$ (Lymar et al. 1996; Denicola et al. 1995, 1996). Development of antibodies that recognize nitrated tyrosyl residues (3-nitrotyrosine) has revealed the extent of nitration under a variety of disease conditions (Ye et al. 1996; Greenacre and Ischiropoulos 2001; Radi et al. 2001).

Caution should be exercised when attributing the cytotoxic actions of activated macrophages expressing iNOS to $ONOO^-$ formation (Ischiropoulos et al. 1992). The 10^9 M^{-1} s^{-1} rate constant for the $NO+O_2^-$ reaction signifies that these species will react with each other at near diffusion control (Huie and Padmaja 1993; Kissner et al. 1997). The stoichiometry of the reactants must also be taken into consideration. Maximal oxidation and nitration mediated by $ONOO^-$ are observed when the rate of NO synthesis is equivalent to the rate of O_2^- production. A tip in balance in favor of either reactant results in a rapid decline or abatement of oxidation and nitration due to reactions secondary to $ONOO^-$ formation; for instance, reaction between ONOOH and NO (Rubbo et al. 1994; Miles et al. 1996; Wink 1997; Jourd'heuil et al. 1999; Espey 2002b,c). In general, exposure to $ONOO^-$ does not elicit cytotoxicity until relatively high bolus concentrations (>200 μM) are applied (Zhu et al. 1992; Espey et al. 2002d). Therefore, a prominent role for $ONOO^-$ as a macrophage effector RNOS would require a substantial and prolonged synthesis of both NO and O_2^- at equivalent rates.

Synchronization of both NO and O_{2-} rates of formation during an immune response would require a precise regulation to produce significant $ONOO^-$. Kinetic modeling of macrophages predict that the limited diffusion of O_2^- will restrict migration of $ONOO^-$ as an effector RNOS (Chen and Deen 2001). Assembly and translocation of the NADPH oxidase complex is a rapid process relative to de novo iNOS expression and protein synthesis, which fosters a temporal dissociation between O_2^- and NO production within an individual activated macrophage (Iyengar et al. 1987; Doyle et al. 1994; Espey et al. 2000b; Pfeiffer et al. 2001). This dichotomy gives rise to either ROS or RNOS formation during distinct phases of macrophage phagocytosis and immune responses, largely exclusive of $ONOO^-$ generation (Vazquez-Torres et al. 2000; Mastroeni et al. 2000; Espey et al. 2000b, 2002b; Pfeiffer et al. 2001). However, numerous studies have suggested that formation of $ONOO^-$ and derived intermediates are essential for killing pathogens in macrophage phagosomes or cytoplasm (Saran and Bors 1994; Nozaki et al. 1997; Linares et al. 2001; Hickman-Davis 2002). These reports are hampered by the lack of pharmacology or analytical methodology specific for $ONOO^-$ versus other ROS and RNOS pathways.

Innate immunity is dependent upon coordination between neutrophil and macrophage populations. Neutrophil infiltration characteristically precedes monocyte recruitment or activation of resident macrophages during acute inflammatory reactions (Metchnikoff 1905). This sequence results in an initial predominance of ROS generation by neutrophils at inflammatory foci. The ROS phase is downregulated by subsequent iNOS catalyzed formation of NO within activated macrophages. RNOS derived from NO autoxidation (see "NO and O_2" below) interfere with assembly of the neutrophil NADPH oxidase complex (Clancy et al. 1992; Fujii et al. 1997; Lee et al. 2000) and diapedesis of neutrophils through the vasculature (Kubes et al. 1991; Granger and Kubes 1994). In this manner, coincident generation of O_2^- and NO by neutrophils and macrophages, respectively, are spatially and temporally segregated minimizing the probability of $ONOO^-$ formation. A similar reciprocal relationship between endothelial NAD(P)H oxidase activation and eNOS in hemodynamic shear stress-induced monocyte chemotactic protein expression has been described (Wung et al. 2001). These data illustrate that $ONOO^-$ may be implicated only within discrete zones where conditions are ideal for contemporaneous and equivalent rates of NO and O_2^- formation. It should be emphasized that NO is a highly effective ROS scavenger (Wink et al. 1995, 2001; Espey et al. 2002d).

A challenge for effective pharmacological intervention of iNOS in macrophages is to identify the timing for specific RNOS participation in the different phases of innate and adaptive immune responses. Although RNOS can compromise host cellular functions, they also play pivotal roles in both eradication of pathogens and abatement of ROS bystander injury.

3.2.2
NO and H_2O_2

Bacteria are rich in proteins containing Fe and Cu that facilitate electron transfer reactions (Salerno 1996). NO can disrupt these proteins (e.g., Fe-sulfur center of ferredoxins) causing the release of redox active metal ions (Drapier 1997; Poole and Hughes 2000). As described above ("Haber-Weiss Cycle"), the bacterial and mammalian cytotoxicity of H_2O_2 involves reduction by either Fe^{2+} or Cu^{1+} to yield either hypervalent metal ions or $^{\cdot}OH$ (Fenton 1894; Clifford and Repine 1982; Halliwell and Gutteridge 1984, 1999). NO-mediated labialization of metalloproteins in *Escherichia coli* bacteria has been shown to potentiate H_2O_2 toxicity more than 1,000-fold (Pacelli et al. 1995). Mobilization of transition metals from the periplasmic space and inner membrane to the genome was evidenced by DNA double-strand breaks.

Catalase is a heme-containing enzyme that catalyzes the catabolism of H_2O_2 into H_2O and O_2 (Chance 1947). NO can limit H_2O_2 catabolism through formation of nitrosylheme-catalase (Brown 1995; Brunelli et al. 2001). This may serve to augment the synergistic bactericidal action of NO and H_2O_2, provided NO is not overtly consumed by excess catalase. Paradoxically, other studies have suggested that virulence may be related to SOD-catalyzed conversion of O_2^- to H_2O_2 (De Groote et al. 1997). Periplasmic Cu,Zn-SOD may protect pathogenic bacteria, such as *E. coli* and *Salmonella typhimurium*, from redox injury (Benov et al. 1995; De Groote et al. 1997). Salmonella deficient in this Cu,Zn-SOD have reduced survival in macrophages and attenuated virulence in mice. These data illustrate that the tandem bactericidal action NO, O_2^- and H_2O_2 are dependent on a complex interrelationship between both host and bacterial systems that interact with ROS and RNOS.

Synergism between NO and H_2O_2 is not limited to bactericidal processes. NO formation by activated macrophages can disrupt transition metal homeostasis in tumor cell targets as well (Hibbs et al. 1984). Viability of a human ovarian cancer cell line exposed to 3-morpholino-sydnonimine, a compound that decomposes to simultaneously form NO and O_2^-, was found to not involve $ONOO^-$ formation. Rather, the mechanism of cytotoxicity was dependent NO and H_2O_2-mediated reduction of trace metals and generation of potent oxidants (e.g., $^{\cdot}OH$; see "Haber-Weiss Cycle" above; Farias-Eisner et al. 1996). In contrast to *Salmonella* and Cu,Zn-SOD, it was found that these cancer cells rely on the glutathione peroxidase-glutathione reductase for H_2O_2 resistance and potential toxicity synergism with NO and reduced metals.

3.2.3
NO_2^- and H_2O_2

Nitrite is an end product of NO autoxidation (Eq. 8; see section "NO and O_2" below; Wink et al. 1993, 2000; Fukuto 1995).

$$4NO + O_2 \rightarrow 2N_2O_3 \rightarrow 4H^+ + 4NO_2^- \quad (8)$$

Marked increases in NO_2^- are observed during immune activation, which correlates well with iNOS induction in macrophages (Stuehr and Marletta 1985; Hibbs et al. 1987a,b; Granger et al. 1991, 1996; Ignarro et al. 1993; Lewis et al. 1995; Grisham et al. 1996; Espey et al. 2000b). Nitrite can be converted to NO_3^- by oxyhemoproteins (Rodkey 1976; Doyle et al. 1985; Ignarro et al. 1993) suggesting that the parenchymal concentration of NO_2^- at sites of inflammation may be much higher than the low micromolar levels (Miranda et al. 2001) detected in plasma. As mentioned above (in "Myeloperoxidase" section), MPO in the presence of H_2O_2 readily catalyzes oxidation of NO_2^- to generate NO_2 (Eq. 9 ; Sampson et al. 1995; Eiserich et al. 1996; van der Vliet et al. 1997; Hazen et al. 1999; Abu-Soud and Hazen 2000; van Dalen et al. 2000; Espey et al. 2002d).

$$2NO_2^- + H_2O_2 + 2H^+ \rightarrow 2NO_2 + 2H_2O \quad (9)$$

Similar to $ONOO^-$, NO_2 is an oxidant and nitrating agent. However, NO_2 produced by MPO/H_2O_2 and NO_2^- elicited substantially greater nitration than an equivalent amount of $ONOO^-$ generated from xanthine oxidase/hypoxanthine (an O_2^- source) and NO (Espey et al. 2002d). Differences in nitration were most pronounced under conditions that required transmembrane diffusion of the RNOS into target cells, consistent with data showing a significant lifetime for NO_2 (Espey et al. 2001, 2002d). Catalytic production of NO_2 by the MPO pathway is much less restricted by the spatial and temporal constraints on $ONOO^-$ formation in that all the components do not need to be generated contemporaneously. Nitration mediated by the MPO/H_2O_2 route may predominate subsequent to macrophage expression of iNOS and NO_2^- accumulation in lesion sites. Under these conditions, cellular or free MPO deposits in combination with episodic formation of H_2O_2 may lead to NO_2 generation over an extended period, which may play a role during inflammatory flares and chronic disease settings. In addition to MPO, protoporphyrin IX (hemin) in the presence of H_2O_2 catalyzes NO_2^- oxidation to form NO_2 resulting in nitration of tyrosyl-containing peptides (Thomas et al. 2002). NO production by activated macrophages may dislodge hemin from hemeproteins and generate copious amounts of NO_2^-. These low molecular weight agents may then move to sites involved in NADPH oxidase activity and H_2O_2 formation causing subsequent nitrative modifications via formation of NO_2.

3.2.4
NO and O_2

In the presence of O_2, the lifetime of NO is inversely proportional to its concentration due to the second order dependency of the autoxidation reaction on NO, which is third order overall (Eq. 10; Ford et al. 1993ε; Wink et al. 1993, 2000; Fukuto 1995).

$$-d[NO]/dt = k_{NO}[NO]^2[O_2] \quad (10)$$

The rate of NO autoxidation is influenced by the solubility of both NO and O_2 (Liu et al. 1998). Both species partition to the hydrophobic phase of membranes (Shaw and Vosper 1977; Subczynski and Hyde 1983), which accelerates the rate of NO autoxidation approximately 300-fold relative to the surrounding aqueous phase (Liu et al. 1998). The rate-limiting step for NO autoxidation in hydrophobic medium is the formation of NO_2, which subsequently reacts with NO to form equilibrium with N_2O_3 (Eqs. 11 and 12; Wink et al. 1993, 2000; Ford et al. 1993; Fukuto 1995).

$$2NO + O_2 \rightarrow 2NO_2 \quad (11)$$

$$NO_2 + NO \leftrightarrow N_2O_3 \quad (12)$$

Indeed, the autoxidation process within the architecture of intact cells was found to be distinct from that which occurs in the aqueous extracellular medium, suggesting that cellular hydrophobic domains in conjunction with scavenger composition and location serve to focus RNOS chemistry to discrete sites during NO formation (Espey et al. 2001b) and emphasize the importance of NO_2 intermediacy as a determinant in the functional outcome of NO biosynthesis (Espey et al. 2002b,c,d).

NO autoxidation becomes more competitive with other NO reaction pathways (e.g., NO+oxyhemeproteins) under conditions of high NO biosynthesis because the rate of autoxidation is proportional to the square of the NO concentration. Consistent with this, macrophages expressing iNOS have been shown to form NO adducts on a variety of nucleophiles via formation of N_2O_3 (Miwa et al. 1987; Iyengar et al. 1987; Kosaka et al. 1989; Wink et al. 1997, 2000; Espey et al. 2001a,b). This process is termed nitrosation, where N_2O_3 acts as a nitrosonium donor (NO^+; Williams 1988). Nitrosation has been implicated in the modification of a variety of proteins containing critical thiol, amine, and hydroxyl residues (Wink et al. 2000; Espey et al. 2000a, 2001). Of particular interest, N_2O_3 formation and nitrosative facility of macrophages was dependent on the route of activation (Espey et al. 2000b). Macrophages stimulated with IFN-γ and LPS contained a twofold greater iNOS protein level compared to macrophages activated with IFN-γ and either TNF-α or IL-1β; however, the difference in nitrosative capacity was greater than 30-fold. These data show that the NO profiles derived from iNOS can be distinct and depend on inductive signal cascades. Formation of NO adducts on amines and thiols can occur during concurrent

O_2^- and NO biosynthesis through either N_2O_3-mediated nitrosation or oxidative nitrosylation (e.g., thyl radical+NO→RSNO) further demonstrating the complex interrelationships between ROS and RNOS and the functional outcomes of macrophage activation (Espey et al. 2002c).

4 Oxygen

An understanding of the dynamic changes in molecular O_2 usage is crucial to deciphering macrophage ROS and RNOS cascades. On the basis of changes in electron paramagnetic resonance spectroscopy signals from spin labels DEPMPO (for O_2^-) and PDT (for O_2), O_2^- generation from activated neutrophils corresponded to 50% of oxygen uptake (Roubaud et al. 1998). Metabolic labeling studies suggest that consumption of O_2 also increases upon stimulation in macrophages, but a smaller percentage is reduced to O_2^- (Baehner and Johnston 1972; Baehner et al. 1975; Reiss and Roos 1978). In severely hypoxic environments, it is possible that oxygen tension may be insufficient to sustain O_2^- formation (Edwards and Lloyd 1988). However, a rebound phenomenon of augmented O_2^- generation may occur upon re-oxygenation (Wilhelm et al. 1997). Generation of N_2O_3 by activated macrophages can nitrosate intracellular glutathione to form *S-nitrosoglutathione* thereby activating bioreductive metabolism via the hexose-monophosphate pathway and glutathione reductase (Albina and Mastrofrancesco 1993; Clancy et al. 1994). Additional changes in O_2 usage by activated macrophages and surrounding cells is manifest by the reversible inhibition of mitochondrial respiration by NO, which has a greater affinity for cytochrome oxidase than O_2 (Brown 1997, 2000; Boveris and Poderoso 2000). NO also interacts within macrophages to stabilize hypoxia-inducible factor and modulate its associated transcription networks (Brüne et al. 2001; Sandau et al. 2001).

5 Conclusion

These examples underlie the concept that participation of ROS and RNOS in disease processes is highly dependent on the state of tissue. Parameters, such as pO_2, may change quickly as waves of different leukocyte populations are engaged within inflammatory lesions over a period of time, subsequently affecting the levels and types of reactive intermediates formed. The action of ROS and RNOS are diverse and dependent on their rates of formation, mobility, interaction with each other, and the composition of the surrounding milieu. Coordination of macrophage-derived ROS and RNOS formation involves signaling from soluble, cell–cell, and matrix cues. These pathways vary greatly among the specialized members of the macrophage lineage. Macrophages have numerous mechanisms that control and balance formation of specific ROS and RNOS toward combating pathogens and preserving host tissue. With an increased awareness of feedback mechanisms, pharmacological strategies aimed at alleviating macrophage-mediated cellular stress and injury will improve.

Acknowledgements. The author thanks Dr. David A. Wink for hisfriendship,mentoring and support.

6 References

Abrahamsson T, Brandt U, Marklund SL, Sjoqvist PO (1992) Vascular bound recombinant extracellular superoxide dismutase type C protects against the detrimental effects of superoxide radicals on endothelium-dependent arterial relaxation. Circ Res 70:264–271

Abu-Soud HM, Hazen SL (2000) Nitric oxide modulates the catalytic activity of myeloperoxidase. J Biol Chem 275:5425–5430

Adak S, Wang Q, Stuehr DJ (2000) Arginine conversion to nitroxide by tetrahydrobiopterin-free neuronal nitric-oxide synthase. Implications for mechanism. J Biol Chem 275:33554–33561

Albina JE, Mastrofrancesco B (1993) Modulation of glucose metabolism in macrophages by products of nitric oxide synthase. Am J Physiol 264:C1594–1599

Alderton WK, Cooper CE, Knowles RG (2001) Nitric oxide synthases: structure, function and inhibition. Biochem J 357:593–615

Amin AR, Attur M, Vyas P, Leszczynska-Piziak J, Levartovsky D, Rediske J, Clancy RM, Vora KA, Abramson SB (1995–96) Expression of nitric oxide synthase in human peripheral blood mononuclear cells and neutrophils. J Inflamm 47:190–205

Annane D, Sanquer S, Sebille V, Faye A, Djuranovic D, Raphael JC, Gajdos P, Bellissant E (2000) Compartmentalised inducible nitric-oxide synthase activity in septic shock. Lancet 355:1143–1148

Babior BM, Kipnes RS, Curnutte JT (1973) Biological defense mechanisms: the production by leukocytes of superoxide, a potential bactericidal agent. J Clin Invest 52:741–744

Babior BM. NADPH oxidase: an update. (1999) Blood 93:1464–76

Baehner RL, Johnston RB Jr (1972) Monocyte function in children with neutropenia and chronic infections. Blood 40:31–41

Baehner RL, Murrmann SK, Davis J, Johnston RB Jr (1975) The role of superoxide anion and hydrogen peroxide in phagocytosis-associated oxidative metabolic reactions. J Clin Invest 56:571–576

Baek KJ, Thiel BA, Lucas S, Stuehr DJ (1993) Macrophage nitric oxide synthase subunits. Purification, characterization and role of prosthetic groups and substrate regulating their association into dimeric enzyme. J Biol Chem 268:21120–21129

Bartberger MD, Fukuto JM, Houk KN (2001) On the acidity and reactivity of HNO in aqueous solution and biological systems. Proc Natl Acad Sci USA 98:2194–2198

Bastian NR, Hibbs JB (1994) Assembly and regulation of NADPH oxidase and nitric oxide synthase. Curr Opin Immunol 6:131–139

Benov L, Chang LY, Day B, Fridovich I (1995) Copper, zinc superoxide dismutase in Escherichia coli: periplasmic localization. Arch Biochem Biophys 319:508–511

Berton G, Gordon S (1983) Modulation of macrophage mannosyl-specific receptors by cultivation on immobilized zymosan. Effects on superoxide-anion release and phagocytosis. Immunology 49:705–715

Berton G, Yan SR, Fumagalli L, Lowell CA (1996) Neutrophil activation by adhesion: mechanisms and pathophysiological implications. Int J Clin Lab Res 26:160–177

Biggar WD and Sturgess JM (1976) Peroxidase activity of alveolar macrophages. Lab Invest 34:31–42

Bodel PT, Nichols BA, Bainton DF (1977) Appearance of peroxidase reactivity within the rough endoplasmic reticulum of blood monocytes after surface adherence. J Exp Med 145:264–274

Bogdan C (2000) The function of nitric oxide in the immune system. Handbook Exp Pharm, Mayer B, Ed., 143:443–492

Bogle RG, Baydoun AR, Pearson JD, Moncada S, Mann GE (1992) L-Arginine transport is increased in macrophages generating nitric oxide. Biochem J 284:15–18

Boveris A, Poderoso JJ (2000) Regulation of oxygen metabolism by nitric oxide. In Nitric Oxide. Biology and Pathobiology. Ignarro LJ, Ed., Academic Press, San Diego, 355–368

Bradley PP, Christensen RD, Rothstein G. (1982) Cellular and extracellular myeloperoxidase in pyogenic inflammation. Blood 60:618–622

Brar SS, Kennedy TP, Sturrock AB, Huecksteadt TP, Quinn MT, Murphy TM, Chitano P, Hoidal JR (2002) NADPH oxidase promotes NF-kappaB activation and proliferation in human airway smooth muscle. Am J Physiol Lung Cell Mol Physiol 282:L782–795

Bromberg Y, Shani E, Joseph G, Gorzalczany Y, Sperling O, Pick E (1994) The GDP-bound form of the small G protein Rac1 p21 is a potent activator of the superoxide-forming NADPH oxidase of macrophages. J Biol Chem 269:7055–7058

Brown GC (1995) Reversible binding and inhibition of catalase by nitric oxide. Eur J Biochem 232:188–191

Brown GC (1997) Nitric oxide inhibition of cytochrome oxidase and mitochondrial respiration: implications for inflammatory, neurodegenerative and ischaemic pathologies. Mol Cell Biochem 174:189–192

Brown GC (2000) Nitric oxide as a competitive inhibitor of oxygen consumption in the mitochondrial respiratory chain. Acta Physiol Scand 168:667–674

Brüne B, von Knethen A, Sandau KB (2001) Transcription factors p53 and HIF-1alpha as targets of nitric oxide. Cell Signal 13:525–533

Brunelli L, Yermilov V, Beckman JS (2001) Modulation of catalase peroxidatic and catalatic activity by nitric oxide. Free Radic Biol Med 30:709–714

Chance B (1947) An intermediate compound in the catalase-hydrogen peroxide reaction. Acta Chem Scand 1: 236–267

Chartrain NA, Geller DA, Koty PP, Sitrin NF, Nussler AK, Hoffman EP, Billiar TR, Hutchinson NI, Mudgett JS (1994) Molecular cloning, structure, and chromosomal localization of the human inducible nitric oxide synthase gene. J Biol Chem 269:6765–6772

Chen B, Deen WM (2001) Analysis of the effects of cell spacing and liquid depth on nitric oxide and its oxidation products in cell cultures. Chem Res Toxicol 14:135–147

Chen Y, Panda K, Stuehr DJ (2002) Control of nitric oxide synthase dimer assembly by a heme-NO-dependent mechanism. Biochemistry 41:4618–4625

Chenais B, Yapo A, Lepoivre M and Tenu J-P (1993) N-hydroxy-L-arginine, a reactional intermediate in nitric oxide biosynthesis, induces cytostasis in human and murine tumor cells. Biochem Biophys Res Commun 196:558–1565

Chisolm GM 3rd, Hazen SL, Fox PL, Cathcart MK (1999) The oxidation of lipoproteins by monocytes-macrophages. Biochemical and biological mechanisms. J Biol Chem 274:25959–25962

Clancy RM, Amin AR, Abramson SB (1998) The role of nitric oxide in inflammation and immunity. Arthritis Rheum 41:1141–1151

Clancy RM, Leszczynska-Piziak J, Abramson SB (1992) Nitric oxide, an endothelial cell relaxation factor, inhibits neutrophil superoxide anion production via a direct action on the NADPH oxidase. J Clin Invest 90:1116–1121

Clancy RM, Levartovsky D, Leszczynska-Piziak J, Yegudin J, Abramson SB (1994) Nitric oxide reacts with intracellular glutathione and activates the hexose monophosphate shunt in human neutrophils: evidence for S-nitrosoglutathione as a bioactive intermediary. Proc Natl Acad Sci USA 91:3680–3684

Clark RA (1999) Activation of the neutrophil respiratory burst oxidase. J Infect Dis 179:S309–317

Clifford DP, Repine JE (1982) Hydrogen peroxide mediated killing of bacteria. Mol Cell Biochem 49:143–149

Closs EI, Scheld J-S, Sharafi M, Förstermann U (2000) Substrate supply for NO synthase in macrophages and endothelial cells: The cationic amino acid transporters (CATs). Mol Pharm 57:68–74

Colton CA, Gbadegesin M, Wink DA, Miranda KM, Espey MG, Vicini S (2001) Nitroxyl anion regulation of the NMDA receptor. J Neurochem 78:1126–1134

Cross AR, Jones OT (1986) The effect of the inhibitor diphenylene iodonium on the superoxide-generating system of neutrophils. Specific labelling of a component polypeptide of the oxidase. Biochem J 237:111–116

Crow JP, Beckman JS (1995) Reactions between nitric oxide, superoxide, and peroxynitrite: footprints of peroxynitrite in vivo. Adv Pharmacol 34:17–43

Daugherty A, Dunn JL, Rateri DL, Heinecke JW (1994) Myeloperoxidase, a catalyst for lipoprotein oxidation, is expressed in human atherosclerotic lesions. J Clin Invest 94:437–444

De Groote MA, Ochsner UA, Shiloh MU, Nathan C, McCord JM, Dinauer MC, Libby SJ, Vazquez-Torres A, Xu Y, Fang FC (1997) Periplasmic superoxide dismutase protects Salmonella from products of phagocyte NADPH-oxidase and nitric oxide synthase. Proc Natl Acad Sci USA 94:13997–14001

DeLeo FR, Allen LA, Apicella M, Nauseef WM (1999) NADPH oxidase activation and assembly during phagocytosis. J. Immunol.163:6732–6740

Denicola A, Freeman BA, Trujillo M, Radi R (1996) Peroxynitrite reaction with carbon dioxide/bicarbonate: kinetics and influence on peroxynitrite-mediated oxidations. Arch Biochem Biophys 333:49–58

Denicola A, Souza JM, Gatti RM, Augusto O, Radi R (1995) Desferrioxamine inhibition of the hydroxyl radical-like reactivity of peroxynitrite: role of the hydroxamic groups. Free Radic Biol Med 19:11–19

Devadas S, Zaritskaya L, Rhee SG, Oberley L, Williams MS (2002) Discrete generation of superoxide and hydrogen peroxide by T cell receptor stimulation: selective regulation of mitogen-activated protein kinase activation and fas ligand expression. J Exp Med 195:59–70

Diaz-Guerra, MJM, Castrillo A, Martin-Sanz P, and Bosca L (1999) Negative regulation by phosphatidylinositol 3-kinase of inducible nitric oxide synthase expression in macrophages. J Immunol 162:6184–6190

Doyle AG, Herbein G, Montaner LJ, Minty AJ, Caput D, Ferrara P, Gordon S (1994) Interleukin-13 alters the activation state of murine macrophages in vitro: comparison with interleukin-4 and interferon-gamma. Eur J Immunol 24:1441–1445

Doyle MP, Herman JG, Dykstra RL (1985) Autocatalytic oxidation of hemoglobin induced by nitrite: activation and chemical inhibition. J Free Radic Biol Med 1:145–153

Drapier JC (1997) Interplay between NO and [Fe-S] clusters: relevance to biological systems. Methods 11:319–329

Espey MG, Miranda KM, Feelisch M, Fukuto J, Grisham MB, Vitek M, Wink DA (2000a) Mechanisms of cell death governed by the balance between nitrosative and oxidative stress. Annal NY Acad Sci 288:209–221

Espey MG, Miranda KM, Pluta RM, Wink DA (2000b) Nitrosative capacity of macrophages is dependent on nitric-oxide synthase induction signals. J Biol Chem 275:11341–11347

Espey MG, Miranda KM, Thomas DD, Wink DA (2001) Distinction between nitrosating mechanisms within human cells and aqueous solution. J Biol Chem 276:3085–3091

Espey MG, Miranda KM, Thomas DD, Wink DA. (2002a) Ingress and reactive chemistry of nitroxyl-derived species within human cells. Free Rad Biol Med 33:827–834

Espey MG, Miranda KM, Thomas DD, Xavier S, Citrin D, Vitek MP, Wink DA (2002b) A chemical perspective on the interplay between NO, reactive oxygen species and reactive nitrogen oxide species. Annal NY Acad Sci 962:195–206
Espey MG, Thomas DD, Miranda KM, Wink DA. (2002c) Focusing of nitric oxide mediated nitrosation and oxidative nitrosylation as a consequence of reaction with superoxide Proc Natl Acad Sci USA 99:11127–11132
Espey MG, Xavier S, Thomas DD, Miranda KM, Wink DA. (2002d) Direct real-time evaluation of nitration with green fluorescent protein in solution and within human cells reveals the impact of nitrogen dioxide vs. peroxynitrite mechanisms. Proc Natl Acad Sci USA 99:3481–3486
Klebanoff SJ (1967) A peroxidase-mediated antimicrobial system in leukocytes. J Clin Invest 46:1078
Klebanoff SJ (1980) Oxygen metabolism and the toxic properties of phagocytes. Ann Intern Med 93:480–489
Edwards SW, Lloyd D (1988) The relationship between superoxide generation, cytochrome b and oxygen in activated neutrophils. FEBS Lett 227:39–42
Eiserich JP, Hristova M, Cross CE, Jones AD, Freeman BA, Halliwell B, van der Vliet A (1998) Formation of nitric oxide-derived inflammatory oxidants by myeloperoxidase in neutrophils. Nature 391:393–397
Eissa NT, Yuan JW, Haggerty CM, Choo EK, Palmer CD, Moss J (1998) Cloning and characterization of human inducible nitric oxide synthase splice variants: a domain, encoded by exons 8 and 9, is critical for dimerization. Proc Natl Acad Sci USA 95:7625–7630
Farias-Eisner R, Chaudhuri G, Aeberhard E, Fukuto JM (1996) The chemistry and tumoricidal activity of nitric oxide/hydrogen peroxide and the implications to cell resistance/susceptibility. J Biol Chem 271:6144–6151
Fenton, HJH (1894) Oxidation of tartaric acid in the presence of iron. J Chem Soc 65:899–910
Ford PC, Wink DA, Stanbury DM (1993) Autoxidation kinetics of aqueous nitric oxide. FEBS Lett 326:1–3
Forman HJ, Torres M (2001a) Signaling by the respiratory burst in macrophages. IUBMB Life 51:365–71
Forman HJ, Torres M (2001b) Redox signaling in macrophages. Mol Aspects Med. 22:189–216
Fridovich I (1995) Superoxide radical and superoxide dismutases. Annu Rev Biochem 64:97–112
Fridovich I (1997) Superoxide anion radical (O2-.), superoxide dismutases, and related matters. J Biol Chem. 272:18515–7
Fujii H, Ichimori K, Hoshiai K, Nakazawa H (1997) Nitric oxide inactivates NADPH oxidase in pig neutrophils by inhibiting its assembling process. J Biol Chem 272:32773–32778
Fukuto JM (1995) Chemistry of nitric oxide: biologically relevant aspects. Adv Pharmacol 34:1–15
Gamaley IA, Kirpichnikova KM, Klyubin IV (1994) Activation of murine macrophages by hydrogen peroxide. Cell Signal 6:949–957
Ganster RW, Geller DA (2000) Molecular regulation of inducible nitric oxide synthase. In Nitric Oxide, Ignarro LJ, Ed., Academic Press, San Diego, 129–156
Garcia RC, Segal AW (1988) Phosphorylation of the subunits of cytochrome b-245 upon triggering of the respiratory burst of human neutrophils and macrophages. Biochem J 252:901–904
Geller DA, Billiar TR (1998) Molecular biology of nitric oxide synthases. Cancer Metastasis Rev 17:7–23
Geller DA, Lowenstein CJ, Shapiro RA, Nussler AK, Silvio MD, Wang SC, Nakayama DK, Simmons RL, Snyder SH, Billiar TR (1993) Molecular cloning and expression of in-

ducible nitric oxide synthase from human hepatocytes. Proc Natl Acad Sci USA 90: 3491–3495

Gobert AP, McGee DJ, Akhtar M, Mendz GL, Newton JC, Cheng Y, Mobley HL, Wilson KT (2001) Helicobacter pylori arginase inhibits nitric oxide production by eukaryotic cells: a strategy for bacterial survival. Proc Natl Acad Sci USA 98:13844–13849

Granger DL, Hibbs JB Jr, Broadnax LM (1991) Urinary nitrate excretion in relation to murine macrophage activation. Influence of dietary L-arginine and oral NG-monomethyl-L-arginine.J Immunol 146:1294–1302

Granger DL, Taintor RR, Boockvar KS, Hibbs JB Jr. (1996) Measurement of nitrate and nitrite in biological samples using nitrate reductase and Griess reaction. Methods Enzymol 268:142–51

Granger DN, Kubes P (1994) The microcirculation and inflammation: modulation of leukocyte-endothelial cell adhesion. J Leukoc Biol 55:662–675

Greenacre SA, Ischiropoulos H (2001) Tyrosine nitration: localisation, quantification, consequences for protein function and signal transduction. Free Radic Res 34:541–581

Griendling, KK, Sorescu D, Ushio-Fukai M (2000) NAD(P)H oxidase Role in cardiovascular biology and disease. Circ Res 86:494–501

Grisham MB, Johnson GG, Lancaster JR Jr (1996) Quantitation of nitrate and nitrite in extracellular fluids. Methods Enzymol 268:237–246

Guo FH, De Raeve HR, Rice TW, Stuehr DJ, Thunnissen FB, Erzurum SC (1995) Continuous nitric oxide synthesis by inducible nitric oxide synthase in normal human airway epithelium in vivo. Proc Natl Acad Sci USA 92:7809–7813

Haber F, Weiss J (1934) The catalytic decomposition of hydrogen peroxide by iron salts. Proc Royal Soc London, Series A, 147:332–351

Halliwell B, Gutteridge JMC (1984) Oxygen toxicity, oxygen radicals, transition-metals and disease. Biochem J 219:1–14

Halliwell B, Gutteridge JMC (1990) The antioxidants of human extracellular fluids. Arch Biochem Biophys 280:1–8

Halliwell B, Gutteridge JMC (1999) Free Radicals in Biology and Medicine, Oxford University Press

Hampton M, Vissers M, Keenan J, Winterbourn C (2002) Oxidant-mediated phosphatidylserine exposure and macrophage uptake of activated neutrophils: possible impairment in chronic granulomatous disease. J Leukoc Biol 71:775–781

Hazen SL (2000) Oxidation and atherosclerosis. Free Radic Biol Med 28:1683–1684

Hazen SL, Hsu FF, Duffin K, Heinecke JW (1996) Molecular chlorine generated by the myeloperoxidase-hydrogen peroxide-chloride system of phagocytes converts low density lipoprotein cholesterol into a family of chlorinated sterols. J Biol Chem 271:23080–23088

Hazen SL, Zhang R, Shen Z, Wu W, Podrez EA, MacPherson JC, Schmitt D, Mitra SN, Mukhopadhyay C, Chen Y, Cohen PA, Hoff HF, Abu-Soud HM (1999) Formation of nitric oxide-derived oxidants by myeloperoxidase in monocytes: pathways for monocyte-mediated protein nitration and lipid peroxidation in vivo. Circ Res. 85:950–958

Heinecke JW, Li W, Daehnke HL 3rd, Goldstein, JA (1993) Dityrosine, a specific marker of oxidation, is synthesized by the myeloperoxidase-hydrogen peroxide system of human neutrophils and macrophages J Biol Chem 268:4069–4077

Hibbs JB Jr, Taintor RR, Vavrin Z (1984) Iron depletion: possible cause of tumor cell cytotoxicity induced by activated macrophages. Biochem Biophys Res Commun 123:716–723

Hibbs JB Jr, Taintor RR, Vavrin Z (1987a) Macrophage cytotoxicity: role for L-arginine deiminase activity and imino nitrogen oxidation to nitrite. Science 235:473–476

Hibbs JB Jr, Taintor RR, Vavrin Z, Rachlin EM (1988) Nitric oxide: a cytotoxic activated macrophage effector molecule. Biochem Biophys Res Commun 157:87–94

Hibbs JB Jr, Vavrin Z, Taintor RR (1987b) L-Arginine is required for expression of the activated macrophage effector mechanism causing selective metabolic inhibition in target cells. J Immunol 138:550–565
Hickman-Davis JM, O'Reilly P, Davis IC, Peti-Peterdi J, Davis G, Young KR, Devlin RB, Matalon S (2002) Killing of Klebsiella pneumoniae by human alveolar macrophages. Am J Physiol Lung Cell Mol Physiol 282:L944–56
Hobbs AJ, Fukuto JM, Ignarro LJ (1994) Formation of free nitric oxide from L-arginine by nitric oxide synthase: direct enhancement of generation by superoxide dismutase. Proc Natl Acad Sci USA 91:10992–10996
Huie RE, Padmaja S (1993) The reaction of NO with superoxide. Free Radical Res Commun 18:195–199
Hughes MN (1999) Relationships between nitric oxide, nitroxyl ion, nitrosonium cation and peroxynitrite. Biochim Biophys Acta 1411:263–272
Hume DA, Gordon S, Thornalley PJ, Bannister JV (1983) The production of oxygen-centered radicals by bacillus-Calmette-Guerin-activated macrophages. An electron paramagnetic resonance study of the response to phorbol myristate acetate. Biochim Biophys Acta 763:245–250
Ignarro LJ, Fukuto JM, Griscavage JM, Rogers NE, Byrns RE (1993) Oxidation of nitric oxide in aqueous solution to nitrite but not nitrate: comparison with enzymatically formed nitric oxide from L-arginine. Proc Natl Acad Sci USA 90:8103–8107
Ischiropoulos H, Zhu L, Beckman JS (1992) Peroxynitrite formation from macrophage-derived nitric oxide. Arch Biochem Biophys 298:446–451
Iyengar R, Stuehr DJ, Marletta MA. (1987) Macrophage synthesis of nitrite, nitrate, and N-nitrosamines: precursors and role of the respiratory burst. Proc Natl Acad Sci USA 84:6369–6373
Jourd'heuil D, Miranda KM, Kim SM, Espey MG, Vodovotz Y, Laroux S, Mai CT, Miles AM, Grisham MB, Wink DA. (1999) The oxidative and nitrosative chemistry of the nitric oxide/superoxide reaction in the presence of bicarbonate. Arch Biochem Biophys 365:92–100
Kettle AJ, Winterbourn CC (1997) Myeloperoxidase: A key regulator of neutrophil oxidant production. Redox Rep 3:3–15
Kissner R, Nauser T, Bugnon P, Lye PG, Koppenol WH (1997) Formation and properties of peroxynitrite as studied by laser flash photolysis, high-pressure stopped-flow technique, and pulse radiolysis. Chem Res Toxicol 10:1285–1292
Koppenol WH (1985) The reaction of ferrous EDTA with hydrogen peroxide: evidence against hydroxyl radical formation. J Free Radic Biol Med 1:281–5
Koppenol WH, Moreno JJ, Pryor WA, Ischiropoulos H, Beckman JS (1992) Peroxynitrite, a cloaked oxidant formed by nitric oxide and superoxide. Chem Res Toxicol 5:834–842
Kosaka H, Wishnok JS, Miwa M, Leaf CD, Tannenbaum SR (1989) Nitrosation by stimulated macrophages. Inhibitors, enhancers and substrates Carcinogenesis 10:563–566
Kubes P, Suzuki M, Granger DN (1991) Nitric oxide: an endogenous modulator of leukocyte adhesion. Proc Natl Acad Sci USA 88:4651–4655
Lee, J.R., and G.A. Koretzky (1998) Production of reactive oxygen intermediates following CD40 ligation correlates with c-Jun N-terminal kinase activation and IL-6 secretion in murine B lymphocytes. Eur J Immunol 28:41884197
Lee C, Miura K, Liu X, Zweier JL (2000) Biphasic regulation of leukocyte superoxide generation by nitric oxide and peroxynitrite. J Biol Chem 275:38965–38972
Lefkowitz DL, Lefkowitz SS (2001) Macrophage-neutrophil interaction: a paradigm for chronic inflammation revisited. Immunol Cell Biol 79:502–506
Lefkowitz DL, Mills K, Morgan D, Lefkowitz SS (1992) Macrophage activation and immunomodulation by myeloperoxidase. Proc Soc Exp Biol Med 199:204–210

Lefkowitz DL, Roberts E, Grattendick K, Schwab C, Stuart R, Lincoln J, Allen RC, Moguilevsky N, Bollen A, Lefkowitz SS (2000) The endothelium and cytokine secretion: the role of peroxidases as immunoregulators. Cell Immunol 202:23–30

Leung KP, Goren MB. (1989) Uptake and utilization of human polymorphonuclear leukocyte granule myeloperoxidase by mouse peritoneal macrophages. Cell Tissue Res 257:653–666

Lewis RS, Tamir S, Tannenbaum SR, Deen WM (1995) Kinetic analysis of the fate of nitric oxide synthesized by macrophages in vitro. J Biol Chem 270:29350–29355

Linares E, Giorgio S, Mortara RA, Santos CX, Yamada AT, Augusto O (2001) Role of peroxynitrite in macrophage microbicidal mechanisms in vivo revealed by protein nitration and hydroxylation. Free Radic Biol Med 30:1234–1242

Lincoln JA, Lefkowitz DL, Cain T, Castro A, Mills KC, Lefkowitz SS, Moguilevsky N, Bollen A. (1995) Exogenous myeloperoxidase enhances bacterial phagocytosis and intracellular killing by macrophages. Infect Immun 63:3042–3047

Liu X, Miller MJ, Joshi MS, Thomas DD, Lancaster JR Jr (1998) Accelerated reaction of nitric oxide with O_2 within the hydrophobic interior of biological membranes. Proc Natl Acad Sci USA 95:2175–2179

Lowenstein CJ, Glatt CS, Bredt DS, Snyder SH. (1992) Cloned and expressed macrophage nitric oxide synthase contrasts with the brain enzyme. Proc Natl Acad Sci USA 89:6711–6715

Lymar SV, Hurst JK (1995) Role of compartmentation in promoting toxicity of leukocyte-generated strong oxidants. Chem Res Toxicol 8:833–840

Lymar SV, Jiang Q, Hurst JK (1996) Mechanism of carbon dioxide-catalyzed oxidation of tyrosine by peroxynitrite. Biochemistry 35:7855–7861

Lyons CR, Orloff GJ, Cunningham JM. (1992) Molecular cloning and functional expression of an inducible nitric oxide synthase from a murine macrophage cell line. J Biol Chem 267:6370–6374

Ma XL, Gao F, Liu G-L, Lopez BL, Christopher TA, Fukuto JM, Wink DA, Feelisch M (1999) Opposite effects of nitric oxide and nitroxyl on postischemic myocardial injury. Proc Natl Acad Sci USA 96:14617–14622

MacMicking, J., Xie, Q.W., Nathan, C. 1997. Nitric oxide and macrophage function. Annu. Rev. Immunol 15:323–350

Mannick JB, Asano K, Izumi K, Kieff E, Stamler JS (1994) Nitric oxide produced by human B lymphocytes inhibits apoptosis and Epstein-Barr virus reactivation. Cell 79:1137–1146

Marletta MA, Yoon PS, Iyengar R, Leaf CD, Wishnok JS (1988) Macrophage oxidation of L-arginine to nitrite and nitrate: nitric oxide is an intermediate. Biochemistry 27:8706–8711

Mastroeni, P., Vazquez-Torres, A., Fang, F.C., Xu, Y., Khan, S., Hormaeche, C.E., Dougan, G. 2000. Antimicrobial actions of the NADPH phagocyte oxidase and inducible nitric oxide synthase in experimental salmonellosis. II. Effects on microbial proliferation and host survival in vivo. J Exp Med 192:237–247

Metchnikoff E (1905) In Immunity to infectious diseases. Cambridge University Press, London

Miles AM, Bohle DS, Glassbrenner PA, Hansert B, Wink DA, Grisham MB (1996) Modulation of superoxide-dependent oxidation and hydroxylation reactions by nitric oxide. J Biol Chem 271:40–47

Miranda KM, Espey MG, Wink DA (2001) A rapid, simple spectrophotometric method for simultaneous detection of nitrate and nitrite. Nitric Oxide 5:62–71

Miranda KM, Espey MG, Yamada K, Krishna M, Ludwick N, Kim S, Jourd'heuil D, Grisham MB, Feelisch M, Fukuto JM, Wink DA (2002) Unique oxidative mechanisms for the reactive nitrogen oxide species, nitroxyl anion. J Biol Chem 276:1720–1727

Miranda KM, Yamada K-I, Espey MG, Thomas DD, DeGraff W, Mitchell JB, Krishna MC, Colton CA, Wink DA (2002) Further evidence for distinct reactive intermediates from

nitroxyl and peroxynitrite: Effects of buffer composition on the chemistry of Angeli's salt and synthetic peroxynitrite. Arch Biochem Biophys, in press
Miwa M, Stuehr DJ, Marletta MA, Wishnok JS, Tannenbaum SR (1987) Nitrosation of amines by stimulated macrophages. Carcinogenesis 8:955–958
Morris SM Jr (1999) Arginine synthesis, metabolism and transport: Regulators of nitric oxide synthases. In Cellular and Molecular Biology of Nitric Oxide, Laskin JD, Laskin DL Eds., Dekker, New York, 57–86
Morris SM Jr, Kepka-Lenhart D, Chen LC (1998) Differential regulation of arginases and inducible nitric oxide synthase in murine macrophage cells. Am J Physiol 275:E740–747
Nakagawara A, Nathan CF, Cohn ZA (1981) Hydrogen peroxide metabolism in human monocytes during differentiation in vitro. J Clin Invest 68:1243–1252
Nanda A, Curnutte JT, Grinstein S (1994) Activation of H+ conductance in neutrophils requires assembly of components of the respiratory burst oxidase but not its redox function. J Clin Invest. 93:1770–1775
Nathan C, Srimal S, Faber C, Sanchez E, Kabbash L, Asch A, Gailit J, Wright SD (1989) Cytokine-induced respiratory burst of human neutrophils: dependence on extracellular matrix proteins aand CD1/CD18 integrins. J Cell Biol 109:1341–1349
Nathan C, Xie QW (1994) Regulation of biosynthesis of nitric oxide. J Biol Chem 269:13725–13728
Nauseef WM, Malech HL (1986) Analysis of the peptide subunits of human neutrophil myeloperoxidase. Blood 67:1504–1507
Nauseef WM, Volpp BD, McCormick S, Leidal KG, Clark RA (1991) Assembly of the neutrophil respiratory burst oxidase. Protein kinase C promotes cytoskeletal and membrane association of cytosolic oxidase components. J Biol Chem 266:5911–5917
Nozaki Y, Hasegawa Y, Ichiyama S, Nakashima I, Shimokata K (1997) Mechanism of nitric oxide-dependent killing of Mycobacterium bovis BCG in human alveolar macrophages. Infect Immun 65:3644–3647
Pacelli R, Wink DA, Cook JA, Krishna MC, DeGraff W, Friedman N, Tsokos M, Samuni A, Mitchell JB (1995) Nitric oxide potentiates hydrogen peroxide-induced killing of Escherichia coli. J Exp Med 182:1469–1479
Paolocci N, Saavedra WF, Miranda KM, Martignani C, Isoda T, Hare JM, Espey MG, Fukuto, JM, Feelisch M, Wink DA, Kass DA (2001) Nitroxyl anion exerts redox-sensitive positive cardiac inotropy in vivo by calcitonin gene-related peptide signaling. Proc Natl Acad Sci USA 98:10463–10468
Paul-Eugene N, Mossalayi D, Sarfati M, Yamaoka K, Aubry JP, Bonnefoy JY, Dugas B, Kolb JP (1993) Evidence for a role of Fc epsilon RII/CD23 in the IL-4-induced nitric oxide production by normal human mononuclear phagocytes. Cell Immunol Jul;163(2):314-8
Pfeiffer S, Lass A, Schmidt K, Mayer B. (2001) Protein tyrosine nitration in cytokine-activated murine macrophages. Involvement of a peroxidase/nitrite pathway rather than peroxynitrite. J Biol Chem 276:34051–34058
Pickl WF, Majdic O, Kohl P, Stockl J, Riedl E, Scheinecker C, Bello-Fernandez C, Knapp W (1996) Molecular and functional characteristics of dendritic cells generated from highly purified CD14+ peripheral blood monocytes. J Immunol 157:3850–3859
Prada-Delgado A, Carrasco-Marin E, Bokoch GM, Alvarez-Dominguez C (2001) Interferon-gamma listericidal action is mediated by novel Rab5a functions at the phagosomal environment. J Biol Chem 276:19059–19065
Podrez EA, Abu-Soud HM, Hazen SL. (2000) Myeloperoxidase-generated oxidants and atherosclerosis. Free Radic Biol Med 28:1717–1725
Poole RK, Hughes MN (2000) New functions for the ancient globin family: bacterial responses to nitric oxide and nitrosative stress. Mol Microbiol 36:775–783
Pryor WA, Squadrito GL (1995) The chemistry of peroxynitrite: a product from the reaction of nitric oxide with superoxide. Am J Physiol 268:L699–722

Radi R, Peluffo G, Alvarez MN, Naviliat M, Cayota A. (2001) Unraveling peroxynitrite formation in biological systems. Free Radic Biol Med 30:463–488

Ramos CL, Pou S, Britigan BE, Cohen MS, Rosen GM (1992) Spin trapping evidence for myeloperoxidase-dependent hydroxyl radical formation by human neutrophils and monocytes. J Biol Chem 267:8307–8312

Reeves EP, Lu H, Jacobs HL, Messina CG, Bolsover S, Gabella G, Potma EO, Warley A, Roes J, Segal AW (2002) Killing activity of neutrophils is mediated through activation of proteases by K+ flux. Nature 416:291–297

Reiling N, Ulmer AJ, Duchrow M, Ernst M, Flad HD, Hauschildt S (1994) Nitric oxide synthase: mRNA expression of different isoforms in human monocytes/macrophages. Eur J Immunol 24:1941–1944

Reiss M, Roos D (1978) Differences in oxygen metabolism of phagocytosing monocytes and neutrophils. J Clin Invest 61:480–488

Rhee SG (1999) Redox signaling: hydrogen peroxide as intracellular messenger. Exp Mol Med. 31:53–59

Robinson JM; Badwey JA (1994) Production of active oxygen species by phagocytic leukocytes. Immunol Ser 60:159–178

Robinson JM, Badwey JA. (1995) The NADPH oxidase complex of phagocytic leukocytes: a biochemical and cytochemical view. Histochem Cell Biol 103:163–180

Rodkey FL (1976) A mechanism for the conversion of oxyhemoglobin to methemoglobin by nitrite. Clin Chem. 22:1986–1990

Rodrigues MR, Rodriguez D, Russo M, Campa A. (2002) Macrophage activation includes high intracellular myeloperoxidase activity. Biochem Biophys Res Commun. 292:869–873

Rosen GM, Pou S, Ramos CL, Cohen MS, Britigan BE, Ramos CL, Cohen MS, Britigan BE (1995) Free radicals and phagocytic cells FASEB J 9:200–209

Root RK, Cohen MS (1981) The microbicidal mechanisms of human neutrophils and eosinophils.Rev Infect Dis 3:565–598

Roubaud V, Sankarapandi S, Kuppusamy P, Tordo P, Zweier JL (1998) Quantitative measurement of superoxide generation and oxygen consumption from leukocytes using electron paramagnetic resonance spectroscopy. Anal Biochem 257:210–217

Rubbo H, Radi R, Trujillo M, Telleri R, Kalyanaraman B, Barnes S, Kirk M, Freeman BA. (1994) Nitric oxide regulation of superoxide and peroxynitrite-dependent lipid peroxidation. Formation of novel nitrogen-containing oxidized lipid derivatives. J Biol Chem 269:26066–26075

Salerno JC (1996) Nitric oxide complexes of metalloproteins. In Niric Oxide. Principles and actions. Lancaster JR Jr., Ed., Academic Press, San Diego, 83–110

Salerno JC, Harris DE, Irizarry K, Patel B, Morales AJ, Smith SM, Martasek P, Roman LJ, Masters BS, Jones CL, Weissman BA, Lane P, Liu Q, Gross SS (1997) An autoinhibitory control element defines calcium-regulated isoforms of nitric oxide synthase. J Biol Chem 272:29769–29777

Sandau KB, Fandrey J, Brüne B (2001) Accumulation of HIF-1alpha under the influence of nitric oxide. Blood 97:1009–1015

Sampson JB, Ye Y, Rosen H, Beckman JS (1995) Myeloperoxidase and horseradish peroxidase catalyze tyrosine nitration in proteins from nitrite and hydrogen peroxide. Arch Biochem Biophys 356:207–213

Saran M, Bors W (1994) Signalling by O_2^- and NO: how far can either radical, or any specific reaction product, transmit a message under in vivo conditions? Chem-Biol Interact 90:35–45

Savill JS, Wyllie AH, Henson JE, Walport MJ, Henson PM, Haslett C (1989) Macrophage phagocytosis of aging neutrophils in inflammation. Programmed cell death in the neutrophil leads to its recognition by macrophages. J Clin Invest 83:865–875

Sbarra AJ, Karnovsky ML (1959) The biochemical basis of phagocytosis. 1. Metabolic changes during the ingestion of particles by polymorphonuclear leukocytes. J Biol Chem 234:1355–1362

Schmidt HHHW, Hofmann H, Schindler U, Shutenko ZS, Cunningham DD, Feelisch M No ·NO from NO synthase. (1996) Proc Natl Acad Sci USA 93:14492–14497

Sharara AI, Perkins DJ, Misukonis MA, Chan SU, Dominitz JA, Weinberg JB (1997) Interferon (IFN)-alpha activation of human blood mononuclear cells in vitro and in vivo for nitric oxide synthase (NOS) type 2 mRNA and protein expression: possible relationship of induced NOS2 to the anti-hepatitis C effects of IFN-alpha in vivo. J Exp Med 186:1495–1502

Shaw AW, Vosper AJ (1977) Solubility of nitric oxide in aqueous and nonaqueous solvents. J Chem Soc Faraday Trans 18:1239–1244

Shepherd VL, Hoidal JR (1990) Clearance of neutrophil-derived myeloperoxidase by the macrophage mannose receptor. Am J Respir Cell Mol Biol 2:335–340

Sies H, de Groot H (1992) Role of reactive oxygen species in cell toxicity. Toxicol Lett. 65:547–551

Snell JC, Chernyshev O, Gilbert DL, Colton CA (1997) Polyribonucleotides induce nitric oxide production by human monocyte-derived macrophages. J Leukoc Biol 62:369–373

Steiner, P, Efferen L, Durkin HG, Joseph GK, and Nowakowski M. Nitric oxide production by human alveolar macrophages in pulmonary disease. (1996) Ann NY Acad Sci 797:246–249

Stuehr DJ (1999) Mammalian nitric oxide synthases. Biochim Biophys Acta 1411:217–230

Stuehr DJ, Ghosh S (2000) Enzymology of nitric oxide. Handbook Exp Pharm, Mayer B., Ed., 143:33–70

Stuehr DJ, Marletta MA (1985) Mammalian nitrate biosynthesis: mouse macrophages produce nitrite and nitrate in response to Escherichia coli lipopolysaccharide. Proc Natl Acad Sci USA 82:7737–7742

Subczynski WK, Hyde JS (1983) Concentration of oxygen in lipid bilayers using a spin-label method. Biophys J 41:283–286

Tauber AI, Borregaard N, Simons E, Wright (1983) Chronic granulomatous disease: a syndrome of phagocyte oxidase deficiencies. J. Med (Baltimore) 62:286–309

Taylor BS, Geller DA (2000) Molecular regulation of the human inducible nitric oxide synthase (iNOS) gene. Shock 13:413–424

Thomas DD, Espey MG, Vitek MP, Miranda KM, Wink DA (2002) Protein nitration by nitrite and peroxide is mediated by heme and free iron through Fenton-type chemistry: an alternative to the NO/O_2^- reaction. Proc Natl Acad Sci USA 99:12691–12696

Tsuruta T, Tani K, Shimane M, Ozawa K, Takahashi S, Tsuchimoto D, Takahashi K, Nagata S, Sato N, Asano S. (1996) Effects of myeloid cell growth factors on alkaline phosphatase, myeloperoxidase, defensin and granulocyte colony-stimulating factor receptor mRNA expression in haemopoietic cells of normal individuals and myeloid disorders. Br J Haematol 92:9–22

van Dalen CJ, Winterbourn CC, Senthilmohan R, Kettle AJ (2000) Nitrite as a substrate and inhibitor of myeloperoxidase. Implications for nitration and hypochlorous acid production at sites of inflammation. J Biol Chem 275:11638–11644

van der Vliet A, Eiserich JP, Halliwell B, Cross CE (1997) Formation of reactive nitrogen species during peroxidase-catalyzed oxidation of nitrite. A potential additional mechanism of nitric oxide-dependent toxicity. J Biol Chem 272:7617–7625

Vazquez-Torres A, Fang FC (2001) Salmonella evasion of the NADPH phagocyte oxidase. Microbes Infect 3:1313–1320

Vazquez-Torres A, Jones-Carson J, Mastroeni P, Ischiropoulos H, Fang FC (2000) Antimicrobial actions of the NADPH phagocyte oxidase and inducible nitric oxide synthase in experimental salmonellosis. I. Effects on microbial killing by activated peritoneal macrophages in vitro. J Exp Med 192:227–236

Venema RC, Ju H, Zou R, Ryan JW, Venema VJ (1997) Subunit interactions of endothelial nitric-oxide synthase. Comparisons to the neuronal and inducible nitric-oxide synthase isoforms. J Biol Chem 272:1276–1282
Vitek MP, Snell J, Dawson H, Colton CA (1997) Modulation of nitric oxide production in human macrophages by apolipoprotein-E and amyloid-beta peptide. Biochem Biophys Res Commun 240:391–394
Vodovotz Y (1997) Control of nitric oxide production by transforming growth factor-beta1: mechanistic insights and potential relevance to human disease. Nitric Oxide 1:3–17
Vodovotz Y, Russell D, Xie QW, Bogdan C, Nathan C (1995) Vesicle membrane association of nitric oxide synthase in primary mouse macrophages. J Immunol 154:2914–2925
Webb JL, Harvey MW, Holden DW, Evans TJ (2001) Macrophage nitric oxide synthase associates with cortical actin but is not recruited to phagosomes. Infect Immun 69:6391–400
Weinberg JB (1998) Nitric oxide production and nitric oxide synthase type 2 expression by human mononuclear phagocytes: a review. Mol Med 4:557–591
Wilcox CS (2002) Reactive oxygen species: roles in blood pressure and kidney function. Curr Hypertens Rep 4:160–166
Wilhelm J, Frydrychova M, Hezinova A, Vizek M (1997) Production of hydrogen peroxide by peritoneal macrophages from rats exposed to subacute and chronic hypoxia. Physiol Res 46:35–39
Williams DLH (1988) In Nitrosation, Cambridge Press, Oxford, United Kingdom
Wink DA, Darbyshire JF, Nims RW, Saavedra JE, Ford PC (1993) Reactions of the bioregulatory agent nitric oxide in oxygenated aqueous media: determination of the kinetics for oxidation and nitrosation by intermediates generated in the NO/O2 reaction. Chem Res Toxicol 6:23–27
Wink DA, Cook JA, Kim SY, Vodovotz Y, Pacelli R, Krishna MC, Russo A, Mitchell JB, Jourd'heuil D, Miles AM, Grisham MB. (1997) Superoxide modulates the oxidation and nitrosation of thiols by nitric oxide-derived reactive intermediates. Chemical aspects involved in the balance between oxidative and nitrosative stress. J Biol Chem 272:11147–11151
Wink DA, Cook JA, Krishna MC, Hanbauer I, DeGraff W, Gamson J, Mitchell JB. (1995) Nitric oxide protects against alkyl peroxide-mediated cytotoxicity: further insights into the role nitric oxide plays in oxidative stress. Arch Biochem Biophys 319:402–407
Wink DA, Miranda KM, Espey MG, Pluta RM, Hewett SJ, Colton C, Vitek M, Feelisch M, Grisham MB (2001) Mechanisms of the antioxidant effects of nitric oxide. Antioxid Redox Signal 3:203–213
Wink DA, Miranda KM, Espey MG, Mitchell JB, Grisham MB, Fukuto J, Feelisch M (2000) The chemical biology of NO. Balancing NO with oxidative and nitrosative stress. Handbook of Exp Pharm, Mayer B, Ed., Springer Verlag, Berlin, 143:7–32
Wink DA, Wink CB, Nims RW, Ford PC (1994) Oxidizing intermediates generated in the Fenton reagent: Kinetic arguments against the intermediacy of the hydroxyl radical. Environ Health Perspec 102:11–15
Winterbourn CC, Vissers MC, Kettle AJ (2000) Myeloperoxidase. Curr Opin Hematol 7:53–58
Wung BS, Cheng JJ, Shyue SK, Wang DL (2001) NO modulates monocyte chemotactic protein-1 expression in endothelial cells under cyclic strain.Arterioscler Thromb Vasc Biol 21:1941–1947
Wymann MP, Kernen P, Deranleau DA, Baggiolini M (1989) Respiratory burst oscillations in human neutrophils and their correlation with fluctuations in apparent cell shape. J Biol Chem 264:15829–15834

Xie QW, Cho HJ, Calaycay J, Mumford RA, Swiderek KM, Lee TD, Ding A, Troso T, Nathan C (1992) Cloning and characterization of inducible nitric oxide synthase from mouse macrophages. Science 256:225–228

Ye YZ, Strong M, Huang ZQ, Beckman JS (1996) Antibodies that recognize nitrotyrosine. Methods Enzymol 269:201–209

Zamora R, Billiar TR (2000) Nitric oxide: A true inflammatory mediator. Handbook Exp Pharm, Mayer B, Ed., 143:493–524

Zamora R, Vodovotz Y, Billiar TR (2000) Inducible nitric oxide synthase and inflammatory diseases. Mol Med 6:347–373

Zeng C, Morrison AR (2001) Disruption of the actin cytoskeleton regulates cytokine-induced iNOS expression. Am J Physiol Cell Physiol 281:C932–40

Zhou M, Brown EJ (1993) Leukocyte response integrin and integrin-associated protein act as a signal transduction unit in generation of a phagocyte respiratory burst. J Exp Med. 178:1165–1174

Zhu L, Gunn C, Beckman JS (1992) Bactericidal activity of peroxynitrite. Arch Biochem Biophys 298:452–457

Proteases

J. A. Mahoney

Department of Medicine,
The Johns Hopkins University School of Medicine,
Baltimore, MD 21205, USA, e-mail: jmahoney@jhmi.edu

Abstract The macrophage, as a gatekeeper to both the innate and acquired immune systems, has great potential as a therapeutic target for such diverse human disease states as bacterial and viral infection, autoimmunity, inflammatory diseases, and cancer. The phenotype of macrophages in different tissues varies markedly between tissues. While this characteristic creates technical challenges in terms of isolation and characterization of resident tissue macrophages, it opens the possibility of targeting individual tissue-specific macrophage populations for pharmacologic intervention. The proteases are among the most numerous and abundant of enzyme classes, representing 1%–4% of all proteins encoded by eukaryotic genomes. Proteases are particularly abundant in macrophages, where they are critical players in many key functions of the macrophage, such as degradation of exogenous, potentially pathogenic proteins; digestion of both foreign and self proteins into peptides for presentation by MHC class I and II; and functional regulation of target proteins, for example by removal of a regulatory domain or a transmembrane anchor. This chapter reviews some of the proteases expressed in macrophages, and discusses what functional roles have been shown for, or postulated for, these enzymes. The enzymes discussed here are divided into two main groups: ectoproteases, which cleave amino acids from either end of a protein or peptide, and endoproteases, which cleave proteins at internal sites. Examples are given illustrating the actions of proteases within the macrophage, at the cell surface, and after secretion into the extracellular milieu.

Keywords Aminopeptidase, Angiotensin converting enzyme, Carboxypeptidase, Caspase, Cathepsin, CPVL, Matrix metalloprotease, TNF-? converting enzyme

1 Introduction

The role of macrophages lies at the interface of the innate and adaptive immune systems. Innate immune functions include phagocytosis of both unopsonized and opsonized pathogens, release of toxic free radicals, and secretion of inflammatory mediators such as cytokines, chemokines, and a large variety of other extracellular signaling molecules. Within the adaptive immune response, macrophages process and present antigen to T cells, and are capable of providing both immunogenic and tolerogenic signaling through secretion of cytokines and other soluble mediators. This unique immunomodulatory role of macrophages makes them ideal candidates for pharmacological intervention, with the potential to treat a highly diverse set of human diseases, including bacterial and viral infections, autoimmunity, inflammatory conditions, and cancer. An important hallmark of macrophages is their ability to adapt to their cellular surroundings, leading to extreme phenotypic diversity of macrophages. While this diversity presents the researcher with certain challenges, it also represents a unique opportunity for pharmacological intervention. If the nature of this diversity is understood, it may be possible to treat restricted subsets of macrophages without affecting others, thus greatly increasing drug specificity.

The proteases are among the most numerous and abundant of enzyme classes. The MEROPS database (Rawlings et al. 2002) (http://www.merops.sanger.ac.uk) lists nearly 400 different human proteases for which a chromosomal location has been mapped, representing more than 1% of the human genome. Analysis of the more completely characterized eukaryotic genomes has shown that proteases comprise between 1.7% and 3.9% of expressed genes, suggesting the presence of a large number of unknown or poorly characterized human proteases. Proteases can be subdivided functionally into those that can cleave internal polypeptide sequences (endopeptidases) and those that cleave only from one end of the substrate molecule (ectopeptidases). The ectopeptidases are further divided into enzymes cleaving from the N and C termini, called aminopeptidases and carboxypeptidases, respectively. Proteases perform a wide variety of physiological functions throughout the body, both inside and outside the cell. From a pharmacological standpoint, a comprehensive understanding of protease function is critical, both for identification of new drug targets and to anticipate and ameliorate the side effects associated with protease inhibition. Known protease functions include: (1) intracellular destruction of proteins that are senescent, misfolded, or expressed cyclically; (2) breakdown of foreign, potentially pathogenic proteins; (3) digestion of both foreign and self proteins into peptides for presentation by MHC class I and II; (4) activation and execution of cell death cascades, acting either on itself or on an adjacent target cell; (5) functional activation and/or inactivation of enzymes, bioactive peptides, and many other types of proteins by proteolytic removal of a regulatory domain; (6) release of proteins from the plasma membrane or other membrane-bound compartment by cleavage of a transmembrane domain; and (7) digestion of proteins in the digestive tract for nutritional purposes.

Since the vast array of proteases in the human body display widely variable specificities and inhibitor sensitivities, the value of proteolysis as a pharmacologic target is clear. Targeting proteolytic pathways has led to such drug successes as angiotensin converting enzyme inhibitors for high blood pressure (Douglas 1985) (see below), the plasmin-targeted thrombolytic agents such as streptokinase and urokinase (Reilly 1985), and HIV protease inhibitors as part of combination therapy for HIV infection (Hammer et al. 1997; Gulick et al. 1997). Proteases are particularly abundant in macrophages, where they perform a wide variety of functions. These functions, which will be described in more detail below, include destruction of phagocytosed material, trimming of peptides for presentation by MHC class II molecules, alterations of extracellular matrix components, and a variety of regulatory roles. This chapter will provide an overview of the properties of proteases found in macrophages, and will attempt to highlight some areas of interest for further basic biochemical study and potential pharmacologic intervention.

2
Ectoproteases

Ectoproteases differ from the more abundant endoproteases in that they cleave substrates only at the carboxy or amino terminus. This functional difference has a structural basis: the substrate binding sites of ectoproteases tend to be solvent-accessible at only one end, thus allowing cleavage at the end of a protein but not in the middle of a polypeptide loop. This difference has important implications in design of synthetic inhibitors. Properties of ectoproteases found in macrophages are discussed below. Alternate names are shown in parentheses.

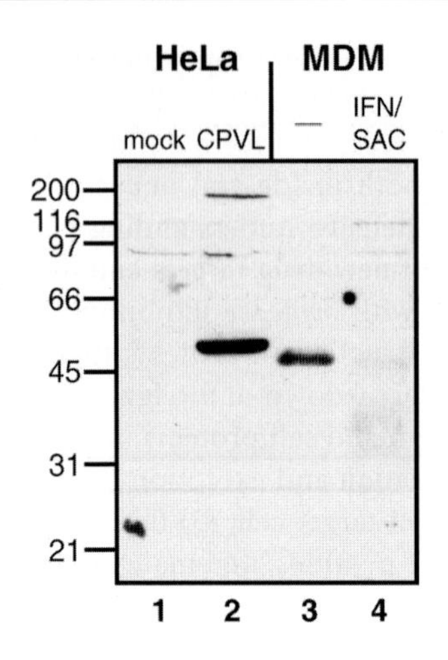

Fig. 1 Regulation of CPVL expression by macrophage activation state. Polyclonal anti-CPVL Western blot of lysates from mock- and CPVL-transfected HeLa cells (*lanes 1 and 2*) shows a major specific band of about 48 kDa. Human monocyte-derived macrophages (*MDM, lanes 3 and 4*) were cultured for approximately 10 days in RPMI 1640 plus 10% fetal bovine serum, either alone or with IFN-γ and *Staphylococcus aureus* cells (*SAC*, a source of lipopolysaccharide). Activation led to complete loss of the immunoreactive band. Equal protein loading was confirmed by Ponceau S staining (not shown). Note that recombinant CPVL migrates slightly more slowly because of an N-terminal epitope tag

2.1 Carboxypeptidase, Vitellogenic-Like (CPVL, CP-Mac)

CPVL (MEROPS ID S10.003) is a 476 amino acid serine carboxypeptidase, discovered as a result of a differential display polymerase chain reaction (PCR) screen for novel macrophage-specific genes (Mahoney et al. 2001). RT-PCR, Northern blot, and Western blot analysis confirm that, among hematopoietic cells, CPVL is indeed restricted to the monocytic lineage. CPVL mRNA was readily detected throughout the lineage, whereas protein expression was absent or low in monocytes and relatively abundant in mature monocyte-derived macrophages. Outside the immune system, however, a wider expression pattern was apparent. High levels of expression, as judged by Northern blot, were apparent in kidney and heart, two organs with few macrophages, while lung and liver, which have much larger macrophage populations, expressed little or no CPVL mRNA. We postulate that CPVL has two distinct expression profiles, one in a subset of tissue macrophages and a second one, presumably representing a separate function, in organs of the cardiovascular system. This pattern is similar to the expression pattern of angiotensin-converting enzyme (ACE, see below), which also shows significant expression in macrophages, heart, and kidney, although ACE is detectable in virtually all organs because of its presence in vascular endothelial cells (Dzau et al. 2001). Moreover, an alternate isoform of ACE is highly expressed in testis. Surveys of the human expressed sequence tags (EST) database suggest that CPVL is also expressed in testis.

The physiological function of CPVL is not currently known. Both primary sequence analysis and pulse-chase experiments (unpublished studies) suggest a lumenal and/or secreted distribution. In preliminary immunocytochemical studies using confocal microscopy, CPVL showed cytoplasmic expression in a vesicular pattern that did not coincide with endosomes, lysosomes, or MHC class II peptide-loading compartments (B. Ntolosi, R. DaSilva, and S. Gordon, personal communication). Protein levels of CPVL are strikingly modulated by inflammatory stimuli: Culture of developing macrophages in the presence of interferon (IFN)-γ and lipopolysaccharide causes a dramatic downregulation of cellular CPVL expression (Fig. 1). While a great deal more work is required to ascertain the importance of this macrophage-restricted protease, we speculate that CPVL may play a role in the macrophage inflammatory response.

2.2
Carboxypeptidase M (CPM)

CPM (MEROPS ID M14.006) is a 439-amino acid glycosylphosphatidyl inositol-linked, membrane-bound metallo-carboxypeptidase (Tan et al. 1989) with preference for removal of lysine or arginine. While it is found in several different cell types, it is strongly expressed in monocytic lineage cells. Just as is the case with CPVL, CPM expression is upregulated during maturation of monocytes to macrophages (Rehli et al. 1995). CPM cleaves the C-terminal basic amino acid from a variety of biologically active peptide substrates, including bradykinin, dynorphin A(1–13), and enkephalins (Skidgel et al. 1989). It has been suggested, though not yet explicitly shown, that CPM would cleave and thus inactivate the anaphylatoxins C3a/C4a/C5a, in the same way as the related liver enzyme carboxypeptidase N (Rehli et al. 2000).

The level of CPM expression in vivo is highly dependent on the activation and/or differentiation state of the cells. Expression in macrophages of secondary lymph organs is low, but macrophage CPM expression during rejection of kidney transplants is much higher, and this elevated expression is inhibited by cyclosporin treatment (Andreesen et al. 1988). Moreover, monocytes from patients with aplastic anemia (Andreesen et al. 1989) or HIV (Andreesen et al. 1990) do not display maturation-induced CPM upregulation, even when cultured with healthy serum. These data suggest that CPM activation may be part of a macrophage inflammatory process. While these types of data show a compelling correlation between CPM levels and macrophage activation, proof of the importance of CPM in macrophage activation will require more understanding of its physiologically relevant substrates.

2.3
CD13 (Aminopeptidase N, Alanine aminopeptidase)

CD13 (MEROPS ID M01.001) is, like CPM, a plasma membrane-bound ectoprotease, consisting of 967 amino acids and one N-terminal transmembrane domain (Olsen et al. 1988). CD13 is ubiquitously expressed, and highly expressed in monocytic and granulocytic cells. As an aminopeptidase, CD13 removes single amino acids from the N termini of proteins and peptides. Like most ectopeptidases, CD13 is capable of removing amino acids from small bioactive peptides. Strikingly, removal of a single amino terminal residue from the chemokine monocyte chemotactic protein (MCP)-1 converts this basophil-activating chemokine into an eosinophil-activating one (Weber et al. 1996), taking advantage of the alternate expression of chemokine receptors with overlapping specificity in these cell types. However, its primary role appears to be the trimming of antigenic peptides bound to MHC molecules (Larsen et al. 1996). Therefore, the relative efficiency of CD13 to perform this function on antigen-presenting cells may significantly affect the balance of epitopes presented to T cells, with important ramifications in autoimmunity.

In addition to these physiologic roles, CD13 is also the receptor for coronaviruses to attach to endothelial cells of the upper respiratory tract (Yeager et al. 1992). Cytomegalovirus also uses CD13 as its receptor, as evidenced by the inhibition of both viral binding and infection with anti-CD13 antibodies in vitro (Soderberg et al. 1993). Cytomegalovirus binding to targets can lead to production of chronic graft-versus-host disease and pathogenic anti-CD13 autoantibodies (Soderberg et al. 1996).

2.4 Lysosomal Protective Protein (LPP, Cathepsin A, PPCA, Lysosomal Carboxypeptidase A)

LPP (MEROPS ID S10.002) is a lysosomally localized serine carboxypeptidase that, along with CPVL and a smooth muscle cell protein called RISC (Chen et al. 2001), make up the only three serine carboxypeptidases known in mammals. LPP was first isolated as the gene mutated in the human lysosomal storage disease galactosialidosis (Galjart et al. 1988), a syndrome caused by instability of lysosomal beta galactosidase and neuraminidase in the absence of a 54-kDa "protective protein," which normally protects these enzymes from degradation in the harsh lysosomal environment. Phenotypes vary by exact mutation, but generally include dwarfism, mental retardation, and a macular cherry-red spot. Sequence analysis showed similarity to serine proteases, and serine carboxypeptidase activity was soon confirmed but, importantly, shown not to be required for the protective function (Galjart et al. 1991).

LPP has three distinct enzymatic activities: an esterase activity, a deamidase activity, and a carboxypeptidase activity (Jackman et al. 1990). The protein is made as a 452 amino acid, 54-kDa precursor, which is then cleaved into 32-kDa and 20-kDa subunits (Pshezhetsky 1998). Mature LPP cleaves a variety of bioactive peptides in vitro, including bradykinin, endothelin I, substance P, and oxytocin (Jackman et al. 1990; Pshezhetsky 1998). However, the physiologically relevant substrates are not known.

Significant progress has recently been made in the treatment of a galactosialidosis model disease in mice (LPP knockout mice), which raises the possibility that this disease may be treatable in humans. D'Azzo and colleagues transplanted LPP knockout mice with bone marrow cells transduced to overexpress LPP under the influence of the colony stimulating factor-1 promoter, which directs expression to monocytes and macrophages. Since bone marrow-derived monocytic lineage cells traffic to essentially all tissues and secrete LPP, resident cells may take up the secreted LPP, thus curing the defect in their own lysosomes. The treated mice showed marked reduction in symptoms and histopathology, with virtually all but the loss of cerebellar Purkinje cells corrected (Hahn et al. 1998). Their most recent model used murine stem cell virus to stably infect bone marrow cells with LPP and green fluorescent protein from a bicistronic vector (Leimig et al. 2002). The treated mice showed marked improvement for many months, including sparing of Purkinje cells. As this impressive work continues on the protective role LPP and its link to galactosialidosis, little information has emerged so far on the physiological role of the enzyme activities of this protein. Lysosomal storage disorders are discussed in detail in Chap. 11.

2.5
Angiotensin Converting Enzyme (ACE, Peptidyl Dipeptidase, Kininase II)

ACE (MEROPS ID X06.001) is a 1,306 amino acid cell surface bound protein containing two independent metalloprotease domains (Soubrier et al. 1988). It is widely expressed in somatic cells, and an alternate form with only the C-terminal protease domain is expressed only in male germ cells. ACE is a dicarboxypeptidase, cleaving two amino acids from the C terminus of the inactive angiotensin I, thus creating the powerful vasoconstrictor angiotensin II. Inhibitors of ACE such as captopril, enalapril, and numerous others have been extremely valuable agents for controlling hypertension (Douglas 1985).

Macrophages express high levels of ACE, and several reports within the past few years have emphasized the pathophysiological importance of macrophage ACE. It has long been known that ACE inhibition is beneficial in the treatment of atherosclerosis. Diet et al. showed that ACE accumulates in atherosclerotic plaques, and that the main source of ACE is foam cells, the characteristic lipid-laden macrophages of atherosclerosis (Diet et al. 1996). Moreover, they showed that differentiation of the monocytic cell line THP-1 into a macrophage phenotype led to an increase in ACE activity, and that increase was potentiated by addition of acetylated LDL. The mechanism of this effect is unknown, but probably involves inhibition of inflammatory mediators such as MCP-1 and interleukin (IL)-12 in the macrophages (Hernandez-Presa et al. 1997; Constantinescu et al. 1998). Finally, treatment of human mononuclear cells with ACE inhibitors in vitro decreased the synthesis of tissue factor, the clotting cascade initiator implicated in arterial thrombosis (Napoleone et al. 2000).

3
Endoproteases

3.1
Cathepsins

The term cathepsins does not refer to a group of proteins related by evolution, but rather by location and function. Cathepsins are a group of lysosomal proteases, most of which are involved in the degradation of phagocytosed or endocytosed products. They can be of any enzyme class, although most are cysteine proteases. Many of the cathepsins are ubiquitously expressed, but as the numbers of known cathepsins increases, some cell type-specific examples are emerging.

Cathepsins B and D (MEROPS IDs C01.060, A01.009) have been implicated in the degradation of apolipoproteins in macrophages (Kuroda et al. 1994). Apolipoproteins from oxidized LDL particles are not digested well by macrophages, potentially leading to accumulation of foam cell macrophages and atherosclerosis. This inhibition of apolipoprotein digestion appears to be mediated by oxidized LDL inhibition of cathepsin B, via an unknown mechanism (Hoppe et al. 1994).

Cathepsin K (MEROPS ID C01.036) is one of the most cell type-specific of this group. While cathepsin K was originally thought to be expressed only on osteoclasts, the macrophage lineage cells responsible for bone resorption (see Chap. 19, this volume), recent evidence suggests that macrophages involved in foreign body responses, such as multinucleated giant cells or epithelioid cells in granulomas, also express it (Buhling et al. 2001). Resident tissue macrophages did not express cathepsin K, whereas cathepsins B and L were expressed on both resident and foreign body-elicited macrophages.

A final example of macrophage-specific cathepsin function comes from the study of processing of the invariant chain, Ii, by antigen-presenting cells. The invariant chain associates with nascent MHC class II molecules, to prevent binding of endogenous antigens. When the class II molecule enters the endosomal compartment, Ii is cleaved by a cathepsin, leaving only the small class II-associated invariant-chain peptide (CLIP) in the MHC groove, to be exchanged for a newly processed antigenic peptide. Chapman and colleagues showed that cathepsin S (MEROPS ID C01.034) is required for Ii cleavage in B cells and dendritic cells (Shi et al. 1999). They went on to show that MHC class II presentation in macrophages of cathepsin S knockout mice was normal, and identified a novel protease, cathepsin F (MEROPS ID C01.018), responsible for this activity in macrophages (Shi et al. 2000).

3.2 Caspase-1 (Interleukin 1β Converting Enzyme, ICE)

Caspases are cysteine endoproteases that cleave after Asp residues, in the context of a four amino acid recognition motif, in a wide variety of protein substrates. Caspases are best known for their involvement in the apoptosis cascade (Earnshaw et al. 1999). However, a subset of caspases are primarily involved in proteolytic release of cytokine precursors from the membrane, for action at a distant site. The best-known example is caspase-1 (MEROPS ID C14.001), which was identified by its ability to release IL-1β from monocytes and macrophages (Thornberry et al. 1992). Mice deficient in caspase-1 cannot release IL-1β, and thus are resistant to septic shock, an IL-1β-dependent process (Li et al. 1995). It was subsequently shown that caspase-1 also catalyzes the release of IL-18, or interferon γ inducing factor (Ghayur et al. 1997; Gu et al. 1997). These data raise the possibility that a specific inhibitor of caspase-1 may be a useful treatment for sepsis. In the meantime, this area has provided an explanation for the regulatory effect of nitric oxide on inflammatory cytokine release. Nitric oxide potently inhibits cysteine proteases by S-nitrosylation of the active site cysteine. Kim et al. showed that activated macrophages treated in vitro with a nitric oxide synthase inhibitor released fourfold more IL-1β than those untreated (Kim et al. 1998). Furthermore, mice deficient in inducible nitric oxide synthase produced more IL-1β and more interferon γ in response to challenge with endotoxin.

3.3 Proteases Secreted to Act on Their Environment

3.3.1 Macrophage Gelatinase (Matrix Metalloprotease 9, MMP9, Gelatinase B)

MMP9 (MEROPS ID M10.009) is a 707 amino acid zinc metalloprotease (Wilhelm et al. 1989) and member of the matrix metalloprotease (MMP) family, a large (>20 different genes discovered) family of zinc metalloproteases responsible for the clearance and remodeling of the extracellular matrix, with downstream effects in areas such as development and wound healing (Nagase and Woessner 1999). The MMPs are highly regulated by gene expression, by synthesis as inactive preproenzymes, and by the presence of inhibitors such as the tissue inhibitors of metalloproteases (TIMPs). Unlike some other groups of proteases, such as the cathepsins, MMPs are generally not expressed in normal tissue, but expression is induced by a variety of stimuli, such as cytokines, growth factors, and others.

MMP9 is expressed on macrophages and neutrophils, and like all gelatinases, degrades a variety of extracellular components such as collagens, elastin, and fibronectin. However, the physiological roles played by this enzyme (and the others in this section; see below) are surprisingly broad, as demonstrated by the results of knockout studies. MMP9 knockout mice showed defects in bone formation and vascularization, caused by a lack of MMP9 in chondroclasts, the multinucleated, bone marrow-derived cells that resorb cartilage (Vu et al. 1998). Other studies showed that MMP9 knockout mice had reduced capacity for outgrowth of oligodendrocyte processes in the developing brain (Oh et al. 1999), and diminished ability for exogenously implanted tumors to metastasize (Itoh et al. 1999). Intriguingly, the tumors did not express MMP9, but rather required MMP9 secreted from host cells for successful metastasis. Finally, MMP9 has been shown to cleave a short amino terminal peptide from the neutrophil chemokine IL-8, and this modified IL-8 was more than tenfold more potent in neutrophil activation and chemotaxis assays (Van den Steen et al. 2000). Therefore, MMP9 has pleiotropic effects because of its matrix proteolytic functions, and moreover, exerts immunostimulatory effects by modification of a chemokine.

3.3.2
Macrophage Metalloelastase (MME, MMP12)

MMP12 (MEROPS ID M10.009) is a 470 amino zinc metalloprotease (Shapiro et al. 1993) in the MMP family. MMP12 degrades elastin and other extracellular matrix components. Consistent with the notion that many MMPs are inducible proteins, MMP12 signal is only detected in placenta (a highly macrophage-enriched tissue) by Northern blotting (Belaaouaj et al. 1995). Experiments with MMP12 promoter constructs indicated that MMP gene expression was induced by LPS in a mouse macrophage cell line, but not in human umbilical vein endothelial cells. Knockout studies showed that MMP12 was necessary for matrix degradation by macrophages in vitro and in vivo. MMP12$^{-/-}$ macrophages had reduced proteolytic activity against insoluble elastin in vitro, and the ability to penetrate Matrigel artificial basement membranes in vitro and in vivo was abolished (Shipley et al. 1996). Strikingly, the knockout conferred complete protection in an experimental model of cigarette smoke-induced emphysema (Hautamaki et al. 1997). Control mice exposed to cigarette smoke for 3 months showed immunohistochemical evidence of MMP12-positive macrophage recruitment to the lungs, and increased mean alveolar air space. Neither effect was detectable with MMP12$^{-/-}$ mice. If macrophage recruitment of MMP12$^{-/-}$ macrophages was artificially induced by adding the chemokine MCP-1, the (MMP12-negative) macrophages were detected in the lung, but there was still no change in mean alveolar air space. These experiments suggest that an MMP12-specific inhibitor has potential therapeutic value in the setting of pathological macrophage recruitment.

3.3.3
Leukocyte Elastase (LE, Neutrophil Elastase)

LE (MEROPS ID S01.131), despite sharing many properties with MMPs 9 and 12, is not a member of the MMP family, but rather is a 218 amino acid serine endoprotease (Sinha et al. 1987). Although LE is expressed in macrophages, it is most highly expressed in neutrophils, where it can be the cause of destructive lung disease (Mitsuhashi et al. 1999). It was recently reported that cystic fibrosis patients have impaired removal of apoptotic inflammatory cells by macrophages. This defect is caused by LE-mediated cleavage of the phosphatidylserine receptor that recognizes apoptotic cells for uptake (Vandivier et al. 2002). The lost capacity to dispose of toxic mediators from apoptotic inflammatory cells may exacerbate the deleterious effects of these cells.

3.4 Proteases Acting at the Cell Surface

3.4.1 TNF-α Converting Enzyme (TACE, ADAM17, CD156b)

Tumor necrosis factor (TNF)-α is a powerful pro-inflammatory cytokine that, like IL-1β, is synthesized in a plasma membrane-bound form, and is then released into the extracellular space by TACE (MEROPS ID M12.217). TACE is an 824 amino acid member of the ADAM (the name is derived from a disintegrin and metalloprotease) family, a group of over 40 proteins containing a disintegrin domain that binds integrins and a metalloprotease domain similar to those in the MMP family (Black et al. 1997; Moss et al. 1997). While TACE is ubiquitously expressed, it is highly expressed in some macrophage populations, where TNF-α is made and secreted. Because TNF-α is often implicated in harmful inflammatory pathways, there has been great interest in devising methods to inactivate TACE. Methods for inhibiting TACE include use of the natural inhibitor TIMP-3 (Amour et al. 1998), creation of a recombinant dominant negative form of the enzyme (Solomon et al. 1999), and development of small molecule inhibitors (Barlaam et al. 1999).

Surprisingly, attempts to create a TACE knockout mouse led to the finding that TACE has a much broader substrate specificity. Mice carrying a mutation in the Zn binding site of TACE had a large number of developmental abnormalities, and most died between embryonic day 17.5 and 1 day after birth (Peschon et al. 1998). Their mutations were reminiscent of those in mice deficient in transforming growth factor (TGF)-α, which is also released from the plasma membrane by a cleavage event. The authors went on to show that TACE is responsible for cleavage of the ectodomains of TGF-α, L-selectin, and TNF receptor p75. Subsequent work by other groups has shown that other proteins of potential therapeutic significance are also cleaved by TACE, including the amyloid protein precursor associated with Alzheimer's disease (Buxbaum et al. 1998), and the receptor for colony stimulating factor-1, the factor critical for commitment of precursor cells to the monocytic lineage (Rovida et al. 2001).

3.4.2 Macrophage Mannose Receptor Secretase

Macrophage mannose receptor (MMR) is a 180-kDa glycoprotein expressed on the plasma membrane of macrophages, some dendritic cells, and a few isolated endothelial cell types (Linehan et al. 2000). Eight C-type lectin domains mediate its functions as a phagocytic and endocytic receptor, recognizing mannose- and fucose-containing structures. A second carbohydrate recognition domain, the N-terminal cysteine-rich (CR) domain, mediates binding to sulfated sugars on ligands such as sialoadhesin and CD45 expressed on marginal zone macrophages and in germinal centers (Martinez-Pomares et al. 1996; Martinez-

Pomares et al. 1999). MMR is released from macrophages in vitro and in vivo by a metalloprotease-type secretase (Martinez-Pomares et al. 1998). Martinez-Pomares and Gordon have proposed that MMR may transport polysaccharide antigens to secondary lymphoid organs for generation of immune responses (Martinez-Pomares and Gordon 1999). To date little is known about the nature of the MMR secretase, except that it is present on macrophages, and is susceptible to hydroxamate-based inhibitors. Given these facts, one may speculate that MMR secretase is, in fact, TACE. However, since most of the over 40 members of the ADAM family share the same basic domain structure as TACE, and as yet have no described function, there is no shortage of good candidates.

4 Summary

Proteases are critical players in many of the central functions of macrophages, including digestion of phagocytosed material, processing of foreign antigens for presentation on MHC class II molecules, tissue remodeling, and regulation of immune responses by activating, inactivating, or releasing from membranes a host of immune-active proteins and peptides. The task of understanding the roles of these enzymes is complicated by the fact that many proteases have multiple and overlapping functions. However, with the appropriate tools and insight, assisted by the arrival of whole genome data sets, the prospects for major advances in understanding the pathophysiology of this system are bright, leading to significant advancements in the treatment of human disease.

Acknowledgements. This work was supported by National Institutes of Health Grant DE12354. I thank Dr. Antony Rosen for his support and encouragement.

5 References

Amour A, Slocombe PM, Webster A, Butler M, Knight CG, Smith BJ, Stephens PE, Shelley C, Hutton M, Knauper V, Docherty AJ, Murphy G (1998) TNF-alpha converting enzyme (TACE) is inhibited by TIMP-3. FEBS Lett 435: 39–44

Andreesen R, Gadd S, Costabel U, Leser HG, Speth V, Cesnik B, Atkins RC (1988) Human macrophage maturation and heterogeneity: restricted expression of late differentiation antigens in situ. Cell Tissue Res 253: 271–279

Andreesen R, Brugger W, Thomssen C, Rehm A, Speck B, Lohr GW (1989) Defective monocyte-to-macrophage maturation in patients with aplastic anemia. Blood 74: 2150–2156

Andreesen R, Brugger W, Kunze R, Stille W, von Briesen H (1990) Defective monocyte to macrophage maturation in human immunodeficiency virus infection. Res Virol 141: 217–224

Barlaam B, Bird TG, Lambert-Van Der Brempt C, Campbell D, Foster SJ, Maciewicz R (1999) New alpha-substituted succinate-based hydroxamic acids as TNFalpha convertase inhibitors. J Med Chem 42: 4890–4908

Belaaouaj A, Shipley JM, Kobayashi DK, Zimonjic DB, Popescu N, Silverman GA, Shapiro SD (1995) Human macrophage metalloelastase. Genomic organization, chromosomal location, gene linkage, and tissue-specific expression. J Biol Chem 270: 14568–14575

Black RA, Rauch CT, Kozlosky CJ, Peschon JJ, Slack JL, Wolfson MF, Castner BJ, Stocking KL, Reddy P, Srinivasan S, Nelson N, Boiani N, Schooley KA, Gerhart M, Davis R, Fitzner JN, Johnson RS, Paxton RJ, March CJ, Cerretti DP (1997) A metalloproteinase disintegrin that releases tumour-necrosis factor- alpha from cells. Nature 385: 729–733

Buhling F, Reisenauer A, Gerber A, Kruger S, Weber E, Bromme D, Roessner A, Ansorge S, Welte T, Rocken C (2001) Cathepsin K–a marker of macrophage differentiation? J Pathol 195: 375–382

Buxbaum JD, Liu KN, Luo Y, Slack JL, Stocking KL, Peschon JJ, Johnson RS, Castner BJ, Cerretti DP, Black RA (1998) Evidence that tumor necrosis factor alpha converting enzyme is involved in regulated alpha-secretase cleavage of the Alzheimer amyloid protein precursor. J Biol Chem 273: 27765–27767

Chen J, Streb JW, Maltby KM, Kitchen CM, Miano JM (2001) Cloning of a novel retinoid-inducible serine carboxypeptidase from vascular smooth muscle cells. J Biol Chem 276: 34175–34181

Constantinescu CS, Goodman DB, Ventura ES (1998) Captopril and lisinopril suppress production of interleukin-12 by human peripheral blood mononuclear cells. Immunol Lett 62: 25–31

Diet F, Pratt RE, Berry GJ, Momose N, Gibbons GH, Dzau VJ (1996) Increased accumulation of tissue ACE in human atherosclerotic coronary artery disease. Circulation 94: 2756–2767

Douglas WW (1985) Polypeptides-Angiotensin, Plasma Kinins, and Others. In: Gilman AG, Goodman LS, Rall TW, Murad F (eds) Goodman and Gilman's the pharmacological basis of therapeutics. MacMillan Publishing Co., New York, 639–659

Dzau VJ, Bernstein K, Celermajer D, Cohen J, Dahlof B, Deanfield J, Diez J, Drexler H, Ferrari R, van Gilst W, Hansson L, Hornig B, Husain A, Johnston C, Lazar H, Lonn E, Luscher T, Mancini J, Mimran A, Pepine C, Rabelink T, Remme W, Ruilope L, Ruzicka M, Schunkert H, Swedberg K, Unger T, Vaughan D, Weber M (2001) The relevance of tissue angiotensin-converting enzyme: manifestations in mechanistic and endpoint data. Am J Cardiol 88: 1L-20L

Earnshaw WC, Martins LM, Kaufmann SH (1999) Mammalian caspases: structure, activation, substrates, and functions during apoptosis. Annu Rev Biochem 68: 383–424

Galjart NJ, Gillemans N, Harris A, van der Horst GT, Verheijen FW, Galjaard H, d'Azzo A (1988) Expression of cDNA encoding the human "protective protein" associated with lysosomal beta-galactosidase and neuraminidase: homology to yeast proteases. Cell 54: 755–764

Galjart NJ, Morreau H, Willemsen R, Gillemans N, Bonten EJ, d'Azzo A (1991) Human lysosomal protective protein has cathepsin A-like activity distinct from its protective function. J Biol Chem 266: 14754–14762

Ghayur T, Banerjee S, Hugunin M, Butler D, Herzog L, Carter A, Quintal L, Sekut L, Talanian R, Paskind M, Wong W, Kamen R, Tracey D, Allen H (1997) Caspase-1 processes IFN-gamma-inducing factor and regulates LPS-induced IFN-gamma production. Nature 386: 619–623

Gu Y, Kuida K, Tsutsui H, Ku G, Hsiao K, Fleming MA, Hayashi N, Higashino K, Okamura H, Nakanishi K, Kurimoto M, Tanimoto T, Flavell RA, Sato V, Harding MW, Livingston DJ, Su MS (1997) Activation of interferon-gamma inducing factor mediated by interleukin- 1beta converting enzyme. Science 275: 206–209

Gulick RM, Mellors JW, Havlir D, Eron JJ, Gonzalez C, McMahon D, Richman DD, Valentine FT, Jonas L, Meibohm A, Emini EA, Chodakewitz JA (1997) Treatment with indinavir, zidovudine, and lamivudine in adults with human immunodeficiency virus infection and prior antiretroviral therapy. N Engl J Med 337: 734–739

Hahn CN, del Pilar MM, Zhou XY, Mann LW, d'Azzo A (1998) Correction of murine galactosialidosis by bone marrow-derived macrophages overexpressing human protective protein/cathepsin A under control of the colony-stimulating factor-1 receptor promoter. Proc Natl Acad Sci U S A 95: 14880–14885

Hammer SM, Squires KE, Hughes MD, Grimes JM, Demeter LM, Currier JS, Eron JJ, Jr., Feinberg JE, Balfour HH, Jr., Deyton LR, Chodakewitz JA, Fischl MA (1997) A controlled trial of two nucleoside analogues plus indinavir in persons with human immunodeficiency virus infection and CD4 cell counts of 200 per cubic millimeter or less. AIDS Clinical Trials Group 320 Study Team. N Engl J Med 337: 725–733

Hautamaki RD, Kobayashi DK, Senior RM, Shapiro SD (1997) Requirement for macrophage elastase for cigarette smoke-induced emphysema in mice. Science 277: 2002–2004

Hernandez-Presa M, Bustos C, Ortego M, Tunon J, Renedo G, Ruiz-Ortega M, Egido J (1997) Angiotensin-converting enzyme inhibition prevents arterial nuclear factor-kappa B activation, monocyte chemoattractant protein-1 expression, and macrophage infiltration in a rabbit model of early accelerated atherosclerosis. Circulation 95: 1532–1541

Hoppe G, O'Neil J, Hoff HF (1994) Inactivation of lysosomal proteases by oxidized low density lipoprotein is partially responsible for its poor degradation by mouse peritoneal macrophages. J Clin Invest 94: 1506–1512

Itoh T, Tanioka M, Matsuda H, Nishimoto H, Yoshioka T, Suzuki R, Uehira M (1999) Experimental metastasis is suppressed in MMP-9-deficient mice. Clin Exp Metastasis 17: 177–181

Jackman HL, Tan FL, Tamei H, Beurling-Harbury C, Li XY, Skidgel RA, Erdos EG (1990) A peptidase in human platelets that deamidates tachykinins. Probable identity with the lysosomal "protective protein". J Biol Chem 265: 11265–11272

Kim YM, Talanian RV, Li J, Billiar TR (1998) Nitric oxide prevents IL-1beta and IFN-gamma-inducing factor (IL-18) release from macrophages by inhibiting caspase-1 (IL-1beta-converting enzyme). J Immunol 161: 4122–4128

Kuroda T, Yoshinari M, Okamura K, Okazawa K, Ikenoue H, Sato K, Fujishima M (1994) Effects of lysosomal protease inhibitors on the degradation of acetylated low density lipoprotein in cultured rat peritoneal macrophages. J Atheroscler Thromb 1: 41–44

Larsen SL, Pedersen LO, Buus S, Stryhn A (1996) T cell responses affected by aminopeptidase N (CD13)-mediated trimming of major histocompatibility complex class II-bound peptides. J Exp Med 184: 183–189

Leimig T, Mann L, Martin MP, Bonten E, Persons D, Knowles J, Allay JA, Cunningham J, Nienhuis AW, Smeyne R, d'Azzo A (2002) Functional amelioration of murine galactosialidosis by genetically modified bone marrow hematopoietic progenitor cells. Blood 99: 3169–3178

Li P, Allen H, Banerjee S, Franklin S, Herzog L, Johnston C, McDowell J, Paskind M, Rodman L, Salfeld J, . (1995) Mice deficient in IL-1 beta-converting enzyme are defective in production of mature IL-1 beta and resistant to endotoxic shock. Cell 80: 401–411

Linehan SA, Martinez-Pomares L, Gordon S (2000) Mannose receptor and scavenger receptor: two macrophage pattern recognition receptors with diverse functions in tissue homeostasis and host defense. Adv Exp Med Biol 479: 1–14

Mahoney JA, Ntolosi B, DaSilva RP, Gordon S, McKnight AJ (2001) Cloning and characterization of CPVL, a novel serine carboxypeptidase, from human macrophages. Genomics 72: 243–251

Martinez-Pomares L, Gordon S (1999) Potential role of the mannose receptor in antigen transport. Immunol Lett 65: 9–13

Martinez-Pomares L, Kosco-Vilbois M, Darley E, Tree P, Herren S, Bonnefoy JY, Gordon S (1996) Fc chimeric protein containing the cysteine-rich domain of the murine man-

nose receptor binds to macrophages from splenic marginal zone and lymph node subcapsular sinus and to germinal centers. J Exp Med 184: 1927–1937
Martinez-Pomares L, Crocker PR, Da Silva R, Holmes N, Colominas C, Rudd P, Dwek R, Gordon S (1999) Cell-specific glycoforms of sialoadhesin and CD45 are counter-receptors for the cysteine-rich domain of the mannose receptor. J Biol Chem 274: 35211–35218
Martinez-Pomares L, Mahoney JA, Kaposzta R, Linehan SA, Stahl PD, Gordon S (1998) A functional soluble form of the murine mannose receptor is produced by macrophages in vitro and is present in mouse serum. J Biol Chem 273: 23376–23380
Mitsuhashi H, Nonaka T, Hamamura I, Kishimoto T, Muratani E, Fujii K (1999) Pharmacological activities of TEI-8362, a novel inhibitor of human neutrophil elastase. Br J Pharmacol 126: 1147–1152
Moss ML, Jin SL, Milla ME, Bickett DM, Burkhart W, Carter HL, Chen WJ, Clay WC, Didsbury JR, Hassler D, Hoffman CR, Kost TA, Lambert MH, Leesnitzer MA, McCauley P, McGeehan G, Mitchell J, Moyer M, Pahel G, Rocque W, Overton LK, Schoenen F, Seaton T, Su JL, Becherer JD, . (1997) Cloning of a disintegrin metalloproteinase that processes precursor tumour-necrosis factor-alpha. Nature 385: 733–736
Nagase H, Woessner JF, Jr. (1999) Matrix metalloproteinases. J Biol Chem 274: 21491–21494
Napoleone E, Di Santo A, Camera M, Tremoli E, Lorenzet R (2000) Angiotensin-converting enzyme inhibitors downregulate tissue factor synthesis in monocytes. Circ Res 86: 139–143
Oh LY, Larsen PH, Krekoski CA, Edwards DR, Donovan F, Werb Z, Yong VW (1999) Matrix metalloproteinase-9/gelatinase B is required for process outgrowth by oligodendrocytes. J Neurosci 19: 8464–8475
Olsen J, Cowell GM, Konigshofer E, Danielsen EM, Moller J, Laustsen L, Hansen OC, Welinder KG, Engberg J, Hunziker W, . (1988) Complete amino acid sequence of human intestinal aminopeptidase N as deduced from cloned cDNA. FEBS Lett 238: 307–314
Peschon JJ, Slack JL, Reddy P, Stocking KL, Sunnarborg SW, Lee DC, Russell WE, Castner BJ, Johnson RS, Fitzner JN, Boyce RW, Nelson N, Kozlosky CJ, Wolfson MF, Rauch CT, Cerretti DP, Paxton RJ, March CJ, Black RA (1998) An essential role for ectodomain shedding in mammalian development. Science 282: 1281–1284
Pshezhetsky AV (1998) Lysosomal carboxypeptidase A. In: Barrett AJ, Rawlings ND, Woessner JF (eds) Handbook of proteolytic enzymes. Academic Press, San Diego, 393–398
Rawlings ND, O'Brien E, Barrett AJ (2002) MEROPS: the protease database. Nucleic Acids Res 30: 343–346
Rehli M, Krause SW, Kreutz M, Andreesen R (1995) Carboxypeptidase M is identical to the MAX.1 antigen and its expression is associated with monocyte to macrophage differentiation. J Biol Chem 270: 15644–15649
Rehli M, Krause SW, Andreesen R (2000) The membrane-bound ectopeptidase CPM as a marker of macrophage maturation in vitro and in vivo. Adv Exp Med Biol 477: 205–216
Reilly RA (1985) Anticoagulant, antithrombotic, and thrombolytic drugs. In: Gilman AG, Goodman LS, Rall TW, Murad F (eds) Goodman and Gilman's the pharmacological basis of therapeutics. MacMillan Publishing Co., New York, 1338–1362
Rovida E, Paccagnini A, Del Rosso M, Peschon J, Dello SP (2001) TNF-alpha-converting enzyme cleaves the macrophage colony-stimulating factor receptor in macrophages undergoing activation. J Immunol 166: 1583–1589
Shapiro SD, Kobayashi DK, Ley TJ (1993) Cloning and characterization of a unique elastolytic metalloproteinase produced by human alveolar macrophages. J Biol Chem 268: 23824–23829

Shi GP, Villadangos JA, Dranoff G, Small C, Gu L, Haley KJ, Riese R, Ploegh HL, Chapman HA (1999) Cathepsin S required for normal MHC class II peptide loading and germinal center development. Immunity 10: 197–206

Shi GP, Bryant RA, Riese R, Verhelst S, Driessen C, Li Z, Bromme D, Ploegh HL, Chapman HA (2000) Role for cathepsin F in invariant chain processing and major histocompatibility complex class II peptide loading by macrophages. J Exp Med 191: 1177–1186

Shipley JM, Wesselschmidt RL, Kobayashi DK, Ley TJ, Shapiro SD (1996) Metalloelastase is required for macrophage-mediated proteolysis and matrix invasion in mice. Proc Natl Acad Sci U S A 93: 3942–3946

Sinha S, Watorek W, Karr S, Giles J, Bode W, Travis J (1987) Primary structure of human neutrophil elastase. Proc Natl Acad Sci U S A 84: 2228–2232

Skidgel RA, Davis RM, Tan F (1989) Human carboxypeptidase M. Purification and characterization of a membrane-bound carboxypeptidase that cleaves peptide hormones. J Biol Chem 264: 2236–2241

Soderberg C, Giugni TD, Zaia JA, Larsson S, Wahlberg JM, Moller E (1993) CD13 (human aminopeptidase N) mediates human cytomegalovirus infection. J Virol 67: 6576–6585

Soderberg C, Larsson S, Rozell BL, Sumitran-Karuppan S, Ljungman P, Moller E (1996) Cytomegalovirus-induced CD13-specific autoimmunity–a possible cause of chronic graft-vs-host disease. Transplantation 61: 600–609

Solomon KA, Pesti N, Wu G, Newton RC (1999) Cutting edge: a dominant negative form of TNF-alpha converting enzyme inhibits proTNF and TNFRII secretion. J Immunol 163: 4105–4108

Soubrier F, Alhenc-Gelas F, Hubert C, Allegrini J, John M, Tregear G, Corvol P (1988) Two putative active centers in human angiotensin I-converting enzyme revealed by molecular cloning. Proc Natl Acad Sci U S A 85: 9386–9390

Tan F, Chan SJ, Steiner DF, Schilling JW, Skidgel RA (1989) Molecular cloning and sequencing of the cDNA for human membrane-bound carboxypeptidase M. Comparison with carboxypeptidases A, B, H, and N. J Biol Chem 264: 13165–13170

Thornberry NA, Bull HG, Calaycay JR, Chapman KT, Howard AD, Kostura MJ, Miller DK, Molineaux SM, Weidner JR, Aunins J, . (1992) A novel heterodimeric cysteine protease is required for interleukin-1 beta processing in monocytes. Nature 356: 768–774

Van den Steen PE, Proost P, Wuyts A, Van Damme J, Opdenakker G (2000) Neutrophil gelatinase B potentiates interleukin-8 tenfold by aminoterminal processing, whereas it degrades CTAP-III, PF-4, and GRO- alpha and leaves RANTES and MCP-2 intact. Blood 96: 2673–2681

Vandivier RW, Fadok VA, Hoffmann PR, Bratton DL, Penvari C, Brown KK, Brain JD, Accurso FJ, Henson PM (2002) Elastase-mediated phosphatidylserine receptor cleavage impairs apoptotic cell clearance in cystic fibrosis and bronchiectasis. J Clin Invest 109: 661–670

Vu TH, Shipley JM, Bergers G, Berger JE, Helms JA, Hanahan D, Shapiro SD, Senior RM, Werb Z (1998) MMP-9/gelatinase B is a key regulator of growth plate angiogenesis and apoptosis of hypertrophic chondrocytes. Cell 93: 411–422

Weber M, Uguccioni M, Baggiolini M, Clark-Lewis I, Dahinden CA (1996) Deletion of the NH2-terminal residue converts monocyte chemotactic protein 1 from an activator of basophil mediator release to an eosinophil chemoattractant. J Exp Med 183: 681–685

Wilhelm SM, Collier IE, Marmer BL, Eisen AZ, Grant GA, Goldberg GI (1989) SV40-transformed human lung fibroblasts secrete a 92-kDa type IV collagenase which is identical to that secreted by normal human macrophages. J Biol Chem 264: 17213–17221

Yeager CL, Ashmun RA, Williams RK, Cardellichio CB, Shapiro LH, Look AT, Holmes KV (1992) Human aminopeptidase N is a receptor for human coronavirus 229E. Nature 357: 420–422

Targeting the Chemokine System

Z. Johnson · A. Frauenschuh · A. E. I. Proudfoot

Serono Pharmaceutical Research Institute,
14 Chemin des Aulx, Plan les Ouates, 1228 Geneva, Switzerland
e-mail: amanda.proudfoot@serono.com

1 Introduction to the Chemokine System

The chemokine superfamily consists of small, basic, heparin-binding proteins that play a pivotal role in basal trafficking as well as in activation and recruitment of leukocytes from the circulation to sites of inflammation. The chemokines are a subset of the cytokine family and are distinguished from other cytokines in that they activate seven-transmembrane (7TM) G protein-coupled receptors. They are a large family, with approximately 50 members identified to date, and for which 19 receptors have been described. The chemokine family is divided structurally into four subfamilies CXC, CC, CX_3C and C, based on the position of the amino terminal cysteine residues. The majority of chemokines fall into the CXC or CC groups (also referred to as α and β subclasses respectively), and hence have been the most extensively studied. The known chemokine/receptor pairs are depicted in Fig. 1, which also indicates a second division based on the recent advances in chemokine biology—chemokines are either expressed constitutively and control basal trafficking or homing, or are inducible,

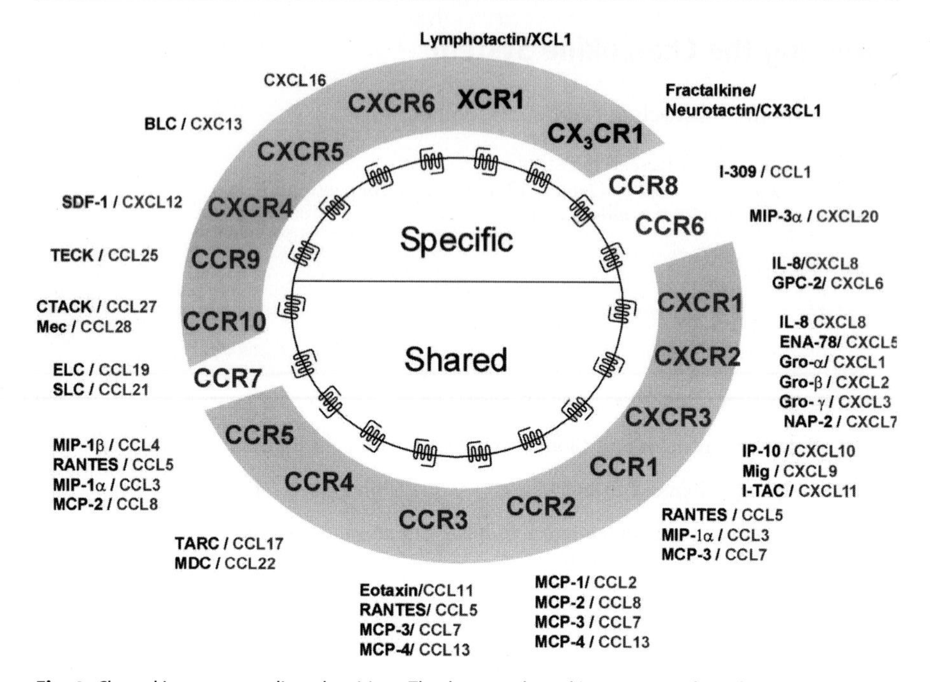

Fig. 1 Chemokine receptor–ligand pairing. The known chemokine receptor–ligand pairs are shown. Both the common names for the chemokines, as well as their systematic nomenclature are given. Receptors that bind a single ligand are classified as specific, while those that bind more than one ligand are classified as shared. Receptors which are constitutively expressed are shaded in *grey*, while those that are inducible are shaded in *orange*. This division must not be considered as absolute, since the receptors that are not shaded have overlapping properties, for example CCR6 is downregulated, while CCR7 is upregulated during the maturation process of dendritic cells, and CCR8 has been implicated in allergic inflammation, although it is expressed constitutively in the thymus. The main point to be made is that the receptors classified as inducible have been shown to play a role in inflammation

and are involved in inflammation. Chemokines were generally named according to the function that was identified such as monocyte chemoattractant protein (MCP) or neutrophil activating peptide (NAP) but since many chemokines were concomitantly identified in more than one laboratory, a single sequence was attributed more than one name. Therefore a systematic nomenclature was recently adopted (Zlotnik and Yoshie 2000) and both common and systematic names are shown in Fig. 1.

Chemokines do not necessarily have a high level of homology at the level of primary amino acid sequence, but they have a very highly conserved monomeric three-dimensional structural fold, which is conferred on them by the canonical 4-cysteine motif, which the majority possess. Their monomeric fold is superimposable for all chemokines whose structures have been solved to date (Fig. 2A), independent of the subfamily to which they belong. However, many chemokines form dimers, and the dimeric structure of CXC chemokines is very different from that of the CC chemokines as shown in Fig. 2. The CXC subclass

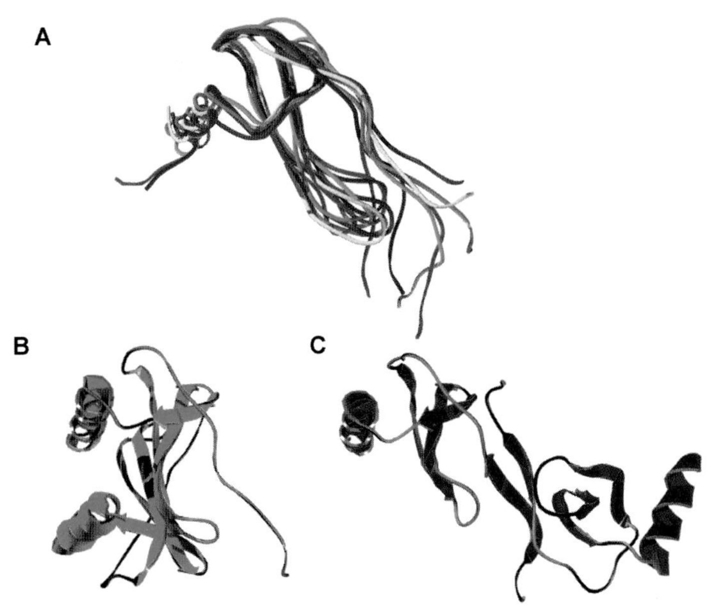

Fig. 2A–C The monomeric and dimeric topology of chemokines. **A** The superposition of seven monomeric chemokine structures. The CC chemokines are shown in *blue*: RANTES/CCL5, *dark blue*, MCP-1/CCL2, *medium blue* and MIP-1β, *light blue*; the CXC chemokines are shown in *green*: PF4/CXCL4, *dark green*, IL-8/CXCL8, *medium green* and NAP-2/CXCL7, *light green*, and the CX_3C chemokine, fractalkine in shown in *red*. **B** The dimeric structure of the CXC chemokine IL-8/CXCL8. **C** The dimeric structure of the CC chemokine RANTES/CCL5

form compact dimers, with the two carboxy helices lining a groove formed by 4 beta sheets, rather similar to the MHC class II groove, while the CC chemokines form elongated cylindrical dimers through interactions of their amino termini. It is widely debated whether the dimeric structure is in fact physiologically relevant, since dimers form at micromolar concentrations, whereas chemokines are active at nanomolar concentrations. While it is well documented that the monomeric forms of chemokines that are known to dimerize are fully active in vitro (Rajarathnam et al. 1994; Paavola et al. 1998; Laurence et al. 2000), it is not known whether in vivo they may dimerize on the receptor. Certain chemokines such as RANTES/CCL5 are also known to form higher order aggregates, which have been shown to be important both in inflammation (Appay et al. 1999) and in inhibition of HIV infectivity (Wagner et al. 1998).

As mentioned above, chemokines are involved in both basal trafficking and inflammatory cell recruitment. Figure 1 highlights an interesting contrast between the receptors involved in basal trafficking (controlled by homeostatic chemokines), which tend to be specific, and those involved in inflammatory cell trafficking, which are shared by several chemokines. As is the rule for all such rules, this division is not strict, since CCR10 is considered to be constitutive, but is involved in certain skin inflammatory syndromes (Reiss et al. 2001;

Homey et al. 2002) while its ligand CTACK/CCL27 is upregulated by pro-inflammatory cytokines (Homey et al. 2002). The role of the homeostatic chemokines is to maintain the "normal" physiologic trafficking of cells for routine immunosurveillance requiring antigen sampling in the secondary lymphoid organs. Mice in which their genes have been deleted have demonstrated the importance of these receptors in development. Knockout mice deficient in either the CXCR4 receptor (Zou et al. 1998a,b; Ma et al. 1998; Tachibana et al. 1998) or SDF-1 (Nagasawa et al. 1996) die perinatally and the embryos show conspicuous defects in the hematopoietic and nervous systems. CCR7 and CXCR5 play important roles in homing of T cells and B cells respectively to secondary lymphoid tissue such as lymph nodes, and spleen, and both $CCR7^{-/-}$ and $CXCR5^{-/-}$ mice show severe defects in secondary lymphoid tissue architecture (Forster et al. 1996, 1999).

In inflammatory situations, the expression of inducible or inflammatory chemokines is upregulated. Control of expression of these chemokines is under the temporal control of pro-inflammatory cytokines. Pro-inflammatory cytokines such as interleukin (IL)-1β, tumor necrosis factor (TNF)-α, and interferon (IFN)-γ alone or in combination induce chemokine expression at sites of inflammation in nonlymphoid tissue (Segerer et al. 2000). In fact, the 3 ligands for CXCR3 were all identified upon upregulation with IFN-γ, hence their common names: IP-10, interferon-inducible protein 10 (Luster et al. 1988); MIG, monokine induced by interferon-γ (Farber 1990; Cole et al. 1998) and I-TAC/CXCL-5, interferon-inducible T-cell alpha chemoattractant (Rani et al. 1996). Activated cells, which possess appropriate chemokine receptors, upregulated during effector cell generation (Moser and Loetscher 2001) are then attracted to the site of inflammation. This is in contrast with resting peripheral cells, which express only homeostatic chemokine receptors. Different cell populations characterize different inflammatory responses, and the predominance of one cell type over another at a site of inflammation is probably dictated by the chemokines expressed at that particular site of inflammation/injury and the receptor expression pattern on the leukocytes.

The fact that chemokines interact with 7TM receptors not only differentiates them from other cytokines, but also renders their receptors tractable targets for therapeutic intervention. Screening for small-molecule inhibitors of cytokine receptors has been rather unsuccessful, whereas a huge proportion of drugs currently on the market act on members of the 7TM superfamily. There is, however, a feature of the chemokine family which may detract from their suitability as therapeutic targets, namely, the redundancy shown in Fig. 1, since multiple chemokines can bind certain receptors, and several chemokines can bind to more than one receptor. However, the levels of control in vivo such as temporal and spatial expression patterns, differential glycosaminoglycan (GAG) binding, differential receptor trafficking patterns as well as the different signaling pathways are far from fully delineated.

The interaction of chemokines on target cells is mediated through seven-transmembrane G protein-coupled receptors, which are usually of the Gi- Gi/Go type (Murphy 1996). Following activation, which has been reported to result in dimerization (Mellado et al. 2001), chemokine receptors become either partially or totally desensitized to repeated stimulation—this process may be important in helping the cell sense a chemotactic gradient, so that it may move through the stroma to the inflammatory site. Binding of a chemokine to its receptor induces a conformational change in the cytoplasmic tail, which promotes the activation of the signaling cascades. Several intracellular signaling pathways, which are believed to influence each other through crosstalk, are induced and lead to cellular adhesion, migration, degranulation, and gene expression (Jiang et al. 1997; Szabo et al. 1998; Katanaev 2001). Some main components of the signaling events include release of intracellular Ca^{2+} via the phospholipase C (PLCβ) pathway and the activation of several isoforms of the protein kinase (PKC) family and phosphoinositide 3-kinase (PI3Kγ). One early event stimulated by ligand binding is the recruitment and activation of a heterotrimeric G protein complex leading to dissociation of the Gα subunit from the membrane-anchored G$\beta\gamma$-heterodimer (Gether and Kobilka 1998). The G$\beta\gamma$ subunit released transduces signaling events that lead to chemotaxis in motile cells in a Gα-independent manner (Parent and Devreotes 1999). Another G$\beta\gamma$-activated pathway involves PI-3 kinase γ (PI3Kγ) where PI3Kγ activation induces the production of phosphoinositil triphosphate (PIP3) (Jiang et al. 1997) and is involved in mitogen-activated protein kinase (MAPK) activation (Bondeva et al. 1998). Both effects are implicated in the chemotactic response. However, other factors mediated through small guanosine triphosphate (GTP)-binding proteins of the Rho-family, whose linkage to the G protein-coupled receptors is not fully understood, have been proved to be involved in chemotaxis (Hart et al. 1998). Chemokines also activate another set of cytoplasmic protein kinases, Janus kinases (JAKs), which phosphorylate signal transducers and activators of transcription (STATs) proteins, which affect gene expression (Mellado et al. 1998; Wong and Fish 1998). While the link between chemotaxis and PI3Kγ has been clearly established (Hirsch et al. 2000), the biological responses mediated by the JAK/STAT signaling pathways remain to be elucidated. As more insight is gained into this complex network, potential targets for therapeutic intervention should be discovered as has been postulated for an intracellular CCR2 receptor antagonist (Yokochi et al. 2001).

2 Cellular Recruitment and Chemotaxis Assays

In order for circulating leukocytes to reach sites of inflammation, they must cross the endothelial cell barrier. Leukocyte transmigration usually occurs at post-capillary venules, and is a multistep process first described by Butcher (1991) in which chemokines play at least two major roles, as illustrated in Fig. 3. Leukocytes first slow down by a selectin-mediated rolling, enabling the initial encounter with chemokines, which are presented on the endothelial cell surface by their immobilization on proteoglycans—introducing the second important intramolecular interaction in chemokine biology their interaction with GAGs.

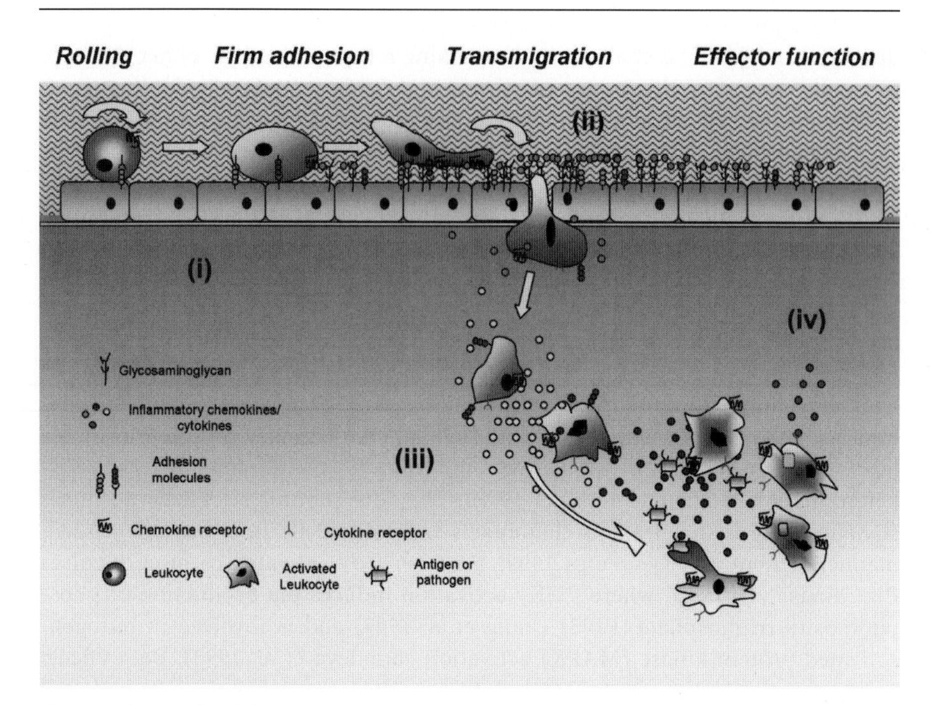

Fig. 3 Orchestrated cellular recruitment mediated by chemokines. Leukocytes circulating in the bloodstream, enter the underlying tissue through the endothelial cell layer (*i*), a process called transendothelial migration. (*ii*) The leukocyte is first slowed down by interactions between selectin/mucin molecules, which cause it to roll along the endothelium. Close to sites of inflammation, the endothelial layer is activated by pro-inflammatory cytokines, resulting in the upregulation of expression of adhesion molecules. Chemokines are secreted and are immobilized on the surface by binding to glycosaminoglycans (GAGs). The interaction of leukocytes with these immobilized chemokines changes the affinity of surface integrins, resulting in firm arrest of the leukocyte, followed by transmigration through the endothelium. (*iii*) Once the migrating cell has crossed the endothelial layer, it follows the haptotactic gradient of chemokines immobilized on the stroma and matrix of interstitial cells. (*iv*) When the cell reaches the site of inflammation, pro-inflammatory cytokines and chemokines induce distinct effector functions

Chemokine/receptor binding on leukocytes causes integrin activation resulting in increased avidity, and thus firm adhesion with adhesion molecules on the endothelial cell surface (Springer 1990; Dustin and Springer 1991). Once firm adhesion to the endothelial surface has been established, the leukocyte transmigrates from the lumen to the site of inflammation in the tissue by a process of haptotaxis (movement along a solid phase gradient). Chemokines bind GAGs, which are present in the extracellular matrix, to create such a gradient or a "guideline" that the cells may use to navigate towards the site of inflammation, where they exert their effects. It is probable that, in vivo, several gradients of different chemokines are present in the tissue, separated temporally and spatially [reviewed in Moser and Loetscher (2001) and Devalaraja and Richmond (1999)]. Specificity introduced to the system both spatially and temporally has been demonstrated by Foxman et al. (1997). In this elegant study, the authors demonstrate that neutrophils navigate towards agonists in a spatially defined and sequential manner.

Recently, the first example of reverse cell migration induced by a chemokine has been reported (Poznansky et al. 2000). This follows the observation that while there are high levels of SDF-1 (a chemoattractant for mature T lymphocytes) in bone marrow and thymic tissue, there is a paucity of this cell type in these areas. This may be explained by a theory put forward by Zlatopolskiy and Laurence (2001) who propose that differential signaling mechanisms control forward or reverse cell migration in response to a chemokine, depending on the critical concentration of that chemokine.

The ability of chemokines to recruit cells has been mimicked in vitro by chemotaxis assays. The classical in vitro assay of the chemotactic function of chemokines has been a system using the Boyden chamber or a similar method of cell chemotaxis [methods reviewed in Wilkinson (1998)]. Briefly, these experiments involve the monitoring of cell migration across a permeable membrane in the positive direction of the agent that is being tested. In these experiments, the chemotactic gradient is created artificially by increasing the chemokine concentration in the lower wells. While this system is widely used as a primary test for the chemotactic ability of a molecule, as with all in vitro methods, it is only an imitation of the in vivo situation. More sophisticated chemotaxis assays have also been described whereby the cell must migrate across an endothelial cell layer and, to make the system more physiological, a shear flow can be applied (Cinamon et al. 2001). However, the best demonstration of the chemoattractant properties should be those in vivo.

Experiments using a flow chamber and endothelial cells to study the differential roles of chemokine in monocyte arrest and extravasation have shown that while GRO-α/CXCR2 is important in transforming monocytes from a rolling state to one of firm arrest, MCP-1/CCL2 is involved in the subsequent shape change and transmigration step (Weber et al. 1999). It is noteworthy that in these studies the effects of MCP-1/CCL2 could only be seen under flow conditions, suggesting that flow may play an important role in the transmigration step, an observation which was also made by another group (Cinamon et al. 2001).

While it would appear simple to monitor cellular recruitment by injecting the chemokine of interest into an animal, in vivo chemotaxis has not been widely used. However, certain methods have been reported where the chemokine is administered into a "pocket" in an animal. The pocket may be a naturally occurring space in the animal; for example, the chemokine can be applied to the lung via the intra tracheal route (Hisada et al. 1999), or into the peritoneal cavity by injection (Z. Johnson and A.E.I. Proudfoot, unpublished observations) or implantation of a chemokine-soaked sponge (Fine et al. 2000). In other situations the pocket may be artificial—for example, by creating an air pouch on the animal's back (Takano et al. 1999). Cellular recruitment can also be induced by the administration of a chemokine by intra dermal injection and MCP-1/CCL2, MCP-2/CCL8, and MCP-3/CCL7 in rabbits have all been shown to recruit monocytes to the site of injection in the skin (Van Damme et al. 1992). Eotaxin, which was identified by biochemical means as the eosinophil recruitment factor from the BAL of sensitized guinea pigs, when injected into guinea pig skin induces an accumulation of eosinophils as would be expected (Jose et al. 1994). Interestingly, however, RANTES/CCL5, which is also an eosinophil chemoattractant in vitro, will only attract an inflammatory cell infiltrate of eosinophils and monocytes following injection into the skin of dogs previously subjected to helminth infection (Meurer et al. 1993). The ability of RANTES/CCL5 to recruit eosinophils only in sensitized subjects was further demonstrated in man where RANTES/CCL5 injection into skin of non-atopic subjects did not elicit an inflammatory response, compared with the predominantly eosinophilic infiltrate observed following RANTES/CCL5 injection in atopic patients (Beck et al. 1997). Intradermal injection of the prototypic CXC chemokine IL-8/CXCL8 in rabbits induces plasma exudation which is dependent on specific recruitment of neutrophils (Colditz et al. 1989; Foster et al. 1989; Rampart et al. 1989; Colditz et al. 1990), and the same exclusive recruitment of neutrophils induced by IL-8/CXCL8 has been confirmed in man (Leonard et al. 1991; Swensson et al. 1991).

In addition to the injection of specific chemokines in vivo, chemotactic agents such as thioglycollate or the bacterial cell wall component lipopolysaccharide (LPS) have been shown to produce a pronounced and specific cellular infiltrate following injection in vivo. LPS induces a rapid initial recruitment of neutrophils followed later by monocytes into the host tissue (Evans et al. 1989; Ghosh et al. 1993), which has been shown to be partially mediated by localized production of chemokines, including MCP-1/CCL2, RANTES/CCL5, and MIP-1α/CCL3 and -β/CCL4 (Kopydlowski et al. 1999). Injection of aged thioglycollate medium has been used as a method for eliciting peritoneal neutrophils or macrophages for many years, with neutrophil recruitment occurring at 4 h (Rodrick et al. 1982) while 3–4 days after injection the predominant cell type in the peritoneal cavity are macrophages (Mishell 1980). Macrophage recruitment after thioglycollate administration involves MCP-1/CCR2, since both the MCP-1$^{-/-}$ and the CCR2$^{-/-}$ mice show impaired responses in this model (Boring et al. 1997; Kurihara et al. 1997; Kuziel et al. 1997), and administration of an anti-CCR2 antibody also significantly inhibits cell recruitment in this model (Mack et al. 2001).

3
Macrophage Chemokine and Receptor Expression Involved in Disease

The aim of this chapter is to concentrate on the effects of chemokines on the monocyte/macrophage leukocyte phenotype in disease, which is the central theme of this book. Several chemokine receptors are expressed on monocytes which are responsible for the rapid and directed migration of these cells from the circulation into tissue, where they mature into phagocytic macrophages involved in cell-mediated host defense against infection. It is very important to note that macrophages have a different chemokine-receptor expression pattern from circulating monocytes (Kaufmann et al. 2001). A good example of differential expression is shown by the RANTES receptors. Circulating monocytes express high levels of CCR1, but very low levels of CCR5 and CCR3. Thus RANTES/CCL5 induces a robust response of monocyte chemotaxis, while eotaxin/CCL11 and MIP-1β/CCL4, specific ligands for CCR3 and CCR5 respectively, induce only moderate responses (Proudfoot et al. 1999) Furthermore, the amino terminally modified variant AOP-RANTES, which is a potent agonist of CCR5, but not of CCR1 (Proudfoot et al. 1999) is not able to induce monocyte chemotaxis (Simmons et al. 1997). On the contrary, AOP-RANTES induces a robust calcium response on monocytes that have acquired the macrophage phenotype in vitro, while it is not able to elicit this response in freshly isolated monocytes (Proudfoot et al. 1999), indicative of the altered expression pattern of chemokine receptors on monocytes and macrophages.

The levels of chemokines and their receptors in macrophage-mediated pathologies such as rheumatoid arthritis (RA); multiple sclerosis (MS) and arteriosclerosis have been extensively studied. In RA, macrophages are believed to play a pivotal role in nefarious activities such as joint destruction, and the synovial fluid from RA patients has been shown to contain a variety of chemokines which attract monocytes, including macrophage inflammatory protein (MIP)-1α/CCL3 (Hosaka et al. 1994; Koch et al. 1994), MIP-1β/CCL4 (Koch et al. 1995), MCP-1/CCL2 (Koch et al. 1992; Villiger et al. 1992) and RANTES/CCL5 (Rathanaswami et al. 1993; Volin et al. 1998). A study of the chemokine receptors expressed in the three leukocyte-trafficking compartments of RA (peripheral blood, synovial fluid and synovial tissue) showed that CCR1 and CCR2 are highly expressed on normal and RA peripheral blood monocytes, but are expressed at low levels on these cells in the synovial fluid, suggesting that they are involved in the recruitment process. Other receptors such as CCR3, CCR4, and CCR5 were found to be upregulated on peripheral blood mononuclear cells (PBMC) from arthritic patients compared to normal samples (Katschke et al. 2001). These results suggest that differential chemokine-receptor expression patterns play important roles in monocyte/macrophage recruitment and retention in disease.

In another chronic autoimmune disease in which macrophages are believed to play a major role, MS, the chemokines IP-10/CXCL10, Mig/CXCL9, which act via CXCR3, and RANTES/CCL5, which acts via CCR1, CCR3 and CCR5 were all shown to be upregulated in the cerebrospinal fluid (CSF) of patients during an MS attack, whereas MCP-1/CCL2 levels, which acts via CCR2, were decreased (Sorensen et al. 1999). The study of the receptor levels in the CSF and on cells present in MS lesions revealed that CXCR3 is elevated in the CSF compared with the level of expression in the peripheral blood (Sorensen et al. 1999). Other studies have shown that CCR5 receptor expression is increased on T cells and macrophages both in the CSF and in MS lesions (Balashov et al. 1999; Sorensen et al. 1999; Strunk et al. 2000). Further evidence that CCR5 plays an important role in the progression and pathology of MS comes from the observation that MS patients that are heterozygous for the CCR5Δ32 allele, which codes a nonfunctional CCR5, have prolonged disease-free intervals between MS attacks compared with individuals expressing wildtype CCR5 (Sellebjerg et al. 2000). Recently the same phenomenon of decreased CCR1 expression in recruited cells that was observed in the synovial fluid of RA patients was seen in MS lesions, whereas CCR5 expression is increased (Trebst and Ransohoff 2001).

The macrophage is known to play a major role in arteriosclerosis, and the observation of high levels of one of the major monocyte/macrophage chemoattractant, MCP-1/CCL2, in atherosclerotic plaques taken from patient samples is not surprising (Nelken et al. 1991). The role of this chemokine and its receptor CCR2 is well borne out by their deletion in mice as is described below. In the remainder of this chapter, we will describe the approaches that would interfere with the role of chemokines as potential strategies to interfere with the inflammatory process in disease, and review the current status of chemokine anti-inflammatory therapeutics.

4
Proof of Concept of Interference with the Chemokine System as Anti-inflammatory Therapies

Before describing the therapeutic strategies that could be applied in man, it is worthwhile to briefly review the evidence that has validated the role of the chemokine system and macrophage biology in animal models using genetic manipulation.

The first genetic manipulation approach is the creation of transgenic mice, an overexpression of a specific chemokine. The theoretical advantage of this approach is that high levels of the chemokine may be maintained at a specific site for a sustained period, perhaps mimicking the effects of an inflammatory response. Surprisingly, the targeted overexpression of MCP-1/CCL2 to the pancreas under the control of the insulin promoter showed that, while extensive monocytic infiltration was induced, this was not paralleled with inflammation, leading to the important observation that cellular recruitment alone is not sufficient to create an inflammatory response, but that a second activation or danger signal is required, as has been suggested by the work of Matzinger (1994). Similarly, overexpression of MCP-1/CCL2 in alveolar cells, so that the chemokine was secreted into the bronchoalveolar space, resulted in an increase in monocytes and lymphocytes in the bronchoalveolar lavage, again without an accompanying inflammation (Gunn et al. 1997)

The alternative genetic approach is to generate knockout mice by gene deletion. In contrast to deletion of constitutive receptors and ligands, which have shown striking phenotypes, inflammatory receptor, and ligand-knockout mice have no unusual phenotype under normal conditions, but do show phenotypes when subjected to inflammatory stress. A good example is the MIP-1α/CCL3 knockout, which only demonstrated a phenotype in infection with Coxsackie virus (Cook et al. 1995). Knockout mice in which the gene for MCP-1/CCL2 has been deleted have been far more informative. These mice showed a fourfold reduction in recruitment of macrophages in response to thioglycollate administered peritoneally, even though the exact mechanism of cellular recruitment remains to be elucidated (Gosling et al. 1999). Further evidence for the role of MCP-1/CCL2 in monocyte recruitment was shown in a delayed-type hypersensitivity (DTH) model. In the MCP-1$^{-/-}$ mouse, an impairment of macrophage accumulation in DTH lesions was observed compared with wildtype mice, despite the fact that the swelling response was similar in both wildtype and gene-deleted mice. Perhaps the most convincing data generated from these mice is in a mouse model of MS, experimental autoimmune encephalomyelitis (EAE). In this study, MCP-1$^{-/-}$ mice were shown to be significantly resistant to EAE following active immunization, with a corresponding impairment of recruitment of macrophages to the CNS (Huang et al. 2001). These experiments demonstrated a very important fact with respect to redundancy. While MCP-1/CCL2 acts only on CCR2, it is not the only chemokine to activate this receptor; yet the deletion of this chemokine mirrored the effect of deleting the receptor itself. In other words, the other CCR2 ligands were not able to compensate. CCR2$^{-/-}$ mice have shown similar macrophage recruitment defects, where macrophage recruitment in response to peritoneal thioglycollate administration is severely impaired. In murine models of arteriosclerosis, both MCP-1/CCL2 and CCR2$^{-/-}$ mice have unequivocally demonstrated the importance of this receptor/ligand pair (Gosling et al. 1999)

Knockout mice do not always confirm the role for certain ligands or receptors, since deletion of MIP-1α/CCL3 and CCR5 showed that these mice remained fully susceptible to EAE (Tran et al. 2000). However, MIP-1α/CCL3 knockout mice show a decrease in cuprizone-induced demyelination (McMahon et al. 2001). These results are in contrast to the neutralization of MIP-1α as described below, and with the strikingly high levels of expression of CCR5 in MS lesions, we believe that interpretations with knockout mice should be taken with care, as compensatory mechanisms may occur in gene-deleted animals—especially in a system as “redundant” as the chemokine system.

5
Anti-inflammatory Strategies

There are several approaches that could be undertaken to inhibit the chemokine-mediated recruitment of monocyte/macrophages into inflammatory sites, which are summarized in Fig. 4. The most frequently applied strategy is that of preventing the interaction between the chemokine and its receptor, and to this end several approaches have been adopted.

The use of neutralizing monoclonal antibodies, principally against the chemokines themselves, has been used extensively in animal models of disease, although surprisingly, few are being developed for therapeutic use perhaps a reflection of the belief that orally available small molecule receptor inhibitors would supersede the use of antibodies. However, published results prove that the use of neutralizing antibodies against specific chemokines can successfully

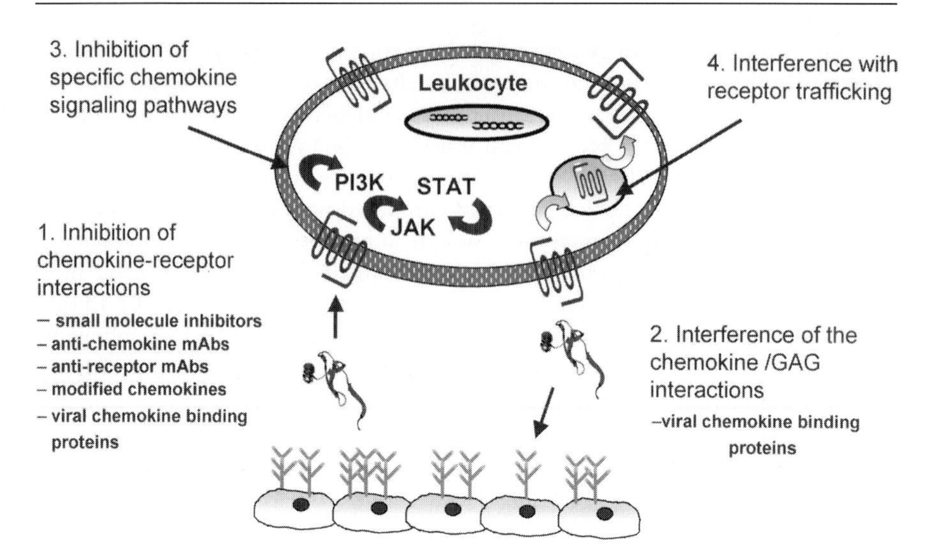

Fig. 4 The possibilities of intervention points in the chemokine system to prevent inflammation. Four possible strategies are depicted. (*1*) The most frequent strategy applied to date, which is the prevention of chemokine/receptor interactions. (*2*) Chemokines are believed to interact with cell surface glycosaminoglycans in order to form the chemotactic gradient. (*3*) Chemokines may have specific signalling pathways. (*4*) Receptor endocytotic pathways have been shown to differ for different chemokine receptors

block specific inflammation in animal models. The blockade of several different CC chemokines in T helper (Th)2 inflammation using a murine model of airways inflammation induced by ovalbumin sensitization revealed that there is a coordinated action of chemokines in the inflammatory process which orchestrates the complete response. Eotaxin/CCL2 and MCP-1/CCL2 were shown to be important for lung inflammation (eosinophil and monocyte infiltration respectively) and bronchial hyperreactivity (BHR), whereas neutralizing MIP-1α/CCL3 had little effect. Use of a neutralizing antibody against the CXC chemokine IL-8/CXCL8 was also shown to be effective at blocking reperfusion-associated lung injury in a rabbit model (Sekido et al. 1993).

Similar efficacious treatments using this approach have been demonstrated in Th1 disease models. In collagen-induced arthritis in rodents, treatment with an anti-MCP-1/CCL2 mAb improved disease by significantly reducing the number of infiltrating macrophages into the lesions, correlating with a reduction in ankle swelling by 30% (Ogata et al. 1997). A neutralizing anti-RANTES/CCL5 polyclonal antibody similarly improved adjuvant induced arthritis by significantly reducing the cellular infiltrate, maintaining joint integrity and preventing bone destruction (Barnes et al. 1998). In EAE, macrophage accumulation has been shown to correspond with levels of MCP-1/CCL2 at the site of inflammation (Ransohoff and Tani 1998). Administration of an anti-MCP-1/CCL2 antibody prevented relapses in rodent EAE (Karpus and Kennedy 1997), suggesting that the macrophage plays a role in this process. On the other hand, anti-MIP-1α/CCL3 treatment prevented the disease onset in an acute model, but not in relapsing and remitting EAE (Karpus and Kennedy 1997). A mAb against rat MCP-1/CCL2 was also used to demonstrate that MCP-1 is responsible for infiltration of monocytes in bleomycin-induced lung injury in rats (Sakanashi et al. 1994). An antibody to MCP-1/CCL2 was able to inhibit monocyte migration across an in vitro model of the blood brain barrier (BBB) by 85%, which supports the role of MCP-1/CCL2 and its receptor CCR2 in recruiting monocytes into the brain (Weiss et al. 1998).

Neutralizing antibodies to rodent chemokine receptors are not widely available, introducing a limiting factor to correlate data obtained in rodent models to receptor antagonism in human disease. Recently, mAbs against murine CCR2 and CCR5 have been described (Mack et al. 2001), which hopefully will provide more information as to the efficacy of blocking these receptors in macrophage-mediated inflammation. While Eotaxin/CCL11-mediated eosinophil recruitment in the guinea pig has been shown to be blocked by a mAb against guinea pig CCR3 (Sabroe et al. 1998), there is no report of its efficacy in airways inflammation.

A second approach to preventing ligand-mediated receptor activation is to use a modified ligand that retains high-affinity binding to the receptor but has lost the ability to induce signaling, and is thus a receptor antagonist. The amino terminal region has been shown by many studies to be crucial for receptor activation, and therefore such receptor antagonists have been generated by the modification of this region, either by truncation or elongation. There are several examples described in vitro of the former approach with IL-8, (Clark-Lewis et al. 1991), RANTES/CCL5 (Gong et al. 1996) and MCP-1/CCL2 (Zhang and Rollins 1995; Kim et al. 1996), but not many in vivo. However, the truncated MCP-1/CCL2 analog (9–76)-MCP-1/CCL2 was shown to have very good efficacy in the murine model of arthritis in MRL-*lpr* mice; its therapeutic administration reduced disease symptoms (Gong and Clark-Lewis 1995). The elongation approach has been extensively studied in vivo with the RANTES analog, Met-RANTES (Proudfoot et al. 1996), where it has shown efficacy in a variety of models which implicate monocyte/involvement such as collagen-induced arthritis (Plater-Zyberk et al. 1997), glomerular crescentic nephritis (Lloyd et al. 1997), organ transplant (Grone et al. 1999), and colitis (Ajuebor et al. 2001).

While the power of biological therapeutics should not be ignored, the dogma of the pharmaceutical industry is based on the superiority of orally bioavailable small molecule inhibitors. Chemokine receptors, which belong to the GPCR class, are ideal targets. However, the majority of these receptors have small ligands, and while chemokines are small proteins, they are significantly larger compared to ligands such as histamine, adrenaline, or dopamine. Moreover, the fact that they bind several ligands, and that ligand binding can be allotropic because the ligands do not necessarily bind to the same site (Cox et al. 2001), renders the search for small molecules more challenging. However, the combination of focused high-throughput screening and medicinal chemistry has led to the development of highly-potent small molecule chemokine receptor antagonists, for which the number in the patent literature is now over 200—no mean feat since this receptor class was only identified just over a decade ago. Certain small molecule antagonists lack receptor selectivity, which may in fact be advantageous, as has been suggested for the dual CCR1/CCR3 inhibitor (Sabroe et al. 2000). However, specific small molecules have been described in the peer-reviewed literature, such as the CCR1 inhibitor BX471, CXCR2 inhibitors, and the CCR5 inhibitors, TAK779 and SCH-C, although TAK779 also has significant inhibitory properties for the closely related receptor, CCR2. To date the only anti-inflammatory in vivo data published are for the CCR1 antagonist BX471, where the compound has been shown to be efficacious in rat heart transplant rejection by reducing RANTES-mediated monocyte adhesion on inflamed endothelium (Horuk et al. 2001), in renal fibrosis by reducing macrophage and lymphocyte infiltration into the kidney (Anders et al. 2002), and in a rat model of EAE (Liang et al. 2000). The CCR5 compound TAK779 demonstrates an important feature in that it is a non-competitive inhibitor, since it binds to a cavity in the transmembrane domains rather than to the ligand-binding site(s) (Dragic et al. 2000). This is particularly important in view of the fact that certain receptors such as CXCR3 have distinct ligand-binding sites (Cox et al. 2001), which makes the identification of competitive inhibitors impossible.

It should be noted that while the published data on small molecule inhibitors are still very limited, since clinical trials have started for some of these inhibitors, numerous conference presentations indicate that there is considerable progress.

6
The Role of the Macrophage in Infection: HIV

In this chapter we have focused on the role of chemokines and their receptors in inflammation, but the importance of these proteins and the macrophage with respect to HIV cannot be ignored. It is widely believed that the macrophage is the cell involved in viral transmission, although primary transmission through infection of T lymphocytes and the dendritic cell cannot be excluded. For many years it was known that the high-affinity interaction with CD4 (Dalgleish et al. 1984) was not sufficient for infection of host cells, but it was only in 1996 that chemokine co-receptors were identified as the essential co-receptors (reviewed in Berger 1997). The initial binding of HIV to CD4 on the target cell surface leads to a conformational change in gp120, a viral envelope surface glycoprotein, which then interacts with the chemokine receptor, inducing a second conformational change which exposes another viral protein gp41, also known as the fusion peptide, that is inserted into the host membrane, resulting in membrane fusion and virus entry into the cell. It had been shown a few months before the discovery of the chemokine receptors as co-receptors that three chemokines, RANTES/CCL5, MIP-1α/CCL3, and MIP-1β/CCL4 were able to reduce HIV infection of cells (Cocchi et al. 1995). This immediately indicated that a blockade of the chemokine receptors could be an anti-HIV therapeutic strategy. This strategy was borne out by the finding that homozygous individuals with a mutation in the CCR5 gene (Δ32-CCR5), which prevents the cell surface expression of the receptor, are resistant to infection by HIV (Samson et al. 1996). It is still not completely clear why transmission is almost exclusively mediated by CCR5-using HIV-1 strains, also called R5 strains, and not by CXCR4 or X4 strains, since macrophages express all the required components, CD4, CCR5, as well as CXCR4.

The establishment of the principle that infection could be prevented with chemokine receptor antagonists caused a huge effort by the pharmaceutical industry to find small-molecule inhibitors. CCR5 has proved more tractable than CXCR4, with nanomolar compounds described in the literature (Baba et al. 1999; Strizki et al. 2001). SCH-C has the advantage of having very good oral bioavailability, allowing an in vivo proof of concept in severe combined immunodeficiency (SCID) mice, since studies in non-human primates were excluded, owing to the high degree of specificity of the compound for the human receptor (Strizki et al. 2001). The start of clinical trials with this molecule has been announced at conferences. Many more chemical series have been granted patent filing, with the total number now being over 50. On the other hand, only one CXCR4 inhibitor series has been described in the literature (Donzella et al. 1998), which has the handicap of not being very "drug-like," with a molecular weight over 1,000 Da.

Studies with modified chemokines as potential anti-HIV therapies have provided very interesting information as to the biology behind HIV inhibition, but their clinical development programs are not yet known. Chemokines could inhibit HIV infection by two mechanisms. First, steric hindrance through acting as pure competitive or non-competitive inhibitors as is the case for TAK779, or alternatively by inducing endocytosis of the receptor, thereby stripping the essential co-receptor from the cell surface. While the first mechanism obviously plays an important role, the efficacy of the second mechanism was demonstrated by a chemically modified RANTES/CCL5 variant, created by the chemical coupling of a pentacarbon alkyl chain to the N-terminus of the chemokine, and hence named aminooxy pentane RANTES (AOP-RANTES) (Simmons et al. 1997). This analog was shown to be far more potent than the truncated RANTES receptor antagonist (9–68)-RANTES (Arenzana-Seisdedos et al. 1996) and Met-RANTES (Simmons et al. 1997) and was subsequently shown not to be a CCR5 antagonist but rather a superagonist of this receptor in that it caused enhanced phosphorylation of CCR5 (Oppermann et al. 1999) as well as enhanced receptor downmodulation properties (Mack et al. 1998). But more surprising was its property to prevent recycling of functional receptors (Mack et al. 1998). Although the mechanism of this phenomenon is not yet fully elucidated, it appears that while it does allow receptor recycling, the receptor is immediately re-internalized (Signoret et al. 2000), thereby rendering it inaccessible to the HIV gp120 interaction. The efficacy of preventing cell-surface receptor expression has been elegantly demonstrated using another approach by the genetic addition of the endoplasmic retrieval peptide KDEL to RANTES/CCL5, MIP-1αCCL3, or SDF-1/CXCL5 (Chen et al. 1998) to create so-called intrakines, which bind to newly synthesized receptors CCR5 or CXCR4, respectively, and inhibit their transportation to the cell surface.

7 Can We Learn Strategic Therapeutic Approaches from Pathogens?

Pathogens have evolved over the millennia in an attempt to evade the host's immune response. Examples of "silent" (non-viral) pathogens are leeches and ticks that are able to feed on their host in the absence of an inflammatory or allergic response. Anti-cytokine and chemokine activities have been identified in tick saliva (Gillespie et al. 2001; Hajnicka et al. 2001) although the molecular identities of these molecules are as yet unknown. However, the best documented pathogens are viruses whose rapid mutational capacities could allow them to become immunologically silent or to devise other methods to escape the host response. Several mammalian viral species have been found to express arrays of immunomodulating molecules, such as chemokines, chemokine receptors, or chemokine-binding proteins, which are used to increase survival and dissemination in vivo and/or decrease the recognition of the virus by its host. Virally expressed chemokines may enhance dissemination by either inducing the recruitment of host cells (McFadden and Murphy 2000) or inducing a Th2-shift of the host immune response (Weber et al. 2001). Human herpes virus-8 encoded viral MIP-II binds potently to both CC and CXC receptors and can efficiently block HIV infection (Zhou et al. 2001). In fact, this molecule crosses one of the barriers of chemokine receptor specificity in that it is able to inhibit both R5 and X4 HIV viral infection (Kledal et al. 1997), a feat that man has yet to achieve. In a rat model for glomerular nephritis, the in vivo administration of vMIP-II attenuates the disease (Chen et al. 1998) and studies on ischemic brain injury have shown that it might also be useful in the therapeutic intervention of stroke (Takami et al. 2001). A soluble chemokine receptor antagonist, MC148R, is expressed by the molluscum contagiosum virus that might become useful as an anti-inflammatory molecule, as it has been shown to be highly specific for CCR8 and hence for inhibiting monocyte invasion and dendritic cell function (Luttichau et al. 2000).

Poxviruses have revealed another interesting class of chemokine inhibitors in the guise of molecules that bind and neutralize chemokines, but which have no mammalian homologs. These binding proteins fall into different classes (Lusso 2000). The first, belonging to the T1 or p35 type (Lalani and McFadden 1997; Stine et al. 1999) show promiscuous binding to several beta-chemokines and are able to inhibit in vitro chemokine activities, thereby interfering with chemokine receptor interactions (Lalani and McFadden 1997; Alcami et al. 1998). Another chemokine-binding protein, M3, encoded by the murine γ-herpesvirus-68, binds members of all four classes of chemokines (Bridgeman et al. 2001). Its expression prevents the recruitment of T cells, but not of B cells and neutrophils

(Parry et al. 2000), which seems to have an evolutionary advantage as the virus replicates in B cells. Another class is represented by the myxoma viral protein, M-T7, which is unable to inhibit in vitro activities despite the fact that it interacts in vitro with a broad range of CC and CXC chemokines. Its mode of action is believed to be through interaction with their heparin-binding domains (Lalani et al. 1997), thereby inhibiting the formation of a chemokine gradient on the endothelial surface and adds weight to the approach suggested in Fig. 3 as a possible anti-inflammatory strategy. The administration of this protein in animal models of inflammation has validated this concept (Liu et al. 2000). Lastly, functional chemokine receptor-like molecules are expressed by some viruses. The receptor US-28 is expressed by the human cytomegalovirus, presumably to help its dissemination in the human host by mediating migration of infected smooth-muscle cells, a fact that can be linked to the acceleration of vascular disease (Streblow et al. 1999, 2001). It is certain that with more structural information and biological data becoming available, viruses could perhaps teach us how to create molecules with, for example, the property described above of vMIP-II, the only molecule that can block both R5 and X4 HIV infection.

8
Conclusions

There is sufficient evidence, some of it summarized in this chapter, that despite the redundancy apparent in the chemokine system, that interference with the system has a significant therapeutic potential. The therapeutic areas encompass those involving macrophage function, including many inflammatory pathologies, as well as infectious disease such as AIDS, where the macrophage is believed to play a pivotal role. It is, perhaps, in the latter area where chemokine antagonists are most advanced, where small-molecule inhibitors of the principal co-receptor, CCR5, are already in clinical trials.

We have described some of the many results obtained with an approach used by that of the biotechnology field: protein therapeutics. Both neutralizing antibodies and modified chemokines in the guise of receptor antagonists have shown to have efficacy. While it is well accepted that an orally available small-molecule inhibitor is the preferred molecule of choice, antibody therapies have a well-proven track record in the clinic. Hopefully biological therapeutics will find their niche as anti-chemokine strategies, in the form of antibodies, receptor antagonists, or pathogen-derived chemokine-binding proteins.

To date the focus has been on receptor antagonists, but as we have illustrated in Fig. 3, we believe that there are many other approaches, some of which are not even depicted in this illustration. The signaling pathway are far from being fully delineated, and may provide attractive targets which interfere not only with chemotaxis, but with other aspects of chemokine biology such as cellular activation, differentiation, angiogenesis, and metastasis, to name but a few. The approaches to block chemokine action could also be more indirect, such as heterologous receptor desensitization, interference with chemokine gene transduction by gene therapy, or prevention of receptor expression for the prevention of HIV infection using subtle strategies such as trapping in the endoplasmic reticulum. In summary, the relatively new area of immunology opened up by the field of chemokine biology promises many novel therapeutic approaches.

9 References

Ajuebor MN, Hogaboam CM, Kunkel SL, Proudfoot AE, Wallace JL (2001) The chemokine RANTES is a crucial mediator of the progression from acute to chronic colitis in the rat. J Immunol 166:552–558

Alcami A, Symons JA, Collins PD, Williams TJ, Smith GL (1998) Blockade of chemokine activity by a soluble chemokine binding protein from vaccinia virus. J Immunol 160:624–633

Anders HJ, Vielhauer V, Frink M, Linde Y, Cohen CD, Blattner SM, Kretzler M, Strutz F, Mack M, Grone HJ, Onuffer J, Horuk R, Nelson PJ, Schlondorff D (2002) A chemokine receptor CCR-1 antagonist reduces renal fibrosis after unilateral ureter ligation. J Clin Invest 109:251–259

Appay V, Brown A, Cribbes S, Randle E, Czaplewski LG (1999) Aggregation of RANTES is responsible for its inflammatory properties. Characterization of nonaggregating, noninflammatory RANTES mutants. J Biol Chem 274:27505–27512

Arenzana-Seisdedos F, Virelizier JL, Rousset D, Clark-Lewis I, Loetscher P, Moser B, Baggiolini M (1996) HIV blocked by chemokine antagonist. Nature 383:400

Baba M, Nishimura O, Kanzaki N, Okamoto M, Sawada H, Iizawa Y, Shiraishi M, Aramaki Y, Okonogi K, Ogawa Y, Meguro K, Fujino M (1999) A small-molecule, nonpeptide CCR5 antagonist with highly potent and selective anti-HIV-1 activity. Proc Natl Acad Sci U S A 96:5698–5703

Balashov KE, Rottman JB, Weiner HL, Hancock WW (1999) CCR5(+) and CXCR3(+) T cells are increased in multiple sclerosis and their ligands MIP-1alpha and IP-10 are expressed in demyelinating brain lesions. Proc Natl Acad Sci U S A 96:6873–6878

Barnes DA, Tse J, Kaufhold M, Owen M, Hesselgesser J, Strieter R, Horuk R, Perez HD (1998) Polyclonal antibody directed against human RANTES ameliorates disease in the Lewis rat adjuvant-induced arthritis model. J Clin Invest 101:2910–2919

Beck LA, Dalke S, Leiferman KM, Bickel CA, Hamilton R, Rosen H, Bochner BS, Schleimer RP (1997) Cutaneous injection of RANTES causes eosinophil recruitment: comparison of nonallergic and allergic human subjects. J Immunol 159:2962–2972

Berger EA. HIV entry and tropism: the chemokine receptor connection. AIDS 11, S3-S16. 1997

Bondeva T, Pirola L, Bulgarelli-Leva G, Rubio I, Wetzker R, Wymann MP (1998) Bifurcation of lipid and protein kinase signals of PI3Kgamma to the protein kinases PKB and MAPK. Science 282:293–296

Boring L, Gosling J, Chensue SW, Kunkel SL, Farese RV, Jr., Broxmeyer HE, Charo IF (1997) Impaired monocyte migration and reduced type 1 (Th1) cytokine responses in C-C chemokine receptor 2 knockout mice. J Clin Invest 100:2552–2561

Bridgeman A, Stevenson PG, Simas JP, Efstathiou S (2001) A secreted chemokine binding protein encoded by murine gammaherpesvirus-68 is necessary for the establishment of a normal latent load. J Exp Med 194:301–312

Butcher EC (1991) Leukocyte-endothelial cell recognition: three (or more) steps to specificity and diversity. Cell 67:1033–1036

Chen S, Bacon KB, Li L, Garcia GE, Xia Y, Lo D, Thompson DA, Siani MA, Yamamoto T, Harrison JK, Feng L (1998) In vivo inhibition of CC and CX3C chemokine-induced leukocyte infiltration and attenuation of glomerulonephritis in Wistar-Kyoto (WKY) rats by vMIP-II. J Exp Med 188:193–198

Cinamon G, Shinder V, Alon R (2001) Shear forces promote lymphocyte migration across vascular endothelium bearing apical chemokines. Nat Immunol 2:515–522

Clark-Lewis I, Schumacher C, Baggiolini M, Moser B (1991) Structure-activity relationships of interleukin-8 determined using chemically synthesized analogs. Critical role of NH2-terminal residues and evidence for uncoupling of neutrophil chemotaxis, exocytosis, and receptor binding activities. J Biol Chem 266:23128–23134

Cocchi F, DeVico AL, Garzino-Demo A, Arya SK, Gallo RC, Lusso P (1995) Identification of RANTES, MIP-1 alpha, and MIP-1 beta as the major HIV- suppressive factors produced by CD8+ T cells. Science 270:1811–1815

Colditz I, Zwahlen R, Dewald B, Baggiolini M (1989) In vivo inflammatory activity of neutrophil-activating factor, a novel chemotactic peptide derived from human monocytes. Am J Pathol 134:755–760

Colditz IG, Zwahlen RD, Baggiolini M (1990) Neutrophil accumulation and plasma leakage induced in vivo by neutrophil-activating peptide-1. J Leukoc Biol 48:129–137

Cole KE, Strick CA, Paradis TJ, Ogborne KT, Loetscher M, Gladue RP, Lin W, Boyd JG, Moser B, Wood DE, Sahagan BG, Neote K (1998) Interferon-inducible T Cell Alpha Chemoattractant (I-TAC): A Novel Non-ELR CXC Chemokine with Potent Activity on Activated T Cells through Selective High Affinity Binding to CXCR3. J Exp Med 187:2009–2021

Cook DN, Beck MA, Coffman TM, Kirby SL, Sheridan JF, Pragnell IB, Smithies O (1995) Requirement of MIP-1 alpha for an inflammatory response to viral infection. Science 269:1583–1585

Cox MA, Jenh CH, Gonsiorek W, Fine J, Narula SK, Zavodny PJ, Hipkin, RW. (2001) Human interferon-inducible 10-kDa protein and human interferon-inducible T cell alpha chemoattractant are allotopic ligands for human CXCR3: Differential binding to receptor states. Mol Pharmacol. 59:707–715

Dalgleish AG, Beverley PC, Clapham PR, Crawford DH, Greaves MF, Weiss RA (1984) The CD4 (T4) antigen is an essential component of the receptor for the AIDS retrovirus. Nature 312:763–767

Devalaraja MN, Richmond A (1999) Multiple chemotactic factors: fine control or redundancy? Trends Pharmacol Sci 20:151–156

Donzella GA, Schols D, Lin SW, Este JA, Nagashima KA, Maddon PJ, Allaway GP, Sakmar TP, Henson G, De Clercq E, Moore JP (1998) AMD3100, a small molecule inhibitor of HIV-1 entry via the CXCR4 co-receptor. Nat Med 4:72–77

Dragic T, Trkola A, Thompson DA, Cormier EG, Kajumo FA, Maxwell E, Lin SW, Ying W, Smith SO, Sakmar TP, Moore JP (2000) A binding pocket for a small molecule inhibitor of HIV-1 entry within the transmembrane helices of CCR5. Proc Natl Acad Sci U S A 97:5639–5644

Dustin ML, Springer TA (1991) Role of lymphocyte adhesion receptors in transient interactions and cell locomotion. Annu Rev Immunol 9:27–66

Evans GF, Snyder YM, Butler LD, Zuckerman SH (1989) Differential expression of interleukin-1 and tumor necrosis factor in murine septic shock models. Circ Shock 29:279–290

Farber JM (1990) A Macrophage mRNA Selectively Induced by {gamma}-Interferon Encodes a Member of the Platelet Factor 4 Family of Cytokines. PNAS 87:5238–5242

Fine JS, Jackson JV, Rojas-Triana A, Bober LA (2000) Evaluation of chemokine- and phlogistin-mediated leukocyte chemotaxis using an in vivo sponge model. Inflammation 24:331–346

Forster R, Mattis AE, Kremmer E, Wolf E, Brem G, Lipp M (1996) A putative chemokine receptor, BLR1, directs B cell migration to defined lymphoid organs and specific anatomic compartments of the spleen. Cell 87:1037–1047

Forster R, Schubel A, Breitfeld D, Kremmer E, Renner-Muller I, Wolf E, Lipp M (1999) CCR7 coordinates the primary immune response by establishing functional microenvironments in secondary lymphoid organs. Cell 99:23–33
Foster SJ, Aked DM, Schroder JM, Christophers E (1989) Acute inflammatory effects of a monocyte-derived neutrophil-activating peptide in rabbit skin. Immunology 67:181–183
Foxman EF, Campbell JJ, Butcher EC (1997) Multistep navigation and the combinatorial control of leukocyte chemotaxis. J Cell Biol 139:1349–1360
Gether U, Kobilka BK (1998) G protein-coupled receptors. II. Mechanism of agonist activation. J Biol Chem 273:17979–17982
Ghosh S, Latimer RD, Gray BM, Harwood RJ, Oduro A (1993) Endotoxin-induced organ injury. Crit Care Med 21:S19-S24
Gillespie RD, Dolan MC, Piesman J, Titus RG (2001) Identification of an IL-2 binding protein in the saliva of the Lyme disease vector tick, Ixodes scapularis. J Immunol 166:4319–4326
Gong JH, Clark-Lewis I (1995) Antagonists of monocyte chemoattractant protein 1 identified by modification of functionally critical NH2-terminal residues. J Exp Med 181:631–640
Gong JH, Uguccioni M, Dewald B, Baggiolini M, Clark-Lewis I (1996) RANTES and MCP-3 antagonists bind multiple chemokine receptors. J Biol Chem 271:10521–10527
Gosling J, Slaymaker S, Gu L, Tseng S, Zlot CH, Young SG, Rollins BJ, Charo IF (1999) MCP-1 deficiency reduces susceptibility to atherosclerosis in mice that overexpress human apolipoprotein B. J Clin Invest 103:773–778
Grone HJ, Weber C, Weber KC, Grone EF, Rabelink T, Klier CM, Wells TC, Proudfoot AE, Schlondorff D, Nelson PJ (1999) Met-RANTES reduces vascular and tubular damage during acute renal transplant rejection: blocking monocyte arrest and recruitment. FASEB J 13:1371–1383
Gunn MD, Nelken NA, Liao X, Williams LT (1997) Monocyte chemoattractant protein-1 is sufficient for the chemotaxis of monocytes and lymphocytes in transgenic mice but requires an additional stimulus for inflammatory activation. J Immunol 158:376–383
Hajnicka V, Kocakova P, Slavikova M, Slovak M, Gasperik J, Fuchsberger N, Nuttall PA (2001) Anti-interleukin-8 activity of tick salivary gland extracts. Parasite Immunol 23:483–489
Hart MJ, Jiang X, Kozasa T, Roscoe W, Singer WD, Gilman AG, Sternweis PC, Bollag G (1998) Direct stimulation of the guanine nucleotide exchange activity of p115 RhoGEF by Galpha13. Science 280:2112–2114
Hirsch E, Katanaev VL, Garlanda C, Azzolino O, Pirola L, Silengo L, Sozzani S, Mantovani A, Altruda F, Wymann MP (2000) Central role for G protein-coupled phosphoinositide 3-kinase gamma in inflammation. Science 287:1049–1053
Hisada T, Hellewell PG, Teixeira MM, Malm MG, Salmon M, Huang TJ, Chung KF (1999) alpha4 integrin-dependent eotaxin induction of bronchial hyperresponsiveness and eosinophil migration in interleukin-5 transgenic mice. Am J Respir Cell Mol Biol 20:992–1000
Homey B, Alenius H, Muller A, Soto H, Bowman EP, Yuan W, McEvoy L, Lauerma AI, Assmann T, Bunemann E, Lehto M, Wolff H, Yen D, Marxhausen H, To W, Sedgwick J, Ruzicka T, Lehmann P, Zlotnik A (2002) CCL27–CCR10 interactions regulate T cell–mediated skin inflammation. Nat Med 8:157–165
Horuk R, Clayberger C, Krensky AM, Wang Z, Grone HJ, Weber C, Weber KS, Nelson PJ, May K, Rosser M, Dunning L, Liang M, Buckman B, Ghannam A, Ng HP, Islam I, Bauman JG, Wei GP, Monahan S, Xu W, Snider RM, Morrissey MM, Hesselgesser J, Perez HD (2001) A non-peptide functional antagonist of the CCR1 chemokine receptor is effective in rat heart transplant rejection. J Biol Chem 276:4199–4204

Hosaka S, Akahoshi T, Wada C, Kondo H (1994) Expression of the chemokine superfamily in rheumatoid arthritis. Clin Exp Immunol 97:451–457
Huang DR, Wang J, Kivisakk P, Rollins BJ, Ransohoff RM (2001) Absence of monocyte chemoattractant protein 1 in mice leads to decreased local macrophage recruitment and antigen-specific T helper cell type 1 immune response in experimental autoimmune encephalomyelitis. J Exp Med 193:713–726
Jiang H, Kuang Y, Wu Y, Xie W, Simon MI, Wu D (1997) Roles of phospholipase C beta2 in chemoattractant-elicited responses. Proc Natl Acad Sci U S A 94:7971–7975
Jose PJ, Griffiths-Johnson DA, Collins PD, Walsh DT, Moqbel R, Totty NF, Truong O, Hsuan JJ, Williams TJ (1994) Eotaxin: a potent eosinophil chemoattractant cytokine detected in a guinea pig model of allergic airways inflammation. J Exp Med 179:881–887
Karpus WJ, Kennedy KJ (1997) MIP-1alpha and MCP-1 differentially regulate acute and relapsing autoimmune encephalomyelitis as well as Th1/Th2 lymphocyte differentiation. J Leukoc Biol 62:681–687
Katanaev VL (2001) Signal transduction in neutrophil chemotaxis. Biochemistry 66:351–368
Katschke KJ, Jr., Rottman JB, Ruth JH, Qin S, Wu L, LaRosa G, Ponath P, Park CC, Pope RM, Koch AE (2001) Differential expression of chemokine receptors on peripheral blood, synovial fluid, and synovial tissue monocytes/macrophages in rheumatoid arthritis. Arthritis Rheum 44:1022–1032
Kaufmann A, Salentin R, Gemsa D, Sprenger H (2001) Increase of CCR1 and CCR5 expression and enhanced functional response to MIP-1 alpha during differentiation of human monocytes to macrophages. J Leukoc Biol 69:248–252
Kim KS, Rajarathnam K, Clark-Lewis I, Sykes BD (1996) Structural characterization of a monomeric chemokine: monocyte chemoattractant protein-3. FEBS Lett 395:277–282
Kledal TN, Rosenkilde MM, Coulin F, Simmons G, Johnsen AH, Alouani S, Power CA, Luttichau HR, Gerstoft J, Clapham PR, Clarklewis I, Wells TNC, Schwartz TW (1997) A broad-spectrum chemokine antagonist encoded by kaposis sarcoma- associated herpesvirus. Science 277:1656–1659
Koch AE, Kunkel SL, Harlow LA, Johnson B, Evanoff HL, Haines GK, Burdick MD, Pope RM, Strieter RM (1992) Enhanced production of monocyte chemoattractant protein-1 in rheumatoid arthritis. J Clin Invest 90:772–779
Koch AE, Kunkel SL, Harlow LA, Mazarakis DD, Haines GK, Burdick MD, Pope RM, Strieter RM (1994) Macrophage inflammatory protein-1 alpha. A novel chemotactic cytokine for macrophages in rheumatoid arthritis. J Clin Invest 93:921–928
Koch AE, Kunkel SL, Shah MR, Fu R, Mazarakis DD, Haines GK, Burdick MD, Pope RM, Strieter RM (1995) Macrophage inflammatory protein-1 beta: a C-C chemokine in osteoarthritis. Clin Immunol Immunopathol 77:307–314
Kopydlowski KM, Salkowski CA, Cody MJ, van Rooijen N, Major J, Hamilton TA, Vogel SN (1999) Regulation of macrophage chemokine expression by lipopolysaccharide in vitro and in vivo. J Immunol 163:1537–1544
Kurihara T, Warr G, Loy J, Bravo R (1997) Defects in macrophage recruitment and host defense in mice lacking the CCR2 chemokine receptor. J Exp Med 186:1757–1762
Kuziel WA, Morgan SJ, Dawson TC, Griffin S, Smithies O, Ley K, Maeda N (1997) Severe reduction in leukocyte adhesion and monocyte extravasation in mice deficient in CC chemokine receptor 2. Proc Natl Acad Sci U S A 94:12053–12058
Lalani AS, Graham K, Mossman K, Rajarathnam K, Clark-Lewis I, Kelvin D, McFadden G (1997) The purified myxoma virus gamma interferon receptor homolog M-T7 interacts with the heparin-binding domains of chemokines. J Virol 71:4356–4363
Lalani AS, McFadden G (1997) Secreted poxvirus chemokine binding proteins. J Leukoc Biol 62:570–576

Laurence JS, Blanpain C, Burgner JW, Parmentier M, Liwang PJ (2000) CC chemokine MIP-1 beta can function as a monomer and depends on Phe13 for receptor binding. Biochemistry 39:3401–3409

Leonard EJ, Yoshimura T, Tanaka S, Raffeld M (1991) Neutrophil recruitment by intradermally injected neutrophil attractant/activation protein-1. J Invest Dermatol 96:690–694

Liang M, Mallari C, Rosser M, Ng HP, May K, Monahan S, Bauman JG, Islam I, Ghannam A, Buckman B, Shaw K, Wei GP, Xu W, Zhao Z, Ho E, Shen J, Oanh H, Subramanyam B, Vergona R, Taub D, Dunning L, Harvey S, Snider RM, Hesselgesser J, Morrissey MM, Perez HD (2000) Identification and characterization of a potent, selective, and orally active antagonist of the CC chemokine receptor-1. J Biol Chem 275:19000–19008

Liu L, Lalani A, Dai E, Seet B, Macauley C, Singh R, Fan L, McFadden G, Lucas A (2000) The viral anti-inflammatory chemokine-binding protein M-T7 reduces intimal hyperplasia after vascular injury. J Clin Invest 105:1613–1621

Lloyd CM, Dorf ME, Proudfoot A, Salant DJ, Gutierrez-Ramos JC (1997) Role of MCP-1 and RANTES in inflammation and progression to fibrosis during murine crescentic nephritis. J Leukoc Biol 62:676–680

Lusso P (2000) Chemokines and viruses: the dearest enemies. Virology 273:228–240

Luster AD, Weinshank RL, Feinman R, Ravetch JV (1988) Molecular and biochemical characterization of a novel gamma-interferon-inducible protein. J Biol Chem 263:12036–12043

Luttichau HR, Stine J, Boesen TP, Johnsen AH, Chantry D, Gerstoft J, Schwartz TW (2000) A highly selective CC chemokine receptor (CCR)8 antagonist encoded by the poxvirus molluscum contagiosum. J Exp Med 191:171–180

Ma Q, Jones D, Borghesani PR, Segal RA, Nagasawa T, Kishimoto T, Bronson RT, Springer TA (1998) Impaired B-lymphopoiesis, myelopoiesis, and derailed cerebellar neuron migration in C. Proc Natl Acad Sci U S A 95:9448–9453

Mack M, Cihak J, Simonis C, Luckow B, Proudfoot AE, Plachy J, Bruhl H, Frink M, Anders HJ, Vielhauer V, Pfirstinger J, Stangassinger M, Schlondorff D (2001) Expression and characterization of the chemokine receptors CCR2 and CCR5 in mice. J Immunol 166:4697–4704

Mack M, Luckow B, Nelson PJ, Cihak J, Simmons G, Clapham PR, Signoret N, Marsh M, Stangassinger M, Borlat F, Wells TNC, Schlondorff D, Proudfoot AEI (1998) Aminooxypentane-RANTES induces CCR5 internalization but inhibits recycling: a novel inhibitory mechanism of HIV infectivity. J Exp Med 187:1215–1224

Matzinger P (1994) Tolerance, danger, and the extended family. Annu Rev Immunol 12:991–1045

McFadden G, Murphy PM (2000) Host-related immunomodulators encoded by poxviruses and herpesviruses. Curr Opin Microbiol 3:371–378

McMahon EJ, Cook DN, Suzuki K, Matsushima GK (2001) Absence of macrophage-inflammatory protein-1alpha delays central nervous system demyelination in the presence of an intact blood-brain barrier. J Immunol 167:2964–2971

Mellado M, Rodriguez-Frade JM, Aragay A, del Real G, Martin AM, Vila-Coro AJ, Serrano A, Mayor F, Jr., Martinez A (1998) The chemokine monocyte chemotactic protein 1 triggers Janus kinase 2 activation and tyrosine phosphorylation of the CCR2B receptor. J Immunol 161:805–813

Mellado M, Rodriguez-Frade JM, Manes S, Martinez A (2001) Chemokine signaling and functional responses: the role of receptor dimerization and TK pathway activation. Annu Rev Immunol 19:397–421

Meurer R, Van Riper G, Feeney W, Cunningham P, Hora D, Jr., Springer MS, MacIntyre DE, Rosen H (1993) Formation of eosinophilic and monocytic intradermal inflammatory sites in the dog by injection of human RANTES but not human monocyte chemoattractant protein 1, human macrophage inflammatory protein 1 alpha, or human interleukin 8. J Exp Med 178:1913–1921

Mishell B.B. (1980) Preparation of Mouse Cell Suspensions. In: Mishell B.B., Shiigi S.M. (eds) Selected Methods In Cellular Immunology. W.H. Freeman and Company, New York, pp. 6–8

Moser B, Loetscher P (2001) Lymphocyte traffic control by chemokines. Nat Immunol 2:123–128

Murphy PM (1996) Chemokine receptors: structure, function and role in microbial pathogenesis. Cytokine Growth Factor Rev 7:47–64

Nagasawa T, Hirota S, Tachibana K, Takakura N, Nishikawa S, Kitamura Y, Yoshida N, Kikutani H, Kishimoto T (1996) Defects of B-cell lymphopoiesis and bone-marrow myelopoiesis in mice lacking the CXC chemokine PBSF/SDF-1. Nature 382:635–638

Nelken NA, Coughlin SR, Gordon D, Wilcox JN (1991) Monocyte chemoattractant protein-1 in human atheromatous plaques. J Clin Invest 88:1121–1127

Ogata H, Takeya M, Yoshimura T, Takagi K, Takahashi K (1997) The role of monocyte chemoattractant protein-1 (MCP-1) in the pathogenesis of collagen-induced arthritis in rats. J Pathol 182:106–114

Oppermann M, Mack M, Proudfoot AE, Olbrich H (1999) Differential effects of CC chemokines on CC chemokine receptor 5 (CCR5) phosphorylation and identification of phosphorylation sites on the CCR5 carboxyl terminus. J Biol Chem 274:8875–8885

Paavola CD, Hemmerich S, Grunberger D, Polsky I, Bloom A, Freedman R, Mulkins M, Bhakta S, McCarley D, Wiesent L, Wong B, Jarnagin K, Handel TM (1998) Monomeric monocyte chemoattractant protein-1 (MCP-1) binds and activates the MCP-1 receptor CCR2B. J Biol Chem 273:33157–33165

Parent CA, Devreotes PN (1999) A cell's sense of direction. Science 284:765–770

Parry CM, Simas JP, Smith VP, Stewart CA, Minson AC, Efstathiou S, Alcami A (2000) A broad spectrum secreted chemokine binding protein encoded by a herpesvirus. J Exp Med 191:573–578

Plater-Zyberk C, Hoogewerf AJ, Proudfoot AE, Power CA, Wells TN (1997) Effect of a CC chemokine receptor antagonist on collagen induced arthritis in DBA/1 mice. Immunol Lett 57:117–120

Poznansky MC, Olszak IT, Foxall R, Evans RH, Luster AD, Scadden DT (2000) Active movement of T cells away from a chemokine. Nat Med 6:543–548

Proudfoot AE, Buser R, Borlat F, Alouani S, Soler D, Offord RE, Schroder JM, Power CA, Wells TN (1999) Amino-terminally modified RANTES analogues demonstrate differential effects on RANTES receptors. J Biol Chem 274:32478–32485

Proudfoot AEI, Power CA, Hoogewerf AJ, Montjovent MO, Borlat F, Offord RE, Wells TNC (1996) Extension of recombinant human RANTES by the retention of the initiating methionine produces a potent antagonist. J Biol Chem 271:2599–2603

Rajarathnam K, Sykes BD, Kay CM, Dewald B, Geiser T, Baggiolini M, Clark-Lewis I (1994) Neutrophil activation by monomeric interleukin-8. Science 264:90–92

Rampart M, Van Damme J, Zonnekeyn L, Herman AG (1989) Granulocyte chemotactic protein/interleukin-8 induces plasma leakage and neutrophil accumulation in rabbit skin. Am J Pathol 135:21–25

Rani MR, Foster GR, Leung S, Leaman D, Stark GR, Ransohoff RM (1996) Characterization of beta-R1, a gene that is selectively induced by interferon beta (IFN-beta) compared with IFN-alpha. J Biol Chem 271:22878–22884

Ransohoff RM, Tani M (1998) Do chemokines mediate leukocyte recruitment in post-traumatic CNS inflammation? Trends Neurosci 21:154–159

Rathanaswami P, Hachicha M, Sadick M, Schall TJ, McColl SR (1993) Expression of the cytokine RANTES in human rheumatoid synovial fibroblasts. Differential regulation of RANTES and interleukin-8 genes by inflammatory cytokines. J Biol Chem 268:5834–5839

Reiss Y, Proudfoot AE, Power CA, Campbell JJ, Butcher EC (2001) CC chemokine receptor (CCR)4 and the CCR10 ligand cutaneous T cell-attracting chemokine (CTACK) in lymphocyte trafficking to inflamed skin. J Exp Med 194:1541–1547

Rodrick ML, Lamster IB, Sonis ST, Pender SG, Kolodkin AB, Fitzgerald JE, Wilson RE (1982) Effects of supernatants of polymorphonuclear neutrophils recruited by different inflammatory substances on mitogen responses of lymphocytes. Inflammation 6:1–11

Sabroe I, Conroy DM, Gerard NP, Li Y, Collins PD, Post TW, Jose PJ, Williams TJ, Gerard CJ, Ponath PD (1998) Cloning and characterization of the guinea pig eosinophil eotaxin receptor, C-C chemokine receptor-3: blockade using a monoclonal antibody in vivo. J Immunol 161:6139–6147

Sabroe I, Peck MJ, Van Keulen BJ, Jorritsma A, Simmons G, Clapham PR, Williams TJ, Pease JE (2000) A small molecule antagonist of chemokine receptors CCR1 and CCR3. Potent inhibition of eosinophil function and CCR3-mediated HIV-1 entry. J Biol Chem 275:25985–25992

Sakanashi Y, Takeya M, Yoshimura T, Feng L, Morioka T, Takahashi K (1994) Kinetics of macrophage subpopulations and expression of monocyte chemoattractant protein-1 (MCP-1) in bleomycin-induced lung injury of rats studied by a novel monoclonal antibody against rat MCP-1. J Leukoc Biol 56:741–750

Samson M, Libert F, Doranz BJ, Rucker J, Liesnard C, Farber CM, Saragosti S, Lapoumeroulie C, Cognaux J, Forceille C, Muyldermans G, Verhofstede C, Burtonboy G, Georges M, Imai T, Rana S, Yi Y, Smyth RJ, Collman RG, Doms RW, Vassart G, Parmentier M (1996) Resistance to HIV-1 infection in caucasian individuals bearing mutant alleles of the CCR-5 chemokine receptor gene. Nature 382:722–725

Segerer S, Nelson PJ, Schlondorff D (2000) Chemokines, chemokine receptors, and renal disease: from basic science to pathophysiologic and therapeutic studies. J Am Soc Nephrol 11:152–176

Sekido N, Mukaida N, Harada A, Nakanishi I, Watanabe Y, Matsushima K (1993) Prevention of lung reperfusion injury in rabbits by a monoclonal antibody against interleukin-8. Nature 365:654–657

Sellebjerg F, Madsen HO, Jensen CV, Jensen J, Garred P (2000) CCR5 delta32, matrix metalloproteinase-9 and disease activity in multiple sclerosis. J Neuroimmunol 102:98–106

Signoret N, Pelchen-Matthews A, Mack M, Proudfoot AEI, Marsh M (2000) Endocytosis and recycling of the HIV coreceptor CCR5. J Cell Biol. 151:1281–1293

Simmons G, Clapham PR, Picard L, Offord RE, Rosenkilde, MM, Schwartz TW, Buser R, Wells TNC, Proudfoot AEI (1997) Potent inhibition of HIV-1 infectivity in macrophages and lymphocytes by a novel CCR5 antagonist. Science 276:276–279

Sorensen TL, Tani M, Jensen J, Pierce V, Lucchinetti C, Folcik VA, Qin S, Rottman J, Sellebjerg F, Strieter RM, Frederiksen JL, Ransohoff RM (1999) Expression of specific chemokines and chemokine receptors in the central nervous system of multiple sclerosis patients. J Clin Invest 103:807–815

Springer TA (1990) Adhesion receptors of the immune system. Nature 346:425–434

Stine JT, Chantry D, Gray P (1999) Virally encoded chemokines and chemokine receptors: genetic embezzlement of host DNA. Chem Immunol 72:161–180

Streblow DN, Orloff SL, Nelson JA (2001) Do pathogens accelerate atherosclerosis? J Nutr 131:2798S-2804S

Streblow DN, Soderberg-Naucler C, Vieira J, Smith P, Wakabayashi E, Ruchti F, Mattison K, Altschuler Y, Nelson JA (1999) The human cytomegalovirus chemokine receptor US28 mediates vascular smooth muscle cell migration. Cell 99:511–520

Strizki JM, Xu S, Wagner NE, Wojcik L, Liu J, Hou Y, Endres M, Palani A, Shapiro S, Clader JW, Greenlee WJ, Tagat JR, McCombie S, Cox K, Fawzi AB, Chou CC, Pugliese-Sivo C, Davies L, Moreno ME, Ho DD, Trkola A, Stoddart CA, Moore JP, Reyes GR, Baroudy BM (2001) SCH-C (SCH 351125), an orally bioavailable, small molecule antagonist of the chemokine receptor CCR5, is a potent inhibitor of HIV-1 infection in vitro and in vivo. Proc Natl Acad Sci U S A 98:12718–12723

Strunk T, Bubel S, Mascher B, Schlenke P, Kirchner H, Wandinger KP (2000) Increased numbers of CCR5+ interferon-gamma- and tumor necrosis factor- alpha-secreting T lymphocytes in multiple sclerosis patients. Ann Neurol 47:269–273

Swensson O, Schubert C, Christophers E, Schroder JM (1991) Inflammatory properties of neutrophil-activating protein-1/interleukin 8 (NAP-1/IL-8) in human skin: a light- and electronmicroscopic study. J Invest Dermatol 96:682–689

Szabo C, Scott GS, Virag L, Egnaczyk G, Salzman AL, Shanley TP, Hasko G (1998) Suppression of macrophage inflammatory protein (MIP)-1alpha production and collagen-induced arthritis by adenosine receptor agonists. Br J Pharmacol 125:379–387

Tachibana K, Hirota S, Iizasa H, Yoshida H, Kawabata K, Kataoka Y, Kitamura Y, Matsushima K, Yoshida N, Nishikawa S, Kishimoto T, Nagasawa T (1998) The chemokine receptor CXCR4 is essential for vascularization of the gastrointestinal tract. Nature 393:591–594

Takami S, Minami M, Nagata I, Namura S, Satoh M (2001) Chemokine receptor antagonist peptide, viral MIP-II, protects the brain against focal cerebral ischemia in mice. J Cereb Blood Flow Metab 21:1430–1435

Takano K, Al Mokdad M, Shibata F, Tsuchiya H, Nakagawa H (1999) Rat macrophage inflammatory protein-1alpha, a CC chemokine, acts as a neutrophil chemoattractant in vitro and in vivo. Inflammation 23:411–424

Tran EH, Kuziel WA, Owens T (2000) Induction of experimental autoimmune encephalomyelitis in C57BL/6 mice deficient in either the chemokine macrophage inflammatory protein-1alpha or its CCR5 receptor. Eur J Immunol 30:1410–1415

Trebst C, Ransohoff RM (2001) Investigating chemokines and chemokine receptors in patients with multiple sclerosis: opportunities and challenges. Arch Neurol 58:1975–1980

Van Damme J, Proost P, Lenaerts JP, Opdenakker G (1992) Structural and functional identification of two human, tumor-derived monocyte chemotactic proteins (MCP-2 and MCP-3) belonging to the chemokine family. J Exp Med 176:59–65

Villiger PM, Terkeltaub R, Lotz M (1992) Monocyte chemoattractant protein-1 (MCP-1) expression in human articular cartilage. Induction by peptide regulatory factors and differential effects of dexamethasone and retinoic acid. J Clin Invest 90:488–496

Volin MV, Shah MR, Tokuhira M, Haines GK, Woods JM, Koch AE (1998) RANTES expression and contribution to monocyte chemotaxis in arthritis. Clin Immunol Immunopathol 89:44–53

Wagner L, Yang OO, Garcia-Zepeda EA, Ge Y, Kalams SA, Walker BD, Pasternack MS, Luster AD (1998) b-Chemokines are released from HIV-1-specific cytolytic granules complexed to potreoglycans. Nature 391:908–911

Weber KS, Grone HJ, Rocken M, Klier C, Gu S, Wank R, Proudfoot AE, Nelson PJ, Weber C (2001) Selective recruitment of Th2-type cells and evasion from a cytotoxic immune response mediated by viral macrophage inhibitory protein-II. Eur J Immunol 31:2458–2466

Weber KS, von Hundelshausen P, Clark-Lewis I, Weber PC, Weber C (1999) Differential immobilization and hierarchical involvement of chemokines in monocyte arrest and transmigration on inflamed endothelium in shear flow. Eur J Immunol 29:700–712

Weiss JM, Downie SA, Lyman WD, Berman JW (1998) Astrocyte-derived monocyte-chemoattractant protein-1 directs the transmigration of leukocytes across a model of the human blood-brain barrier. J Immunol 161:6896–6903

Wilkinson PC (1998) Assays of leukocyte locomotion and chemotaxis. J Immunol Methods 216:139–153

Wong M, Fish EN (1998) RANTES and MIP-1alpha activate stats in T cells. J Biol Chem 273:309–314

Yokochi S, Hashimoto H, Ishiwata Y, Shimokawa H, Haino M, Terashima Y, Matsushima K (2001) An anti-inflammatory drug, propagermanium, may target GPI-anchored proteins associated with an MCP-1 receptor, CCR2. J Interferon Cytokine Res 21:389–398
Zhang Y, Rollins BJ (1995) A dominant negative inhibitor indicates that monocyte chemoattractant protein 1 functions as a dimer. Mol Cell Biol 15:4851–4855
Zhou N, Luo Z, Luo J, Liu D, Hall JW, Pomerantz RJ, Huang Z (2001) Structural and functional characterization of human CXCR4 as a chemokine receptor and HIV-1 co-receptor by mutagenesis and molecular modeling studies. J Biol Chem 276:42826–42833
Zlatopolskiy A, Laurence J (2001) 'Reverse gear' cellular movement mediated by chemokines. Immunol Cell Biol 79:340–344
Zlotnik A, Yoshie O (2000) Chemokines: a new classification system and their role in immunity. Immunity 12:121–127
Zou YR, Kottmann AH, Kuroda M, Taniuchi I, Littman DR (1998a) Function of the chemokine receptor CXCR4 in haematopoiesis and in cerebellar development. Nature 393:595–599

Antimicrobial Peptides

T. Ganz · R. I. Lehrer

School of Medicine,
University of California,
Los Angeles, CA 90095, USA
e-mail: tganz@mednet.ucla.edu

Abstract Antimicrobial peptides, mainly defensins and cathelicidins, are abundant components of granulocytes, Paneth cells of the small intestine, inflamed epithelia, and rabbit alveolar macrophages. There is increasing evidence that in these settings antimicrobial peptides and larger proteins contribute to microbicidal activity and other host defense functions. However, antimicrobial peptides and antimicrobial proteins (with the exception of lysozyme) are present at most in small amounts in most types of macrophages. In some cases, antimicrobial peptides may be difficult to detect at the protein level because macrophages lack granules, the large preformed storage compartment for antimicrobial peptides of granulocytes and Paneth cells, and may instead synthesize antimicrobial substances continually or on demand. Alternatively, histones, other nucleoproteins, or as yet unrecognized polypeptides or nonprotein components may contribute to oxygen-independent killing in macrophages.

Keywords Cathelicidins, Defensins, Histones, Host defense

1 Overview

Antimicrobial peptides are polypeptides of mass less than 10 kDa. They are found in a host-defense context and manifest antimicrobial activity in vitro when tested under conditions found in their proposed sites of action. Antimicrobial peptides may act alone or in concert with other host-derived molecules. In the last 20 years, hundreds of different antimicrobial peptides have been discovered in the phagocytes, epithelia, and secretions of vertebrates, and in the blood cells and tissues of invertebrates. The peptides comprise one end of a continuum of gene-encoded antimicrobial molecules that range in size from the large protein complexes that form complement pores, through intermediate size (10–100 kDa) proteins that can target and disrupt microbial membranes or sequester essential micronutrients.

It is important to acknowledge "up front" that most available information about antimicrobial peptides comes *not* from studies of macrophages, but from other cells—mainly polymorphonuclear leukocytes (PMN) and epithelial cells. Epithelia, most prominently the epidermis, synthesize antimicrobial peptides constitutively or in response to signals associated with injury, inflammation, or infection. In the PMN of many species, antimicrobial peptides are synthesized constitutively and in such abundance that they constitute the predominant polypeptides in crude cell extracts. Polymorphonuclear leukocytes store their antimicrobial peptides in cytoplasmic granules of several types. Some of these subcellular organelles (e.g., the primary or azurophil granules of human PMN) preferentially deliver their contents to phagosomes, while others (e.g., the secondary or specific granules of human PMN) release their contents mostly to the cell surface and external milieu. In the inflamed epidermis, the site of antimicrobial peptide deposition is not yet well characterized. In either case, the presence of high concentrations of stored antimicrobial peptides has facilitated their biochemical detection, recovery, and analysis. In contrast, with the prominent exception of rabbit alveolar macrophages, the macrophages examined to date have not contained high concentrations of antimicrobial peptides other than histones or fragments thereof.

Nevertheless, even macrophages from mice, an animal whose PMN are relatively deficient in antimicrobial peptides (Eisenhauer and Lehrer 1992) clearly possess antimicrobial mechanisms other than the production of reactive oxygen or nitrogen products (Shiloh et al. 1999). Several possible scenarios have been considered. First, there may a substantial contribution to antimicrobial activity from phagosome acidification, nutrient depletion (e.g., depletion of iron or tryptophan), or other mechanisms not dependent on the delivery of high concentrations of antimicrobial substances into the phagosome. Second, sufficient amounts of antimicrobial substances may be delivered to the phagosome by active synthesis coupled with vesicular transport, even in the absence of a storage compartment. Lysozyme (muramidase), a peptidoglycan-degrading enzyme found in most macrophages, is more often secreted than retained by murine peritoneal macrophages. Third, an external source of antimicrobial substances, including granule proteins and peptides from PMN or histones from phagocytized and degraded nuclear material could contribute to killing in vivo. Finally, as yet unrecognized antimicrobial substances, including autologous histones and histone fragments may contribute to phagocytic killing in macrophages. We will first discuss the biology of antimicrobial peptides in mammalian polymorphonuclear leukocytes before reviewing the more limited information available in the diverse types of animal macrophages.

2
Antimicrobial Peptides in PMN

Within each PMN there are several thousand membrane-bounded cytoplasmic structures—"granules"—whose contents include cationic polypeptides complexed to an anionic proteoglycan matrix (Olsson 1969; Parmley et al. 1986). The granules are selectively recovered from broken PMN by differential centrifugation, and their contents can be extracted by acidic solvents. The extract typically consists of four groups of proteins (Modrzakowski et al. 1979; Greenwald and Ganz 1987) that can be separated by molecular weight. In the highest molecular weight fraction (>50 kDa), myeloperoxidase and lactoferrin are the predominant proteins, accompanied in some species by the somewhat less abundant B/PI (bactericidal/permeability inducing protein). In human PMN, the next fraction consists of ~30-kDa serine proteases (elastase, cathepsin G, proteinase 3) accompanied by azurocidin/CAP37, their enzymatically inactive homolog. The third fraction (~14 kDa) is predominantly lysozyme, but also contains the pro-forms of antimicrobial cathelicidin peptides. The last fraction (3–4 kDa), which is especially abundant in human and rabbit PMNs, contains defensins.

From the cellular content of the various antimicrobial polypeptides and the volume occupied by their cytoplasmic granules, it can be estimated that each of the major peptides and proteins exists at multi-milligram/ml concentrations within these organelles. Electron micrographs of phagocytic PMN suggest that the granule contents are deposited onto the surfaces of ingested microbes with relatively little dilution.

3
Defensins

Mammalian defensins (Ganz et al. 1985a; Selsted et al. 1985) are a family of genetically related peptides that possess a framework of six cysteines with three characteristic disulfide linkages. Based on the spacing of the cysteines and the pattern of their connections, defensins are further subdivided into α-, β-, and θ-defensins. All three families have been found in PMN. α-Defensins occur in PMNs from humans and other primates, rats, hamsters, guinea pigs, and rabbits. β-Defensins were found in cattle (Selsted et al. 1996) and fowl (Harwig et al. 1994), and θ-defensins were identified in rhesus monkeys (Tang et al. 1999). The PMN of mice, pigs, sheep, and horses apparently lack defensins altogether. In Paneth cells, a specialized granule-rich epithelial cell implicated in the defense of small-intestinal crypts, only α-defensins (Ouellette and Lualdi 1990; Selsted et al. 1992a; Bevins et al. 1996) have been well documented in humans, mice, and rabbits, while in other epithelial cells β-defensins (Diamond et al. 1991; Bensch et al. 1995; Harder et al. 1997) predominate.

Post-phagocytic degranulation and nicotinamide adenine dinucleotide phosphate (NADPH) oxidase activation provide the phagocytic vacuoles of PMN with myeloperoxidase and hydrogen peroxide—the catalytic machinery and material needed to iodinate proteins. If radioactive iodine is provided, the contents of phagocytic vacuoles are rapidly radioiodinated. Such phagocytic vacuoles can be recovered from broken cells and analyzed by electrophoresis and autoradiography. In the phagocytic vacuoles of human PMN that ingested *Salmonella typhimurium*, defensins were the predominant iodinated polypeptide (Joiner et al. 1989).

In vitro, defensins display concentration-dependent antibacterial and antifungal activity, and several α-defensins were shown to inactivate herpes simplex virus (Daher et al. 1986). Their potency depends on the primary sequence and cationic charge of the peptide, the microbial target, and its metabolic state, and the composition of the test medium. In particular, the activity of defensins against gram-negative bacteria and many fungi is competitively inhibited by increasing concentrations of physiologic cations such as sodium, potassium, and especially calcium or magnesium, as well as by several serum proteins (Lehrer et al. 1988; Panyutich and Ganz 1991; Panyutich et al. 1994, 1995). Although the available information on the ionic composition of the phagosomal fluid is limited, the concentration of defensins in the phagocytic vacuole is so high that inhibitory ionic factors are likely to be overwhelmed. High defensin concentrations also occur at epithelial surfaces that interface with the external environment, as recently reported for β-defensin 1 in the porcine tongue (Shi et al. 1999).

A similar argument also holds for rabbit alveolar macrophages, phagocytes that contain macrophage cationic peptides (MCP)-1 and -2 (Patterson Delafield et al. 1980). These peptides are structurally (Selsted et al. 1983) identical to NP-1 and -2, the two most cationic defensins of rabbit PMN (Selsted et al. 1984). That MCP-1 and -2 are produced by alveolar macrophages was shown in metabolic labeling experiments with ^{35}S-cysteine (Ganz et al. 1985b) and by their high levels of MCP-1 and -2 mRNA (Ganz et al. 1989). The concentration of MCPs in alveolar macrophages increases during postnatal maturation of rabbit lungs (Ganz et al. 1985b), and is further increased by Freund's adjuvant-induced lung inflammation (Lehrer et al. 1981), so that the amounts of MCPs per cell are similar to the corresponding defensins in rabbit neutrophils. Surprisingly, rabbit peritoneal macrophages lack defensins (Ganz et al. 1989), even when induced by Freund's adjuvant. Thus, the expression of defensins appears to be a marker of alveolar macrophage differentiation in the rabbit. With the exception of rabbit alveolar macrophages, no other macrophage in any animal species has yet been shown to contain defensins in amounts comparable to human or rabbit PMN.

Defensin mRNA expression (predominantly bovine β-defensins 4 and 5) has also been detected in bovine alveolar macrophages by polymerase chain reaction (PCR) and Northern blots (Ryan et al. 1998), but cellular concentrations or secretion rates for defensin peptides have not yet been published. Duits et al. recently reported that chimpanzee alveolar macrophages contained mRNA for β-defensin-1 (Duits et al. 2000). In the 1980s, we did not detect defensins in human alveolar macrophage extracts by electrophoretic methods, nor by immunostaining with antibodies against human neutrophil α-defensins HNP1–3. However, human monocytes from peripheral blood do contain detectable HNP1–3 peptides as shown by co-immunostaining with CD14 and anti-HNP1–3 antibody (Agerberth et al. 2000). The concentration and eventual fate of these defensins has not yet been determined.

The recent discovery of cyclic (θ) minidefensins in the leukocytes of a primate, *Macaca mulatta* (the Rhesus monkey), is remarkable for many reasons (Tang et al. 1999; Leonova et al. 2001). These peptides have only 18 residues, including six cysteines that form a three-rung disulfide ladder between its two antiparallel β-sheets. The only known cyclic peptides of animal origin, each minidefensin derives from two peptide precursors (demidefensins), each of which contributes nine total residues to the mature peptide. Although the cellular machinery responsible for splicing and trimming the precursors remains to be defined, it is operational in human leukocytes (Tang et al. 1999). No cyclic minidefensin peptides are known to exist in humans; however, normal human bone marrow expresses an mRNA homologous to the rhesus minidefensin precursors. A synthetic cyclic minidefensin (retrocyclin) whose sequence was based on this mRNA, was remarkably effective in protecting human cells from in vitro infection by HIV-1 (Cole et al. 2002).

4 Cathelicidins

Cathelicidins are a family of mammalian antimicrobial peptides that share a conserved cathelin domain with approximately 100 residues (Zanetti et al. 1995). The active peptide is located C-terminally to this domain, and is in most cases released by proteolysis during the secretion of the peptides from PMN. Like defensins, the C-terminal peptides are broadly antimicrobial but (Turner et al. 1998; Ganz et al. 2000) appear to be less sensitive than defensins to the inhibitory effects of salt and serum. Thus, cathelicidins may have evolved to act in extracellular spaces, where salt and serum components are present in abundance. The function of the cathelin domain is not known with certainty, but it could prevent premature intracellular activation of the potentially cytotoxic peptides, and target the peptide to its subcellular compartment—the specific (or secondary) granules of PMN (Sorensen et al. 1997). The C-terminal peptides are highly varied, and include α-helical peptides, β-sheet peptides, and other structural forms. The conservation of the cathelin domain encoded by exons 1–3 and the extreme variability of the C-terminal peptide encoded by exon 4 have suggested that this gene family evolved by exon (domain) swapping (Zhao et al. 1995). Cathelicidins are highly abundant and represented by multiple genes in cattle (Romeo et al. 1988; Gennaro et al. 1989; Selsted et al. 1992b) and pigs (Agerberth et al. 1991; Kokryakov et al. 1993; Boman et al. 1993; Storici and Zanetti 1993; Zanetti et al. 1994; Tossi et al. 1995). The number of genes in murine (Gallo et al. 1997), rabbit (Ooi et al. 1990; Zarember et al. 1997; Larrick et al. 1993) and human (Cowland et al. 1995; Larrick et al. 1995; Gudmundsson et al. 1996) is considerably smaller (1–3), but these peptides are relatively abundant. The human cathelicidin LL-37 has been detected in monocytes by immunostaining (Agerberth et al. 2000), but the location and concentration of the peptide in these cells is not known. Cathelicidins have not yet been reported in macrophages, but it is not clear if adequate attempts have been made to detect them.

5
Other Antimicrobial Peptides in Macrophages

The same extractive methods and assays that detected defensins and cathelicidins in PMN of many animal species and in alveolar macrophages of rabbits have also been applied to resting and interferon-γ induced murine macrophages and macrophage-like cell lines. In these studies, antimicrobial peptides similar to those of PMN were not detected, but histones and a ribosomal protein were responsible for the antimicrobial activity of cell extracts (Hiemstra et al. 1993, 1999). Histones in particular have a long history of showing up in screens for antimicrobial substances (Hirsch 1958) and perhaps it is finally time to pay attention to them.

Macrophages have an impressive lysosome system and are keen practitioners of phagosome–lysosome fusion. It is clear that many pathogens that survive within macrophages actively prevent phagosome–lysosome fusion, as shown first for *Mycobacterium tuberculosis* (Armstrong and Hart 1971; Hart 1979). Other organisms now known to subvert phagosome–lysosome fusion in macrophages include *Brucella* spp., *Legionella pneumophila*, *Salmonella typhimurium* and the protozoan, *Toxoplasma gondii*. Interference with phagolysosomal fusion almost certainly prevents some antimicrobial substances from reaching the phagosomes but the nature of these substances is not known.

Although histones are generally thought of as “nuclear” proteins, they are “born” (synthesized) in the cytoplasm, travel to and live in the nucleus, and “die” (undergo proteolysis) within lysosomes (Odaka and Mizuochi 1999). Consequently, the lysosomal compartment of macrophages may contain a brew that is rich in histones and other polycationic nucleoproteins. Exposure to such a mixture could be noxious for many, if not all, ingested bacteria. It still remains to be determined whether histones and other cationic “nuclear” proteins reach sufficient concentrations in phagolysosomes to contribute to the killing of ingested microbes. If so, it would also be of interest to ascertain whether these proteins enter phagolysosomal compartments during recycling of internal nuclear material, phagocytosis of extrinsic cellular material, or after de novo synthesis induced by cytokines or phagocytosis. As histones and other cationic nucleoproteins have dedicated transport systems for nuclear delivery (Jakel et al. 2002), it would also be of interest to learn if similar pathways exist to deliver them to the phagolysosomal compartment.

6 Summary and Conclusions

Macrophages employ multiple antimicrobial pathways to kill or inhibit phagocytized microbes and microbes that have evolved to parasitize macrophages. Although the pathways that generate reactive oxygen and nitrogen intermediates are similar to those in PMN, there are substantial differences in the oxygen-independent antimicrobial mechanisms. With the possible exception of their juvenile forms (i.e., monocytes) most macrophages lack an extensive primary storage compartment analogous to the granules found in PMNs. Without a storehouse of dedicated antimicrobial peptides and proteins, macrophages appear to rely on alternative mechanisms linked to phagolysosomal fusion to render the environment around ingested bacteria antimicrobial. What effector molecules do macrophages use to kill ingested organisms? If they are polypeptides, then intact and cleaved portions of histones and other highly cationic "nuclear" proteins are the prime suspects. If they are not polypeptides, then an important chapter in macrophage biology (unfortunately, not this one) remains to be written.

7 References

Agerberth B, Charo J, Werr J, Olsson B, Idali F, Lindbom L, Kiessling R, Jornvall H, Wigzell H, Gudmundsson GH (2000) The human antimicrobial and chemotactic peptides LL-37 and alpha-defensins are expressed by specific lymphocyte and monocyte populations. Blood 96:3086–3093

Agerberth B, Lee JY, Bergman T, Carlquist M, Boman HG, Mutt V, Jornvall H (1991) Amino acid sequence of PR-39. Isolation from pig intestine of a new member of the family of proline-arginine-rich antibacterial peptides. Eur J Biochem 202:849–854

Armstrong JA, Hart PD (1971) Response of cultured macrophages to Mycobacterium tuberculosis, with observations on fusion of lysosomes with phagosomes. J Exp Med 134:713–740

Bensch KW, Raida M, Magert HJ, Schulz-Knappe P, Forssmann WG (1995) hBD-1: a novel beta-defensin from human plasma. FEBS Lett 368:331–335

Bevins CL, Jones DE, Dutra A, Schaffzin J, Muenke M (1996) Human enteric defensin genes: chromosomal map position and a model for possible evolutionary relationships. Genomics 31:95–106

Boman HG, Agerberth B, Boman A (1993) Mechanisms of action on Escherichia coli of cecropin P1 and PR- 39, two antibacterial peptides from pig intestine. Infect Immun 61:2978–2984

Cole AM, Hong T, Boo LM, Nguyen T, Zhao C, Bristol G, Zack JA, Waring AJ, Yang OO, Lehrer RI (2002) Retrocyclin: a primate peptide that protects cells from infection by T- and M-tropic strains of HIV-1. Proc Natl Acad Sci USA 99:1813–1818

Cowland JB, Johnsen AH, Borregaard N (1995) hCAP-18, a cathelin/pro-bactenecin-like protein of human neutrophil specific granules. FEBS Lett 368:173–176

Daher KA, Selsted ME, Lehrer RI (1986) Direct inactivation of viruses by human granulocyte defensins. J Virol 60:1068–1074

Diamond G, Zasloff M, Eck H, Brasseur M, Maloy WL, Bevins CL (1991) Tracheal antimicrobial peptide, a cysteine-rich peptide from mammalian tracheal mucosa: peptide isolation and cloning of a cDNA. Proc Natl Acad Sci USA 88:3952–3956

Duits LA, Langermans JA, Paltansing S, van der ST, Vervenne RA, Frost PA, Hiemstra PS, Thomas AW, Nibbering PH (2000) Expression of beta-defensin-1 in chimpanzee (Pan troglodytes) airways. J Med Primatol 29:318–323

Eisenhauer PB, Lehrer RI (1992) Mouse neutrophils lack defensins. Infect Immun 60:3446–3447

Gallo RL, Kim KJ, Bernfield M, Kozak CA, Zanetti M, Merluzzi L, Gennaro R (1997) Identification of CRAMP, a cathelin-related antimicrobial peptide expressed in the embryonic and adult mouse. J Biol Chem 272:13088–13093

Ganz T, Bellm L, Lehrer RI (2000) Protegrins: new antibiotics of mammalian origin. Expert Opin Investig Drugs 9:1731–1742

Ganz T, Rayner JR, Valore EV, Tumolo A, Talmadge K, Fuller F (1989) The structure of the rabbit macrophage defensin genes and their organ-specific expression. J Immunol 143:1358–1365

Ganz T, Selsted ME, Szklarek D, Harwig SS, Daher K, Bainton DF, Lehrer RI (1985a) Defensins. Natural peptide antibiotics of human neutrophils. J Clin Invest 76:1427–1435

Ganz T, Sherman MP, Selsted ME, Lehrer RI (1985b) Newborn rabbit alveolar macrophages are deficient in two microbicidal cationic peptides, MCP-1 and MCP-2. Am Rev Respir Dis 132:901–904

Gennaro R, Skerlavaj B, Romeo D (1989) Purification, composition, and activity of two bactenecins, antibacterial peptides of bovine neutrophils. Infect Immun 57:3142–3146

Greenwald GI, Ganz T (1987) Defensins mediate the microbicidal activity of human neutrophil granule extract against Acinetobacter calcoaceticus. Infect Immun 55:1365–1368

Gudmundsson GH, Agerberth B, Odeberg J, Bergman T, Olsson B, Salcedo R (1996) The human gene FALL39 and processing of the cathelin precursor to the antibacterial peptide LL-37 in granulocytes. Eur J Biochem 238:325–332

Harder J, Bartels J, Christophers E, Schroeder J-M (1997) A peptide antibiotic from human skin. Nature 387:861–862

Hart PD (1979) Phagosome-lysosome fusion in macrophages: a hinge in the intracellular fate of ingested microorganisms? Front Biol 48:409–423

Harwig SS, Swiderek KM, Kokryakov VN, Lee TD, Lehrer RI (1994) Primary structure of gallinacin-1, an antimicrobial beta-defensin from chicken leukocytes. In: Crabb JW (ed) Techniques in Protein Chemistry V. Academic Press, San Diego, pp. 81–88

Hiemstra PS, Eisenhauer PB, Harwig SS, van den Barselaar MT, van Furth R, Lehrer RI (1993) Antimicrobial proteins of murine macrophages. Infect Immun 61:3038–3046

Hiemstra PS, van den Barselaar MT, Roest M, Nibbering PH, van Furth R (1999) Ubiquicidin, a novel murine microbicidal protein present in the cytosolic fraction of macrophages. J Leukoc Biol 66:423–428

Hirsch JG (1958) Bactericidal action of histone. J Exp Med 108:925–944

Jakel S, Mingot JM, Schwarzmaier P, Hartmann E, Gorlich D (2002) Importins fulfil a dual function as nuclear import receptors and cytoplasmic chaperones for exposed basic domains. EMBO J 21:377–386
Joiner KA, Ganz T, Albert J, Rotrosen D (1989) The opsonizing ligand on Salmonella typhimurium influences incorporation of specific, but not azurophil, granule constituents into neutrophil phagosomes. J Cell Biol 109:2771–2782
Kokryakov VN, Harwig SS, Panyutich EA, Shevchenko AA, Aleshina GM, Shamova OV, Korneva HA, Lehrer RI (1993) Protegrins: leukocyte antimicrobial peptides that combine features of corticostatic defensins and tachyplesins. FEBS Lett 327:231–236
Larrick JW, Hirata M, Balint RF, Lee J, Zhong J, Wright SC (1995) Human CAP18: a novel antimicrobial lipopolysaccharide-binding protein. Infect Immun 63:1291–1297
Larrick JW, Hirata M, Shimomoura Y, Yoshida M, Zheng H, Zhong J, Wright SC (1993) Antimicrobial activity of rabbit CAP18-derived peptides. Antimicrob Agents Chemother 37:2534–2539
Lehrer RI, Ganz T, Szklarek D, Selsted ME (1988) Modulation of the in vitro candidacidal activity of human neutrophil defensins by target cell metabolism and divalent cations. J Clin Invest 81:1829–1835
Lehrer RI, Szklarek D, Selsted ME, Fleischmann J (1981) Increased content of microbicidal cationic peptides in rabbit alveolar macrophages elicited by complete Freund adjuvant. Infect Immun 33:775–778
Leonova L, Kokryakov VN, Aleshina G, Hong T, Nguyen T, Zhao C, Waring AJ, Lehrer RI (2001) Circular minidefensins and posttranslational generation of molecular diversity. J Leukoc Biol 70:461–464
Modrzakowski MC, Cooney MH, Martin LE, Spitznagel JK (1979) Bactericidal activity of fractionated granule contents from human polymorphonuclear leukocytes. Infect Immun 23:587–591
Odaka C, Mizuochi T (1999) Role of macrophage lysosomal enzymes in the degradation of nucleosomes of apoptotic cells. J Immunol 163:5346–5352
Olsson I (1969) Chondroitin sulfate proteolycan of human leukocytes. Biochim Biophys Acta 177:241–249
Ooi CE, Weiss J, Levy O, Elsbach P (1990) Isolation of two isoforms of a novel 15-kDa protein from rabbit polymorphonuclear leukocytes that modulate the antibacterial actions of other leukocyte proteins. J Biol Chem 265:15956–15962
Ouellette AJ, Lualdi JC (1990) A novel mouse gene family coding for cationic, cysteine-rich peptides. Regulation in small intestine and cells of myeloid origin. J Biol Chem 265:9831–9837
Panyutich A, Ganz T (1991) Activated alpha 2-macroglobulin is a principal defensin-binding protein. Am J Respir Cell Mol Biol 5:101–106
Panyutich AV, Hiemstra PS, Van Wetering S, Ganz T (1995) Human neutrophil defensin and serpins form complexes and inactivate each other. Am J Respir Cell Mol Biol 12:351–357
Panyutich AV, Szold O, Poon PH, Tseng Y, Ganz T (1994) Identification of defensin binding to C1 complement. FEBS Lett 356:169–173
Parmley RT, Doran T, Boyd RL, Gilbert C (1986) Unmasking and redistribution of lysosomal sulfated glycoconjugates in phagocytic polymorphonuclear leukocytes. J Histochem Cytochem 34:1701–1707
Patterson Delafield J, Martinez RJ, Lehrer RI (1980) Microbicidal cationic proteins in rabbit alveolar macrophages: a potential host defense mechanism. Infect Immun 30:180–192
Romeo D, Skerlavaj B, Bolognesi M, Gennaro R (1988) Structure and bactericidal activity of an antibiotic dodecapeptide purified from bovine neutrophils. J Biol Chem 263:9573–9575
Ryan LK, Rhodes J, Bhat M, Diamond G (1998) Expression of beta-defensin genes in bovine alveolar macrophages. Infect Immun 66:878–881

Selsted ME, Brown DM, DeLange RJ, Lehrer RI (1983) Primary structures of MCP-1 and MCP-2, natural peptide antibiotics of rabbit lung macrophages. J Biol Chem 258:14485–14489

Selsted ME, Harwig SS, Ganz T, Schilling JW, Lehrer RI (1985) Primary structures of three human neutrophil defensins. J Clin Invest 76:1436–1439

Selsted ME, Miller SI, Henschen AH, Ouellette AJ (1992a) Enteric defensins: antibiotic peptide components of intestinal host defense. J Cell Biol 118:929–936

Selsted ME, Novotny MJ, Morris WL, Tang YQ, Smith W, Cullor JS (1992b) Indolicidin, a novel bactericidal tridecapeptide amide from neutrophils. J Biol Chem 267:4292–4295

Selsted ME, Szklarek D, Lehrer RI (1984) Purification and antibacterial activity of antimicrobial peptides of rabbit granulocytes. Infect Immun 45:150–154

Selsted ME, Tang YQ, Morris WL, McGuire PA, Novotny MJ, Smith W, Henschen AH, Cullor JS (1996) Purification, primary structures, and antibacterial activities of beta-defensins, a new family of antimicrobial peptides from bovine neutrophils. J Biol Chem 271:16430

Shi J, Zhang G, Wu H, Ross C, Blecha F, Ganz T (1999) Porcine epithelial beta-defensin 1 is expressed in the dorsal tongue at antimicrobial concentrations. Infect Immun 67:3121–3127

Shiloh MU, MacMicking JD, Nicholson S, Brause JE, Potter S, Marino M, Fang F, Dinauer M, Nathan C (1999) Phenotype of mice and macrophages deficient in both phagocyte oxidase and inducible nitric oxide synthase. Immunity 10:29–38

Sorensen O, Arnljots K, Cowland JB, Bainton DF, Borregaard N (1997) The human antibacterial cathelicidin, hCAP-18, is synthesized in myelocytes and metamyelocytes and localized to specific granules in neutrophils. Blood 90:2796–2803

Storici P, Zanetti M (1993) A cDNA derived from pig bone marrow cells predicts a sequence identical to the intestinal antibacterial peptide PR-39. Biochem Biophys Res Commun 196:1058–1065

Tang YQ, Yuan J, Osapay G, Osapay K, Tran D, Miller CJ, Ouellette AJ, Selsted ME (1999) A cyclic antimicrobial peptide produced in primate leukocytes by the ligation of two truncated alpha-defensins [see comments]. Science 286:498–502

Tossi A, Scocchi M, Zanetti M, Storici P, Gennaro R (1995) PMAP-37, a novel antibacterial peptide from pig myeloid cells. cDNA cloning, chemical synthesis and activity. Eur J Biochem 228:941–946

Turner J, Cho Y, Dinh NN, Waring AJ, Lehrer RI (1998) Activities of LL-37, a cathelin-associated antimicrobial peptide of human neutrophils. Antimicrob Agents Chemother 42:2206–2214

Zanetti M, Gennaro R, Romeo D (1995) Cathelicidins: a novel protein family with a common proregion and a variable C-terminal antimicrobial domain. FEBS Lett 374:1–5

Zanetti M, Storici P, Tossi A, Scocchi M, Gennaro R (1994) Molecular cloning and chemical synthesis of a novel antibacterial peptide derived from pig myeloid cells. J Biol Chem 269:7855–8

Zarember K, Elsbach P, Shin-Kim K, Weiss J (1997) p15 s (15-kD antimicrobial proteins) are stored in the secondary granules of Rabbit granulocytes: implications for antibacterial synergy with the bactericidal/permeability-increasing protein in inflammatory fluids. Blood 89:672–679

Zhao C, Ganz T, Lehrer RI (1995) Structures of genes for two cathelin-associated antimicrobial peptides: prophenin-2 and PR-39. FEBS Lett 376:130–134

Macrophage Phospholipid Products

Questions on the Synthesis, Secretion and Biologic Effects of Phospholipid Mediators

P. M. Henson

Program in Cell Biology, Department of Pediatrics,
National Jewish Medical and Research Center,
1400 Jackson Street, Denver, CO 80206, USA, e-mail: hensonp@njc.org

Abstract This chapter focuses on a number of elements in the pharmacology, biochemistry, and biology of three families of extracellular phospholipid mediators as they relate to macrophages. In particular, it raises questions that are unique to phospholipids as mediators, including mechanisms of secretion, membrane association, and presentation to their receptors. A number of similarities between the different groups of phospholipids is emphasized, even though the lysophosphatides and PAFs act through seven transmembrane G protein-linked receptors, and phosphatidylserine does not. The phospholipid mediators discussed herein have very broad, and highly potent biologic activities, providing a real motive for appropriate pharmacological manipulation, both in the macrophage and more generally.

Keywords Eicosanoids, Lipid mediators, Lysophosphatides, Macrophages, PAF, Phosphatidylserine, Phospholipids, Receptors

1 Questions on the Synthesis, Secretion and Biologic Effects of Phospholipid Mediators

Macrophages both produce and respond to a host of different types of lipids with an equally bewildering array of physiologic and/or pathologic consequences. The pharmacology of lipid products in relation to macrophage function is thus an immense subject with many potential points of therapeutic attack, from the enzymology of lipid synthesizing and metabolizing enzymes to manipulation of the relevant intracellular and extracellular binding proteins or receptors. Therefore, this chapter will emphasize conceptual issues in lipid pharmacology rather than a catalog of active molecules and their receptors. It will also focus primarily on extracellularly acting phospholipids, in particular the PAF or platelet activating family of phospholipids, the family of lysophospholipid growth factors, sometimes called PLGFs (phospholipid growth-factors), and phosphatidylserines. Each family has broad biologic effects acting via one or more specific cell-surface receptors but with additional complexity related to intracellular metabolism and effects as well as its ability to insert into, and cross, lipid membranes.

The first two groups are generally thought to act in the fluid environment on cell-surface receptors, while PS may mediate its action while remaining on the presenting-cell membrane. However, as we shall see, each of these families may have the potential to act from either location. Surprisingly, although macrophages are known to respond to these bioactive phospholipids, their effects and signaling pathways have not received the degree of investigation that one might have expected.

Phospholipids are the main constituents of cell membranes. Their key structural elements are a glycerol backbone, one or two hydrophobic acyl, alkyl or alky-1-enyl side chains and a phosphate-linked polar head group, usually serine, ethanolamine, choline, glycerol, or various inositides. They are oriented in the membrane lipid bilayer with the head groups facing the extracellular medium or the cytoplasmic contents and the hydrophobic domains comprising the two leaflets of the lipid bilayer, admixed with other lipidic molecules such as cholesterol and the transmembrane domains of proteins. The phospholipid component of cell membranes is now recognized as being far from an inert matrix for active proteins to float in. Rather it is both a dynamic, highly active environment for such proteins as well as itself providing a critical supply of mediators and signaling molecules for extra- and intra-cellular regulation. These latter functions imply specific binding sites on proteins for phospholipids, a subject of ever-widening interest. Such binding may also be to specific receptors, i.e., leads to signaling activation and cellular responses. In addition, phospholipids can be released/secreted to the outside of the cell and there signal to other cells in the environment via such receptors on the outer cell surface.

Important to our understanding of the complex roles played by membrane phospholipids is the recognition that they are not uniformly distributed in the membrane, either within the plane of the membrane or between its inner and outer leaflet. There is a significant asymmetry between the two leaflets, with phosphatidylcholines (PCs) primarily found in the outer leaflet along with the lipid family of sphingomyelins (SM) or glycosphingolipids. The anionic phosphatidylserine and ethanolamine (PS and PE) are predominantly located in the inner leaflet. As discussed below, this asymmetry will turn out to have important implications in membrane function, lipid secretion and uptake, and cell recognition and stimulation.

Given the size of the subject, the bibliography is representational rather than exhaustive. In many cases, reviews will be cited instead of the original manuscripts, since there may be large numbers of these. Accordingly, I wish to tender my apologies in advance for any seeming omissions. The intent in this chapter is to raise questions, point to potential directions, and suggest future pharmacologic approaches to macrophage lipid mediator production and response.

1.1 The Questions

Detailed consideration of macrophage production and response to bioactive phospholipids raises a number of questions many of which relate to lipid mediators in general and certainly to those belonging to the family of phospholipids. These include: (1) location of the synthesizing enzymes within the cell, (2) mechanisms of transport to the plasma membrane, (3) translocation across the plasma membrane, (4) liberation of the phospholipids from the source cell into the environment, (5) special issues relating to presentation of amphipathic lipids to receptors on the cell membrane of the responding cells, (6) the possibility of direct receptor engagement on the responding cell with surface-membrane phospholipid expressed on the cell of origin and (7) the implications of phospholipid uptake and further metabolism, biosynthesis, and even re-secretion by responding cells such as macrophages.

1.2 The Players

Three examples of externally active, bioactive phospholipid families will form the focus of this discussion: (1) PAF and its analogs, including oxidized PCs of appropriate structure to bind the PAF receptor; (2) Lysophospholipids sometimes called lysophospholipid growth factors and including lyso-phosphatidic acids (LPA), lysosphingophospholipids such as sphingosine-1-phosphate (SIP, sometimes known as SPP) and lysophosphatidylcholines (LPC); (3) Phosphatidylserine (PS). In addition, lipoproteins may also contain members of these potentially active phospholipids. A key point in considering any of these bioactive phospholipids is that all of them can have highly varied structures and in truth represent families of molecules. For the lysophospholipids the *sn1* group can vary. For PAFs and PS, substituents occur in both the *sn1* and *sn2* position. In addition, the structure of PAFs, i.e., *sn1* alkyl or acyl PCs with a short chain *sn2* acyl group (usually acetate) is not strikingly different from lysophospholipids.

As mentioned above, most phospholipids can and do interact with proteins and, as a consequence, exhibit biologic responses, whether these proteins are considered receptors or not. The key distinction here is extracellular action on cell-surface receptors. An additional complexity for phospholipids is their ability to cross the membrane into the cell, there to act on intracellular responding proteins, or to be metabolized into products that have such activities. Elements of this process will only be discussed here in the context of transcellular biosynthesis of the bioactive phospholipids mentioned above. Another huge, and largely unfathomed, possibility is that intracellularly active phospholipids, such as the products of sphingomyelin metabolism (ceramides, sphingosine, sphingosine-1-P, etc.) or the myriad phosphatidylinositol metabolites may gain access to the outside of the cell and act on nearby or adherent cells either on external receptors if present, or more likely, by internalization and interaction with intracellular binding proteins/receptors. Largely ignored so far in this spectrum of potential biologic effects are phosphatidylethanolamines, phosphatidylglycerols, or phospholipids with variations in the *sn1* linkage such as plasmalogens. It seems highly unlikely that, in the long run, these will prove be devoid of extracellular effects and biologic activities.

1.3 Platelet Activating Factor(s) and Eicosanoids

Cleavage of the *sn2* arachidonyl group from PC generates two products, each with enormous potential for further metabolism and biologic function. The hydrolysis is mediated by phospholipases A2, of which the most important for generation of arachidonate products, also known as eicosanoids, is the type IV, arachidonyl-specific cytoplasmic or cPLA2. The critical role played by this upstream enzyme is exemplified by the phenotypic effects seen in genetically deficient mice (Klivenyi et al. 1998; Fujishima et al. 1999) and in the great interest in its modulation shown at various times by the pharmaceutical industry. The relevance for this chapter is that translocation and activation of cPLA2 is a first step in generation of the myriad eicosanoids, i.e., the subsequent products of arachidonate metabolism as well as being one source for lysophosphatides and synthesis of the phospholipid mediator, platelet activating factor (PAF). These latter are of importance as candidate mediators for a wide range of biologic processes, from fertilization and parturition to inflammation and brain function. PAF has additional historical interest in being one of the first phospholipid mediators to be described and shown to have a distinct receptor.

Eicosanoids are produced following oxidation of the released arachidonic acid by cyclo-oxygenases (COX enzymes or PGH synthases), lipoxygenases, and P450 family enzymes. PAFs are derived from the other product of *sn2* fatty acid hydrolysis, the LPC, by the action of a specific acetyltransferase which inserts an ester-linked acetyl group in place of the original arachidonate. This pathway is sometimes known as the remodeling pathway to distinguish if from de novo synthesis of PAF, the latter being a constitutive process and of less importance in bulk production of the mediator. The initial PC substrate for the $cPLA_2$ can have either an acyl- or alkyl-linked *sn1* fatty acid yielding either acyl or alkyl PAF. The susceptibility of the *sn1* acyl group to lysophospholipase (Nakagawa et al. 1992) or phospholipase A1 activity (including from the cPLA2 itself, which has an additional activity in this regard) makes the alkyl forms more stable and may explain why they were the first to be identified. Increasingly in recent years, it has been recognized that there is an alternative pathway for stimulated PAF production involving the activity of a CoA-independent transacylase (Blank et al. 1995; Winkler et al. 1995). This process may be of particular relevance in the more persistent production of PAF seen in macrophages (Shamsuddin et al. 1997; Svetlov et al. 1997) but, nevertheless does require an initial source of lysophosphatide as recipient of the transacylase process.

Importantly, included in the PAF family are a variety of oxidation products of PCs, since one effect of oxidation of *sn2* unsaturated fatty acids is chain-shortening, i.e., creating molecules with short chain acyl groups that therefore exhibit similar structure and function to PAF (Marathe et al. 1999; Marathe et al. 2000; Marathe et al. 2001) reacting with the same receptor and being susceptible to the same PAF acetylhydrolase. Indeed it has been questioned whether one of the raisons d'être for the plasma acetylhydrolase is to deal with inappropriate concentrations of such oxidized phospholipids, which, for example are found in preparations of oxidized low-density lipoprotein (LDL) (Marathe et al. 2001). The ability of macrophages to mount an oxidative burst is suspected of enhancing this mode of PC modification, i.e., leading to an alternative, non-enzymatic generation of PAF-like activities.

As indicated, macrophages synthesize and "secrete" PAFs (Ninio et al. 1982; Dentan et al. 1996). However, they also synthesize and secrete significant amounts of PAF acetylhydrolase (Ninio et al. 1982; Elstad et al. 1989; Stafforini et al. 1990) an enzyme that specifically hydrolyses short chain *sn2* fatty acids (including acetate), thereby returning the PAF to lyso PAF. Because of the importance of the alkyl forms of PAF, the term lysoPAF has often been used even though the molecule is directly equivalent to lysoPC containing an *sn1* alkyl linkage. This can then be remodeled, either by re-acetylation back to PAF or by acylation with other, longer chain fatty acids. The presence in macrophage and macrophage supernatants of this enzyme (Ninio et al. 1982; Stafforini et al. 1990) may have led to underestimates in early investigations of PAF production by this cell type. Macrophages are a significant source of plasma acetylhydrolase (Howard and Olson 2000), an enzyme that circulates in association with lipoproteins and has been shown to have significant anti-inflammatory effects in vivo (Stafforini et al. 1987; Prescott 1997; Stafforini et al. 1999). The enzyme is also expressed by macrophages in atherosclerotic plaques (Hakkinen et al. 1999).

1.3.1 Actions of PAFs

PAFs have an extremely broad range of activities and are thought to act primarily through a specific, seven transmembrane, G protein-linked receptor (GPCR), the PAFR (see Henson 2000a,b). To date, only one such extracellular membrane receptor has been identified, although many pharmacological studies have suggested that all the actions of PAF are difficult to reconcile with a single receptor. In addition, as we shall see, possible intracellular effects of this group of molecules needs to be considered, and here too, the likelihood of additional binding molecules or receptors within the cell has been discussed (see Henson 2000a,b; Yamada et al. 1999). Effects on macrophages are complex. Mononuclear phagocytes express the GPCR PAF receptor. Most investigations have suggested the induction of proinflammatory responses including inflammatory mediator synthesis and secretion, as well as effects on motility and weak chemotactic activity for monocytes. One of the intriguing effects of PAFs on inflammatory cells is their ability to induce priming. Here, PAF increases the responsiveness of the cell to other stimuli, even though it does not (or only weakly) initiate the response itself. This priming effect is also seen in macrophages (Kucey et al. 1991; Bautista and Spitzer 1992; Waga et al. 1993; Rose et al. 1995; Bozza et al. 1996; Yamaguchi et al. 1999).

On the other hand, in the context of interaction with apoptotic cells (see below), macrophages appear to respond to PAF with a potential anti-inflammatory effect (Fadok et al. 1998a). Since this was blocked by a PAF receptor antagonist, a potentially complex signaling from this receptor might be suggested. However, one or both of two alternatives may explain these diverse effects. As mentioned, other cell surface receptors for this molecule may exist. As discussed below, PAF may also be translocated across the membrane and there gain access to intracellular "receptors" or binding proteins. One such candidate is the peroxisome proliferator-activated receptors (PPAR) family of nuclear receptors, particularly PPARγ This has been shown to bind to, and be activated by, oxidized alkyl PC moieties from oxidized LDL (Davies et al. 2001) and could conceivably respond to other members of the "PAF" family. There is separate evidence that PPARγ can mediate anti-inflammatory responses in macrophages (Delerive et al. 2001; Alleva et al. 2002) suggesting a potentially fruitful area for future investigation.

1.4 Lysophosphatides

LPC and LPA contain a glycerol backbone whereas SIP and related family members are lysosphingophospholipids. They all act through closely related, seven transmembrane G protein-coupled receptors including a family of 8 or more EDG receptors for LPA and SIP (Chun et al. 2002) and a newly described G2A receptor for LPC (Kabarowski et al. 2001) or GPR4 receptor for sphingosylphosphorylcholine (Zhu et al. 2001). These molecules are found in significant amounts in the circulation, in association with albumin, other binding proteins, or lipoproteins, and in levels that can vary in different disease states.

Stimulated production of LPA (Moolenaar 1995, 2000) can arise from at least three pathways. In one, phospholipids are hydrolyzed to phosphatidic acid (PA) by phospholipases D (Exton 2000, 2002) with subsequent hydrolysis of the *sn2* acyl group by phospholipases A_2. When the source is membrane vesicles, presumably sPA_2s are responsible (Goetzl and Lynch 2000). In a second pathway, as in activated platelets, diacylglycerol kinase phosphorylates DAG (derived from the action of phospholipase C) to PA with subsequent PLA_2 action to generate LPA. This is released into the medium (see below). The third pathway involves oxidative chain-shortening effects (e.g., on LDL) similar to that seen for non-enzymatic production of PAF-like molecules (Goetzl and Lynch 2000). Similar pathways are operative for the generation of sphingosine-1-phosphate, SIP, starting with sphingomyelinase action on membrane sphingomyelin (Hannun and Bell 1993), but require the final action of sphingosine kinases, very widely distributed, stimulatable, intracellular enzymes located in both the cytosol and membrane fractions of cells (Liu et al. 2000). To no great surprise, the intracellular levels of SIP are highly regulated, in part by the actions of phosphohydrolases (Le et al. 2002).

LPC can be produced by the action of any of the many phospholipases A_2 that can act on PC. In the cell, this can include the calcium dependent $cPLA_2$ that is so important for eicosanoid generation as well as calcium-independent enzymes. However, as already discussed, the presence of reacylating enzymes would be expected in general to result in rapid removal of the LPC. In fact, the relatively high levels of acetyl-CoA in the cell and the presence of an acetyltransferase (see above) has even led to the suggestion that intracellular PAF production might be, in part, a mechanism for rapid removal of potentially toxic LPC. (LPC acts as a potent detergent at higher concentrations.) Extracellular $sPLA_2$s from multiple sources (including from infectious organisms) acting on PC in membrane vesicles, damaged cells, lipoproteins, etc. can all initiate LPC production. Macrophage-derived $sPLA_2$ would be a likely source in many circumstances including, for example, in LPS-induced LPC generation in the lung (Arbibe et al. 1998).

As expected from the discussions above, LPC has also been suggested to arise from oxidation of PC-containing *sn2* unsaturated acyl groups, for example of LDL. A cautionary note here is the distinction between chain shortening, leading to generation of short-chain *sn2* acyl groups, which would act like PAF on the PAF receptor, versus the generation of true LPC. In even more general terms, a clear comparison between the PAF and lysophosphatide family of lipids in terms of production and receptor-driven effects has still to be completed. To this point, the two literatures have tended to remain separate.

Initially considered as having possible ionophore-like activity, LPA soon became recognized as a ligand for receptor-mediated calcium mobilization (Moolenaar 2000) and a host of other cellular responses including cell replication, cytoskeletal regulation, and movement or inflammatory mediator generation. This broad spectrum of activities extends to SIP and LPC (see Moolenaar 2000; Tigyi et al. 2000; Hla et al. 2001; Graler and Goetzl 2002; Spiegel and Milstien 2002). These molecules also exhibit a wide spectrum of intracellular roles (see for example: Hla et al. 1999). Actions on the immune system and inflammation have not received as much attention as on tissue cells but are likely to be important (see Graler and Goetzl 2002). LPC has been suggested to have varying effects on lymphocytes, and genetic deletion of the LPC receptor G2A resulted in progressive inflammation and a lupus-like syndrome (Le et al. 2001).

Roles for these mediators in macrophage function are significantly understudied. However, macrophages do express a number of the EDG receptors (Goetzl et al. 2000a; Hornuss et al. 2001; Lee et al. 2002). LPC has been reported to have chemotactic activity on monocytes (Quinn et al. 1988) and the broad abilities of the lysophospholipid mediators to affect cell movement (Spiegel et al. 2002b) (in part through actions on the *rho* family of GTPases) suggest at least one likely group of effects.

LPA and SIP likely act as proinflammatory stimuli and appear able to enhance inflammatory mediator production (Lee et al. 2002). On the other hand, LPA activation of adenylate cyclase has been reported in RAW cells (Lin et al. 1999) raising the possibility of more complex effects. LPA induces calcium mobilization in microglial cells (Moller et al. 2001). It has also been suggested to function as a macrophage survival stimulus acting through phosphoinositol (PI)3 kinase and presumably Akt (Koh et al. 1998). These authors raise the possibility that LPA may explain some of the known survival effects of serum (plasma). LPC can also activate PKCs in macrophages (Prokazova et al. 1998), enhances FcR-mediated phagocytosis (Morito et al. 2000), and when injected into the spinal cord, initiates macrophage accumulation (Ousman and David 2000), although whether by direct or indirect effects is not clear. In general, the full spectrum of activities, likely autocrine and paracrine effects, regulatory processes, and in vivo biologic relevance for the macrophage remain to be determined.

1.5
Phosphatidylserine

PS is normally found in intracellular membranes and on the inner leaflet of the plasma membrane. As discussed below, when it becomes expressed on the outer leaflet, especially in apoptotic cells, it serves as a recognition signal for removal of such cells. A specific receptor (PSR) has been described that recognizes the polar head group of PS in a stereospecific fashion and which mediates this activity (Fadok et al. 1998b, 2000, 2001a,b). On the other hand, PS is recognized by a host of proteins, both within and without the cell. For example, many of the intracellular signaling proteins, such as most of the PKCs use PS as a co-factor. Many of the annexin family of molecules also bind PS, and annexin V has become a standard marker for detecting the phospholipid on the cell membrane. In the extracellular environment, coagulation factors V and X bind and use PS to accelerate clotting. Molecules such as GAS-6, β glycoprotein 1, MFG-E8, members of the collectin family (see below) and probably many others are known to interact with PS. Scavenger receptors can also bind anionic phospholipids, including PS. In many cases, unlike the PS receptor mentioned above, these binding proteins do not show high specificity for PS and will also bind PE, PI, or PA. On the other hand, this does not diminish their potential role in recognizing and responding to PS in the environment or on cell surfaces.

A significant source of PS in the extracellular environment are membrane vesicles. Activation of platelets results in external PS expression and concomitant assembly and activation of the coagulation cascade. It is now recognized that liberated vesicles are the major source of this activity. Most cell types, including mononuclear phagocytes, actively release vesicles and do so to a greater degree when activated. As discussed below, we suspect a high degree of PS expression on such structures. Membrane fragmentation during cell death (necrosis or post-apoptotic cytolysis) also results in vesicle formation and, without the normal mechanisms for regulating phospholipid asymmetry (see below) they are likely to exhibit PS on their surfaces.

Macrophages can expose PS on their surface during stimulation, phagocytosis, and apoptosis. They express the PS receptor (Fadok et al. 1992, 1993, 1998b, 2000) and respond to PS binding of this by ingestion (for example of the apoptotic cell or membrane vesicle) through, what we suggest is a process of stimulated macropinocytosis (Hoffmann et al. 2001). Ligation of this receptor also induces the generation of anti-inflammatory molecules and signals and blocks the production of proinflammatory chemokines, cytokines, growth factors, and eicosanoids (Fadok et al. 1998a).

In summary, then, macrophages make and respond to PAFs, LPCs, and LPAs. They can express PS on their surfaces, probably release PS-expressing vesicles as well as recognize and respond to PS on other cells and vesicles. They also can both make and respond to a wide variety of eicosanoid types, including prostanoids, HPETES and HETES, leukotrienes (LTB4 and sulfidopeptide leukotrienes), lipoxins, and isoprostanes. This clearly shows the breadth of expression of both the appropriate synthesizing enzymes as well as of surface receptors for the products; although, not surprisingly, different macrophage populations may express different patterns of these and may upregulate or downregulate them to different degrees to different stimuli.

As noted, there are a number of additional common features exhibited by these families of bioactive phospholipids. They each have significant intracellular signaling and co-factor roles as well as exhibiting specific extracellular receptors. They can be found in lipoproteins and may therefore play important additional roles in atherogenesis. Intriguingly, antibodies against these phospholipids (LPCs, PS, etc.) are not uncommon, particularly in autoimmune disease.

2 Sites of Synthesis

Phospholipid synthesis is segregated to different sites within the cell depending on the lipid species being generated (Voelker 2000). There is also a large body of work addressing membrane biosynthesis, which does not need to be considered particularly here. Most of the plasma membrane PS in the resting cell ends up on the inner leaflet. More to the point for our discussions is the site at which the biologically active forms of the externally acting phospholipids are generated.

It is now recognized that many of the key synthetic enzymes for these mediators translocate to, or are located at, the nuclear membrane, endoplasmic reticulum, or Golgi. This includes the initiating $cPLA_2$, as well as the PGH synthases involved in prostanoid generation and the 5-lipoxygenase (5-LO) required for synthesis of leukotrienes. The acetyltransferase involved in PAF synthesis is also thought to be found at these intracellular sites (Record and Snyder 1990; Samples et al. 1999). The co-localization of these various enzymes would certainly make for enhanced efficiency of product generation, particularly if the lipid substrates and intermediates are inserted into, or bound to, hydrophobic membrane domains at this one site. On the other hand, location on what is presumed to be the cytoplasmic face of nuclear or endoplasmic reticulum membrane does raise questions about transport of products and intermediates from this site to the plasma membrane, not to mention passage across this to the outside environment (see below).

Secretary phospholipases A_2 can, under specialized conditions, act on phospholipids of the external membrane to generate lysophosphatides and macrophages synthesize and secrete $sPLA_2$s including those of groups II, V and X. (Berger et al. 1999; Morioka et al. 2000; Jaross et al. 2002) Some of these show preference for PS or PE, but others may have broader specificity. In a number of cell types, including macrophages, there is also evidence for complex functional coupling between $sPLA_2$ and $cPLA_2$ (Balsinde et al. 1998) although this may not apply to all types of macrophage or stimuli (Dieter et al. 2002). Whether these externally acting enzymes can generate bioactive LPA or LPC [or lyso-PAF for later internalization and conversion to PAF at the endoplasm reticulum (ER)] is not clear but must certainly remain a possibility. Intracellular PLA_2s must be seen as likely candidates for the generation of the bioactive lysophosphatides and these may include those with varying *sn2* fatty acid specificities, implying varying degrees of accompanying arachidonate release and eicosanoid synthesis. One poorly studied or considered question is a possible source of bioactive phospholipids from intracellular organelles (granules, vesicles, etc.). Once again, the issue of local generation of active phospholipids at intracellular membranes must raise questions of access to the plasma membrane and to the outside environment.

3 How Do Intracellularly Generated Bioactive Phospholipids Gain Access to the Plasma Membrane and Extracellular Environment?

The bulk of cell phospholipid is found in membrane bilayers, in intracellular membranes and organelles or in the plasma membrane. However, in order to supply biologically active phospholipids to the external environment, most of these molecules must transit one or more membrane bilayers. Initially synthesized in the ER, often modified enzymatically subsequently, the lipid elements of membrane assembly are a subject of detailed ongoing investigation. Transit across the cytoplasm to the cell membrane is clearly required not only for maintenance of the membrane itself but also for "secretion" of the bioactive phospholipid mediators. Access of these to the external environment may come from three main processes. In one, conventional secretion involving vesicle (or granule) transport to the plasma membrane with discharge of contents would cause release of any contained phospholipids. It would also lead to surface expression of phospholipids inserted into the intraluminal membrane leaflet of the vesicle or granule.

In a second process, phospholipids may be transported from synthesis or modification sites within the cell to the inner surface of the plasma membrane, probably in physical association with transport proteins. Insertion into this inner leaflet would leave the phospholipid available for regulated transmembrane movement (translocation) to the outer leaflet, a process sometimes called "flop" to contrast with inward movement across the membrane ("flip") or bidirectional movement (flip-flop or "scrambling"). Fusion of transport vesicles with the inner surface of plasma membranes would also supply the inner leaflet of the latter with new phospholipids that could later be translocated to the outer surface.

A third potential mechanism for supply of extracellular phospholipids is more complex and could include elements of each of the former. Many cell types in vitro release membrane vesicles from their surface, both spontaneously and in response to stimuli. There is also increasing evidence for the presence of such vesicles in the circulation, and such vesicles may contain oxidized phospholipids with biologic effects on monocytes (Huber et al. 2002). Stimulated monocytes, for example, release large numbers of such vesicles and these have been shown to contain protein mediators of inflammation (MacKenzie et al. 2001). Such vesicles may also supply externally expressed bioactive phospholipids to receptors on other cells. The mechanisms for vesicle release are only now beginning to be studied. Teleologically, over-supply of plasma membrane from secretory events would need to be regulated to maintain cell size and surface area. This could be achieved both by re-internalization of the extra membrane or its pinching off and release of vesicles to the environment. It seems likely that the vesicles could also act as stimuli to adjacent or even distant cells.

As far as PAF and lysophosphatide mediators is concerned, the exact mechanisms of such transport are not known. Intracellular PAF binding proteins have been described in macrophages and could serve as transporters (Banks et al. 1988; Lumb et al. 1990). The phosphoinositol transfer proteins (PITPs) also carry single molecules of PC (as well as sphingomyelin and phosphatidic acid) and exchange them for membrane-resident phospholipid molecules (Li et al. 2002). They might serve similar roles for the bioactive phospholipids under discussion. PAF or lyso-PAF appear able to move efficiently across the cytoplasm in both directions, since radiolabeled compounds can be seen in the nuclear membrane by autoradiography very soon after addition to the outside of the cell (F.H. Chilton, P.M. Henson and R.C. Murphy, unpublished observations). This could certainly fit a transport protein model. On the other hand, the molecules are relatively hydrophilic, so simple diffusion cannot be ruled out. The possible presence of PAF, or lysophosphatides, in intracellular transport vesicles has not received much attention, but there seems no intrinsic reasons why this could not also serve as a route for export.

4
How Do Bioactive Phospholipids Cross the Plasma Membrane?

This question of how bioactive phospholipids cross the plasma membrane is important for both secretion and surface expression of active phospholipids as well as for uptake and then re-expression in transcellular biosynthetic processes. Studies with PAF have shown that this molecule can be actively transported across the plasma membrane in both directions. At issue for such a process is insertion into one leaflet of the bilayer followed by a mechanism for flip or flop (inward or outward) to the other leaflet. Compounding factors are the possibility that the phospholipid might be metabolized at the membrane and/or may also be interacting with proteinaceous membrane receptors. PAF proved useful as a tool for understanding membrane phospholipid movement, in part because the *sn1* ether linkage is resistant to phospholipase cleavage. Short-chain *sn2* analogs that were biologically active but also resistant to hydrolysis were additionally helpful (Bratton et al. 1991, 1992; Bratton 1994).

Phosphatidylcholines with long-chain fatty acids are only poorly moved across the membrane bilayer, but those with shorter *sn2*-linked groups moved more rapidly (Zhou et al. 1997). However, in both cases, translocation is enhanced after activation of the cells. In the case of PAF, this activation can arise from ligation of the PAF receptor itself, so that stimulation through this receptor can enhance non-selective uptake of PAF or other phospholipids that have become inserted into one or other membrane leaflet. While uptake of PAF may occur through internalization of the ligated receptor, this appears to represent only a small proportion of total uptake; most presumably occurs through the action of flippases. Thus, blockade of the specific PAF receptor but ligation of unrelated G protein-linked receptors can induce just as much uptake.

4.1 Phospholipid Scrambling

The physical processes involved in phospholipid translocation across the plasma membrane are unknown. However, a group of phospholipid "scramblases" appear to be required (Williamson et al. 1995, 2001; Comfurius et al. 1996; Zhou et al. 1997; Frasch et al. 2000; Sims and Wiedmer 2001). The name derives from their ability to move phospholipids bidirectionally across the membrane, thereby reducing the inherent phospholipid asymmetry between the inner and outer leaflets, i.e., scrambling the membrane. The activity seems to be significantly independent of phospholipid head group and, as suggested above, probably acts more efficiently on molecules with shorter *sn2* substituents. It has been suggested that scramblase activation requires PKC-dependent phosphorylation and the presence of calcium for optimal activity (Frasch et al. 2000). There may also be a tyrosine phosphorylation event involved (Sun et al. 2001). Scramblases are a family of type 2 transmembrane proteins whose exact contribution to the phospholipid movement remains to be determined. It is hard to see how they can directly mediate this process. Assembly into multimers in the membrane might lead to local disorder in the lipid bilayer to enhance movement. More likely, other membrane proteins are involved. Possible candidates for such are members of the ATP cassette family of proteins already known to participate in phospholipid and cholesterol import or export.

Regulation of membrane phospholipid distribution is suggested to be important and requires some concerted investigation. The lipid environment is increasingly seen to play a critical role in the function of membrane proteins, receptors, etc., and the activity of phospholipid translocation processes is bound to contribute to this and the overall cell functions. A recently emphasized case in point is the recognition of proteases that appear to function within the hydrophobic domain of the lipid bilayer (Wolfe and Selkoe 2002). Activation of many, or even most, cell types (i.e., activation of PKCs and mobilization of calcium) leads to transient membrane phospholipid scrambling. The relationship of this to defined membrane domains (e.g., cholesterol-rich regions, rafts, caveolae, etc.) remains to be clarified, although early evidence suggests ties between rafts and scrambling (Kunzelmann et al. 2002).

Since resting cells, including macrophages, maintain membrane phospholipid asymmetry, casual activation of scramblase and scrambling must generally be transient, implying both a cessation of the scramblase activation and a rectification process to return any externalized PS or PE back to the inner leaflet. Although different cell stimuli induce scrambling at different rates and for different lengths of time [G protein receptors, fast and shorter, or growth factors such as macrophage colony-stimulating factor (M-CSF), slow and longer], the process is usually time limited. Although not formally shown at this point, cessation of scrambling may be attributed to the action of phosphatases on the scramblase, inactivation of the PKCs, and/or rectification of the increased calcium levels. However, cessation of the actual scrambling is not enough to maintain homeostasis, the translocated phospholipids are actively returned to their original state. This is achieved by an aminophospholipid translocase. Its activity requires ATP and is blocked by high calcium. Various candidates for this activity have been implicated. A possible aminophospholipid translocase (a member of the P-type ATPase family) is widely distributed from plants to man (Chen et al. 1999; Daleke and Lyles 2000; Ding et al. 2000; Gomes et al. 2000) and although the validity of this candidate has been challenged, more recent data in *Arabidopsis* does add support to an aminophospholipid translocase role for these P-type ATPases (Gomes et al. 2000).

While alterations in phospholipid scrambling have been shown to be important in maintaining or altering phospholipid asymmetry, in surface expression of PS, and in "secretion" of PAF, there has been little attention to a role for release of the lysophosphatide mediators. The comment is made in Moolenaar (2000) that "Precisely how LPA and SIP are released into the extracellular environment remains to be elucidated." We are suggesting that translocation across the membrane mediated by activated phospholipid scramblase as outlined for PAF is an important first step. This would be followed by partitioning onto carrier proteins such as albumin or lipoproteins (see below); this may represent a key release mechanism for these phospholipids as it appears to be also for PAFs.

4.2 Membrane Scrambling and Phosphatidylserine Expression in Apoptosis

By contrast, permanent membrane scrambling is seen in cells that are undergoing the process of apoptosis. In this case PKCδ is cleaved and rendered permanently active by the action of caspase 3, increased calcium levels are maintained, and the aminophospholipid translocase activity is inhibited probably as a consequence of the decreased ATP and increased calcium. This process leads to permanent expression of phosphatidylserine on the apoptotic cell surface now known to be important for apoptotic cell recognition and removal as well as for the biologic consequences of this (see below). As mentioned above, while much emphasis has been placed on PS in this circumstance, scrambling of the membrane phospholipids in apoptosis is likely to have other profound effects on the cell surface. This would range from alterations in other phospholipids (internalization of sphingomyelin, externalization of PE), alterations in lipid and protein orientation and distribution in the membrane, effects on rafts, etc. It also probably contributes to other effects seen in apoptosis, such as membrane blebbing, cell shrinking, binding of recognition molecules such as collectins, thrombospondin, β-glycoprotein 1, GAS-6 (Scott et al. 2001), MFG-E8, (Hanayama et al. 2002). Expression of PS on apoptotic membrane blebs enhances local pro-coagulant activity (Casciola et al. 1996). The increased membrane scrambling that occurs during apoptosis would also be expected to move any internal phospholipid mediators to the outer membrane, which could include PAF and the lysophosphatides. Apoptosis is often accompanied by PLA_2 activation, which could increase the supply of such mediators.

Macrophages are not at all unique in their apoptotic mechanisms or manifestations. Although not studied as much as many other cell types, they too express PS on the membrane and are certainly removed as efficiently as other cells when they become apoptotic in vivo.

4.3
Why May Membrane Phospholipid Translocation Be Important?

The maintenance of phospholipid asymmetry is suggested to play a major role in membrane functions, with regard to the structural and biochemical effects of the lipids themselves as well as for attached or embedded proteins. The role of surface PS expression during macrophage activation and phagocytosis is not yet delineated but is expected to contribute to the processes in some fashion. Intriguingly, such PS exposure has been reported on the macrophage during uptake of apoptotic cells, i.e., during a process that involves PSR recognition of PS also on the target (Marguet et al. 1999; Callahan et al. 2000).

As argued herein, we also suggest an important role for phospholipid scrambling in supplying new phospholipids to the extracellular environment, for access to other cells, or "release" into the surroundings. However, the bidirectionality of the process also suggests a role in uptake of these same phospholipids, especially the lysophosphatides and PAFs with the opportunities for intracellular metabolism, inactivation, or conversion, or even intracellular biologic effects.

Whether there are any pharmaceutical possibilities in altering scrambling and/or aminophospholipid translocase activity during apoptosis is as yet unclear. However, as we begin to understand the processes better, as well as their consequences, such potential may become apparent.

5
How Are Phospholipids Released from the Membrane into the Aqueus Extracellular Milieu?

Bioactive phospholipids that have reached the external membrane leaflet of the originating cell may act on other cells in the environment by release into the medium, blood, or interstitial fluid or because they are directly recognized by receptors on the responding cell. Here we will consider the release aspects.

5.1 Carrier Proteins

Even relatively hydrophilic phospholipids such as LPC may not easily partition from the membrane into an aqueus environment without some help, and the longer the *sn2* fatty acid, the more difficult this process will be. Extracellular binding proteins, particularly albumin, are more probably involved in all these process. Serum albumins have a high capacity for lipid binding on multiple sites and with varying affinities. This makes them ideal as carriers for biologically active lipids. A specific example is the confusion seen in the early studies of a platelet-activating material released from IgE-stimulated blood or leukocytes when the experiments were carried out (as they originally were) in protein-free medium. The inclusion of albumin in the medium immediately enhanced the activity, its stability, and biologic effects, not to mention setting the stage for its isolation and characterization.

However, while simple to state, this issue adds enormous complexity to an understanding of the effects and activities of phospholipid mediators, especially in vivo. Seldom are in vitro experiments carried out in 100% plasma. Addition of 2% or even 10% plasma, serum, or albumin hardly mimics real-life conditions. Albumin in vivo will already have bound lipids so that the surface phospholipids of interest will have to displace these before being "extracted" from the membrane. Partition onto albumin will depend then on many factors: the protein concentration, relative affinity of the phospholipid for given binding sites, whatever phospholipids are already on these sites as well as the hydrophilicity of the phospholipid in question (after all, the albumin itself is a globular, water-soluble molecule). An ability of albumin to physically bind to the cell membrane would contribute to this process. While many studies have shown how difficult it is to completely free isolated cells of albumin, this may be in part due to internalization in microvesicles as well as surface attachment. Might such attachment itself be due to phospholipid binding (see below)?

In vitro stimulation of macrophages, harvesting the supernatant and subsequent examination of active phospholipid effects on responding cells will be critically dependent on all these issues.

In order to further illustrate this point, standard assays for phospholipid translocation across the plasma membrane involve incubation of the cells with lipid (usually supplied on albumin, in micelles or liposomes) to allow insertion into the outer leaflet and translocation. Then, in order to detect intracellular phospholipid, any material remaining on/in the outer leaflet is back-extracted with lipid-free albumin. Not usually considered in such studies is how much of the preexisting extracellular leaflet phospholipid is also extracted by such a procedure. This is probably not of critical importance when investigating translocation in this way. However, the issue does raise intriguing questions about ongoing phospholipid exchange from membrane to albumin and back in tissue culture or even in vivo. When macrophages or other cells express PS on their surface during activation (transient) or apoptosis (permanent), is any of this portioned onto albumin to become potentially available for PS receptors on other cells?

This discussion has focused on albumin in part because of its high capacity and low specificity for lipid binding as well as its high concentration. LPA and LPC in plasma are found bound to albumin. However, there may well be other, more specific carriers in plasma or tissues that play more selective roles. One might wonder, for example, whether the acute phase reactant, C-reactive protein (CRP), which is known to bind phosphorylcholine, could "extract" and carry LPC from a producing cell to one with LPC receptors. Gelsolin has also been suggested as a carrier for lysophosphatides (Goetzl et al. 2000b), particularly perhaps after cellular damage or inflammation. Clearly, the other major source of carriers for bioactive phospholipids are the lipoproteins. The issues of partitioning from membrane to carrier apply here as well and once again from an experimental perspective, few investigators carry out mediator experiments in the presence of physiological levels of lipoproteins. Plasma phospholipid transfer protein (PLTP) exchanges phospholipids between lipoproteins (van Haperen et al. 2000; van Tol 2002), but whether it is important in movement and distribution of the bioactive phospholipids that we are discussing is not yet clear. The presence of PAF acetylhydrolase in lipoproteins raises the intriguing likelihood that it serves here to limit the presence of PAF or bioactive, oxidized, PCs in this circulating pool. An implication might be to confine the effects of highly active PAFs to local sites.

5.2 Vesicles

As discussed above, the other way in which bioactive phospholipids could be "released" from cells is as components of vesicles. The presence of bioactive phospholipids in the membranes of such vesicles would also be expected to allow stimulation of their cognate receptors (see below). Whether vesicles released from monocytes or macrophages do indeed contain such phospholipids remains to be determined. Vesicles and apoptotic bodies liberated during apoptosis also express PS. It seems highly likely that vesicles liberated from activated macrophages do the same and that these could possibly, in the right circumstances, provide a potential stimulus to cells via the PS receptor. Vesicles containing PAF and/or lysophosphatides might be able to directly stimulate the PAF or EDG receptors on target cells.

5.3 Retention or "Storage" of Bioactive Phospholipids

An early observation in the study of PAF synthesis was that in most cell types, relatively little PAF was actually released into the medium. This cellular "retention" has led to questions about intracellular activities of the molecules. We suspect that it reflects in part the combined need for the newly synthesized molecules to be transported across the cytoplasm, translocated across the plasma membrane, and partitioned onto carriers in the environment. In like fashion, the lysophosphatide mediators are also sometimes seen retained or "stored" within cells (e.g., platelets) (Yatomi et al. 1997; Goetzl and Lynch 2000). Whether these molecules are ever stored in vesicles or granules for later export is not yet clear.

6 How Do Phospholipids Stimulate Surface Receptors?

6.1 Presentation

A number of special issues arise in consideration of the way in which a phospholipid might interact with its surface protein receptor.

6.1.1 Albumin and Carrier Proteins

Solubility in the aqueous environment is just as important here as in release from the cell of origin. It is probably no accident that the phospholipid mediators under discussion (lysophosphatides, PAF, etc.) have significant hydrophilic properties in comparison with the bulk of their membrane counterparts. However, it seems likely that even here, presentation on carrier proteins, especially albumin, is the norm. How much is ever free in the aqueous milieu is questionable. This point has certainly been emphasized (and challenged) for interaction of PAF with its receptor (Clay et al. 1990; Grigoriadis et al. 1992). We suggest that in biologic fluids, PAF is bound to one of four potential binding sites on albumin and exchanges from these to the receptor which has a higher affinity for the molecule. Once again, at least one implication from this would be potential competition for PAF on the albumin by other lipids, since the binding sites are not specific. In the circulation, the concentration of albumin (lipid-binding capacity) is high, but in the tissues and in the immediate environment of the macrophage membrane this may not always be the case. In vitro stimulation experiments are seldom carried out in whole plasma. The degree to which LPA or LPC are presented to their receptors on albumin or other potential carriers is unknown, but it seems highly likely from their related structures that the same issues apply.

6.1.2 Membrane Insertion

If not on albumin, the PAF is susceptible to metabolism by plasma PAF acetylhydrolase (probably following partition into lipoprotein particles) and incubation with plasma leads to rapid inactivation. In the presence of cells, the bioactive phospholipids also rapidly and effectively insert into the outer leaflet of plasma membranes. As we have already noted, if the cell is actively scrambling its membrane, the material will then be flipped to the inside with consequences that can include metabolism, remodeling, and even possible direct effects on intracellular signaling pathways. Transcellular metabolism of PAF, lyso-PAF or analogs appears to result from uptake subsequent to membrane insertion rather than following receptor binding. However, this question has not been easy to resolve. PAF receptor antagonists can certainly prevent internalization through the receptor and do not block initial insertion into the outer leaflet of the cell membrane. On the other hand, stimulation of the cell through ligation of the receptor enhances internalization (see above) and often activates or alters the synthetic or metabolic processes within the cell so that antagonist blockade of transcellular metabolism can occur at this step. Examination of transcellular metabolism (see Sect. 7) in the presence of specific PAF receptor antagonists but also with alternative stimuli (i.e., active on unrelated receptors) has helped answer these questions.

Insertion into the membrane too provides a competitive site for the phospholipid mediator interaction, although the affinity will be much lower than that for the receptor. The additional question is whether membrane inserted PAF or lysophosphatides can act on their surface receptors from this site or must be "released" or rebound to albumin to have this effect. Can the membrane act as a store for immediately adjacent (and relatively protected) ligand that contributes to the overall effects of such phospholipid ligands on the receptor?

It is clear that these considerations might play havoc with conventional analysis of receptor–ligand interactions. It is unlikely that these types of molecules are ever in true monomeric form. If not in membranes or on proteins, they tend to interact with themselves or other lipids to form micelles and indeed, in all too many in vitro experiments, appropriate attention to the critical micellar concentration (i.e., the physical form of the ligand) appears to be lacking. Another confounding feature is that insertion of these hydrophilic (amphipathic) phospholipids itself induces physical changes in the membrane, in high enough concentrations, even causing cell lysis. This demands experimental studies with low concentrations of ligand. However, when a cell is itself releasing lysophosphatides to a nearby recipient in vivo, or even in vitro, the local concentrations are not usually known and, especially if released in vesicles, could reach levels with possible physical effects on portions of the responding cell membrane.

6.2 Membrane Presentation

At issue here is whether phospholipid "mediators" in the outer leaflet of cells can, from this site, interact with cognate receptors on another cell. A special case would be presentation from membrane vesicles to receptors on the responding cell.

6.2.1
Phosphatidylserine

Specific recognition of, and response to, PS on the surface of apoptotic cells represents a clear case of this type of membrane presentation. A receptor for PS has been identified (Fadok et al. 2000) that recognizes the polar head group of the phospholipid in a stereospecific fashion. Numerous other extracellular proteins and surface "receptors" are also known to bind PS (see above). In particular in macrophages, these include a variety of scavenger receptors. The extracellular, soluble, PS-binding molecules may serve to cross-link apoptotic cells or membrane vesicles to macrophage surfaces and, thereby, induce or enhance uptake and removal.

We have suggested that the PS receptor plays particularly important roles in such removal. Its blockade or removal prevents much of the apoptotic cell uptake into macrophages in vitro and preliminary evidence supports such a role in vivo as well (Hoffmann et al. 2001; Huynh et al. 2002). More indication of the in vivo importance of this receptor will come from studies of its knockout, which is not yet available. On the other hand, blockade of the PS by attachment of annexin V both in vitro and in vivo has been shown to prevent uptake of apoptotic cells and some of the consequences of this (Bennett et al. 1995; Blankenberg et al. 1998; Stach et al. 2000). Interaction of apoptotic cells with, and uptake into, macrophages does not induce the proinflammatory consequences usually associated with phagocytosis. Rather it results in an active suppression of proinflammatory mediator production (Fadok et al. 1998a). This appears in part due to selective induction of anti-inflammatory mediators such transforming growth factor (TGF)-β. Under some circumstances there is also increased interleukin (IL)-10 production (Voll et al. 1997; Fadok et al. 2001c). These effects appear to be due to the PS receptor since in its absence, or blockade, the anti-inflammatory response is no longer seen. Direct ligation of the PSR with antibody or PS liposomes induces these molecules, is anti-inflammatory and, in vivo, can hasten resolution of an inflammatory response.

Intriguingly, engagement of the PS receptor suppresses proinflammatory eicosanoids (e.g., thromboxanes) and enhances production of PGE_2 and PGI_2 (Fadok et al. 1998a; and W. Vandivier, unpublished observations). In addition, some of the in vitro suppression of proinflammatory mediator production was blocked by COX inhibitors (Fadok et al. 1998a). Selective effects of the PSR on eicosanoid biosynthetic enzymes are implicated. Increasingly, evidence is appearing of subversion of these apoptotic cell recognition systems and their anti-inflammatory consequences by parasites. Thus *Leishmania* may use the PS receptor to interact with macrophages and alter inflammatory reactions (de Freitas Balanco et al. 2001) and *Trypanosoma cruzi*, the PGE_2 production to evade the inflammatory response (Freire-de-Lima et al. 2000). In addition, and so far unexplained, pharmacologic blockade of the PAF receptor also reduced the anti-inflammatory effects of apoptotic cell uptake by macrophages, and this could be mimicked by direct addition of PAF to the cells. While production of

PAF from macrophages interacting with apoptotic cells or PSR ligands could not be detected, this may have been because of low levels of free mediator and/or its attachment to any of the membranes in the system. A potential dual pro- or anti-inflammatory role for PAF and its receptor on macrophages is raised by these observations but requires much more investigation to prove.

At issue here is how PS in the membrane leaflet is bound by the receptor on a responding cell. At least the receptor recognizes the polar head group of the PS and this is facing the aqueous, extracellular environment. (The potential role of other substituents of the PS molecule in PSR binding are at this point unknown.) On the other hand, it would not seem easy for a protein in a cell membrane surrounded by surface carbohydrates, etc. (the glycocalyx) to gain access to the head group of a phospholipid on the cell being recognized also surrounded by surface structures. In general, cell membranes are mutually repulsive unless specific adhesion molecules are engaged. We have suggested a two-part process in recognition and removal of apoptotic cells in which adhesion or tethering ligands play a key role in bringing the two players in close apposition and, we propose, in allowing appropriate engagement of the PS receptor (Hoffmann et al. 2001). Either intrinsically, or because of these geographical constraints, it would appear that the PSR is of low effective affinity, although whether this is true in any real Michaelis Menton sense is completely unknown. Local high densities of PS on membrane blebs or on free vesicles would also serve to enhance the potential for PS receptor activation.

The potential difficulty for PS receptor activation in "normal" circumstances may have biologic implications in that it would mean that transient PS exposure on activated cells would not initiate responses in adjacent macrophages unless a number of other factors also came into play, including the presence of key tethering ligands and local high concentrations of "aggregated" PS on the target cell. Soluble PS-binding proteins acting as bridge molecules might have less of a problem gaining access to the surface phospholipids but again may need other factors or local high surface densities to be effective back on the responding cell.

6.2.2 Surface Effects for PAF and Lysophosphatides

How much any of these potential constraints apply to the other bioactive phospholipids is not at all clear. Their short (or absent) *sn2* fatty acids and higher hydrophilicity means that liberation from the membrane is easier (see above). Numerous studies with PAF analogs and antagonists suggest that the active site in the receptor can "see" each portion of the molecule, i.e., the phosphorylcholine, short chain *sn2* substituent, and the *sn1* group, an alkyl link in this position being more effective. The clear implication is that ultimately the PAF that acts on the receptor is completely free of the membrane of origin or carrier molecule. Can the receptor "extract" the PAF from a cell membrane? Is membrane-associated PAF bioactive? In this regard, an important series of experiments has addressed the effects of endothelial PAF on monocyte adhesion and stimulation (Patel et al. 1993; Zimmerman et al. 1996). Here PAF on the endothelial membrane can stimulate the monocytes. Adhesion molecules are also involved, perhaps to enhance binding to the PAF receptors. What is not clear is whether the PAF is actually membrane-inserted in this circumstance or is bound to an endothelial surface protein serving to "present" the molecule to the responding monocytes.

Whether similar phenomena and constraints occur in the case of SIP, LPA, LPC, or even PS is not yet clear, but seem quite likely.

7 Metabolism Versus Stimulation

Interaction of PAF with macrophages may have additional effects because of the ability of these cells to produce PAF acetylhydrolase. The product of this enzymatic activity is LPC (with either an alkyl or acyl *sn1* substituent depending on the type of PAF). This could occur within the cell or in the local extracellular environment. While such inter-conversion between these phospholipid mediators has not received much attention, transcellular metabolism and biosynthesis has been well documented for eicosanoids. The process adds complexity to our overall understanding of lipid mediators.

7.1 Transcellular Metabolism and Biosynthesis

The terms transcellular metabolism and biosynthesis are used to describe the production of intermediates in the eicosanoid, PAF, or lysophosphatide pathways by one cell, followed by their secretion or release to the outside, their subsequent uptake by other cells in the environment for further metabolism to mature mediators, which then are themselves secreted from the secondary cells for subsequent action (by ligation of specific receptors) on nearby cells.

The best example of this phenomenon is the synthesis of LTA_4 from arachidonate by 5-lipoxygenase (5-LO) on the nuclear membrane in cells such as neutrophils that do not contain the appropriate LTC_4 synthase to complete the synthesis of sulfidopeptide leukotrienes. In the presence of platelets, endothelial cells, or macrophages which do have this latter enzyme, the LTA_4 is efficiently taken up, metabolized to LTC_4 (Maclouf et al. 1996; Fradin et al. 1989; Maclouf et al. 1989; Sala et al. 2000) and then secreted into the environment to act on cells with appropriate receptors for these sulfidopeptide leukotrienes, including the macrophages themselves (Grimminger et al. 1991; Fukai et al. 1996).

Activation of macrophages and other inflammatory cells has been shown not only to result in synthesis and release of PAF, but in most cases where this is carefully examined, of LPC, and/or of lyso-PAF (i.e., alkyl-lyso-PC) as well. Uptake of these [possibly by scrambling, possibly via receptor engagement (Ohshima et al. 2002)] is rapid and efficient and can result in esterification by the PAF acetyltransferase, i.e., the synthesis in this secondary cell of PAF or reacylation with longer chain fatty acids to add to the PC pool (Ohshima et al. 2002). Since macrophages contain and secrete high levels of PAF acetylhydrolase, any PAF they synthesize will always be subject to hydrolysis, reuptake, and re-esterification. The final balance of products is therefore likely to be highly dynamic, depending significantly not only on the initial production of lyso-PAF, the availability of acetyl-CoA, and levels of active acetyltransferase, but also on the extracellular environment. Although not much studied to date, similar effects may be seen with the lysophosphatides

8 Summary

This chapter has addressed a number of elements in the pharmacology, biochemistry, and biology of extracellular phospholipid mediators. It has focused on special features of phospholipids as mediators including mechanisms of secretion, membrane association, and presentation to their receptors. A number of similarities between the different groups of phospholipids were emphasized, even though the lysophosphatides and PAFs act through seven transmembrane, G protein-linked receptors and PS does not. One of the main points perhaps is that because of these special features, the pharmacology may be significantly more complex. On the other hand, these molecules are highly active, which provides real motive for appropriate pharmacological manipulation probably most easily at the level of the specific receptors. The phospholipid mediators discussed herein have very broad biologic activities. Some of this may be because of the numerous receptors (e.g., the EDG family). With others, the breadth cannot be accounted for in receptor heterogeneity (e.g., PAF) and more likely lies at the level of cell response variation and or multiple signaling pathways. Overall, with the possible exception of the PS receptor (study of which itself is in its infancy) the macrophage has not been a major focus for investigation of either these mediators' production or responses. This should be rectified.

Acknowledgements. This work was supported by National Institutes of Health grant HL34303.

9 References

Alleva, D.G., E.B. Johnson, F.M. Lio, S.A. Boehme, P.J. Conlon, and P.D. Crowe. 2002. Regulation of murine macrophage proinflammatory and anti-inflammatory cytokines by ligands for peroxisome proliferator-activated receptor-gamma: counter-regulatory activity by IFN-gamma. J Leukoc Biol. 71:677–685

Arbibe, L., K. Koumanov, D. Vial, C. Rougeot, G. Faure, N. Havet, S. Longacre, B.B. Vargaftig, G. Bereziat, D.R. Voelker, C. Wolf, and L. Touqui. 1998. Generation of lysophospholipids from surfactant in acute lung injury is mediated by type-II phospholipase A2 and inhibited by a direct surfactant protein A-phospholipase A2 protein interaction. J Clin Invest. 102:1152–1160

Balsinde, J., M.A. Balboa, and E.A. Dennis. 1998. Functional coupling between secretory phospholipase A2 and cyclooxygenase-2 and its regulation by cytosolic group IV phospholipase A2. Proc Natl Acad Sci U S A. 95:7951–7956

Banks, J.B., R.L. Wykle, J.T. O'Flaherty, and R.H. Lumb. 1988. Evidence for protein-catalyzed transfer of platelet activating factor by macrophage cytosol. Biochim Biophys Acta. 961:48–52

Bautista, A.P., and J.J. Spitzer. 1992. Platelet activating factor stimulates and primes the liver, Kupffer cells and neutrophils to release superoxide anion. Free Radic Res Commun. 17:195–209

Bennett, M.R., D.F. Gibson, S.M. Schwartz, and J.F. Tait. 1995. Binding and phagocytosis of apoptotic vascular smooth muscle cells is mediated in part by exposure of phosphatidylserine. Circ Res. 77:1136–1142

Berger, A., N. Havet, D. Vial, L. Arbibe, C. Dumarey, M.L. Watson, and L. Touqui. 1999. Dioleylphosphatidylglycerol inhibits the expression of type II phospholipase A2 in macrophages. Am J Respir Crit Care Med. 159:613–618

Blank, M.L., Z.L. Smith, V. Fitzgerald, and F. Snyder. 1995. The CoA-independent transacylase in PAF biosynthesis: tissue distribution and molecular species selectivity. Biochim Biophys Acta. 1254:295–301

Blankenberg, F.G., P.D. Katsikis, J.F. Tait, R.E. Davis, L. Naumovski, K. Ohtsuki, S. Kopiwoda, M.J. Abrams, M. Darkes, R.C. Robbins, H.T. Maecker, and H.W. Strauss. 1998. In vivo detection and imaging of phosphatidylserine expression during programmed cell death. Proc Natl Acad Sci U S A. 95:6349–6354

Bozza, P.T., J.L. Payne, J.L. Goulet, and P.F. Weller. 1996. Mechanisms of platelet-activating factor-induced lipid body formation: requisite roles for 5-lipoxygenase and de novo protein synthesis in the compartmentalization of neutrophil lipids. J Exp Med. 183:1515–1525

Bratton, D.L. 1994. Polyamine inhibition of transbilayer movement of plasma membrane phospholipids in the erythrocyte ghost. J Biol Chem. 269:22517–22523

Bratton, D.L., E. Dreyer, J.M. Kailey, V.A. Fadok, K.L. Clay, and P.M. Henson. 1992. The mechanism of internalization of platelet-activating factor in activated human neutrophils. Enhanced transbilayer movement across the plasma membrane. J Immunol. 148:514–523

Bratton, D.L., J.M. Kailey, K.L. Clay, and P.M. Henson. 1991. A model for the extracellular release of PAF: the influence of plasma membrane phospholipid asymmetry. Biochim Biophys Acta. 1062:24–34

Callahan, M.K., P. Williamson, and R.A. Schlegel. 2000. Surface expression of phosphatidylserine on macrophages is required for phagocytosis of apoptotic thymocytes. Cell Death Differ. 7:645–653

Casciola-Rosen, L., A. Rosen, M. Petri, and M. Schlissel. 1996. Surface blebs on apoptotic cells are sites of enhanced procoagulant activity: implications for coagulation events and antigenic spread in systemic lupus erythematosus. Proc Natl Acad Sci USA. 93:1624–1629

Chen, C.Y., M.F. Ingram, P.H. Rosal, and T.R. Graham. 1999. Role for Drs2p, a P-type ATPase and potential aminophospholipid translocase, in yeast late Golgi function. J Cell Biol. 147:1223–1236

Chun, J., E.J. Goetzl, T. Hla, Y. Igarashi, K.R. Lynch, W. Moolenaar, S. Pyne, and G. Tigyi. 2002. International Union of Pharmacology. XXXIV. Lysophospholipid Receptor Nomenclature. Pharmacol Rev. 54:265–269

Clay, K.L., C. Johnson, and P. Henson. 1990. Binding of platelet activating factor to albumin. Biochim Biophys Acta. 1046:309–314

Comfurius, P., P. Williamson, E.F. Smeets, R.A. Schlegel, E.M. Bevers, and R.F. Zwaal. 1996. Reconstitution of phospholipid scramblase activity from human blood platelets. Biochemistry. 35:7631–7634

Daleke, D.L., and J.V. Lyles. 2000. Identification and purification of aminophospholipid flippases. Biochim Biophys Acta. 1486:108–127

Davies, S.S., A.V. Pontsler, G.K. Marathe, K.A. Harrison, R.C. Murphy, J.C. Hinshaw, G.D. Prestwich, A.S. Hilaire, S.M. Prescott, G.A. Zimmerman, and T.M. McIntyre. 2001. Oxidized alkyl phospholipids are specific, high affinity peroxisome proliferator-activated receptor gamma ligands and agonists. J Biol Chem. 276:16015–16023

de Freitas Balanco, J.M., M.E. Moreira, A. Bonomo, P.T. Bozza, G. Amarante-Mendes, C. Pirmez, and M.A. Barcinski. 2001. Apoptotic mimicry by an obligate intracellular parasite downregulates macrophage microbicidal activity. Curr Biol. 11:1870–1873

Delerive, P., J.C. Fruchart, and B. Staels. 2001. Peroxisome proliferator-activated receptors in inflammation control. J Endocrinol. 169:453–459

Dentan, C., P. Lesnik, M.J. Chapman, and E. Ninio. 1996. Phagocytic activation induces formation of platelet-activating factor in human monocyte-derived macrophages and in macrophage-derived foam cells. Relevance to the inflammatory reaction in atherogenesis. Eur J Biochem. 236:48–55

Dieter, P., A. Kolada, S. Kamionka, A. Schadow, and M. Kaszkin. 2002. Lipopolysaccharide-induced release of arachidonic acid and prostaglandins in liver macrophages: regulation by Group IV cytosolic phospholipase A2, but not by Group V and Group IIA secretory phospholipase A2. Cell Signal. 14:199–204

Ding, J., Z. Wu, B.P. Crider, Y. Ma, X. Li, C. Slaughter, L. Gong, and X.S. Xie. 2000. Identification and functional expression of four isoforms of ATPase II, the putative aminophospholipid translocase. Effect of isoform variation on the ATPase activity and phospholipid specificity. J Biol Chem. 275:23378–23386

Elstad, M.R., D.M. Stafforini, T.M. McIntyre, S.M. Prescott, and G.A. Zimmerman. 1989. Platelet-activating factor acetylhydrolase increases during macrophage differentiation. A novel mechanism that regulates accumulation of platelet-activating factor. J Biol Chem. 264:8467–8470

Exton, J.H. 2000. Phospholipase D. Ann N Y Acad Sci. 905:61–68

Exton, J.H. 2002. Phospholipase D-structure, regulation and function. Rev Physiol Biochem Pharmacol. 144:1–94

Fadok, V.A., A. de Cathelineau, D.L. Daleke, P.M. Henson, and D.L. Bratton. 2001b. Loss of phospholipid asymmetry and surface exposure of phosphatidylserine is required for phagocytosis of apoptotic cells by macrophages and fibroblasts. J. Biol. Chem. 276:1071–1077

Fadok, V.A., D.J. Laszlo, P.W. Noble, L. Weinstein, D.W. Riches, and P.M. Henson. 1993. Particle digestibility is required for induction of the phosphatidylserine recognition mechanism used by murine macrophages to phagocytose apoptotic cells. J Immunol. 151:4274–4285

Fadok, V.A., D.L. Bratton, A. Konowal, P.W. Freed, J.Y. Westcott, and P.M. Henson. 1998a. Macrophages that have ingested apoptotic cells in vitro inhibit proinflammatory cytokine production through autocrine/paracrine mechanisms involving TGF-beta, PGE2, and PAF. J Clin Invest. 101:890–898

Fadok, V.A., D.L. Bratton, and P.M. Henson. 2001a. Phagocyte receptors for apoptotic cells: recognition, uptake, and consequences. J Clin Invest. 108:957–962

Fadok, V.A., D.L. Bratton, D.M. Rose, A. Pearson, R.A. Ezekewitz, and P.M. Henson. 2000. A receptor for phosphatidylserine-specific clearance of apoptotic cells [see comments]. Nature. 405:85–90

Fadok, V.A., D.L. Bratton, L. Guthrie, and P.M. Henson. 2001c. Differential effects of apoptotic versus lysed cells on macrophage production of cytokines: role of proteases. J Immunol. 166:6847–6854

Fadok, V.A., D.L. Bratton, S.C. Frasch, M.L. Warner, and P.M. Henson. 1998b. The role of phosphatidylserine in recognition of apoptotic cells by phagocytes [see comments]. Cell Death Differ. 5:551–562

Fadok, V.A., J.S. Savill, C. Haslett, D.L. Bratton, D.E. Doherty, P.A. Campbell, and P.M. Henson. 1992. Different populations of macrophages use either the vitronectin receptor or the phosphatidylserine receptor to recognize and remove apoptotic cells. J Immunol. 149:4029–4035

Fradin, A., J.A. Zirrolli, J. Maclouf, L. Vausbinder, P.M. Henson, and R.C. Murphy. 1989. Platelet-activating factor and leukotriene biosynthesis in whole blood. A model for the study of transcellular arachidonate metabolism. J Immunol. 143:3680–3685

Frasch, S.C., P.M. Henson, J.M. Kailey, D.A. Richter, M.S. Janes, V.A. Fadok, and D.L. Bratton. 2000. Regulation of phospholipid scramblase activity during apoptosis and cell activation by protein kinase Cdelta. J Biol Chem. 275:23065–23073

Freire-de-Lima, C.G., D.O. Nascimento, M.B. Soares, P.T. Bozza, H.C. Castro-Faria-Neto, F.G. de Mello, G.A. DosReis, and M.F. Lopes. 2000. Uptake of apoptotic cells drives the growth of a pathogenic trypanosome in macrophages. Nature. 403:199–203

Fujishima, H., R.O. Sanchez Mejia, C.O. Bingham, 3rd, B.K. Lam, A. Sapirstein, J.V. Bonventre, K.F. Austen, and J.P. Arm. 1999. Cytosolic phospholipase A2 is essential for both the immediate and the delayed phases of eicosanoid generation in mouse bone marrow-derived mast cells. Proc Natl Acad Sci U S A. 96:4803–4807

Fukai, F., Y. Suzuki, Y. Nishizawa, and T. Katayama. 1996. Transcellular biosynthesis of cysteinyl leukotrienes by Kupffer cell- hepatocyte cooperation in rat liver. Cell Biol Int. 20:423–428

Goetzl, E.J., and K.R. Lynch. 2000. Preface: the omnific lysophospholipid growth factors. Ann N Y Acad Sci. 905:xi-xiv

Goetzl, E.J., H. Lee, T. Azuma, T.P. Stossel, C.W. Turck, and J.S. Karliner. 2000b. Gelsolin binding and cellular presentation of lysophosphatidic acid. J Biol Chem. 275:14573–14578

Goetzl, E.J., Y. Kong, and J.K. Voice. 2000a. Cutting edge: differential constitutive expression of functional receptors for lysophosphatidic acid by human blood lymphocytes. J Immunol. 164:4996–4999

Gomes, E., M.K. Jakobsen, K.B. Axelsen, M. Geisler, and M.G. Palmgren. 2000. Chilling tolerance in Arabidopsis involves ALA1, a member of a new family of putative aminophospholipid translocases. Plant Cell. 12:2441–2454

Graler, M.H., and E.J. Goetzl. 2002. Lysophospholipids and their G protein-coupled receptors in inflammation and immunity. Biochim Biophys Acta. 1582:168–174

Grigoriadis, G., and A.G. Stewart. 1992. Albumin inhibits platelet-activating factor (PAF)-induced responses in platelets and macrophages: implications for the biologically active form of PAF. Br J Pharmacol. 107:73–77

Grimminger, F., U. Sibelius, and W. Seeger. 1991. Amplification of LTB4 generation in AM-PMN cocultures: transcellular 5- lipoxygenase metabolism. Am J Physiol. 261:L195–203

Hakkinen, T., J.S. Luoma, M.O. Hiltunen, C.H. Macphee, K.J. Milliner, L. Patel, S.Q. Rice, D.G. Tew, K. Karkola, and S. Yla-Herttuala. 1999. Lipoprotein-associated phospholipase A(2), platelet-activating factor acetylhydrolase, is expressed by macrophages in human and rabbit atherosclerotic lesions. Arterioscler Thromb Vasc Biol. 19:2909–2917

Hanayama, R., M. Tanaka, K. Miwa, A. Shinohara, A. Iwamatsu, and S. Nagata. 2002. Identification of a factor that links apoptotic cells to phagocytes. Nature. 417:182–187

Hannun, Y.A., and R.M. Bell. 1993. The sphingomyelin cycle: a prototypic sphingolipid signaling pathway. Adv Lipid Res. 25:27–41

Henson. 2000a. PAF. In Cykokine reference manual, Ed. J, J, Openheim

Henson. 2000b. PAF receptors. In Cykokine reference manual, Ed. J, J, Openheim

Hla, T., M.J. Lee, N. Ancellin, C.H. Liu, S. Thangada, B.D. Thompson, and M. Kluk. 1999. Sphingosine-1-phosphate: extracellular mediator or intracellular second messenger? Biochem Pharmacol. 58:201–207

Hla, T., M.J. Lee, N. Ancellin, J.H. Paik, and M.J. Kluk. 2001. Lysophospholipids–receptor revelations. Science. 294:1875–1878

Hoffmann, P.R., A.M. deCathelineau, C.A. Ogden, Y. Leverrier, D.L. Bratton, D.L. Daleke, A.J. Ridley, V.A. Fadok, and P.M. Henson. 2001. Phosphatidylserine (PS) induces PS receptor-mediated macropinocytosis and promotes clearance of apoptotic cells. J Cell Biol. 155:649–659

Hornuss, C., R. Hammermann, M. Fuhrmann, U.R. Juergens, and K. Racke. 2001. Human and rat alveolar macrophages express multiple EDG receptors. Eur J Pharmacol. 429:303–308

Howard, K.M., and M.S. Olson. 2000. The expression and localization of plasma platelet-activating factor acetylhydrolase in endotoxemic rats. J Biol Chem. 275:19891–19896

Huber, J., A. Vales, G. Mitulovic, M. Blumer, R. Schmid, J.L. Witztum, B.R. Binder, and N. Leitinger. 2002. Oxidized membrane vesicles and blebs from apoptotic cells contain biologically active oxidized phospholipids that induce monocyte- endothelial interactions. Arterioscler Thromb Vasc Biol. 22:101–107

Huynh, M.L., V.A. Fadok, and P.M. Henson. 2002. Phosphatidylserine-dependent ingestion of apoptotic cells promotes TGF- beta1 secretion and the resolution of inflammation. J Clin Invest. 109:41–50

Jaross, W., R. Eckey, and M. Menschikowski. 2002. Biological effects of secretory phospholipase A2 group IIA on lipoproteins and in atherogenesis. Eur J Clin Invest. 32:383–393

Kabarowski, J.H., K. Zhu, L.Q. Le, O.N. Witte, and Y. Xu. 2001. Lysophosphatidylcholine as a ligand for the immunoregulatory receptor G2A. Science. 293:702–705

Klivenyi, P., M.F. Beal, R.J. Ferrante, O.A. Andreassen, M. Wermer, M.R. Chin, and J.V. Bonventre. 1998. Mice deficient in group IV cytosolic phospholipase A2 are resistant to MPTP neurotoxicity. J Neurochem. 71:2634–2637

Koh, J.S., W. Lieberthal, S. Heydrick, and J.S. Levine. 1998. Lysophosphatidic acid is a major serum noncytokine survival factor for murine macrophages which acts via the phosphatidylinositol 3-kinase signaling pathway. J Clin Invest. 102:716–727

Kucey, D.S., E.I. Kubicki, and O.D. Rotstein. 1991. Platelet-activating factor primes endotoxin-stimulated macrophage procoagulant activity. J Surg Res. 50:436–441

Kunzelmann-Marche, C., J.M. Freyssinet, and M.C. Martinez. 2002. Loss of plasma membrane phospholipid asymmetry requires raft integrity. Role of transient receptor potential channels and ERK pathway. J Biol Chem. 277:19876–19881

Le Stunff, H., C. Peterson, H. Liu, S. Milstien, and S. Spiegel. 2002. Sphingosine-1-phosphate and lipid phosphohydrolases. Biochim Biophys Acta. 1582:8–17

Le, L.Q., J.H. Kabarowski, Z. Weng, A.B. Satterthwaite, E.T. Harvill, E.R. Jensen, J.F. Miller, and O.N. Witte. 2001. Mice lacking the orphan G protein-coupled receptor G2A develop a late- onset autoimmune syndrome. Immunity. 14:561–571

Lee, H., J.J. Liao, M. Graeler, M.C. Huang, and E.J. Goetzl. 2002. Lysophospholipid regulation of mononuclear phagocytes. Biochim Biophys Acta. 1582:175–177

Li, H., J.M. Tremblay, L.R. Yarbrough, and G.M. Helmkamp, Jr. 2002. Both isoforms of mammalian phosphatidylinositol transfer protein are capable of binding and transporting sphingomyelin. Biochim Biophys Acta. 1580:67–76

Lin, W.W., S.H. Chang, and S.M. Wang. 1999. Roles of atypical protein kinase C in lysophosphatidic acid-induced type II adenylyl cyclase activation in RAW 264.7 macrophages. Br J Pharmacol. 128:1189–1198

Liu, H., M. Sugiura, V.E. Nava, L.C. Edsall, K. Kono, S. Poulton, S. Milstien, T. Kohama, and S. Spiegel. 2000. Molecular cloning and functional characterization of a novel mammalian sphingosine kinase type 2 isoform. J Biol Chem. 275:19513–19520

Lumb, R.H., M. Record, G. Ribbes, G.L. Pool, F. Terce, and H. Chap. 1990. PAF-acether transfer activity in HL-60 cells is induced during differentiation. Biochem Biophys Res Commun. 171:548–554

MacKenzie, A., H.L. Wilson, E. Kiss-Toth, S.K. Dower, R.A. North, and A. Surprenant. 2001. Rapid secretion of interleukin-1beta by microvesicle shedding. Immunity. 15:825–835

Maclouf, J., A. Sala, G. Rossoni, F. Berti, R. Muller-Peddinghaus, and G. Folco. 1996. Consequences of transcellular biosynthesis of leukotriene C4 on organ function. Haemostasis. 26 Suppl 4:28–36

Maclouf, J., R.C. Murphy, and P.M. Henson. 1989. Platelets and endothelial cells contribute to the production of LTC4 by transcellular metabolism with neutrophils. Adv Prostaglandin Thromboxane Leukot Res. 19:259–262

Marathe, G.K., K.A. Harrison, R.C. Murphy, S.M. Prescott, G.A. Zimmerman, and T.M. McIntyre. 2000. Bioactive phospholipid oxidation products. Free Radic Biol Med. 28:1762–1770

Marathe, G.K., S.M. Prescott, G.A. Zimmerman, and T.M. McIntyre. 2001. Oxidized LDL contains inflammatory PAF-like phospholipids. Trends Cardiovasc Med. 11:139–142

Marathe, G.K., S.S. Davies, K.A. Harrison, A.R. Silva, R.C. Murphy, H. Castro-Faria-Neto, S.M. Prescott, G.A. Zimmerman, and T.M. McIntyre. 1999. Inflammatory platelet-activating factor-like phospholipids in oxidized low density lipoproteins are fragmented alkyl phosphatidylcholines. J Biol Chem. 274:28395–28404

Marguet, D., M.F. Luciani, A. Moynault, P. Williamson, and G. Chimini. 1999. Engulfment of apoptotic cells involves the redistribution of membrane phosphatidylserine on phagocyte and prey. Nat Cell Biol. 1:454–456

Moller, T., J.J. Contos, D.B. Musante, J. Chun, and B.R. Ransom. 2001. Expression and function of lysophosphatidic acid receptors in cultured rodent microglial cells. J Biol Chem. 276:25946–25952

Moolenaar, W.H. 1995. Lysophosphatidic acid, a multifunctional phospholipid messenger. J Biol Chem. 270:12949–12952

Moolenaar, W.H. 2000. Development of our current understanding of bioactive lysophospholipids. Ann N Y Acad Sci. 905:1–10

Morioka, Y., A. Saiga, Y. Yokota, N. Suzuki, M. Ikeda, T. Ono, K. Nakano, N. Fujii, J. Ishizaki, H. Arita, and K. Hanasaki. 2000. Mouse group X secretory phospholipase A2 induces a potent release of arachidonic acid from spleen cells and acts as a ligand for the phospholipase A2 receptor. Arch Biochem Biophys. 381:31–42

Morito, T., K. Oishi, M. Yamamoto, and K. Matsumoto. 2000. Biphasic regulation of Fc-receptor mediated phagocytosis of rabbit alveolar macrophages by surfactant phospholipids. Tohoku J Exp Med. 190:15–22

Nakagawa, Y., M. Sugai, K. Karasawa, A. Tokumura, H. Tsukatani, M. Setaka, and S. Nojima. 1992. Possible influence of lysophospholipase on the production of 1-acyl-2-acetylglycerophosphocholine in macrophages. Biochim Biophys Acta. 1126:277–285

Ninio, E., J.M. Mencia-Huerta, F. Heymans, and J. Benveniste. 1982. Biosynthesis of platelet-activating factor. I. Evidence for an acetyl- transferase activity in murine macrophages. Biochim Biophys Acta. 710:23–31

Ohshima, N., S. Ishii, T. Izumi, and T. Shimizu. 2002. Receptor-dependent metabolism of platelet-activating factor in murine macrophages. J Biol Chem. 277:9722–9727

Ousman, S.S., and S. David. 2000. Lysophosphatidylcholine induces rapid recruitment and activation of macrophages in the adult mouse spinal cord. Glia. 30:92–104

Patel, K.D., E. Lorant, D.A. Jones, M. Prescott, T.M. McIntyre, and G.A. Zimmerman. 1993. Juxtacrine interactions of endothelial cells with leukocytes: tethering and signaling molecules. Behring Inst Mitt:144–164

Prescott, S.M. 1997. Inflammatory actions of platelet-activating factor: control by PAF acetylhydrolase. J Investig Allergol Clin Immunol. 7:416

Prokazova, N.V., N.D. Zvezdina, and A.A. Korotaeva. 1998. Effect of lysophosphatidylcholine on transmembrane signal transduction. Biochemistry (Mosc). 63:31–37

Quinn, M.T., S. Parthasarathy, and D. Steinberg. 1988. Lysophosphatidylcholine: a chemotactic factor for human monocytes and its potential role in atherogenesis. Proc Natl Acad Sci U S A. 85:2805–2809

Record, M., and F. Snyder. 1990. Intracellular location of acetyltransferase in the remodeling pathway of PAF biosynthesis in undifferentiated human leukemic cells (HL-60). J Lipid Mediat. 2:1–8

Rose, D.M., V.A. Fadok, D.W. Riches, K.L. Clay, and P.M. Henson. 1995. Autocrine/paracrine involvement of platelet-activating factor and transforming growth factor-beta in the induction of phosphatidylserine recognition by murine macrophages. J Immunol. 155:5819–5825

Sala, A., G. Rossoni, F. Berti, C. Buccellati, A. Bonazzi, J. Maclouf, and G. Folco. 2000. Monoclonal anti-CD18 antibody prevents transcellular biosynthesis of cysteinyl leukotrienes in vitro and in vivo and protects against leukotriene-dependent increase in coronary vascular resistance and myocardial stiffness. Circulation. 101:1436–1440

Samples, B.L., G.L. Pool, and R.H. Lumb. 1999. Subcellular localization of enzyme activities involved in the metabolism of platelet-activating factor in rainbow trout leukocytes. Biochim Biophys Acta. 1437:357–366

Scott, R.S., E.J. McMahon, S.M. Pop, E.A. Reap, R. Caricchio, P.L. Cohen, H.S. Earp, and G.K. Matsushima. 2001. Phagocytosis and clearance of apoptotic cells is mediated by MER. Nature. 411:207–211

Shamsuddin, M., E. Chen, J. Anderson, and L.J. Smith. 1997. Regulation of leukotriene and platelet-activating factor synthesis in human alveolar macrophages. J Lab Clin Med. 130:615–626

Sims, P.J., and T. Wiedmer. 2001. Unraveling the mysteries of phospholipid scrambling. Thromb Haemost. 86:266–275

Spiegel, S., and S. Milstien. 2002. Sphingosine-1-phosphate a key cell signaling molecule. J Biol Chem. In Press

Spiegel, S., D. English, and S. Milstien. 2002b. Sphingosine 1-phosphate signaling: providing cells with a sense of direction. Trends Cell Biol. 12:236–242

Stach, C.M., X. Turnay, R.E. Voll, P.M. Kern, W. Kolowos, T.D. Beyer, J.R. Kalden, and M. Herrmann. 2000. Treatment with annexin V increases immunogenicity of apoptotic human T- cells in Balb/c mice. Cell Death Differ. 7:911–915

Stafforini, D.M., M.R. Elstad, T.M. McIntyre, G.A. Zimmerman, and S.M. Prescott. 1990. Human macrophages secret platelet-activating factor acetylhydrolase. J Biol Chem. 265:9682–9687

Stafforini, D.M., T. Numao, A. Tsodikov, D. Vaitkus, T. Fukuda, N. Watanabe, N. Fueki, T.M. McIntyre, G.A. Zimmerman, S. Makino, and S.M. Prescott. 1999. Deficiency of platelet-activating factor acetylhydrolase is a severity factor for asthma. J Clin Invest. 103:989–997

Stafforini, D.M., T.M. McIntyre, M.E. Carter, and S.M. Prescott. 1987. Human plasma platelet-activating factor acetylhydrolase. Association with lipoprotein particles and role in the degradation of platelet- activating factor. J Biol Chem. 262:4215–4222

Sun, J., J. Zhao, M.A. Schwartz, J.Y. Wang, T. Wiedmer, and P.J. Sims. 2001. c-Abl tyrosine kinase binds and phosphorylates phospholipid scramblase 1. J Biol Chem. 276:28984–28990

Svetlov, S.I., H. Liu, W. Chao, and M.S. Olson. 1997. Regulation of platelet-activating factor (PAF) biosynthesis via coenzyme A-independent transacylase in the macrophage cell line IC-21 stimulated with lipopolysaccharide. Biochim Biophys Acta. 1346:120–130

Tigyi, G., D.J. Fischer, D. Baker, D.A. Wang, J. Yue, N. Nusser, T. Virag, V. Zsiros, K. Liliom, D. Miller, and A. Parrill. 2000. Pharmacological characterization of phospholipid growth-factor receptors. Ann N Y Acad Sci. 905:34–53

van Haperen, R., A. van Tol, P. Vermeulen, M. Jauhiainen, T. van Gent, P. van den Berg, S. Ehnholm, F. Grosveld, A. van der Kamp, and R. de Crom. 2000. Human plasma phos-

pholipid transfer protein increases the antiatherogenic potential of high density lipoproteins in transgenic mice. Arterioscler Thromb Vasc Biol. 20:1082–1088
van Tol, A. 2002. Phospholipid transfer protein. Curr Opin Lipidol. 13:135–139
Voelker, D.R. 2000. Interorganelle transport of aminoglycerophospholipids. Biochim Biophys Acta. 1486:97–107
Voll, R.E., M. Herrmann, E.A. Roth, C. Stach, J.R. Kalden, and I. Girkontaite. 1997. Immunosuppressive effects of apoptotic cells. Nature. 390:350–351
Waga, I., M. Nakamura, Z. Honda, I. Ferby, S. Toyoshima, S. Ishiguro, and T. Shimizu. 1993. Two distinct signal transduction pathways for the activation of guinea- pig macrophages and neutrophils by endotoxin. Biochem Biophys Res Commun. 197:465–472
Williamson, P., A. Christie, T. Kohlin, R.A. Schlegel, P. Comfurius, M. Harmsma, R.F. Zwaal, and E.M. Bevers. 2001. Phospholipid scramblase activation pathways in lymphocytes. Biochemistry. 40:8065–8072
Williamson, P., E.M. Bevers, E.F. Smeets, P. Comfurius, R.A. Schlegel, and R.F. Zwaal. 1995. Continuous analysis of the mechanism of activated transbilayer lipid movement in platelets. Biochemistry. 34:10448–10455
Winkler, J.D., A.N. Fonteh, C.M. Sung, L. Huang, M. Chabot-Fletcher, L.A. Marshall, and F.H. Chilton. 1995. Inhibition of CoA-independent transacylase reduces inflammatory lipid mediators. Adv Prostaglandin Thromboxane Leukot Res. 23:89–91
Wolfe, M.S., and D.J. Selkoe. 2002. Biochemistry. Intramembrane proteases–mixing oil and water. Science. 296:2156–2157
Yamada, M., A. Tanimoto, G. Ichinowatari, H. Yaginuma, and K. Ohuchi. 1999. Possible participation of intracellular platelet-activating factor in tumor necrosis factor-alpha production by rat peritoneal macrophages. Eur J Pharmacol. 374:341–350
Yamaguchi, Y., F. Matsumura, J. Liang, K. Okabe, T. Matsuda, H. Ohshiro, K. Ishihara, E. Akizuki, S. Yamada, and M. Ogawa. 1999. Platelet-activating factor antagonist (TCV-309) attenuates the priming effects of bronchoalveolar macrophages in cerulein-induced pancreatitis rats. Pancreas. 18:355–363
Yatomi, Y., Y. Igarashi, L. Yang, N. Hisano, R. Qi, N. Asazuma, K. Satoh, Y. Ozaki, and S. Kume. 1997. Sphingosine 1-phosphate, a bioactive sphingolipid abundantly stored in platelets, is a normal constituent of human plasma and serum. J Biochem (Tokyo). 121:969–973
Zhou, Q., J. Zhao, J.G. Stout, R.A. Luhm, T. Wiedmer, and P.J. Sims. 1997. Molecular cloning of human plasma membrane phospholipid scramblase. A protein mediating transbilayer movement of plasma membrane phospholipids. J Biol Chem. 272:18240–18244
Zhu, K., L.M. Baudhuin, G. Hong, F.S. Williams, K.L. Cristina, J.H. Kabarowski, O.N. Witte, and Y. Xu. 2001. Sphingosylphosphorylcholine and lysophosphatidylcholine are ligands for the G protein-coupled receptor GPR4. J Biol Chem. 276:41325–41335
Zimmerman, G.A., M.R. Elstad, D.E. Lorant, T.M. McLntyre, S.M. Prescott, M.K. Topham, A.S. Weyrich, and R.E. Whatley. 1996. Platelet-activating factor (PAF): signalling and adhesion in cell-cell interactions. Adv Exp Med Biol. 416:297–304

Part 3
Modulation of Specialised Macrophage Activities

Dendritic Cells Versus Macrophages as Antigen-Presenting Cells: Common and Unique Features

S. Vuckovic · D. N. J. Hart

Mater Medical Research Institute,
Raymond Terrace, 4101 South Brisbane, Queensland, Australia
e-mail: dhart@mmri.mater.org.au

Abstract Dendritic cells (DC) and macrophages contribute to both the innate and adaptive immune responses. It is becoming clear that DC and macrophages can be derived from common precursors, and that monocytes differentiate into DC under defined experimental conditions. Multiple types of DC and macrophages exist with different functional roles. Both immature DC and macrophages have significant phagocytic ability and are recruited by chemokines and cytokines to inflammatory sites. Upon encountering antigen or inflammatory stimuli, DC and macrophages become activated and responsible for several distinct non-specific and specific immunological functions. Most importantly, different stimuli, i.e. different pathogen-associated molecular patterns trigger different DC outcomes. Thus, the different DC subsets regulate the processing/delivery of antigen and provide a variety of costimulatory surface molecules, soluble cytokines and chemokines. DC are uniquely capable of activating primary immunity. This has driven the use of DC for tumour immunotherapy.

Keywords Dendritic cells, Macrophages, Differentiation, Sentinel function, Primary immunity, Immunotherapy

1 Introduction

Dendritic cells (DC) and macrophages belong to a family of antigen-presenting cells (APC). DC show certain similarities with macrophages, such as their differentiation pathway and sentinel functions but also have unique properties, such as efficient induction and maintenance of primary immune responses. Whilst the differentiation pathway that generates human DC in vivo remains unknown, DC and macrophages can be generated from common precursors under defined experimental conditions. As efficient sentinels, DC, like macrophages, are capable of recognizing danger signals derived from pathogens, taking up, processing and presenting antigen. However, DC differ from macrophages in key functions, notably the processing and delivery of antigen and the provision of costimulatory and accessory signals. These unique properties enable DC to function as effective APC, uniquely able to initiate primary immune responses and regulate adaptive immune responses. Because of their capacity to induce and maintain primary immune responses, DC are attractive vehicles for tumour immunotherapy.

This review points out similarities between DC and macrophages in their differentiation pathways and sentinel functions, as well as unique features of DC such as initiation and regulation of primary immune responses. Much of this work depends on the ability to discriminate DC populations with monoclonal antibodies. The currently limited but increasingly important field of DC differentiation antigens has been reviewed elsewhere (Hart et al. 2001, 2002).

2 Differentiation Pathways of DC and Their Relation to Macrophages

Human DC are found in almost all organs and represent a heterogeneous cell population. Based on phenotype and current views as to DC development, the different DC subsets are now universally subdivided into a minimum of two subsets. The myeloid-derived DC may include further subsets, but these certainly include the archetypical DC originally identified in mouse spleen. The lymphoid, plasmacytoid or confusingly, even the monocytoid DC is considered a very distinct subset, distinguished to a certain extent by its morphology but most effectively by its type I interferon (IFN)-producing capability. Myeloid-derived DC include Langerhans cells (LC), interstitial DC (e.g. dermal DC) and CD11c^{+} DC isolated from lymphoid tissues. Their precursors are included in the CD11c^{+} CD123^{dim} blood DC subset. In addition, DC derived from monocytes (Mo-DC) are also considered as myeloid-derived DC. Plasmacytoid blood CD123^{hi}CD11c^{-} DC, are the most probable candidates of lymphoid-derived DC and they are identified in thymus and other lymphoid organs. Their entry into the latter via high endothelial venules is a key distinguishing feature from myeloid DC, which enter lymph nodes via the afferent lymphatics.

Some subsets of myeloid-derived DC, dermal DC, LC and Mo-DC share common precursors with macrophages. $CD34^{+}CD1a^{-}$ precursors could generate DC when cultured with interleukin (IL)-7, tumour necrosis factor (TNF)-α, stem cell factor (SCF) and FLT3L. These precursors could also generate macrophages when cultured with macrophage colony-stimulating factor (M-CSF) (Dalloul et al. 1999). $CD34^{+}$ precursor cells obtained from cord blood or bone marrow differentiate into DC when cultured with granulocyte-macrophage (GM)-CSF and TNF-α (Caux et al. 1996; Caux et al. 1997). In this culture system, differentiation appears to occur via two independent, immature DC intermediates, defined by their exclusive expression of CD14 and CD1a. When cultured with GM-CSF and TNF-α, $CD14^{+}CD1a^{-}$ intermediates generate E-cadherin-mature DC, with a dermal or lymphoid-organ DC phenotype. They could also generate macrophages, when cultured with M-CSF. In contrast, $CD14^{-}CD1a^{+}$ intermediates generate E-cadherin^{+}langerin^{+} LC-like DC. Differentiation of LC can also be achieved from the $CD14^{+}CD1a^{-}$ intermediates, in culture with transforming growth factor (TGF)-β (Jaksits et al. 1999). Similarly, LC-like DC can be generated by culturing monocytes (Geissmann et al. 1998) or blood $CD11c^{+}CD1a^{+}$DC with GM-CSF, IL-4 and TGF-β (Ito et al. 1999).

Peripheral blood monocytes cultured with the cytokine combination of GM-CSF and IL-4 (Sallusto and Lanzavecchia 1994) or GM-CSF and IL-13 (Piemonti et al. 1995; Allavena et al. 1998) differentiate into immature Mo-DC. Differentiation of monocytes into DC occurs during the transendothelial migration of monocytes (Randolph et al. 1998) and takes place in the lymph nodes (Randolph et al. 1999). Interestingly, mouse monocytes, which migrate in vivo, differentiate into DC-like cells that have high late bead uptake capacity but lack CD11c (Randolph et al. 1999), the key DC subset marker. It appears that the minority $CD16^{+}$ monocyte subset may be at least partially committed to DC differentiation (Randolph et al. 2002). Peripheral blood monocytes can be made to differentiate into macrophages, when cultured with M-CSF or GM-CSF (Clark and Kamen 1987; Metcalf 1989).

Cytokines and undefined serum component(s) control the balance between the differentiation of monocytes into DC and into macrophages. The cytokine IL-6 exerts inhibitory effects on DC development and promotes differentiation of monocytes to macrophages after addition to GM-CSF and IL-4 cultures (Mitani et al. 2000). This action of IL-6 can be abrogated by TNF-α, lipopolysaccharide (LPS), IL-1β, CD40L and TGF-β1. Furthermore, certain immunosuppressive cytokines, such as IL-10 prevent the differentiation of monocytes into DC but not into macrophages (Allavena et al. 1998). The results obtained from serum-containing and serum-free culture experiments show that the humoral factor(s) in serum promote differentiation of monocytes into macrophages rather than into DC, when cultured with GM-CSF and IL-4 (Cao et al. 2000).

Notch receptors are conserved transmembrane receptors, which play a central role in regulating cell decision of bipotent precursors. Notch receptors expressed by bipotential progenitors are activated by neighbouring cells bearing Notch ligands, leading to differentiation of Notch-expressing cells along a lineage-specific pathway. Interaction between the transmembrane receptor Notch and its ligand Delta-1 balance the differentiation of blood monocytes towards DC but not towards macrophages (Ohishi et al. 2001).

Immature Mo-DC exposed to TNF-α, LPS, CD40L or prostaglandin E2 (PGE2), following culture with GM-CSF and IL-4, acquire a final commitment towards mature DC expressing a high level of MHC class I and II, the costimulatory molecules CD80 and CD86, and the DC differentiation/activation molecules CMRF-44 (Vuckovic et al. 1998), CMRF56 (Hock et al. 1999) and CD83 (Zhou and Tedder 1996). In addition, type I IFN also promotes differentiation of monocyte to mature DC in culture with granulocyte (G)-CSF or with GM-CSF and IL-4 (Santini et al. 2000; Huang et al. 2001).

At least some subsets of myeloid-derived DC appear capable of differentiating into macrophages, again indicating a developmental link between DC and macrophages. Immature Mo-DC derived after culture with GM-CSF and IL-4 have limited DC commitment and acquire macrophage features after removing cytokines, unless stimulated to differentiate to mature Mo-DC (Palucka et al. 1998). A subset of DC found in the blood, which has a low expression of CD11c molecules, acquired macrophage features following exposure to M-CSF (Robinson et al. 1999). Curiously, the plasmacytoid blood CD123^{hi} DC appear to have CD14^{+} CD16^{+} precursors (Ho et al. 2002).

Differentiation of DC from either CD34^{+} precursor cells or peripheral blood monocytes and the developmental link between DC and macrophages observed in vitro requires in vivo confirmation that they represent physiological counterparts. Further research will address physiological mechanisms governing the differentiation of DC and the link between DC and macrophages. We have previously suggested that the preformed surveillance DC provide primary immune activation (Vuckovic et al. 1998). Recruitment of monocyte precursors and differentiation into DC may represent an inflammatory boost pathway for antigen presentation. Experiments can be designed to address this hypothesis.

3
DC and Macrophages in Pathogen Recognition

DC and macrophages function as sentinels for the cognate and innate immune systems. In the peripheral tissues, DC are found in an immature form characterized by their ability to recognize pathogen-associated molecular patterns (PAMP) shared by large groups of pathogens. Immature DC recognize PAMP through phylogenetically conserved Toll-like receptor (TLR) family members. In humans, six TLR homologues have been reported (Rock et al. 1998) and at least four others have been identified (Bowie and O'Neill 2000). All are type I integral membrane receptors with extracellular leucine-rich repeats and a cytoplasmic portion that is homologous to the signalling domain of the IL-1R. In mice, gene knockout studies indicate that TLR2 is required for gram-positive responses such as peptidoglycan (PGN) (Takeuchi et al. 1999) and outer membrane protein A (OmpA) (Jeannin et al. 2000), and TLR4 is essential for gram-negative responses, including bacterial toxin LPS (Poltorak et al. 1998). TLR3 recognizes double-stranded RNA (dsRNA) or a synthetic dsRNA analogue poly(inosinic acid)–poly(cytidylic acid) [poly (I:C)], a molecular pattern associated with viral infection (Alexopoulou et al. 2001). TLR9 mediates immune responses to unmethylated CpG dinucleotides in the bacterial DNA (Hemmi et al. 2000). The function of other TLRs has yet to be defined.

Whilst several TLR are found on DC, the TLR are more abundantly expressed on macrophages. Mo-DC express TLR1, 2, 3 and 4 whereas TLR5 is barely detectable (Kadowaki et al. 2001; Visintin et al. 2001). Plasmacytoid blood $CD123^{hi}$ DC express TLR7 and 9 (Kadowaki et al. 2001). Macrophages express an abundance of TLR including TLR1, 2, 4, 5 and 8 (Visintin et al. 2001). It is of interest that TLR2 and 4 are required for responses to a number of PAMP and are expressed on both immature Mo-DC and macrophages, which perhaps reflects their similar functions. In accordance with TLR expression patterns, Mo-DC respond to the TLR2-ligand PGN, and plasmacytoid $CD123^{hi}$ blood DC respond to the TLR9-ligand unmethylated CpG dinucleotides in bacterial DNA (Hartmann et al. 1999; Kadowaki et al. 2001). Macrophages respond to microbial molecules known to trigger signalling via TLR2, TLR3 and TLR4, such as PGN, poly (I:C) and LPS, respectively.

Signalling through TLR drives DC and macrophages to produce proinflammatory cytokines. Signalling through TLR2 stimulates Mo-DC to produce large amounts of TNF-α in response to PGN. In contrast, signalling through TLR9 stimulates plasmacytoid blood $CD123^{hi}$ DC to produce IFN-α/β, during antibacterial immune responses. Signalling through TLR2, TLR3 and TLR4 empowers macrophages with the ability to produce large amounts of TNF-α and IL-6 during gram-positive or gram-negative immune responses (Kadowaki et al. 2001).

Recognition of PAMP drives the maturation of DC and progressive downregulation of TLR. This coincides with a functional switch from sentinel to antigen-presenting function. Mature Mo-DC lack any TLR. This results in a loss of responsiveness to LPS and a loss of TNF-α-producing capability. Upon maturation, plasmacytoid blood $CD123^{hi}$ DC downregulate expression of TLR7 and TLR9, lose their ability to produce IFN-α/β and acquire the ability to present antigen to T cells (Kadowaki et al. 2001; Visintin et al. 2001).

The presence of only a limited set of TLR on DC suggests that they might have a restricted ability to recognize PAMP and subsequently have lesser functional plasticity in response to pathogens, compared to macrophages. The recognition of PAMP by innate receptors on DC probably provides stress signals required for antigen targeting and cross-presentation by DC. The recognition of PAMP by macrophages enhances the elimination of bacteria. The possibility of cellular cooperation, in which macrophages provide antigen and relevant signals to DC was raised some time ago (McKenzie et al. 1989).

4 DC and Macrophages Differ in Their Ability to Cross-Present Exogenous Antigens

Initial investigations of DC and macrophages focused on the uptake of exogenous antigens for processing and presentation in the context of MHC class II molecules to $CD4^{+}$T lymphocytes.

The generation of cytotoxic effector T lymphocytes (CTL) responses to tumours (Berard et al. 2000), viruses (Sigal et al. 1999), bacterial antigens (Lenz et al. 2000), graft tissue (Bevan 1976) and even self-antigens (Kurts et al. 1996) also requires the presentation of exogenous antigen by MHC class I molecules on the surface of APC, a process termed cross-priming or cross-presentation. Both Mo-DC and macrophages can take up exogenous antigens in the form of soluble proteins, particulate antigen and cell-associated antigens derived from apoptotic or necrotic cells, and then process and cross-present them in the context of the MHC class I molecules (Rock et al. 1993; Shen et al. 1997). Other reports have indicated that cross-presentation is a specific property of DC (Mitchell et al. 1998; Rodriguez et al. 1999).

DC and macrophages use scavenger receptor-mediated endocytosis to take up exogenous antigen derived from apoptotic cells. Mo-DC use $\alpha_V\beta5$ integrin in cooperation with CD36 and thrombospondin to form a molecular bridge to their apoptotic target (Albert et al. 1998). Macrophages use $\alpha_V\beta3$ integrin, CD36, thrombospondin and the phosphatidylserine-binding protein, the cognate receptor for externalized phosphatidylserine on apoptotic cells (Savill et al. 1992; Fadok et al. 2000). Uptake of soluble proteins by phagocytosis or macropinocytosis (Kovacsovics-Bankowski et al. 1993; Norbury et al. 1995) also leads to cross-presentation but requires higher antigen concentration and could be less relevant in vivo than receptor-mediated endocytosis.

DC and macrophages also express an extensive range of both type I and type II C type lectins that appear to have a role in antigen uptake. The lectin molecules DEC-205 (CD205), MMR (CD206), DC-SIGN (CD209) and BDCA-2 are differentially expressed on DC, Mo-DC and macrophages (Kato et al. 2000; Osugi et al. 2002). DEC-205 delivers antigen deep into the endocytic pathway. Other lectins such as DC-SIGN (Geijtenbeek et al. 2000) and BDCA-2 (Dzionek et al. 2001) are internalized, resulting in effective antigen processing.

Mo-DC are able to prime naïve T cells, and induce CTL responses to antigen derived from apoptotic cells and clear pathogens responsible for the induction of apoptotic cell death (Albert et al. 1998). Macrophages are more efficient at taking up antigens derived from apoptotic cells than Mo-DC, but they degrade rather than cross-present the ingested antigens and subsequently suppress inflammatory responses (Voll et al. 1997; Fadok et al. 1998). This could be explained by the direct transport of internalized antigens from endosome to the cytosol, which exists in DC but not in macrophages (Rodriguez et al. 1999). Direct entry of exogenous antigen into the cytosol results in antigen introduction into the classical transporter associated with the antigen processing (TAP)-dependent MHC class I presentation pathway. Noncytosolic pathways are related to endocytic processing by macrophages and involve loading of peptides on post-Golgi MHC class I molecules (Yewdell et al. 1999).

Exogenous antigens chaperoned by heat shock proteins (HSP) are released into the extracellular milieu during necrotic cell death. A wide array of antigenic peptides are chaperoned by HSP such as tumour-antigenic peptides (Ishii et al. 1999), viral epitopes (Greenstone et al. 1998) or corresponding epitopes from ovalbumin or β-galactosidase-transfected cells (Arnold et al. 1995). They are chaperoned by different HSP including gp96, hsp90 and hsp70. DC and macrophages use the HSP receptor, CD91 to take up HSP-peptide complexes (Binder et al. 2000; Basu et al. 2001). Interactions of HSP with CD91 induce expression of costimulatory molecules on the DC and stimulate both DC and macrophages to secrete cytokines such as TNF-α, GM-CSF and IL-12.

Both DC and macrophages are able to cross-present HSP–peptide complexes in the context of MHC class I molecules and induce antigen-specific CTL responses. Presentation of HSP–peptide complexes occurs exclusively in tissue but not in blood as a result of localized necrotic cell death. Extremely small quantities of peptides (nanograms or picograms) are sufficient to induce CTL responses. Peptides alone or chaperoned by non-HSP proteins such as albumin do not induce CTL responses (Anderson and Srivastava 2000).

Demonstration in vitro that a cell can cross-present antigen does not prove that cell is responsible for the special function of cross-presentation in vivo. There is the paucity of evidence concerning the identity of APC involved in the latter process. den Hann and colleagues identified, for the first time, the APC in lymphoid tissue involved in cross-presentation (den Haan et al. 2000) and showed that mouse $CD8^+$ DC but not $CD8^-$ DC, cross-present antigen in the spleen. The number of cross-presenting APC appears to be very low, and only 1% of these cross-present cell-associated antigen to $CD8^+$ T cells. A similarly low percentage of activated DC is found in human lymphoid tissue (Summers et al. 2000). Rat $CD4^-$ DC containing apoptotic cell remnants have been found to migrate to the T-cell areas of lymph nodes (Huang et al. 2000). These rat $CD4^-$ DC may be the rat equivalent of the mouse $CD8^+$ DC and may be involved in cross-presentation. Macrophage-like cells with cross-presenting function have been isolated from tumours (Ostrand-Rosenberg et al. 1999). The role of distinct types of DC and macrophages in cross-presentation in vivo needs further investigation.

5 The Differentiation of DC

The differentiation (maturation) of DC reduces the high rate of antigen uptake, increases the secretion of the cytokines and chemokines needed for the migration of DC and induces the expression of the antigen presenting and costimulatory molecules required to enhance antigen presentation and initiate an immune response. Oligonucleotide array and proteomics studies indicate the programmed expression of many genes during DC differentiation, most of which are likely to enable the differentiating and migrating DC to respond to microenvironmental regulatory signals.

Most studies show the changes in gene and protein expression that occur during differentiation/maturation of immature Mo-DC into mature Mo-DC (Dietz et al. 2000; Le Naour et al. 2001). This differentiation is accompanied by changes in the expression of several genes encoding proteins involved in cell adhesion and motility. The adhesion molecules, galectin 2, galectin 3, CD11a/LFA-1α, ninjurin 1, macmarcks, syndecan 2, CD44E and presenilin 1, are all downregulated. A truncated form of Cadherin-8 is downregulated. Differentiation of DC induces a switch from galectin 3 (high in immature DC) to galectin 9 (high in mature DC). Expression of secreted proteins involved in cell mortality, autotaxin-t and semaphoring E, macrophage capping protein and vimentin, are upregulated. The concomitant decrease in expression of integrins and other cell adhesion molecules plus the increase in expression of genes involved in cell motility almost certainly contributes to the enhanced migration properties of mature DC compared with immature DC (Barratt-Boyes et al. 2000).

The differentiation of DC is accompanied by marked changes in the expression of cytokines and chemokine expression as well as their receptors. Several genes encoding proinflammatory cytokines and their receptors, such as pointerleukin 1β, TNF-α, CD163, C5a anaphylatoxin receptor, IL-6R, and TNFR, are downregulated. The chemokines, which act as potent neutrophil chemoattractants and activators including CTAPIII, MIP2-α, MIP-2β, ENA78, PF4 and IL-8, are downregulated. Genes encoding anti-inflammatory proteins such as cyclophilin C and TSG-6, are upregulated. The differentiation of DC is also accompanied by the upregulation of osteopontin, a key cytokine involved in T-cell activation (Ashkar et al. 2000). Mac-2-binding protein, an adhesion molecule involved in natural killer (NK) and lymphokine-activated killer (LAK) cell activation and secretion of IL-2 (Ullrich et al. 1994), is upregulated. Upregulation of TGF-α is also observed during DC differentiation. Among leukocytes, only activated macrophages secrete TGF-α. TGF-α secreted by DC may participate in wound healing and repair (Schultz et al. 1991), tumourigenesis (DiGiovanni et al. 1994) and/or providing support for DC homing.

Differentiated DC express IL-7, IL-15 and their appropriate ligands, which stimulate T-cell expansion (Dietz et al. 2000). Such expression of cytokines and their cognate ligands may be analogous to the expression by differentiated DC of IL-12 together with IL-12R, another potent T-cell stimulus (Grohmann et al. 1998). Differentiation of DC is accompanied by increasing levels of CCR7, TARC and STCP-1, all of which are involved in chemotaxis and are needed to target cells into the lymph nodes. Another gene transcribed selectively during DC differentiation is indoleamine 2,3-deoxygenase (IDO). IDO degrades tryptophan required for T-cell proliferation and subsequently suppresses T-cell proliferation (Mellor and Munn 1999). DC that transcribe IDO can reduce local levels of available tryptophan by the action of IDO and protect themselves from the activated cytotoxic T cells they stimulate. The identification of another putative enzyme DCAL (Dekker et al. 2002) induced during DC differentiation indicates there is much more to be learnt about these processes.

Several HSP that participate in antigen processing and presentation, hsp73, hsp27, and calreticulin are also regulated during DC differentiation. Hsp73 binds specifically to the cell surface of monocytes and DC lines, is internalized spontaneously by receptor-mediated endocytosis (Arnold-Schild et al. 1999) and is upregulated during DC differentiation. The role of hsp27 upregulation during DC differentiation is less clear. Increased hsp27 promotes resistance of monocytes to apoptotic cell death (Samali and Cotter 1996). In contrast to hsp73 and hsp27, calreticulin is downregulated during DC differentiation due to post-translational modification. Calreticulin participates in the assembly of MHC class I with peptide and β2-microglobulin in the endoplasmic reticulum, a process required for the presentation of antigenic peptides to cytotoxic T lymphocytes at the cell surface (Krause and Michalak 1997). Proteomic analysis of DC identified a truncated form of calreticulin 32 present only in DC (Le Naour et al. 2001). This form contains the P-domain, a site of chaperone activity, the C-domain, which contains the endoplasmic reticulum retrieval sequence, but lacks the N-domain. The function of calreticulin 32 in mature DC is under investigation.

6 DC Regulate the Adaptive Immune Response

Mature DC provide a permissive environment for inducing immune responses. Both myeloid DC and plasmacytoid CD123^{hi} DC can induce T helper (Th)1 and Th2 immune responses, and despite initial suggestions, there is no stable DC phenotype or subset distinction, which polarizes distinct, Th1 or Th2 immune responses. The type and magnitude of Th immune responses is dependent on the differentiation/activation status of DC regulated by the type of differentiating stimulus, duration of DC activation and the DC–T cell ratio.

Following exposure to CD40L, LPS (Vieira et al. 2000) or dsRNA (Verdijk et al. 1999), mature DC produce IL-12 and consequently drive Th1 responses. Blood myeloid CD11c^{+} DC can generate higher numbers of Th1 effector cells than Mo-DC obtained from the same donors (Osugi et al. 2002). In contrast, PGE2 promotes differentiation of mature DC that produce low levels of IL-12 and drive Th2 immune responses (Kalinski et al. 1998). Type I IFN could promote mature DC with the ability to induce Th1 or Th2 responses, depending on the cytokine combined with type I IFN. Mature DC derived in culture with type I IFN and GM-CSF produce IL-15 and promote Th1 immune responses (Santini et al. 2000). In contrast, mature DC derived in culture with type I IFN, GM-CSF and IL-4, produce IL-10 and favour Th2 immune responses (Huang et al. 2001). Yssel's medium supplemented with LPS or IFN-γ promotes differentiation of mature DC that produce low levels of IL-12, increased levels of IL-10 and direct differentiation of Th cells towards the Th0/Th2 responses (Chang et al. 2000).

DC taken at early time points after induction of maturation induce Th1 responses. DC that are "exhausted" or "polarized" as a results of prolonged activation, lose the ability to produce IL-12 and induce Th2 responses (Langenkamp et al. 2000). The DC:T-cell ratio in these experiments influences outcome, e.g. a high DC:T-cell ratio promotes Th1 and a low DC:T-cell ratio promotes Th2 responses (Tanaka et al. 2000).

Human plasmacytoid CD123^{hi} DC could induce Th1 or Th2 immune responses depending on the type of differentiation factors. When cultured with IL-3, they preferentially promote Th2 immune responses, whereas activated with viruses they prime naïve T cells to produce IFN-γ and IL-10 (Cella et al. 2000; Kadowaki et al. 2000).

Effector CD4$^+$ T cells induced by mature DC are required for recruitment of other effectors such as macrophages and eosinophils and for the induction of CD8$^+$ T-cell mediated CTL responses to cross-presented exogenous antigens. In addition to T-cell activation, it is important to note that cross-presentation can also induce T-cell tolerance. In lymph nodes, in the absence of effector CD4$^+$ T cells, memory CD8$^+$ T cells divided and were subsequently deleted leading to tolerance. In contrast, in the presence of CD4$^+$ T cells, effector IFN-γ-producing CTL occurred (Albert et al. 2001). The contribution of CD4$^+$ T cells can be replaced by CD40 crosslinking and inflammatory cytokines. Macrophages are not capable of generating IFN-γ-producing CTL even in the presence of CD40 crosslinking (Albert et al. 1998) and T cells exposed in this way remain immunologically ignorant. Endogenous antigen-loading via a classical MHC class I pathway allows both macrophages and DC to trigger production of IFN-γ-producing CTL.

7
New Approaches in DC Immunotherapy

Because of their ability to cross-present exogenous antigen and induce and maintain efficient primary immune responses, DC are the main cellular vehicle for clinical trials of vaccine strategies aimed to initiate CTL responses to tumours and pathogens (www.mmri.mater.org.au). In this context, macrophages are not attractive candidates for use in immunotherapy because of their failure to induce primary immune responses.

Despite the variety of strategies that have induced tumour-specific immune responses, the optimal DC-based strategies for human trials still remain to be determined. The most commonly used, clinically approved, approach is based on loading empty MHC class I molecules on DC with exogenous peptides. However, this is limited by peptide restriction to a given HLA type, induction of CTL responses only and limited patient responses to defined tumour antigen. Indeed, many DC-based immunotherapy protocols in human cancer have shown limited efficacy (Nestle et al. 2001), challenging research for improved strategy. Which type of DC preparation to use and how to administer it remain major issues.

As an alternative, apoptotic or necrotic tumour cells can be provided to DC for cross-presentation of tumour antigen. The use of apoptotic or necrotic killed tumour cells as a source of tumour antigen should provide both MHC class I and class II epitopes, leading to diverse immune responses involving polyclonal CTL and helper $CD4^+$ T cells. Moreover, helper $CD4^+$ T cells are able to recruit other effectors such as macrophages and eosinophils. DC loaded with killed allogeneic melanoma cells are able to induce differentiation of naïve T cells into CTL that are specific for a broad spectrum of shared melanoma antigens (Berard et al. 2000). This demonstration of cross-priming against shared tumour antigens builds the basis for using allogeneic tumour cells lines to deliver tumour antigen to DC for vaccination protocols.

Emerging data suggest that HSP participate in antigen presentation and play a central role in the induction of primary immune responses by DC. Tumour-derived HSP are used to treat autologous tumours in patients with advanced tumours, renal cell carcinoma and metastasis melanoma (Anderson and Srivastava 2000). Results showed that the autologous gp96 vaccine was effective in the adjuvant setting and post-vaccination stabilization of disease and CTL-restricted responses against autologous tumours was demonstrated.

New approaches based on the use of apoptotic or necrotic allogeneic tumour cell lines or HSP prepared from autologous tumours to induce responses against tumour antigens may have the advantages of applicability to many patients regardless of HLA type, as well as the generation of tumour-specific $CD4^+$T responses, which may recruit other effectors such as macrophages and eosinophils. The possibility of administering these subcutaneously as adjuvant killed tumour cells or HSP may even avoid antigen loading or transfection of DC prepared in vitro. These approaches will need to be studied and contrasted with more defined methods of antigen loading DC, which are more attractive to the regulatory authorities.

In summary, we have much to learn about the relationship of DC subsets to macrophages and their potential cooperative interactions. The plethora of emerging molecular data give scientists new opportunities and that data will undoubtedly translate rapidly into clinical applications.

8 References

Albert, ML, Pearce, SF, Francisco, LM, Sauter, B, Roy, P, Silverstein, RL and Bhardwaj, N (1998) Immature dendritic cells phagocytose apoptotic cells bas alphavbeta5 and CD36 and cross-present antigens to cytotoxic T lymphocytes. J Exp med. 188:1359–1368

Albert, ML, Sauter, B and Bhardwaj, N (1998) Dendritic cells acquire antigen from apoptotic cells and induce class I-restricted CTLs. Nature. 392:86–89

Albert, ML, Jegathesan, M and Darnell, RB (2001) Dendritic cell maturation is required for the cross-tolerization of CD8+ T cells. Nat Immunol. 2:1010–1017.

Alexopoulou, L, Holt, AC, Medzhitov, R and Flavell, RA (2001) Recognition of double-stranded RNA and activation of NF-kappaB by Toll- like receptor 3. Nature. 413:732–738.

Allavena, P, Piemonti, L, Longoni, D, Bernasconi, S, Stoppacciaro, A, Ruco, L and Mantovani, A (1998) IL-10 prevents the differentiation of monocytes to dendritic cells but promotes their maturation to macrophages. Eur J Immunol. 28:359–369

Anderson, KM and Srivastava, PK (2000) Heat, heat shock, heat shock proteins and death: a central link in innate and adaptive immune responses. Immunol Lett. 74:35–39

Arnold, D, Faath, S, Rammensee, H and Schild, H (1995) Cross-priming of minor histocompatibility antigen-specific cytotoxic T cells upon immunization with the heat shock protein gp96. J Exp Med. 182:885–889

Arnold-Schild, D, Hanau, D, Spehner, D, Schmid, C, Rammensee, HG, de la Salle, H and Schild, H (1999) Cutting edge: receptor-mediated endocytosis of heat shock proteins by professional antigen-presenting cells. J Immunol. 162:3757–3760.

Ashkar, S, Weber, GF, Panoutsakopoulou, V, Sanchirico, ME, Jansson, M, Zawaideh, S, Rittling, SR, Denhardt, DT, Glimcher, MJ and Cantor, H (2000) Eta-1 (osteopontin): an early component of type-1 (cell-mediated) immunity. Science. 287:860–864

Barratt-Boyes, SM, Zimmer, MI, Harshyne, LA, Meyer, EM, Watkins, SC, Capuano, S, 3rd, Murphey-Corb, M, Falo, LD, Jr. and Donnenberg, AD (2000) Maturation and trafficking of monocyte-derived dendritic cells in monkeys: implications for dendritic cell-based vaccines. J Immunol. 164:2487–2495.

Basu, S, Binder, RJ, Ramalingam, T and Srivastava, PK (2001) CD91 is a common receptor for heat shock proteins gp96, hsp90, hsp70, and calreticulin. Immunity. 14:303–313.

Berard, F, Blanco, P, Davoust, J, Neidhart-Berard, EM, Nouri-Shirazi, M, Taquet, N, Rimoldi, D, Cerottini, JC, Banchereau, J and Palucka, AK (2000) Cross-Priming of Naive CD8 T Cells against Melanoma Antigens Using Dendritic Cells Loaded with Killed Allogeneic Melanoma Cells. J Exp Med. 192:1535–1544

Bevan, MJ (1976) Cross-priming for a secondary cytotoxic response to minor H antigens with H-2 congenic cells which do not cross-react in the cytotoxic assay. J Exp Med. 143:1283–1288

Binder, RJ, Han, DK and Srivastava, PK (2000) CD91: a receptor for heat shock protein gp96. Nat Immunol. 1:151–155

Bowie, A and O'Neill, LA (2000) The interleukin-1 receptor/Toll-like receptor superfamily: signal generators for pro-inflammatory interleukins and microbial products. J Leukoc Biol. 67:508–514

Cao, H, Verge, V, Baron, C, Martinache, C, Leon, A, Scholl, S, Gorin, NC, Salamero, J, Assari, S, Bernard, J and Lopez, M (2000) In vitro generation of dendritic cells from human blood monocytes in experimental conditions compatible for in vivo cell therapy. J Hematother Stem Cell Res. 9:183–194.

Caux, C, Vanbervliet, B, Massacrier, C, Dezutter-Dambuyant, C, de Saint-Vis, B, Jacquet, C, Yoneda, K, Imamura, S, Schmitt, D and Banchereau, J (1996) CD34+ hematopoietic progenitors from human cord blood differentiate along two independent dendritic cell pathways in response to GM-CSF+TNF alpha. J Exp Med. 184:695–706

Caux, C, Massacrier, C, Vanbervliet, B, Dubois, B, Durand, I, Cella, M, Lanzavecchia, A and Banchereau, J (1997) CD34+ hematopoietic progenitors from human cord blood differentiate along two independent dendritic cell pathways in response to granulocyte-macrophage colony-stimulating factor plus tumor necrosis factor alpha: II. Functional analysis. Blood. 90:1458–1470

Cella, M, Facchetti, F, Lanzavecchia, A and Colonna, M (2000) Plasmacytoid dendritic cells activated by influenza virus and CD40L drive a potent TH1 polarization. Nat Immunol. 1:305–310.

Chang, CC, Wright, A and Punnonen, J (2000) Monocyte-derived CD1a+ and CD1a- dendritic cell subsets differ in their cytokine production profiles, susceptibilities to transfection, and capacities to direct Th cell differentiation. J Immunol. 165:3584–3591

Clark, SC and Kamen, R (1987) The human hematopoietic colony-stimulating factors. Science. 236:1229–1237

Dalloul, AH, Patry, C, Salamero, J, Canque, B, Grassi, F and Schmitt, C (1999) Functional and phenotypic analysis of thymic CD34+CD1a- progenitor- derived dendritic cells: predominance of CD1a+ differentiation pathway. J Immunol. 162:5821–5828

Dekker, JW, Budhia, S, Angel, NZ, Cooper, BJ, Clark, GJ, Hart, DN and Kato, M (2002) Identification of an S-adenosylhomocysteine hydrolase-like transcript induced during dendritic cell differentiation. Immunogenetics. 53:993–1001.

den Haan, JM, Lehar, SM and Bevan, MJ (2000) CD8(+) but not CD8(-) dendritic cells cross-prime cytotoxic T cells in vivo. J Exp Med. 192:1685–1696.

Dietz, AB, Bulur, PA, Knutson, GJ, Matasic, R and Vuk-Pavlovic, S (2000) Maturation of human monocyte-derived dendritic cells studied by microarray hybridization. Biochem Biophys Res Commun. 275:731–738

DiGiovanni, J, Rho, O, Xian, W and Beltran, L (1994) Role of the epidermal growth factor receptor and transforming growth factor alpha in mouse skin carcinogenesis. Prog Clin Biol Res. 387:113–138

Dzionek, A, Sohma, Y, Nagafune, J, Cella, M, Colonna, M, Facchetti, F, Gunther, G, Johnston, I, Lanzavecchia, A, Nagasaka, T, Okada, T, Vermi, W, Winkels, G, Yamamoto, T, Zysk, M, Yamaguchi, Y and Schmitz, J (2001) BDCA-2, a novel plasmacytoid dendritic cell-specific type II C-type lectin, mediates antigen capture and is a potent inhibitor of interferon alpha/beta induction. J Exp Med. 194:1823–1834.

Fadok, VA, Bratton, DL, Konowal, A, Freed, PW, Westcott, JY and Henson, PM (1998) Macrophages that have ingested apoptotic cells in vitro inhibit proinflammatory cytokine production through autocrine/paracrine mechanisms involving TGF-beta, PGE2, and PAF. J Clin Invest. 101:890–898

Fadok, VA, Bratton, DL, Rose, DM, Pearson, A, Ezekewitz, RA and Henson, PM (2000) A receptor for phosphatidylserine-specific clearance of apoptotic cells. Nature. 405:85–90

Geijtenbeek, TB, Kwon, DS, Torensma, R, van Vliet, SJ, van Duijnhoven, GC, Middel, J, Cornelissen, IL, Nottet, HS, KewalRamani, VN, Littman, DR, Figdor, CG and van Kooyk, Y (2000) DC-SIGN, a dendritic cell-specific HIV-1-binding protein that enhances trans-infection of T cells. Cell. 100:587–597

Geissmann, F, Prost, C, Monnet, JP, Dy, M, Brousse, N and Hermine, O (1998) Transforming growth factor beta1, in the presence of granulocyte/macrophage colony-stimulating factor and interleukin 4, induces differentiation of human peripheral blood monocytes into dendritic Langerhans cells. J Exp Med. 187:961–966

Greenstone, HL, Nieland, JD, de Visser, KE, De Bruijn, ML, Kirnbauer, R, Roden, RB, Lowy, DR, Kast, WM and Schiller, JT (1998) Chimeric papillomavirus virus-like particles elicit antitumor immunity against the E7 oncoprotein in an HPV16 tumor model. Proc Natl Acad Sci U S A. 95:1800–1805

Grohmann, U, Belladonna, ML, Bianchi, R, Orabona, C, Ayroldi, E, Fioretti, MC and Puccetti, P (1998) IL-12 acts directly on DC to promote nuclear localization of NF-kappaB and primes DC for IL-12 production. Immunity. 9:315–323

Hart DNJ, Clark, GJ, A., MKP, Kato, M, Vuckovic, S, Lopez, JA and Wykes, M (2002) 7th leucocyte differentiation antigen workshop DC section summary. In: Mason. D (eds) Leucocyte Typing VII. Oxford University Press, Oxford University Press, pp 283–293

Hart, DNJ, MacDonald, K, Vuckovic, S and Clark, GJ (2001) Phenotypic characterization of dendritic cells. In: M. J. Lotze and A. W. Thompson (eds) Dendritic cells: Biology and Clinical Application. Academic Press, Academic Press, pp 97–117

Hartmann, G, Weiner, GJ and Krieg, AM (1999) CpG DNA: a potent signal for growth, activation, and maturation of human dendritic cells. Proc Natl Acad Sci USA. 96:9305–9310

Hemmi, H, Takeuchi, O, Kawai, T, Kaisho, T, Sato, S, Sanjo, H, Matsumoto, M, Hoshino, K, Wagner, H, Takeda, K and Akira, S (2000) A Toll-like receptor recognizes bacterial DNA.[In Process Citation]. Nature. 408:740–745

Ho, CS, Munster, D, Pyke, CM, Hart, DN and Lopez, JA (2002) Spontaneous generation and survival of blood dendritic cells in mononuclear cell culture without exogenous cytokines. Blood. 99:2897–2904.

Hock, BD, Fearnley, DB, Boyce, A, McLellan, AD, Sorg, RV, Summers, KL and Hart, DN (1999) Human dendritic cells express a 95 kDa activation/differentiation antigen defined by CMRF-56. Tissue Antigens. 53:320–334.

Huang, FP, Platt, N, Wykes, M, Major, JR, Powell, TJ, Jenkins, CD and MacPherson, GG (2000) A discrete subpopulation of dendritic cells transports apoptotic intestinal epithelial cells to T cell areas of mesenteric lymph nodes. J Exp Med. 191:435–444

Huang, YM, Hussien, Y, Yarilin, D, Xiao, BG, Liu, YJ and Link, H (2001) Interferon-beta induces the development of type 2 dendritic cells. Cytokine. 13:264–271

Ishii, T, Udono, H, Yamano, T, Ohta, H, Uenaka, A, Ono, T, Hizuta, A, Tanaka, N, Srivastava, PK and Nakayama, E (1999) Isolation of MHC class I-restricted tumor antigen peptide and its precursors associated with heat shock proteins hsp70, hsp90, and gp96. J Immunol. 162:1303–1309

Ito, T, Inaba, M, Inaba, K, Toki, J, Sogo, S, Iguchi, T, Adachi, Y, Yamaguchi, K, Amakawa, R, Valladeau, J, Saeland, S, Fukuhara, S and Ikehara, S (1999) A CD1a+/CD11c+ subset of human blood dendritic cells is a direct precursor of Langerhans cells. J Immunol. 163:1409–1419

Jaksits S, Kriehuber, E, Charbonnier, AS, Rappersberger, K, Stingl, G and Maurer, D (1999) CD34+ cell-derived CD14+ precursor cells develop into Langerhans cells in a TGF-beta 1-dependent manner. J Immunol. 163:4869–4877

Jeannin, P, Renno, T, Goetsch, L, Miconnet, I, Aubry, JP, Delneste, Y, Herbault, N, Baussant, T, Magistrelli, G, Soulas, C, Romero, P, Cerottini, JC and Bonnefoy, JY (2000) OmpA targets dendritic cells, induces their maturation and delivers antigen into the MHC class I presentation pathway. Nat Immunol. 1:502–509.

Kadowaki, N, Antonenko, S, Lau, JY and Liu, YJ (2000) Natural interferon alpha/beta-producing cells link innate and adaptive immunity. J Exp Med. 192:219–226

Kadowaki, N, Ho, S, Antonenko, S, Malefyt, RW, Kastelein, RA, Bazan, F and Liu, YJ (2001) Subsets of human dendritic cell precursors express different toll-like receptors and respond to different microbial antigens. J Exp Med. 194:863–869.

Kalinski, P, Schuitemaker, JH, Hilkens, CM and Kapsenberg, ML (1998) Prostaglandin E2 induces the final maturation of IL-12-deficient CD1a+CD83+ dendritic cells: the levels of IL-12 are determined during the final dendritic cell maturation and are resistant to further modulation. J Immunol. 161:2804–2809

Kato, M, Neil, TK, Fearnley, DB, McLellan, AD, Vuckovic, S and Hart, DN (2000) Expression of multilectin receptors and comparative FITC-dextran uptake by human dendritic cells. Int Immunol. 12:1511–1519

Kovacsovics-Bankowski, M, Clark, K, Benacerraf, B and Rock, KL (1993) Efficient major histocompatibility complex class I presentation of exogenous antigen upon phagocytosis by macrophages. Proc Natl Acad Sci U S A. 90:4942–4946

Kurts, C, Heath, WR, Carbone, FR, Allison, J, Miller, JF and Kosaka, H (1996) Constitutive class I-restricted exogenous presentation of self antigens in vivo. J Exp Med. 184:923–930

Langenkamp, A, Messi, M, Lanzavecchia, A and Sallusto, F (2000) Kinetics of dendritic cell activation: impact on priming of TH1, TH2 and nonpolarized T cells. Nat Immunol. 1:311–316

Le Naour, F, Hohenkirk, L, Grolleau, A, Misek, DE, Lescure, P, Geiger, JD, Hanash, S and Beretta, L (2001) Profiling changes in gene expression during differentiation and maturation of monocyte-derived dendritic cells using both oligonucleotide microarrays and proteomics. J Biol Chem. 276:17920–17931.

Lenz, LL, Butz, EA and Bevan, MJ (2000) Requirements for bone marrow-derived antigen-presenting cells in priming cytotoxic T cell responses to intracellular pathogens. J Exp Med. 192:1135–1142

McKenzie, JL, Prickett, TCR and Hart, DNJ (1989) Human dendritic cells stimulate allogeneic T cells in the absence of interleukin 1. Immunology. 67:290-297
Mellor, AL and Munn, DH (1999) Tryptophan catabolism and T-cell tolerance: immunosuppression by starvation? Immunol Today. 20:469–473
Metcalf, D (1989) The molecular control of cell division, differentiation commitment and maturation in haemopoietic cells. Nature. 339:27–30
Mitani, H, Katayama, N, Araki, H, Ohishi, K, Kobayashi, K, Suzuki, H, Nishii, K, Masuya, M, Yasukawa, K, Minami, N and Shiku, H (2000) Activity of interleukin 6 in the differentiation of monocytes to macrophages and dendritic cells. Br J Haematol. 109:288–295
Mitchell, DA, Nair, SK and Gilboa, E (1998) Dendritic cell/macrophage precursors capture exogenous antigen for MHC class I presentation by dendritic cells. Eur J Immunol. 28:1923–1933
Nestle, FO, Banchereau, J and Hart, D (2001) Dendritic cells: On the move from bench to bedside. Nat Med. 7:761–765
Norbury, CC, Hewlett, LJ, Prescott, AR, Shastri, N and Watts, C (1995) Class I MHC presentation of exogenous soluble antigen via macropinocytosis in bone marrow macrophages. Immunity. 3:783–791
Ohishi, K, Varnum-Finney, B, Serda, RE, Anasetti, C and Bernstein, ID (2001) The Notch ligand, Delta-1, inhibits the differentiation of monocytes into macrophages but permits their differentiation into dendritic cells. Blood. 98:1402–1407
Ostrand-Rosenberg, S, Pulaski, BA, Clements, VK, Qi, L, Pipeling, MR and Hanyok, LA (1999) Cell-based vaccines for the stimulation of immunity to metastatic cancers. Immunol Rev. 170:101–114
Osugi, Y, Vuckovic, S and Hart, DN (2002) Myeloid blood CD11c(+) dendritic cells and monocyte-derived dendritic cells differ in their ability to stimulate T lymphocytes. Blood. 100:2858–2866
Palucka, KA, Taquet, N, Sanchez-Chapuis, F and Gluckman, JC (1998) Dendritic cells as the terminal stage of monocyte differentiation. J Immunol. 160:4587–4595.
Piemonti, L, Bernasconi, S, Luini, W, Trobonjaca, Z, Minty, A, Allavena, P and Mantovani, A (1995) IL-13 supports differentiation of dendritic cells from circulating precursors in concert with GM-CSF. Eur.Cytokine.Netw. 6:245–252
Poltorak, A, He, X, Smirnova, I, Liu, MY, Huffel, CV, Du, X, Birdwell, D, Alejos, E, Silva, M, Galanos, C, Freudenberg, M, Ricciardi-Castagnoli, P, Layton, B and Beutler, B (1998) Defective LPS signaling in C3H/HeJ and C57BL/10ScCr mice: mutations in Tlr4 gene. Science. 282:2085–2088
Randolph, GJ, Beaulieu, S, Lebecque, S, Steinman, RM and Muller, WA (1998) Differentiation of monocytes into dendritic cells in a model of transendothelial trafficking. Science. 282:479–482
Randolph, GJ, Inaba, K, Robbiani, DF, Steinman, RM and Muller, WA (1999) Differentiation of phagocytic monocytes into lymph node dendritic cells in vivo. Immunity. 11:753–761
Randolph, GJ, Sanchez-Schmitz, G, Liebman, RM and Schakel, K (2002) The CD16(+) (FcgammaRIII(+)) subset of human monocytes preferentially becomes migratory dendritic cells in a model tissue setting. J Exp Med. 196:517–527
Robinson, SP, Patterson, S, English, N, Davies, D, Knight, SC and Reid, CD (1999) Human peripheral blood contains two distinct lineages of dendritic cells. Eur J Immunol. 29:2769–2778
Rock, KL, Rothstein, L, Gamble, S and Fleischacker, C (1993) Characterization of antigen-presenting cells that present exogenous antigens in association with class I MHC molecules. J Immunol. 150:438–446
Rock, FL, Hardiman, G, Timans, JC, Kastelein, RA and Bazan, JF (1998) A family of human receptors structurally related to Drosophila Toll. Proc Natl Acad Sci. 95:588–593

Rodriguez, A, Regnault, A, Kleijmeer, M, Ricciardi-Castagnoli, P and Amigorena, S (1999) Selective transport of internalized antigens to the cytosol for MHC class I presentation in dendritic cells. Nat Cell Biol. 1:362–368

Sallusto, F and Lanzavecchia, A (1994) Efficient presentation of soluble antigen by cultured human dendritic cells is maintained by granulocyte/macrophage colony- stimulating factor plus interleukin 4 and downregulated by tumour necrosis factor-α. J Exp Med. 179:1109–1118

Samali, A and Cotter, TG (1996) Heat shock proteins increase resistance to apoptosis. Exp Cell Res. 223:163–170

Santini, SM, Lapenta, C, Logozzi, M, Parlato, S, Spada, M, Di Pucchio, T and Belardelli, F (2000) Type I interferon as a powerful adjuvant for monocyte-derived dendritic cell development and activity in vitro and in Hu-PBL-SCID mice. J Exp Med. 191:1777–1788.

Savill, J, Hogg, N, Ren, Y and Haslett, C (1992) Thrombospondin cooperates with CD36 and the vitronectin receptor in macrophage recognition of neutrophils undergoing apoptosis. J Clin Invest. 90:1513–1522

Schultz, G, Rotatori, DS and Clark, W (1991) EGF and TGF-alpha in wound healing and repair. J Cell Biochem. 45:346–352

Shen, Z, Reznikoff, G, Dranoff, G and Rock, KL (1997) Cloned dendritic cells can present exogenous antigens on both MHC class I and class II molecules. J Immunol. 158:2723–2730

Sigal, LJ, Crotty, S, Andino, R and Rock, KL (1999) Cytotoxic T-cell immunity to virus-infected non-haematopoietic cells requires presentation of exogenous antigen. Nature. 398:77–80

Summers, KL, Hock, BD, McKenzie, JL and Hart, DNJ (2000) Further phenotypic characterization of DC subsets in human tonsils. In preparation.

Takeuchi, O, Hoshino, K, Kawai, T, Sanjo, H, Takada, H, Ogawa, T, Takeda, K and Akira, S (1999) Differential roles of TLR2 and TLR4 in recognition of gram-negative and gram-positive bacterial cell wall components. Immunity. 11:443–451

Tanaka, H, Demeure, CE, Rubio, M, Delespesse, G and Sarfati, M (2000) Human monocyte-derived dendritic cells induce naive T cell differentiation into T helper cell type 2 (Th2) or Th1/Th2 effectors. Role of stimulator/responder ratio. J Exp Med. 192:405–412.

Ullrich, A, Sures, I, D'Egidio, M, Jallal, B, Powell, TJ, Herbst, R, Dreps, A, Azam, M, Rubinstein, M, Natoli, C and et al. (1994) The secreted tumor-associated antigen 90 K is a potent immune stimulator. J Biol Chem. 269:18401–18407

Verdijk, RM, Mutis, T, Esendam, B, Kamp, J, Melief, CJ, Brand, A and Goulmy, E (1999) Polyriboinosinic polyribocytidylic acid (poly(I:C)) induces stable maturation of functionally active human dendritic cells. J Immunol. 163:57–61

Vieira, PL, de Jong, EC, Wierenga, EA, Kapsenberg, ML and Kalinski, P (2000) Development of Th1-inducing capacity in myeloid dendritic cells requires environmental instruction. J Immunol. 164:4507–4512

Visintin, A, Mazzoni, A, Spitzer, JH, Wyllie, DH, Dower, SK and Segal, DM (2001) Regulation of Toll-like receptors in human monocytes and dendritic cells. J Immunol. 166:249–255

Voll, RE, Herrmann, M, Roth, EA, Stach, C, Kalden, JR and Girkontaite, I (1997) Immunosuppressive effects of apoptotic cells. Nature. 390:350–351

Vuckovic, S, Fearnley, DB, Mannering, SI, Dekker, J, Whyte, LF and Hart, DN (1998) Generation of CMRF-44+ monocyte-derived dendritic cells: insights into phenotype and function. Exp Hematol. 26:1255–1264

Yewdell, JW, Norbury, CC and Bennink, JR (1999) Mechanisms of exogenous antigen presentation by MHC class I molecules in vitro and in vivo: implications for generating CD8+ T cell responses to infectious agents, tumors, transplants, and vaccines. Adv Immunol. 73:1–77

Zhou, LJ and Tedder, TF (1996) CD14+ blood monocytes can differentiate into functionally mature CD83+ dendritic cells. Proc Natl Acad Sci USA. 93:2588–2592

The Osteoclast

T. J. Chambers

Department of Cellular Pathology,
St. George's Hospital Medical School,
Cranmer Terrace, London, SW17 0RE, UK
e-mail: t.chambers@sghms.ac.uk

Abstract The osteoclast is the cell that resorbs bone. It has been known for many years that it is formed from cells of the mononuclear phagocyte system, and that its formation and function are governed by osteoblastic cells. Recently, the molecular basis for this regulation was identified: osteoblastic cells induce osteoclastic differentiation in immature mononuclear phagocytes through expression of macrophage colony-stimulating factor (M-CSF) and receptor-activator of NFκB ligand (RANKL). Osteoblastic regulation of bone resorption is assisted through secretion of an inhibitor, osteoprotegerin (OPG), a soluble (decoy) receptor for RANKL. Transforming growth factor beta (TGF-β), which is present in bone matrix in large amounts, is also essential for osteoclast formation, at least in vitro. Surprisingly, TNF-α can substitute for and is strongly synergistic with RANKL for osteoclast-induction. TNF-α is widely expressed, and RANKL can also be present in situations that are not associated with osteoclast formation, so that the presence of large quantities of TGF-β in bone matrix might explain why osteoclast formation is essentially confined to the bone microenvironment. In this review, recent data concerning the mechanisms underlying the induction of osteoclastic differentiation and function are described, together with recent findings concerning the mechanisms through which osteoclasts adhere to and resorb bone. Several of these mechanisms are currently being exploited for the development of novel therapies for diseases, such as osteoporosis, that are caused by excessive bone resorption.

Keywords Osteoclast, Osteoblast, RANKL, TNF-?, TGF-?, Bone

Abbreviations

$\alpha_v\beta3$	$\beta3$integrin alpha v beta 3
CTR	calcitonin receptor
M-CSF	macrophage colony-stimulating factor
MMP	matrix metalloproteinase
OPG	osteoprotegerin
RANK	receptor activator for NF-κB
RANKL	ligand for RANK
TRAF	TNF receptor-associated factor
TRAP	tartrate-resistant acid phosphatase
TRANCE	tumour necrosis factor-related activation-induced cytokine

1 Introduction

Bone resorption is crucial to the normal development and maintenance of the skeleton, and for the regulation of plasma calcium levels. In development, deficient resorption leads to osteopetrosis and failure of tooth eruption. In the adult, the continuous physiological remodelling of bone, whereby aged or fatigued bone is removed and replaced by new, is dependent on bone resorption. Excessive bone resorption is the key pathophysiological event underlying several diseases, including malignant hypercalcaemia and postmenopausal osteoporosis, in humans.

It has long been the consensus that osteoclasts are responsible for bone resorption. Their origin, though, has until recently been hotly contested. Their multinuclear and debriding characteristics suggested an origin from macrophages, but kinetic and histodynamic studies argued that they are locally derived, and there was even a popular idea that they could interconvert with osteoblasts, the cells that form bone. There have been striking advances in the last few years that have established a very close relationship between these cells and macrophages, and have illuminated many of the mechanisms involved in their formation, regulation and function.

2 Mechanisms of Bone Resorption

The defining characteristic of the osteoclast is its ability to resorb bone. When osteoclasts are placed on a bone surface in vitro, they make deep excavations with extraordinary speed. Macrophages can digest ingested bone particles but only the osteoclast can dissolve bone by an extracellular mechanism; and it achieves this unaided by other cell types (Chambers et al. 1984a). Actively resorbing osteoclasts establish a circle of close contact with the bone surface, known as the sealing zone, below a peripheral ring of cytoplasm from which organelles are excluded (the clear zone of electron microscopy). This clear zone corresponds to a ring of F-actin, which correlates well with functional activation. The bone-apposed surface of the osteoclast circumscribed within this sealing zone is thrown into deep folds to form the 'ruffled border' seen by electron and light microscopy. In this area protons and acid hydrolases are extruded onto the bone surface. The mineral (calcium hydroxyapatite) and organic (predominantly collagen) components of bone are thereby dissolved. Solubilisation of organic components might continue during transocytic vesicular transport of the released material to the basal surface of the osteoclast, which is usually in contact with a blood vessel (Halleen et al. 1999).

Osteoclasts were originally identified as multinuclear cells, but it is not known why they are multinuclear. In vitro and in vivo, they can remain mononuclear, especially when the density is low, yet resorb bone (Kaye 1984; Fuller and Chambers 1989; Prallet et al. 1992), so that multinuclearity is not essential for their function. Multinuclearity may help in regulation, or improve the efficiency with which they can resorb: the effort expended in maintaining an extracellular resorptive micro-environment will be inversely related to the area of the interface with bone.

Osteoclasts have been shown to secrete protons into the subosteoclastic attachment zone (Baron et al. 1985; Silver et al. 1988). This occurs by targeted secretion of hydrochloric acid, by an H^+-ATPase proton pump, through the ruffled border (Vaananen et al. 1990). Concomitant with the development of the ruffled border, the number of intracellular acid compartments promptly decreases as vesicles containing proton pumps are transported to the surface that becomes the ruffled border. This sequence does not occur in osteopetrotic oc/oc mice, which lack an osteoclast-specific component of the vacuolar proton pump (Nakamura et al. 1997; Brady et al. 1999; Li et al. 1999; Nakamura et al. 1999; Scimeca et al. 2000). Mutation of the same subunit is one of the causes of osteopetrosis in humans (Frattini et al. 2000). The recent finding that vacuolar H^+-ATPase at the ruffled border contains osteoclast-specific subunits has further encouraged development of resorption-inhibitors that inhibit the osteoclastic proton pump (Hernando et al. 1995; Van Hille et al. 1995; Li et al. 1996). Recent data suggest that protons are not laterally contained by the so-called sealing zone, but are neutralised by mineral before loss by lateral diffusion occurs (Stenbeck and Horton 2000).

Protons for the proton pump are produced by cytoplasmic carbonic anhydrase II (CAII). High levels are present in osteoclasts (Gay and Mueller 1974). Inhibition of the enzyme suppresses bone resorption by isolated osteoclasts in vitro (Hall et al. 1991) and inherited deficiency leads to osteopetrosis (Sly et al. 1983). Excess cytoplasmic bicarbonate is removed via the chloride–bicarbonate exchanger located in the basolateral membrane (Hall and Chambers 1989). Correspondingly, there are large numbers of chloride channels in the ruffled border, which allow a flow of chloride anions into the resorption lacuna to maintain electroneutrality (Schlesinger et al. 1997). Loss of a component of this channel (ClC9) leads to osteopetrosis in mice and man (Kornak et al. 2001).

After dissolution of the mineral phase, osteoclastic enzymes destroy the organic bone matrix. This process is sensitive to inhibition by leupeptin and other inhibitors of cysteine proteinases (Delaisse et al. 1987; Fuller and Chambers 1995). The overwhelmingly predominant cysteine proteinase in osteoclasts is cathepsin K, while cathepsins B, L and S are rare (Tezuka et al. 1994; Drake et al. 1996; Kamiya et al. 1998; Ishibashi et al. 2001). The enzyme differs from other cathepsins in its ability to act at a relatively high pH (pH 6), which might facilitate its extracellular activity, and in its ability to cleave native collagen in the triple-helical region (Garnero et al. 1998). This collagenase activity, otherwise seen only in the collagenases of the matrix metalloproteinase family and neutrophil elastase, depends upon the presence of chondroitin 4-sulphate, a component of bone (Li et al. 2000b). Cathepsin K has been immunolocalised to resorption pits (Xia et al. 1999). Mutation of the gene for cathepsin K in humans and mice results in osteopetrosis (Gelb et al. 1996; Saftig et al. 1998). This suggests that cathepsin K is the dominant proteinase responsible for the digestion of bone, with little compensation by other cathepsins.

Ostoclasts also express high levels of MMP-9 (gelatinase B) (Reponen et al. 1994). However, deletion of the gene for MMP-9 causes only a transient disturbance of bone resorption (Vu et al. 1998). Since inhibition of MMPs has no discernible effect on bone resorption by isolated osteoclasts (Fuller and Chambers 1995), the enzyme seems likely to be involved in some other function in the osteoclast, such as migration, or mobilisation of matrix-associated growth factors (Vu et al. 1998).

3
Osteoclast Differentiation and Activation

It was established in the 1970s, by parabiosis, tissue grafting, and bone marrow transplants, that osteoclasts were of haematogenous rather that local origin (see Marks 1983 for review). The mononuclear phagocyte series seemed the most likely candidate as precursors because, like the osteoclast, these cells are specialised for debridement and can fuse to form multinuclear cells in the presence of extracellular foreign bodies. On the basis of this origin, two predictions were made concerning the regulation of bone resorption (Chambers 1980). First, an origin for osteoclasts from inherently wandering cells suggests that their localisation is governed by local bone cells such as osteoblasts and osteocytes; and second, that this control needs to include some form of protection for bone from phagocytic attack as a foreign body, since implantation of unprotected bone provokes this reaction.

It became apparent, though, that while osteoclasts share some immunological markers with macrophages (see Athanasou 1996), they are also distinctive (see Chambers 1989; Helfrich and Horton 1993). Osteoclasts lack many markers characteristic of macrophages (e.g. Fc, C3 receptors) and express very high levels of tartrate-resistant acid phosphatase (TRAP) and 'vitronectin receptors' (integrin $\alpha_v\beta3$); and express calcitonin receptors (CTR), which are absent from macrophages. Most distinctively, osteoclasts ex vivo excavate bone within hours, while macrophages show no excavation whatsoever, even on extended incubation on bone surfaces (Chambers and Horton 1984; Chambers et al. 1984a).

When osteoclasts were extracted and tested, it was found that they were indeed governed by osteoblastic cells: agents known to stimulate osteoclasts in intact bone stimulated isolated osteoclasts only in the presence of osteoblastic cells (Chambers 1982; Chambers 1985; Mcsheehy and Chambers 1986; Thomson et al. 1986; Thomson et al. 1987). Similarly, osteoclast differentiation was found to depend on factors from osteoblastic/bone marrow stromal cells (Takahashi et al. 1988; Hattersley and Chambers 1989), the expression of which is increased by resorptive hormones (Fuller and Chambers 1998; Liu et al. 1998; Matsuzaki et al. 1998; Takeda et al. 1999). This led to the view that the induction of bone resorption depended upon a primary interaction of resorptive stimuli with osteoblastic cells, which responded by expressing factor(s) that induced the differentiation and activation of osteoclasts (Chambers 1992; Suda et al. 1995).

One of the factors produced by osteoblastic cells that supports osteoclast formation is M-CSF. The evidence for this came from the discovery that osteopetrosis in the op/op mouse was caused by a stop codon in the gene for M-CSF (Wiktor-Jedrzejczak et al. 1990; Yoshida et al. 1990). M-CSF is nevertheless not sufficient for osteoclast formation in vitro: an additional, osteoblast-derived factor is also required, or macrophages form by default (Hattersley et al. 1991; Takahashi et al. 1991). Nor is M-CSF essential in vivo: osteoclasts are present in op/op mice, but in reduced number, and the osteopetrosis resolves after a few weeks, perhaps as the resorptive burden decreases. Resolution is accelerated by transgenic over-expression of Bcl-2 in mononuclear phagocytes (Lagasse and Weissman 1997). The main role of M-CSF in osteoclast biology appears to be to enhance the survival and proliferation of precursors, and the survival of mature cells. More recently, it has also been found to induce the expression of RANK, the receptor for the osteoclast-inductive ligand RANKL (see below) (Arai et al. 1999). Flt3 ligand (FL) also induces RANK, and might account for the partial redundancy of M-CSF in osteoclast formation (Lean et al. 2001).

A major consequence of the identification of the role of M-CSF in osteoclast biology was the conclusion that, despite its distinct phenotype, the osteoclast derives from the mononuclear phagocyte system. This was reinforced when mice deleted for *c-fos* were found to be osteopetrotic, with absence of osteoclasts. Without c-Fos, precursors form macrophages, either by default or through arrested development, despite an osteoclast-inductive environment (Grigoriadis et al. 1994).

The osteoblast-derived ligand responsible for osteoclast differentiation and activation was independently discovered by Snow Brand Milk Products and Amgen. Both initially found a soluble inhibitor of osteoclast formation [osteoprotegerin (OPG), a soluble receptor of the tumour necrosis factor (TNF) superfamily] (Simonet et al. 1997; Tsuda et al. 1997). They used this to identify the cognate

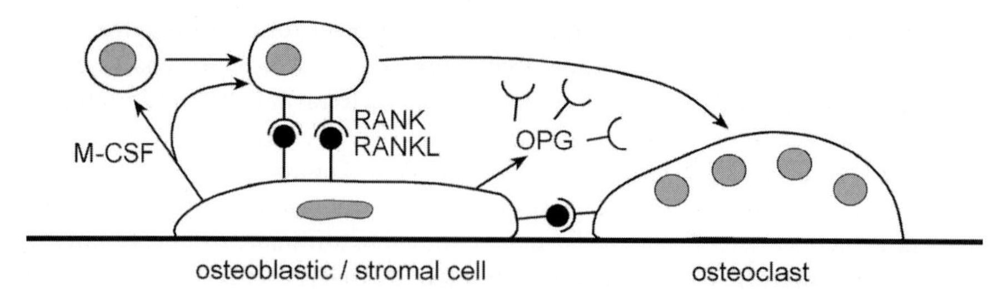

Fig. 1 Induction and regulation of osteoclasts by osteoblastic/stromal cells. Resorptive hormones such as parathyroid hormone, 1,25 dihydroxyvitamin D_3, interleukin-1 and TNF-α induce bone marrow stromal cells and osteoblastic cells to express M-CSF and RANKL. Haemopoietic cells are induced to express RANK by M-CSF or FL, and this interacts with RANKL on the surface of stromal cells to induce osteoclast formation and activity. The stromal cells also produce OPG, the soluble decoy receptor for RANKL. Expression of OPG is regulated in a manner reciprocal to RANKL

ligand, which was identical to the recently discovered TNF superfamily member TRANCE/RANKL (see Suda et al. 1999; Chambers 2000 for reviews).

RANKL is a type I transmembrane protein that (with M-CSF) replaces the need for osteoblastic/stromal cells in the induction and activation of osteoclasts from haemopoietic precursors in vitro (see Fig. 1). Expression of RANKL by osteoblastic cells is upregulated by agents that stimulate bone resorption. Mice in which the gene is deleted have osteopetrosis, caused by complete absence of osteoclasts. RANK is the receptor on osteoclasts and their precursors that interacts with RANKL. Soluble RANK and anti-RANK antibodies suppress osteoclast formation and activity, and deletion of the gene results in complete absence of osteoclasts (see Chambers 2000; Suda et al. 1999 for reviews).

RANK binds TNF receptor-associated factor (TRAF) 1,2,3,5 and 6 (Galibert et al. 1998; Kim et al. 1999). Animals double-mutant for p50/p52 nuclear factor (NF)-κB (Iotsova et al. 1997) or deficient in c-Fos, part of the activator protein (AP)-1 transcription factor complex, have osteopetrosis (Johnson et al. 1992; Wang et al. 1992). TRAF 2,5 and 6 have been shown to activate NF-κB downstream of TNF receptors (see Kim et al. 1999). The same TRAFs activate c-jun N-terminal kinase (JNK), which activates AP-1 (Kim et al. 1999). RANK activates not only NF-κB and JNK (Anderson et al. 1997; Wong et al. 1997; Darnay et al. 1998), but also expression of *c-fos* in osteoclasts (Matsuo et al. 2000). Deletion of the gene for TRAF6 causes osteopetrosis with normal numbers of non-resorptive osteoclasts (Lomaga et al. 1999), or absent osteoclasts (Naito et al. 1999). Many agents that do not induce osteoclast differentiation activate these signals, so RANKL presumably also induces unknown, osteoclast-specific signals.

Although precursors, whether from bone marrow, spleen, blood or peritoneum undergo osteoclastic differentiation when incubated in M-CSF and RANKL, macrophages also form. With continued incubation, only macrophages persist. In semi-solid media, osteoclasts always differentiate in colonies mixed with macrophages. This is unlikely to be the case on bone surfaces. It might be that culture conditions, which are essentially pro-inflammatory (Thyberg 1996; Iyer et al. 1999) (see below) divert some of the precursors to macrophages; or the precursors that provide osteoclasts differ from those used in the above experiments; or there may be additional signals in vivo that ensure osteoclasts form. The proportion of cells that develops into osteoclasts is much greater if precursors are incubated on bone slices, rather than on plastic substrates (Fuller et al. 2000).

Transforming growth factor (TGF)-β is an essential cofactor for osteoclast formation, at least in vitro. It substantially increases the proportion of precursors that become osteoclasts (Sells Galvin et al. 1999; Fuller et al. 2000), and blockade of TGF-β signalling abolishes osteoclast formation, while increasing macrophage numbers (Fuller et al. 2000). Basal osteoclast formation by RANKL in vitro is likely to be due to TGF-β in serum and/or produced by the precursors themselves. In the osteoblast-containing cultures used to analyse osteoclasts before RANKL was discovered, TGF-β had been found to either stimulate or inhibit resorptive cells. Inhibition might reflect negative feedback, because TGF-β, which osteoclasts express, induces production of OPG, the soluble decoy receptor for RANKL, in osteoblastic cells (Murakami et al. 1998; Takai et al. 1998).

In common with many pro-resorptive agents, TNF-α induces osteoblastic cells to stimulate osteoclasts (Thomson et al. 1987), and accordingly induces RANKL expression in osteoblastic cells (Horwood et al. 1998; Hofbauer et al. 1999). Surprisingly (to those who had spent many years looking for such a factor) it also stimulates osteoclast formation and bone resorption in vitro through a direct action on osteoclasts and their precursors (Azuma et al. 2000; Kobayashi et al. 2000; Fuller et al. 2002). This RANKL-like action in vitro appears to be independent of but strongly synergistic with RANKL, and of similar potency (Fuller et al. 2002). However, the role of TNF-α in vivo is less clear: Mice deleted for TNF receptors have normal bone, unlike RANK-deficient animals; and while injection of RANKL cures osteopetrosis in RANKL deficiency, TNF-α does not cure osteopetrosis in RANK-deficient mice (Li et al. 2000a). Even in inflammation, osteoclast formation is dependent on RANKL: It is required for bone loss in experimental arthritis (Kong et al. 1999; Pettit et al. 2001). Osteolysis by TNF-α in vivo might occur through induction of RANKL in osteoblastic cells. Also, extremely low levels of TNF-α are strongly synergistic with RANKL in vitro. Thus, although TNF-α cannot substitute for RANKL, systemic or local TNF-α could promote both RANKL expression and responsiveness, and so increase bone resorption with minimal disturbance to the underlying, RANKL-mediated physiological patterns of osteoclastic resorption.

4 The Osteoclast and the Macrophage

The ability of TNF-α, a ubiquitous macrophage-activating agent, to induce osteoclasts from immature mononuclear phagocytes in vitro, raises several questions concerning the nature of the osteoclast, and its relationship with macrophages. A major effect of TNF-α on macrophages is the activation of bacteriocidal activity (NO/superoxide), which is primed by interferon (IFN)-γ and deactivated by TGF-β. TGF-β by itself enhances phagocytosis and lysosomal enzyme production, and it has been suggested that while IFN-γ induces cytocidal macrophages early in host defence, TGF-β diverts macrophage activity towards debridement in the subsequent healing phase (see Riches 1996; Letterio and Roberts 1998 for reviews). We found that for TNF-α, as for RANKL, TGF-β dramatically increases the proportion of precursors that become osteoclasts, and IFN-γ does the reverse (Fox et al. 2000). The signals through which TGF-β achieves this are unknown, but there is evidence that IFN-γ suppresses osteoclast formation through inhibition of TRAF6 (Takayanagi et al. 2000). Whatever the mechanism, the observations of Fox et al. (2000) imply that lineage in M-CSF-induced precursors is determined by TGF-β/IFN-γ, and merely activated by TNF-α. By analogy, the osteoclast is a lineage determined by TGF-β and activated by TNF-α/RANKL. This suggests a model in which the osteoclast is an alternative and equivalent destiny for macrophage precursors to that of the cytocidal macrophage; it is an activated variant of the debriding macrophage.

In vitro, the osteoclastic phenotype is induced by the particular combination of M-CSF, RANKL/TNF-α and TGF-β. However, while TNF-α is ubiquitous in vivo, osteoclast formation is essentially confined to bone. This might be because the pro-inflammatory cytokine TNF-α and the anti-inflammatory TGF-β do not co-exist at levels sufficient to induce osteoclastic differentiation. Even if they did, because TNF-α is pro-inflammatory, it will normally be accompanied by, or itself induce, inflammatory cytokines such as GM-CSF, IL-4, IL-12, IL-18 and IFN-γ. These are all potent anti-osteoclast agents (Hattersley and Chambers 1990; Lacey et al. 1995; Udagawa et al. 1997; Horwood et al. 2001; Miyamoto et al. 2001). The ability of these cytokines to inhibit osteoclast differentiation might be secondary to their role in macrophage induction/stimulation. In contrast, RANKL is not pro-inflammatory (Fox et al. 2000), and this might facilitate the specific activation of the debridement pathway in macrophages. It might be that RANKL, unlike TNF-α, cures osteoclast deficiency in mice deleted for RANKL because the former does not generate osteoclast-inhibitory cytokines.

This view of the osteoclast, as an activated, non-inflammatory, debriding macrophage, is consistent with the absence of associated inflammatory cells such as neutrophils at sites of resorption. It is noteworthy though that cells with many of the characteristics of osteoclasts, such as multinuclearity, high levels of TRAP and cathepsin K, can be seen in inflammatory sites, particularly where extracellular foreign material is present (Diaz et al. 2000; Buhling et al. 2001). The osteoclast might have evolved as a variant of such defence cells, which became professionalised for the extracellular destruction of a particular substrate of predictable composition.

This does not imply that all multinuclear cells are closely related to osteoclasts. Not only RANKL, but cytokines that strongly inhibit osteoclast formation, such as IFN-γ, IL-3, IL-4 and GM-CSF, induce macrophages to form multinucleate giant cells in vitro. Even those foreign body giant cells that most resemble osteoclasts are probably distinct: If devitalised bone is implanted subcutaneously into mice, it soon becomes covered by multinucleate giant cells, but the bone is not resorbed (Chambers and Horton 1984; Popoff and Marks 1986). Resorption only occurs when osteoblasts appear. Macrophage giant cells cannot be induced to resorb bone by RANKL (Boissy et al. 2001). Multinuclear macrophages may well be as phenotypically diverse as mononuclear macrophages; multinuclearity occurs most readily in immature, 'responsive' macrophages (Möst et al. 1997) and may reflect the intensity of the stimulus or response, rather than its nature.

The osteoclast might then be seen as the equivalent of those specialised macrophage derivatives that populate other tissues under physiologic conditions, such as alveolar macrophages, Kupffer cells and microglia. The question arises, as it does for these cells, as to the extent to which osteoclasts derive from self-sustaining local precursors that were originally established from the blood, or derive continuously from the blood. For osteoclasts, this is completely unknown. If they are replenished from the blood, this does not seem to occur via a typical monocyte, because only a very small proportion (<3%) of monocytes can form osteoclasts in vitro (Quinn et al. 1994; Fuller and Chambers 1998). Osteoclast formation from blood cells also takes a surprisingly long time (>14 days in human), suggesting that origin is from a proliferative subpopulation, rather than from typical monocytes. The resistance of the majority of monocytes to osteoclast differentiation may reflect the resistance to osteoclast formation that develops in bone marrow cells during incubation in M-CSF (Arai et al. 1999; Wani et al. 1999).

Recent observations in a variant of Paget's disease suggest that osteoclasts derive from local precursors. At least some cases of this disease, which is characterised by foci of reckless osteoclasts, were found to be due to an inherited mutation that causes RANK to be overactive (Hughes et al. 2000). This implies that the foci of reckless osteoclasts derive from locally proliferative precursors that have developed a second, activating mutation (if the precursor were haematogenous, reckless osteoclasts would be systemic). Thus, it remains unknown whether in adulthood osteoclasts originate from typical monocytes, or from a subpopulation of immature peripheral blood mononuclear cells, or from a haematogenous cell that establishes a self-sustaining pool of precursors on bone surfaces. The distinction is critical to the design of in vitro models for recruitment mechanisms and precursor responsiveness.

5 How is the Spatial Control of Resorption Achieved?

A striking feature of bone cell biology is the complexity and dynamism of the patterns of osteoclastic localisation associated with bone morphogenesis and restructuring. These patterns depend on the ability of resident bone cells to direct precursors and mature osteoclasts to the sites at which activation of bone resorption is appropriate.

Despite uncertainty regarding the nature of the precursor, some speculations are possible concerning localisation mechanisms. This might occur via direct tethering by M-CSF or RANKL, because both are expressed by bone cells as transmembrane forms. If so, they must be mutually redundant, because systemic administration of (soluble) M-CSF or RANKL to mice deficient in the respective genes for these proteins leads to cure of osteopetrosis with largely normal patterns of osteoclastic localisation. An alternative explanation for the ability of soluble RANKL to localise osteoclasts might be via TNF-α-like induction of adhesion receptors in endothelial cells, for RANKL specifically in bone endothelial cells. However, bone marrow transplants show that RANK is required only on osteoclasts and their precursors for localisation to occur (Li et al. 2000a). While the possibilities above have not been excluded, the most likely model for localisation is one in which bone cells generate patterns of adhesion molecules, such as integrin ligands [especially ligands for the alpha v beta 3 receptor ($\alpha_v\beta3$), which is highly expressed on osteoclasts], in addition to RANKL and M-CSF. An advantage of such a RANKL/M-CSF-independent localisation mechanism would be that it would enable the establishment of a pool of special, non-differentiated osteoclast precursors on bone surfaces.

A second, related question is: Once osteoclasts are localised, how is resorption induced? It might merely be a matter of increased expression of RANKL. RANKL clearly activates resorption behaviour by osteoclasts on bone or dentine slices (Fuller et al. 1998; Burgess et al. 1999). However, it seems unlikely that the same behaviour occurs on other substrates; this would be inappropriate and potentially destructive. It is more likely that there are recognition factors whereby bone allows or assists resorptive behaviour.

The role of the integrin $\alpha_v\beta3$ has received much attention in the context of osteoclast localisation and activation. $\alpha_v\beta3$ is highly expressed in osteoclasts and it is likely to play a role in migration, adhesion and perhaps endocytosis of resorption products (see Väänänen et al. 2000). Antibodies against $\alpha_v\beta3$, or RGD tripeptide-containing peptides such as echistatin and kistrin, are effective inhibitors of bone resorption in vitro and in vivo (Chambers et al. 1986; Horton et al. 1991; Lakkakorpi et al. 1991; Fisher et al. 1993), and resorption is reduced, although still present, in mice deleted for the β_3 gene (McHugh et al. 2000). Osteoclast localisation in vivo, however, appears unaffected by $\alpha_v\beta3$ blockade or deficiency (Masarachia et al. 1998; Yamamoto et al. 1998; McHugh et al. 2000). Surprisingly, $\alpha_v\beta3$ seems to be absent from the sealing zone of close attachment of the resorbing osteoclast to bone (Lakkakorpi et al. 1991; Masarachia et al. 1998), although actin ring formation, and probably migration of osteoclasts over the bone surface, depends on $\alpha_v\beta3$ ligation (Nakamura et al. 1999).

Although $\alpha_v\beta3$ is redundant for osteoclastic localisation, it does greatly assist bone resorption. However, the ubiquitous presence of $\alpha_v\beta3$ ligands makes it unlikely that signalling through this integrin alone is sufficient to induce resorptive behaviour. It seems very likely that there are alternative signals for localisa-

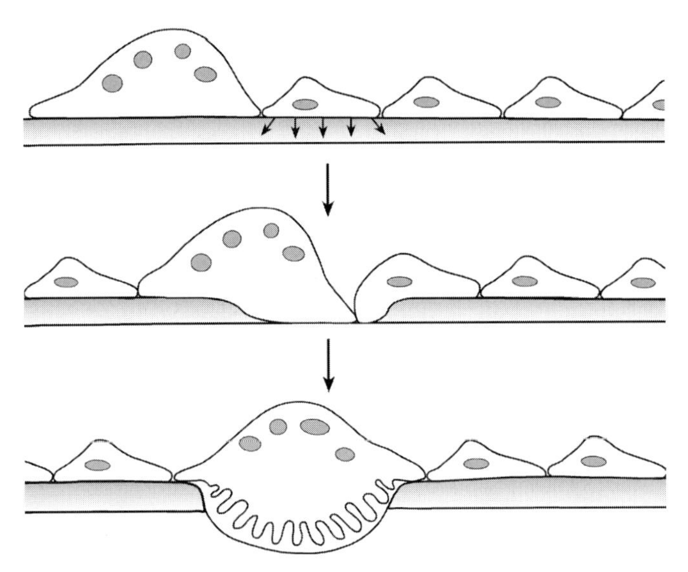

Fig. 2 A model for the mechanism of induction of bone resorption. Osteoclasts undertake resorptive behaviour only when in contact with bone mineral. Hormones and cytokines that stimulate bone resorption also stimulate osteoblastic cells to secrete interstitial collagenase. This exposes osteoclasts to bone mineral, which induces or enables resorptive behaviour. *Shaded area* represents non-mineralised organic material that covers bone surfaces. *Arrows* represent interstitial collagenase

tion, and probable that the substrate provides additional signals to the osteoclasts that resorptive behaviour is appropriate. Bone mineral might be involved in both processes: implantation of inorganic materials such as hydroxyapatite crystals into tissues evokes a foreign body giant cell (multinuclear macrophage) response that leads to sequestration and dissolution of the implanted material. Presumably, therefore, bone has evolved ways to protect itself from such phagocytic attack. One possibility is that the osteoblastic cells on bone surfaces retract to allow resorption (Rodan and Martin 1981). However, in experiments designed to test this idea, surface cells offered only minor protection against resorption (Karsdal et al. 2001). A second possibility is that the non-mineralised layer of organic material that lines bone surfaces represents the protection against phagocytic (or osteoclastic) attack. Then, when bone resorption is required, bone-lining cells remove the organic protective layer to expose bone mineral to 'phagocytic' recognition by the special debrider of bone, the osteoclast (Chambers 1980) (Fig. 2).

This model is consistent with a large body of evidence (see Chambers 1992; Fuller and Chambers 1995 for references). First, while osteoclasts cultured on bone slices require only cysteine proteinases to resorb the bone matrix, resorption in vivo depends upon both cysteine proteinases and interstitial collagenase. Second, interstitial collagenase expression is observed in bone-lining and osteocytic cells immediately adjacent to osteoclasts in bone tissue, but not in osteoclasts (Fuller and Chambers 1995; Zhao et al. 1999); and there is abundant evi-

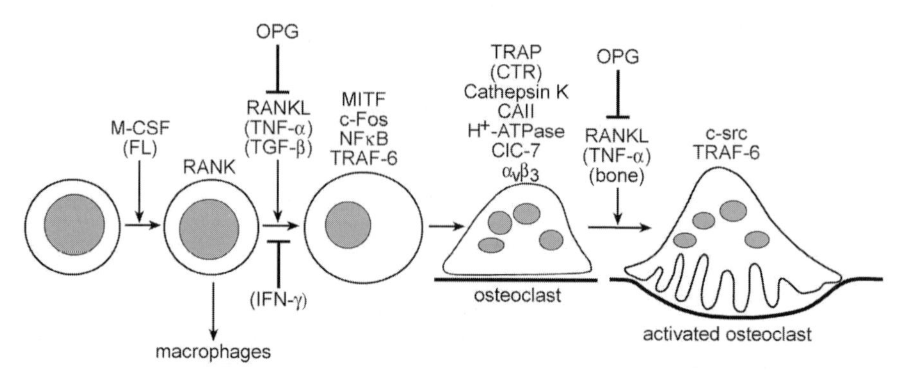

Fig. 3 Molecular mechanisms of osteoclast differentiation and function. An essential role in vivo for all the molecules, except those *in parentheses*, has been demonstrated by gene deletion/mutation. *CAII*, carbonic anhydrase; *ClC7*, chloride transporter; *CTR*, calcitonin receptor; *FL*, flt3 ligand; *MITF*, microphthalmia transcription factor; *TRAP*, tartrate-resistant acid phosphatase. See text for further details

dence that hormones that stimulate resorption strongly stimulate interstitial collagenase expression in osteoblasts (see Chambers 1992 for references). Third, osteoclasts readily resorb bone if mineral is exposed on the bone surface, but do not resorb bone that is unmineralised or has been demineralised (Chambers et al. 1984b; Chambers and Fuller 1985). Last, incubation of native bone surfaces with interstitial collagenase exposes mineral onto the surface, and renders the bone resorption-inductive for osteoclasts (Chambers et al. 1985). The observations provide compelling evidence for a model in which, while osteoclasts can undoubtedly digest all the components of bone including native collagen, unaided by other cell types, the induction of resorptive activity in osteoclasts depends on contact with bone mineral.

6
Pharmacologic Intervention in Bone Disease

Although it is merely the pawn, the osteoclast is invariably the agent, whenever there is excessive bone destruction. The commonest consequence is osteoporosis, but there are many other circumstances in which it is beneficial to suppress excessive bone resorption, such as in the bone destruction that accompanies metastatic and non-metastatic malignancy, rheumatoid arthritis, periodontitis and Paget's disease. The osteoclast has a very specialised function, compared to many other members of the macrophage family. This might be responsible for the wealth of molecules that are unique to or uniquely essential in these cells. The recent remarkable progress in identifying these molecules has provided many new potential targets for the specific and potent inhibition of osteoclastic function (Fig. 3). Many of these, and many other agents such as oestrogens, selective oestrogen receptor modulators (SERMS) and bisphosphonates, that specifically inhibit resorption through less-well understood mechanisms, and many agents beyond, are used in current therapy, or are the subject of drug discovery and development programmes. A detailed treatment of this area is beyond the scope of this article, and the reader is referred elsewhere for reviews (Gowen et al. 2000; Rodan and Martin 2000).

7 References

Anderson DM, Maraskovsky E, Billingsley WL, Dougall WC, Tometsko ME, Roux ER, Teepe MC, DuBose RF, Cosman D, Galibert L (1997) A homologue of the TNF receptor and its ligand enhance T-cell growth and dendritic-cell function. Nature 390: 175–179

Arai F, Miyamoto T, Ohneda O, Inada T, Sudo T, Brasel K, Miyata T, Anderson DM, Suda T (1999) Commitment and differentiation of osteoclast precursor cells by the sequential expression of c-Fms and receptor activator of nuclear factor κB (RANK) receptors. J Exp Med 190: 1741–1754

Athanasou NA (1996) Current concepts review: Cellular biology of bone-resorbing cells. J Bone Joint Surg 78-A: 1096–1112

Azuma Y, Kaji K, Katogi R, Takeshita S, Kudo A (2000) Tumor necrosis factor-α induces differentiation of and bone resorption by osteoclasts. J Biol Chem 275: 4858–4864

Baron R, Neff L, Louvard D, Courtoy PJ (1985) Cell-mediated extracellular acidification and bone resorption: evidence for a low pH in resorbing lacunae and localization of a 100kD lysosomal membrane protein at the osteoclast ruffled border. J Cell Biol 101: 2210–2222

Boissy P, Destaing O, Jurdic P (2001) RANKL induces formation of avian osteoclasts from macrophages but not from macrophage polykaryons. Biochem Biophys Res Comm 288: 340–346

Brady KP, Dushkin H, Fornzler D, Koike T, Magner F, Her H, Gullan S, Segre GV, Green RM, Beier DR (1999) A novel putative transport maps to the osteosclerosis (oc) mutation and is not expressed in the oc mutant mouse. Genomics 56: 254–261

Buhling F, Reisenauer A, Gerber A, Kruger S, Weber E, Bromme D, Roessner A, Ansorge S, Welte T, Rocken C (2001) Cathepsin K—a marker of macrophage differentiation? J Pathol 195: 375–382

Burgess TL, Qian Y-X, Kaufman S, Ring BD, Van G, Capparelli C, Kelly M, Hsu H, Boyle WJ, Dunstan CJ, Hu S, Lacey DL (1999) The ligand for osteoprotegerin (OPGL) directly activates mature osteoclasts. J Cell Biol 145: 527–538

Chambers TJ (1980) The cellular basis of bone resorption. Clin Orthop Rel Res 151: 283–293

Chambers TJ (1982) Osteoblasts release osteoclasts from calcitonin-induced quiescence. J Cell Sci 57: 247–260

Chambers TJ (1985) The pathobiology of the osteoclast. J Clin Pathol 38: 241–252

Chambers TJ (1989) The origin of the osteoclast. In: Peck W (ed(s) Bone and Mineral Research Annual. vol. 6. Elsevier, Amsterdam, p 1–25

Chambers TJ (1992) Regulation of osteoclast development and function. In: Rifkin BR, Gay CV (ed(s) Biology and Physiology of the Osteoclast. CRC Press, Boca Raton, p 105–128

Chambers TJ (2000) Regulation of the differentiation and function of osteoclasts. J Pathol 192: 4–13

Chambers TJ, Darby JA, Fuller K (1985) Mammalian collagenase predisposes bone surfaces to osteoclastic resorption. Cell Tiss Res 241: 671–675

Chambers TJ, Fuller K (1985) Bone cells predispose bone surfaces to resorption by exposure of mineral to osteoclastic contact. J Cell Sci 76: 155–165

Chambers TJ, Fuller K, Darby JA, Pringle JAS, Horton MA (1986) Monoclonal antibodies against osteoclasts inhibit bone resorption *in vitro*. Bone Miner 1: 127–135
Chambers TJ, Horton MA (1984) Failure of cells of the mononuclear phagocyte series to resorb bone. Calcif Tissue Int 36: 556–558
Chambers TJ, Revell PA, Fuller K, Athanasou NA (1984a) Resorption of bone by isolated rabbit osteoclasts. J Cell Sci 66: 383–399
Chambers TJ, Thomson BM, Fuller K (1984b) Effect of substrate composition on bone resorption by rabbit osteoclasts. J Cell Sci 70: 61–71
Darnay BG, Haridas V, Ni J, Moore PA, Aggarwal BB (1998) Characterization of the intracellular domain of receptor activator of NF-kB (RANK). Interaction with tumor necrosis factor receptor-associated factors and activation of NF-κB and c-Jun N-terminal kinase. J Biol Chem 273: 20551–20555
Delaisse J-M, Boyde A, Maconnachie E, Ali NN, Sear CHJ, Eeckhout Y, Vaes G, Jones SJ (1987) The effects of inhibitors of cysteine-proteinases and collagenase on the resorptive activity of isolated osteoclasts. Bone 8: 305–313
Diaz A, Willis AC, Sim RB (2000) Expression of the proteinase specialized in bone resorption, cathepsin K, in granulomatous inflammation. Mol Med 6: 648–659
Drake FH, Dodds RA, James IE, Connor JR, Debouck C, Richardson S, Lee-Rykaczewski E, Coleman L, Rieman D, Barthlow R, Hastings G, Gowen M (1996) Cathepsin K, but not cathepsins B, L, or S, is abundantly expressed in human osteoclasts. J Biol Chem 271: 12511–12516
Fisher JE, Caulfield MP, Sato M, Quaruccio HA, Gould RJ, Garsky VM, Rodan GA, Rosenblatt M (1993) Inhibition of osteoclastic bone resorption in vivo by echistatin, an 'arginyl-glycyl-aspartyl' (RGD)-containing protein. Endocrinology 132: 1411–1413
Fox SW, Fuller K, Bayley KE, Lean JM, Chambers TJ (2000) TGF-β_1 and IFN-γ direct macrophage activation by TNF-α to osteoclastic or cytocidal phenotype. J Immunol 165: 4957–4963
Frattini A, Orchard PJ, Sobacchi C, Giliani S, Abinun M, Mattsson JP, Keeling DJ, Andersson A-K, Wallbrandt P, Zecca L, Notarangelo LD, Vezzoni P, Villa A (2000) Defects in TCIRG1 subunit of the vacuolar proton pump are responsible for a subset of human autosomal recessive osteopetrosis. Nat Gen 25: 343–346
Fuller K, Chambers TJ (1989) Bone matrix stimulates osteoclastic differentiation in cultures of rabbit bone marrow cells. J Bone Miner Res 4: 179–183
Fuller K, Chambers TJ (1995) Localisation of mRNA for collagenase in osteocytic, bone surface and chondrocytic cells but not osteoclasts. J Cell Sci 108: 2221–2230
Fuller K, Chambers TJ (1998) Parathyroid hormone induces bone resorption in human peripheral blood mononuclear cells. Int J Exp Pathol 79: 223–233
Fuller K, Lean JM, Bayley KE, Wani MR, Chambers TJ (2000) A role for TGF-β_1 in osteoclast differentiation and survival. J Cell Sci 113: 2445–2453
Fuller K, Murphy C, Kirstein B, Fox SW, Chambers TJ (2002) TNFα potently activates osteoclasts, through a direct action independent of and strongly synergistic with RANKL. Endocrinology 143: In press
Fuller K, Wong B, Fox S, Choi Y, Chambers TJ (1998) TRANCE is necessary and sufficient for osteoblast-mediated activation of bone resorption in osteoclasts. J Exp Med 188: 997–1001
Galibert L, Tometsko ME, Anderson DM, Cosman D, Dougall WC (1998) The involvement of multiple tumor necrosis factor receptor (TNFR)-associated factors in the signaling mechanisms of receptor activator of NF-kB, a member of the TNFR superfamily. J Biol Chem 273: 34120–34127
Garnero P, Borel O, Byrjalsen I, Ferreras M, Drake FH, McQueney MS, Foged NT, Delmas PD, Delaissé J-M (1998) The collagenolytic activity of cathepsin K is unique among mammalian proteinases. J Biol Chem 273: 32347–32352
Gay CV, Mueller WJ (1974) Carbonic anhydrase and osteoclasts: Localization by labeled inhibitor autoradiography. Science 183: 432–434

Gelb BD, Shi GP, Chapman HA, Desnick RJ (1996) Pycnodysostosis, a lysosomal disease caused by cathepsin K deficiency. Science 273: 1236–1238

Gowen M, Emery JG, Kumar S (2000) Emerging therapies for osteoporosis. Emerging Drugs 5: 1–43

Gowen M, Nedwin GE, Mundy GR (1986) Preferential inhibition of cytokine-stimulated bone resorption by recombinant interferon gamma. J Bone Miner Res 1: 469–474

Grigoriadis AE, Wang Z-Q, Cecchini MG, Hofstetter W, Felix R, Fleisch HA, Wagner EF (1994) c-Fos: a key regulator of osteoclast-macrophage lineage determination and bone remodeling. Science 266: 443–448

Hall TJ, Chambers TJ (1989) Optimal bone resorption by isolated rat osteoclasts requires chloride/bicarbonate exchange. Calcif Tiss Int 45: 378–380

Hall TJ, Higgins W, Tardif C, Chambers TJ (1991) A comparison of the effects of inhibitors of carbonic anhydrase on osteoclastic bone resorption and purified carbonic anhydrase isozyme II. Calcif Tissue Int 49: 328–332

Halleen JM, Räisänen S, Salo JJ, Reddy SV, Roodman GD, Hentunen TA, Lehenkari PP, Kaija H, Vihko P, Väänänen HK (1999) Intracellular fragmentation of bone resorption products by reactive oxygen species generated by tartrate-resistant acid phosphatase. J Biol Chem 274: 22907–22910

Hattersley G, Chambers TJ (1989) Generation of osteoclasts from hemopoietic cells and a multipotential cell line *in vitro*. J Cell Physiol 140: 478–482

Hattersley G, Chambers TJ (1990) Effects of interleukin 3 and of granulocyte-macrophage and macrophage colony stimulating factors on osteoclast differentiation from mouse hemopoietic tissue. J Cell Physiol 142: 201–209

Hattersley G, Owens J, Flanagan AM, Chambers TJ (1991) Macrophage colony stimulating factor (M-CSF) is essential for osteoclast formation *in vitro*. Biochem Biophys Res Comm 177: 526–531

Helfrich MH, Horton MA (1993) Antigens of osteoclasts: phenotypic definition of a specialized hemopoietic cell lineage. In: Horton MA (ed(s) Blood Cell Biochemistry: Macrophages and Related Cells. vol. 5. Plenum Press, New York, p 183–202

Hernando N, Bartkiewicz M, Collin-Osdoby P, Osdoby P, Baron R (1995) Alternative splcing generates a second isoform of the catalytic a subunit of the vacuolar H(+)-ATPase. Proc Natl Acad Sci USA 92: 6087–6091

Hofbauer LC, Lacey DL, Dunstan CR, Spelsberg TC, Riggs BL, Khosla S (1999) Interleukin-1ß and tumor necrosis factor-α, but not interleukin-6, stimulate osteoprotegerin ligand gene expression in human osteoblastic cells. Bone 25: 255–259

Horton MA, Taylor ML, Arnett TR, Helfrich MH (1991) Arg-gly-asp (RGD) peptides and the anti-vitronectin receptor antibody 23C6 inhibit dentine resorption and cell spreading by osteoclasts. Exp Cell Res 195: 368–375

Horwood NJ, Eliott J, Martin TJ, Gillespie MT (1998) Osteotropic agents regulate the expression of osteoclast differentiation factor and osteoprotegerin in osteoblastic stromal cells. Endocrinology 139: 4743–4746

Horwood NJ, Elliott J, Martin TJ, Gillespie MT (2001) IL-12 alone and in synergy with IL-18 inhibits osteoclast formation in vitro. J Immunol 166: 4915–4921

Hughes AE, Ralston SH, Marken J, Bell C, MacPherson H, Wallace RHG, van Hul W, Whyte MP, Nakatsuka K, Hovy L, Anderson DM (2000) Signal peptide mutations in RANK cause familial expansile osteolysis. Nat Gen 24: 45–49

Iotsova V, Caamano J, Loy J, Yang Y, Lewin A, Bravo R (1997) Osteopetrosis in mice lacking NF-KB1 and NF-KB2. Nat Med 3: 1285–1289

Ishibashi O, Inui T, Mori Y, Kurokawa T, Kokubo T, Kumegawa M (2001) Quantification of the expression levels of lysosomal cysteine proteinases in purified human osteoclastic cells by competitive RT-PCR. Calcif Tissue Int 68: 109–116

Iyer VR, Eisen MB, Rosse DT, Schuler G, Moore T, Lee JCF, Trent JM, Staudt LM, Hudson Jr J, Boguski MS, Lashkari D, Shalon D, Botstein D, Brown PO (1999) The transcriptional program in the response of human fibroblasts to serum. Science 283: 83–87

Johnson RS, Spiegelman BM, Papaioannou V (1992) Pleiotropic effects of a null mutation in the *c-fos* proto-oncogene. Cell 71: 577–586

Kamiya T, Kobayashi Y, Kanaoka K, Nakashima T, Kato Y, Mizuno A, Sakai H (1998) Fluorescence microscopic demonstration of cathepsin K activity as the major lysosomal cysteine proteinase in osteoclasts. J Biochem 123: 752–759

Karsdal MA, Fjording MS, Foged NT, Delaissé J-M (2001) Transforming growth factor-β-induced osteoblast elongation regulates osteoclastic bone resorption through a p38 mitogen-activated protein kinase- and matrix metalloproteinase-dependent pathway. J Biol Chem 276: 39350–39358

Kaye M (1984) When is it an osteoclast? J Clin Pathol 37: 398–400

Kim H-H, Lee DE, Shin JN, Lee YS, Jeon YM, Chung C-H, Ni J, Kwon BS, Lee ZH (1999) Receptor activator of NF-kB recruits multiple TRAF family adaptors and activates c-Jun N-terminal kinase. FEBS Lett 443: 297–302

Kobayashi K, Takahashi N, Jimi E, Udagawa N, Takami M, Kotake S, Nakagawa N, Kinosaki M, Yamaguchi K, Shima N, Yasuda H, Morinaga T, Higashio K, Martin TJ, Suda T (2000) Tumor necrosis factor α stimulates osteoclastic differentiation by a mechanism independent of the ODF/RANKL-RANK interaction. J Exp Med 191: 275–285

Kong Y-Y, Felge U, Sarosi I, Bolon B, Tafuri A, Morony S, Capparelli C, Li JI, Elliott R, McCabe S, Wong T, Campagnuolo G, Moran E, Bogoch ER, Van G, Nguyen LT, Ohashi PS, Lacey DL, Fish E, Boyle WJ, Penninger JM (1999) Activated T cells regulate bone loss and joint destruction in adjuvant arthritis through osteoprotegerin ligand. Nature 402: 304–309

Kornak U, Kasper D, Bõsl MR, Kaiser E, Schweizer M, Schulz A, Friedrich W, Delling G, Jentsch T (2001) Loss of the CIC-7 chloride channel leads to osteopetrosis in mice and man. Cell 104: 205–215

Lacey DL, Erdmann JM, Teitelbaum SL, Tan H-L, Ohara J, Shioi A (1995) Interleukin 4, interferon-γ, and prostaglandin E impact the osteoclast cell-forming potential of murine bone marrow macrophages. Endocrinology 136: 2367–2376

Lagasse E, Weissman IL (1997) Enforced expression of Bcl-2 in monocytes rescues macrophages and partially reverses osteopetrosis in *op/op* mice. Cell 89: 1021–1031

Lakkakorpi PT, Horton MA, Helfrich MH, Karhukorpi E-K, Väänänen HK (1991) Vitronectin receptor has a role in bone resorption but does not mediate tight sealing zone attachment of osteoclasts to the bone surface. J Cell Biol 115: 1179–1186

Lean JM, Fuller K, Chambers TJ (2001) Flt3 ligand can substitute for M-CSF in support of osteoclast differentiation and function. BloodIn press

Letterio JJ, Roberts AB (1998) Regulation of immune responses by TGF-ß. Ann Rev Immunol 16: 137–161

Li J, Sarosi I, Yan XQ, Morony S, Capparelli C, Tan HL, McCabe S, Elliott R, Scully S, Van G, Kaufman S, Juan SC, Sun Y, Tarpley J, Martin L, Christensen K, McCabe J, MKostenuik P, Hsu H, Fletcher F, Dunstan CR, Lacey DL, Boyle WJ (2000a) RANK is the intrinsic hematopoietic cell surface receptor that controls osteoclastogenesis and regulation of bone mass and calcium metabolism. Proc Natl Acad Sci USA 97: 1566–1571

Li Y-P, Chen W, Liang Y, Li E, Stashenko P (1999) *Atp6i*-deficient mice exhibit severe osteopetrosis due to loss of osteoclast-mediated extracellular acidification. Nat Gen 23: 447–451

Li Y-P, Chen W, Stashenko P (1996) Molecular cloning and characterization of a putative novel human osteoclast-specific 116-kDa vacuolar proton pump subunit. Biochem Biophys Res Comm 218: 813–821

Li Z, Hou WS, Bromme D (2000b) Collagenolytic activity of cathepsin K is specifically modulated by cartilage-resident chondroitin sulfates. Biochemistry 39: 529–536

Liu BY, Guo J, Lanske B, Divieti P, Kronenberg HM, Bringhurst FR (1998) Conditionally immortalised murine bone marrow stromal cells mediate parathyroid hormone-dependent osteoclastogenesis *in vitro*. Endocrinology 139: 1952–1964

Lomaga MA, Yeh W-C, Sarosi I, Duncan GS, Furlonger C, Ho A, Morony S, Capparelli C, Van G, Kaufman S, van der Heiden A, Itie A, Wakeham A, Khoo W, Sasaki T, Cao Z, Penninger JM, Paige CJ, Lacey DL, Dunstan CR, Boyle WJ, Goeddel DV, Mak TW (1999) TRAF6 deficiency results in osteopetrosis and defective interleukin-1, CD40, and LPS signaling. Genes Develop 13: 1015–1024

Marks SC (1983) The origin of the osteoclast. J Oral Pathol 12: 226–256

Masarachia P, Yamamoto M, Leu C, Rodan G, Duong L (1998) Histomorphometric evidence for echistatin inhibition of bone resorption in mice with secondary hyperparathyroidism. Endocrinology 139: 1401–1410

Matsuo K, Owens JM, Tonko M, Elliott C, Chambers TJ, Wagner EF (2000) *Fosl1* is a transcriptional target of c-Fos during osteoclast differentiation. Nat Gen 24: 184–187

Matsuzaki K, Katayama K, Takahashi Y, Nakamura Y, Udagawa N, Tsurukai T, Nishinakamura R, Toyama Y, Yasbe Y, Hori M, Takahashi N, Suda T (1998) Human osteoclast-like cells are formed from peripheral blood mononuclear cells in a co-culture with SaOS-2 cells transfected with the PTH/PTHrP receptor gene. Endocrinology 140: 925–932

McHugh KP, Hodivala-Dilke K, Zheng MH, Namba N, Lam J, Novack D, Ross FX, Hynes RO, Teitelbaum SL (2000) Mice lacking β_3 integrins are osteosclerotic because of dysfunctional osteoclasts. J Clin Invest 105: 433–440

McSheehy PMJ, Chambers TJ (1986) Osteoblastic cells mediate osteoclastic responsiveness to parathyroid hormone. Endocrinology 118: 824–828

Miyamoto T, Ohneda O, Arai F, Iwamoto K, Okada S, Takagi K, Anderson DM, Suda T (2001) Bifurcation of osteoclasts and dendritic cells from common progenitors. Blood 98: 2544–2554

Möst J, Spötl L, Mayr G, Gasser A, Sarti A, Dierich MP (1997) Formation of multinucleated giant cells in vitro is dependent on the stage of monocyte to macrophage maturation. Blood 89: 662–671

Murakami T, Yamamoto M, Yamamoto M, Ono K, Nishikawa M, Nagata N, Motoyoshi K, Akatsu T (1998) Transforming growth factor-ß1 increases mRNA levels of osteoclastogenesis inhibitory factor in osteoblastic/stromal cells and inhibits the survival of murine osteoclast-like cells. Biochem Biophys Res Comm 252: 747–752

Naito A, Azuma S, Tanaka S, Miyazaki T, Takaki S, Takatsu K, Nakao K, Nakamura K, Katsuki M, Yamamoto T, Inoue J (1999) Severe osteopetrosis, defective interleukin-1 signalling and lymph node organogenesis in TRAF6-deficient mice. Genes Cells 4: 353–362

Nakamura I, Pilkington MF, Lakkakorpi PT, Lipfert L, Sims SM, Dixon SJ, Rodan GA, Duong LT (1999) Role of $\alpha_v\beta3$ integrin in osteoclast migration and formation of the sealing zone. J Cell Sci 112: 3985–3993

Nakamura I, Takahashi N, Udagawa N, Moriyama Y, Kurokawa T, Jimi E, Sasaki T, Suda T (1997) Lack of vacuolar proton ATPase association with the cytoskeleton in osteoclasts of osteosclerotic mice *(oc/oc)* mice. FEBS Lett 401: 207–212

Pettit AR, Ji H, von Stechow D, Muller R, Goldring SR, Choi Y, Benoist C, Gravallese EM (2001) TRANCE/RANKL knockout mice are protected from bone erosion in a serum transfer model of arthritis. Am J Pathol 159: 1689–1699

Popoff SN, Marks Jr SC (1986) Ultrastructure of the giant cell infiltrate of subcutaneously implanted bone particles in rats and mice. Am J Anat 177: 491–503

Prallet B, Male P, Neff L, Baron R (1992) Identification of a functional mononuclear precursor of the osteoclast in chicken medullary bone-marrow cultures. J Bone Miner Res 7: 405–414

Quinn JMW, McGee JOD, Athanasou NA (1994) Cellular and hormonal factors influencing monocyte differentiation to osteoclastic bone-resorbing cells. Endocrinology 134: 2416–2423

Reponen P, Sahlberg C, Muhnaut C, Thesleff I, Tryggvason K (1994) High expression of 92-kD type IV collagenase (gelatinase B) in the osteoclast lineage during mouse development. J Cell Biol 124: 1091–1102

Riches DWH (1996) Macrophage involvement in wound repair, remodeling, and fibrosis. In: Clark RAF (ed(s) The Molecular and Cellular Biology of Wound Repair. Plenum Press, New York, p 95–141

Rodan GA, Martin TJ (1981) The role of osteoblasts in hormonal control of bone resorption. Calcif Tiss Int 33: 349–351

Rodan GA, Martin TJ (2000) Therapeutic approaches to bone diseases. Science 289: 1508–5014

Saftig P, Hunziker E, Wehmeyer O, Jones S, Boyde A, Rommerskirch W, Moritz JD, Schu P, von Figura K (1998) Impaired osteoclastic bone resorption leads to osteopetrosis in cathepsin-K-deficient mice. Proc Natl Acad Sci USA 95: 13453–13458

Schlesinger PH, Blair HC, Teitelbaum SL, Edwards JC (1997) Characterization of the osteoclast ruffled border chloride channel and its role in bone resorption. J Biol Chem 272: 18636–18643

Scimeca J-C, Franchi A, Trojani C, Parrinello H, Grosgeorge J, Robert C, Jaillon O, Poirier C, Gaudray P, Carle GF (2000) The gene encoding the mouse homologue of the human osteoclast-specific 116 kDa V-ATPase subunit bears a deletion in osteosclerotic *(oc/oc)* mutants. Bone 26: 207–213

Sells Galvin RJ, Gatlin CL, Horn JW, Fuson TR (1999) TGF-beta enhances osteoclast differentiation in hematopoietic cell cultures stimulated with RANKL and M-CSF. Biochem Biophys Res Comm 265: 233–239

Silver IA, Murrills RJ, Etherington DJ (1988) Microelectrode studies on the acid microenvironment beneath adherent macrophages and osteoclasts. Exp Cell Res 175: 266–276

Simonet WS, Lacey DL, Dunstan CR, Kelley M, Chang M-S, Lüthy R, Nguyen HQ, Wooden S, Bennett L, Boone T, Shimamoto G, DeRose M, Elliott R, Columbero A, Tan H-L, Trail G, Sullivan J, Davy E, Bucay N, Renshaw-Gegg L, Hughes TM, Hill D, Pattison W, Campbell P, Sander S, Van G, Tarpley J, Derby P, Lee R, Program. AE, Boyle WJ (1997) Osteoprotegerin: a novel secreted protein involved in the regulation of bone density. Cell 89: 309–319

Sly WS, Hewett-Emmett D, Whyte MP, Yu Y-SL, Tashjian RE (1983) Carbonic anhydrase II deficiency identified as the primary defect in the autosomal recessive syndrome of osteopetrosis with renal tubular acidosis and cerebral calcification. Proc Natl Acad Sci USA 80: 2752–2756

Stenbeck G, Horton MA (2000) A new specialized cell-matrix interaction in actively resorbing osteoclasts. J Cell Sci 113: 1577–1587

Suda T, Takahashi N, Martin TJ (1995) Modulation of osteoclast differentiation: update. In: Bikle DD, Negrovilar A (ed(s) Endoc Rev Monographs. vol. 4. Endocrine Society, Bethesda, MD, p 266–270

Suda T, Takahashi N, Udagawa N, Jimi E, Gillespie MT, Martin TJ (1999) Modulation of osteoclast differentiation and function by the new member of the tumor necrosis factor receptor and ligand families. Endocr Rev 20: 345–357

Takahashi N, Akatsu T, Udagawa N, Sasaki T, Yamaguchi A, Moseley JM, Martin TJ, Suda T (1988) Osteoblastic cells are involved in osteoclast formation. Endocrinology 123: 2600–2602

Takahashi N, Udagawa N, Akatsu T, Tanaka H, Isogai Y, Suda T (1991) Deficiency of osteoclasts in osteopetrotic mice is due to a defect in the local microenvironment provided by osteoblastic cells. Endocrinology 128: 1792–1796

Takai H, Kanematsu M, Yano K, Tsuda E, Higashio K, Ikeda K, Watanabe K, Yamada Y (1998) Transforming growth factor-beta stimulates the production of osteoprotegerin/osteoclastogenesis inhibitory factor by bone marrow stromal cells. J Biol Chem 273: 27091–27096

Takayanagi H, Ogasawara K, Hida S, Chiba T, Murata S, Sato K, Takaoka A, Yokochi T, Oda H, Tanaka K, Nakamura K, Taniguchi T (2000) T-cell-mediated regulation of osteoclastogenesis by signalling cross-talk between RANKL and IFN-gamma. Nature 408: 600–605

Takeda S, Yoshizawa T, Nagai Y, Yamato H, Fukumoto S, Sekine K, Kato S, Matsumoto T, Fujita T (1999) Stimulation of osteoclast formation by 1,25-dihydroxyvitamin D requires its binding to vitamin D receptor (VDR) in osteoblastic cells: studies using VDR knockout mice. Endocrinology 140: 1005–1008

Tezuka K-i, Tezuka Y, Maejima A, Sato T, Nemoto K, Kamioka H, Hakeda Y, Kumegawa M (1994) Molecular cloning of a possible cysteine proteinase predominantly expressed in osteoclasts. J Biol Chem 269: 1106–1109

Thomson BM, Mundy GR, Chambers TJ (1987) Tumor necrosis factors α and ß induce osteoblastic cells to stimulate osteoclastic bone resorption. J Immunol 138: 775–779

Thomson BM, Saklatvala J, Chambers TJ (1986) Osteoblasts mediate interleukin 1 stimulation of bone resorption by rat osteoclasts. J Exp Med 164: 104–112

Thyberg J (1996) Differentiated properties and proliferation of arterial smooth muscle cells in culture. Int Rev Cytol 169: 183–265

Tsuda E, Goto M, Mochizuki S, Yano K, Kobayashi F, Morinaga T, Higashio K (1997) Isolation of a novel cytokine from human fibroblasts that specifically inhibits osteoclastogenesis. Biochem Biophys Res Comm 234: 137–142

Udagawa N, Horwood NJ, Elliott J, Mackay A, Owens J, Okamura H, Kurimoto M, Chambers TJ, Martin TJ, Gillespie MT (1997) Interleukin (IL)-18 (interferon gamma inducing factor) is produced by osteoblasts and acts via granulocyte macrophage-colony stimulating factor and not via interferon-gamma to inhibit osteoclast formation. J Exp Med 185: 1005–1012

Väänänen HK, Zhao H, Mulari M, Halleen JM (2000) The cell biology of osteoclast function. J Cell Sci 113: 377–381

Vaananen JK, Karhukorpi EK, Sundquist K, Wallmark B, Hentunen T, Tuukanen J, Lakkakorpi P (1990) Evidence for the presence of a proton pump of the vacuolar H^+-ATPase type in the ruffled borders of osteoclasts. J Cell Biol 111: 1305–1311

Van Hille B, Richener H, Green JR, Bilbe G (1995) The ubiquitous VA68 isoform of subunit A of the vacuolar H(+)-ATPase is highly expressed in human osteoclasts. Biochem Biophys Res Comm 214: 1108–1113

Vu TH, Shipley JM, Bergers G, Berger JE, Helms JA, Hanahan D, Shapiro SD, Senior RM, Werb Z (1998) MMP-9/gelatinase B is a key regulator of growth plate angiogenesis and apoptosis of hypertrophic chondrocytes. Cell 93: 411–422

Wang Z-Q, Ovitt C, Grigoriadis AE, Mohle-Steinlen U, Rüther U, Wagner EF (1992) Bone and haematopoietic defects in mice lacking *c-fos*. Nature 360: 741–745

Wani MR, Fuller K, Kim NS, Choi Y, Chambers T (1999) Prostaglandin E_2 cooperates with TRANCE in osteoclast induction from hemopoietic precursors: synergistic activation of differentiation, cell spreading, and fusion. Endocrinology 140: 1927–1935

Wiktor-Jedrzejczak W, Bartocci A, Ferrante AW, Jr., Ahmed-Ansari A, Sell KW, Pollard JW, Stanley ER (1990) Total absence of colony-stimulating factor 1 in the macrophage-deficient osteopetrotic (op/op) mouse. Proc Natl Acad Sci USA 87: 4828–4832

Wong BR, Rho J, Arron J, Robinson E, Orlinick J, Chao M, Kalachikov S, Cayani E, Bartlett III FS, Frankel WN, Lee SY, Choi Y (1997) TRANCE is a novel ligand of the tumor necrosis factor receptor family that activates c-Jun N-terminal kinase in T cells. J Biol Chem 272: 25190–25194

Xia L, Kilb J, Wex H, Li Z, Lipyansky A, Breuil V, Stein L, Palmer JT, Dempster DW, Bromme D (1999) Localization of rat cathepsin K in osteoclasts and resorption pits: inhibition of bone resorption and cathepsin-K activity by peptidyl vinyl sulfones. Biol Chem 380: 679–687

Yamamoto M, Fisher JE, Gentile M, Seedor JG, Leu C-T, Rodan SB, Rodan GA (1998) The integrin ligand echistatin prevents bone loss in ovariectomized mice and rats. Endocrinology 139: 1411–1419

Yoshida H, Hayashi S-I, Kunisada T, Ogawa M, Nishikawa S, Okamura H, Sudo T, Shultz LD, Nishikawa S-I (1990) The murine mutation osteopetrosis is in the coding region of the macrophage colony stimulating factor gene. Nature 345: 442–444

Zhao W, Byrne MH, Boyce BF, Krane SM (1999) Bone resorption induced by parathyroid hormone is strikingly diminished in collagenase-resistant mutant mice. J Clin Invest 103: 517–524

Macrophages in the Central and Peripheral Nervous System

V. H. Perry

CNS Inflammation Group, School of Biological Sciences,
University of Southampton, Southampton, SO16 7PX, UK
e-mail: vhp@soton.ac.uk

Abstract In the central and peripheral nervous system there are several distinct populations of macrophages. In the brain there are macrophages in the parenchyma and meninges, and there are macrophages associated with the vasculature. The macrophages in the brain parenchyma, the microglia are highly atypical with a distinct morphology and downregulated phenotype. The molecular mechanisms that underpin this unusual phenotype are being unravelled. Microglia rapidly respond to perturbations of their microenvironment and become activated in almost all brain pathologies. There is considerable interest in the possible role that activated microglia may have in brain and spinal cord pathology. The perivascular macrophages abutting the brain vasculature are also highly specialised macrophages and play an important role in communication between systemic inflammation and the brain. There is still much to learn about the role of macrophages in nervous system injury and repair.

Keywords Brain, CNS (central nervous system), Downregulation, Macrophage, Microglia, Neurodegeneration, Perivascular, PNS (peripheral nervous system), Spinal cord

In the central and peripheral nervous system there are large numbers of resident macrophages. These reside in different compartments of the nervous system, and the local microenvironment has a profound effect on their phenotype.

These macrophage populations have different roles in local homeostasis and respond differentially to injury and infection. It is now recognised that inflammation may contribute to the outcome of diverse neurological conditions including stroke, acute traumatic injury, HIV-1 associated dementia and Alzheimer's disease (Perry 1994). In all these conditions it is the macrophage that lies centre stage, and understanding how these cells contribute to brain damage and repair remains a significant challenge. Studies of the phenotype, the regulation of phenotype, and function of these different resident populations will aid in elucidating how these mononuclear phagocyte populations contribute to these diverse disease states.

1 Macrophages of the Central Nervous System

1.1 Microglia

The most abundant macrophage in the central nervous system (CNS) is the microglia. These cells have a distinct morphology (Fig. 1) and a distinct phenotype. The phenotype of the microglia in the normal adult CNS is best described as being in a downregulated or switched-off state (Perry and Gordon 1991). For example, microglia express low levels of major histocompatability complex (MHC) class I and II, low levels of CD45 (Streit et al. 1989) and low or undetectable levels of the scavenger receptor (Bell et al. 1994). The microglia do express F4/80, complement receptor type 3 (CR3) and Fc receptors (Perry et al. 1985). It is likely that the unusual morphology and the atypical phenotype gave rise to the debate as to whether microglia were or were not cells of the mononuclear phagocyte lineage (Ling and Wong 1993). However, numerous studies using bone marrow chimeras and specific panels of antibodies show that these cells are indeed the resident macrophages of the brain parenchyma (see Perry et al. 1993 for review). While there is little doubt that the microglia are of mononuclear phagocyte lineage, the precise origin and time of entry of mononuclear phagocytes into the CNS remains to be clarified (Kaur et al. 2001).

The microglia are present throughout the rostrocaudal axis of the adult CNS. They are more abundant in grey than white matter, and the density and morphology varies depending on their precise location (Lawson et al. 1990). Despite the marked neurochemical differences that exist from one brain region to the next, the phenotype of the parenchymal microglia, as judged by their morphology and expression of cell surface antigens, is rather uniform.

1.1.1 Phenotype

While numerous studies have documented the unusual phenotype of the microglia and the cell surface and cytoplasmic antigens that are, or are not, expressed

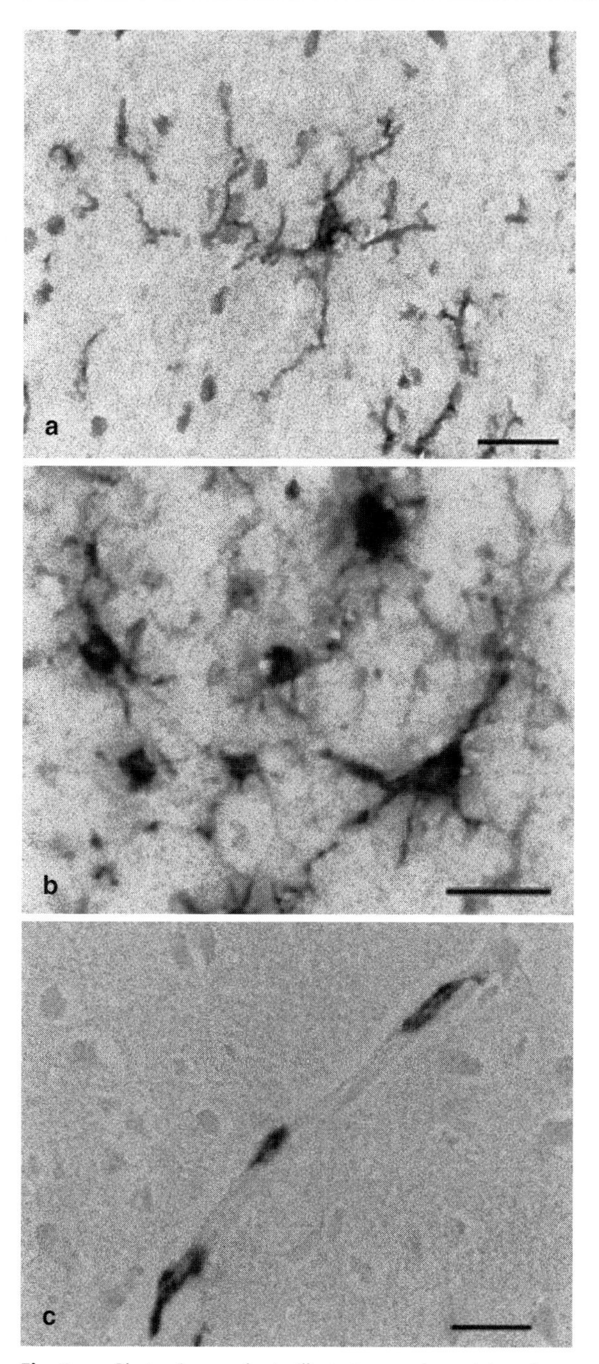

Fig. 1a–c Photomicrographs to illustrate complement type 3 expression on normal microglia (*upper*) and activated microglia (*middle*) in the rat brain. Perivascular macrophages are revealed by the monoclonal antibody ED2 (*lower*). *Scale bar*=20 μm

by these cells, the key issue is to discover the mechanisms that regulate their phenotype. What is it about the CNS microenvironment that so effectively downregulates the macrophage? One approach to this problem has been to investigate the factors that produce a microglia morphology or phenotype in vitro. Not surprisingly given the highly ramified nature of adult microglia (Fig. 1), the isolation of these cells from adult brain has not been routinely carried out, although it is possible (Ford et al. 1996). In the main, those studying microglia in in vitro studies have isolated immature microglia from neonatal brains, a time when the cells are less ramified and greater yields are obtained. However, the cells isolated from neonates, particularly when kept in culture for any length of time, have rather few similarities to microglia of the adult CNS. Indeed they have the phenotype of a typical macrophage, apart from some subtle morphological specializations (Giulian et al. 1995). To investigate the factors that might induce the microglia phenotype, various co-culture systems have been studied. Sievers and colleagues (Schmidtmayer et al. 1994; Sievers et al. 1994) have demonstrated that astrocytes will induce microglia morphology in macrophages derived from brain, spleen or blood and induce some aspects of the phenotype of the microglia. However, the astrocyte-derived cell surface, or secreted molecules, that may be responsible for the phenotype have not been isolated. A number of other systems have also been used to study factors that induce the microglia phenotype in vitro, but these have largely depended on morphological criteria and have not defined precisely the criteria for deciding that a macrophage in vitro actually represents the microglia phenotype in vivo.

At least part of the downregulated phenotype may come about as a consequence of immune regulatory molecules secreted by astrocytes or other cells in the CNS. Low levels of transforming growth factor $\beta 1$ (Kiefer et al. 1995) and interleukin-10 (Strle et al. 2001) are expressed in the normal CNS and they may contribute to microglia downregulation. There is evidence that neuronal activity is involved in the regulation of microglia phenotype, since blockade of neural activity in co-cultures leads to the upregulation of MHC class II expression by microglia (Neumann et al. 1996), an effect shown to be mediated by neurotrophins (Neumann et al. 1998). Recently, it has been shown that the interaction between CD200, expressed on neurons, and CD200R expressed by macrophages is involved in the regulation of microglia phenotype. Microglia in the CNS of mice lacking CD200 were found to have a more activated phenotype and formed focal clusters of microglia, although this was by no means ubiquitous (Hoek et al. 2000). The absence of CD200 also resulted in more rapid induction of experimental allergic encephalomyelitis. Other receptor–ligand interactions of this sort may be critically involved in the microglia phenotype (Barclay et al. 2002). It is known that the extracellular matrix of the CNS is highly atypical, being dominated by proteoglycans (Bandtlow and Zimmermann 2000), and that macrophages are differentially activated depending on the substratum to which they adhere. In an adhesion assay in which macrophages were allowed to adhere to brain sections a novel monoclonal antibody that blocked macrophage adhesion to brain sections but not to spleen suggested the presence of a novel adhesive interaction (Brown et al. 1998).

In addition to the ligand–receptor interactions between microglia and components of the brain microenvironment, which may downregulate these cells, it may also be the absence of stimuli that play a part in the atypical phenotype. In regions of the CNS where the blood–brain barrier is absent and the macrophages are exposed to serum products, such as in the circumventricular organs, the resident brain macrophages are more activated (Perry et al. 1992).

1.1.2
Function

Numerous functions have been ascribed to the microglia, but definitive evidence is largely lacking. During development of the CNS, at least half of the neurons and glia generated in the embryo will not survive into adulthood, and at least some proportion of these cells undergoing apoptosis are phagocytosed by the microglia (Perry et al. 1985). Although it has been suggested that macrophages invading the developing brain may be in some way essential for CNS development, this seems unlikely. A comparison of the distribution of the apoptotic cells and the immature microglia reveals that these macrophages are opportunistic phagocytes rather than attracted to the apoptotic cells (Ashwell 1991). The PU.1-null mouse, which lacks myeloid cells, is viable until at least several weeks postnatally and major abnormalities of the CNS have not been reported (McKercher et al. 1996). This is not to say that there are no abnormalities of the CNS in these mice and results from more detailed studies will be of interest, particularly in the organisation of the fibre tracts where large numbers of macrophages invade the immature brain (Innocenti et al. 1983).

One structure in which the invading macrophages do appear to play an essential developmental role is in the vasculature of the eye (Lang and Bishop 1993). When macrophages were eliminated from the anterior chamber of the eye, the hyaloid artery and pupillary membrane were abnormally persistent (Diez-Rous and Lang 1997). These studies also showed that the macrophages do not just remove the cells of the vasculature but actively induce the endothelial cells to undergo apoptosis. Whether the macrophages are actively involved in remodelling of vasculature elsewhere or other components of the developing CNS remains to be established.

In the normal, healthy adult brain parenchyma, the role of microglia in brain homeostasis is unclear. Although there has been speculation that they may be involved in synaptic modelling, there is little evidence to support this except in the rather special circumstance of the neural lobe. In this structure, resident macrophages phagocytose the endings of hypothalamic magnocellular neurons where they abut the capillaries of the neural lobe (Pow et al. 1989). When the numbers of microglia processes in the neuropil are viewed in the context of the density of synapses, it is hard to envisage these cells playing a significant role in ongoing CNS synaptic plasticity.

The most important role for the microglia is that of first line of defence against injury and infection. It is now well documented that following any injury or pathology of the brain the microglia alter their levels of antigen expression and their morphology (Kreutzberg 1996). The more severe the disturbance, the more radical are the changes in the microglia morphology and phenotype but these cells are typically referred to as activated microglia (Fig. 1, middle). There is considerable interest as to how these cells, or other macrophage populations in the CNS, contribute to the outcome of diverse neurological diseases. In all conditions of acute or chronic neurodegeneration the microglia become activated, but not surprisingly there is no simple relationship between morphological activation and the associated cytokine profile (Walsh et al. 2000; Perry et al. 2002).

One key issue is the role of the microglia in antigen presentation and whether, for example, the upregulation of MHC class II leads to antigen presentation to CD4 T-lymphocytes and propagation of an immune assault on the CNS. There remains some controversy in this area but the weight of the evidence suggests that microglia are rather poor antigen-presenting cells (reviewed in Perry 1998), although other views have been expressed (Aloisi 2001). The normal brain parenchyma lacks dendritic cells and it appears that it is the perivascular macrophages that present antigen to T-lymphocytes patrolling the CNS (see below).

1.2 Perivascular Macrophages

A long overlooked population of CNS macrophages is that closely associated with the vasculature, the perivascular macrophages. These cells lie adjacent to the cerebral endothelial cells behind the blood–brain barrier but separated from the CNS parenchyma by a basement membrane (Graeber 1989) and are present throughout the rostrocaudal axis of the CNS. The perivascular macrophages have a simpler morphology than the microglia (Fig. 1, lower) and also a more activated phenotype. They express readily detectable levels of MHC class I and II CD45 (Streit et al. 1989) and in the normal mouse brain express the scavenger receptor (Mato et al. 1996) which is absent from the normal microglia. In the rat the monoclonal antibody, ED2 was found to be a selective marker of the perivascular macrophages (Graeber et al. 1989).

These cells are not only more activated than the microglia, they also turn over more rapidly. Data from a number of bone marrow chimera studies suggest that a significant proportion is replaced over a period of 3–6 months (Hickey et al. 1992). The important point in this regard is that monocytes are continually trafficking across the normal intact blood–brain barrier, which has implications for understanding how a number of intracellular pathogens, including HIV-1, may enter the CNS.

1.2.1 Function

One might expect from the location of these that cells they are well placed to respond to immune-activating molecules in the blood and are also the first cells that activated T lymphocytes will encounter on crossing the cerebral endothelium. Following a peripheral challenge with endotoxin [lipopolysaccharide (LPS)], to mimic a peripheral infection, it is the perivascular macrophages that first respond by increased synthesis of inhibitory factory $\kappa B\alpha$ and cyclooxygenase-2 (Cox-2) (Nguyen et al. 2002). This sensitivity of the perivascular macrophages relative to the microglia is also seen following intracerebral challenge with LPS or pro-inflammatory cytokines. The perivascular macrophages rapidly upregulate their synthesis of Cox-2 while the microglia do not (Minghetti et al. 1999). These cells play a major role in signalling between the periphery and the brain in the induction of fever and sickness behaviour that accompany systemic infection (Konsman et al. 2002).

The fact that perivascular macrophages constitutively express MHC class II molecules has led to the suggestion that these are the major APCs of the CNS. Studies by Hickey and Kimura (1988) show that these cells have the capacity to present antigen to encephalitogenic $CD4^+$ T lymphocytes. The selective isolation and study of perivascular and meningeal macrophages, as distinct from microglia, also shows that the $CD45^{high}$ population are competent APCs while the $CD45^{low}$ microglia are not (Ford et al. 1996). The observations raise the important question as to whether the perivascular macrophages are akin to, or are, dendritic cells of the CNS with the capacity to migrate from the brain compartment to lymphoid organs. At the present time there is little evidence to support the idea that there are dendritic cells in the perivascular space, or within the parenchyma of the brain. Two lines of evidence suggest that dendritic cells are present in the meninges and choroid plexus but not in the perivascular space or brain parenchyma.

In the rat the monoclonal antibody 0X62 recognises an integrin restricted to a population of dendritic cells and $\gamma\delta$ T lymphocytes (Brenan and Puklavec 1992). Using this antibody it has been shown that there are $OX62^+$/MHC class II^+ in the meninges and choroid plexus (Matyszak and Perry 1996; McMenamin 1999) but they are absent from the brain parenchyma. However, the presence or absence of any single antigen is not sufficient to define a cell as a dendritic cell, it is the functional capacity that is the key. Functional studies in vivo show that there are no dendritic cells in the brain parenchyma.

The microinjection of heat-killed mycobacterium bacillus Calmette-Guérin (BCG) into the ventricles, or on to the surface of the brain, gives rise to a typical overt delayed-type hypersensitivity (DTH) response. However, when the BCG is delivered in such a manner as to restrict it to the brain parenchyma it may reside there for many months undetected by the immune system (Matyszak and Perry 1995). Similar experiments have been performed using influenza virus (Stevenson et al. 1997). These simple experiments demonstrate that neither perivascular macrophages nor microglia are able to phagocytose the BCG or virus and with these antigens then migrate to the lymphoid organs to initiate a primary immune response. In contrast, it has been shown that soluble antigen when delivered to the brain parenchyma, and with the appropriate precautions being taken to limit the delivery to the parenchyma, will rapidly drain via the perivascular spaces to the cervical lymph nodes (Cserr and Knopf 1992). The soluble antigen draining to the periphery results in an effective antibody response to the delivered antigen (Gordon et al. 1992). The functional significance of the drainage of soluble antigens, some of which are likely to be potentially immunogenic CNS antigens, is of interest in the maintenance of tolerance to CNS antigens.

1.3 Macrophages of the Meninges and Choroid Plexus

In the meninges, the membranes covering the brain, and within the stroma of the choroid plexus there are large numbers of macrophages. Although these macrophages lie outside the CNS itself, it is clear that these cells have the potential to play a significant part in immunological reactions in the CNS. These macrophages have a more activated phenotype than the microglia and perivascular cells and there are also some dendritic cells (Matyszak et al. 1992; Matyszak and Perry 1996; McMenamin 1999). These macrophage populations are involved in both innate and acquired immune reactions (see above) in the CNS.

2 Macrophages of the Peripheral Nervous System

Within the endoneurium of peripheral nerves there are resident macrophages which constitute about 5% of the total cell population. There are also large numbers of macrophages in the membranes covering the nerves and peripheral ganglia (Braun et al. 1993). It is not known whether these macrophage populations play any significant role in peripheral nervous system (PNS) homeostasis, but they are likely involved in the response to nerve injury. There has been considerable interest as to whether the monocytes that invade a peripheral nerve after nerve injury play a part in the regeneration response (Lazarov-Spiegler et al. 1998). However, data show that macrophages phagocytose the myelin and axon debris in the distal segment of the injured nerve but play a rather minor role in the regeneration of the peripheral nerve fibres; it is the peripheral glial cell, the Schwann cell, that is key to successful regeneration (Hughes and Perry 1999).

3 References

Aloisi F (2001) Immune function of microglia. Glia 36:165–79
Ashwell K W S (1991) The distribution of microglia and cell death in the fetal rat forebrain. Dev Brain Res 58:1–12
Bandtlow CE and Zimmermann DR (2000) Proteoglycans in the developing brain: new conceptual insights for old proteins. Physiol Rev 80:1267–1290
Barclay AN, Wright, GJ, Brooke G and Brown MH (2002) CD200 and membrane protein interactions in the control of myeloid cells. Trends Immunol 23:285–290
Bell MD, Lopez-Gonzalez R, Lawson L, Hughes D, Fraser I, Gordon S and Perry VH (1994) Upregulation of the macrophage scavenger receptor in response to different forms of injury in the CNS. J Neurocytol 23: 605–613
Braun J S, Kaisling B and Le Hir M (1993) Cellular components of the immune barrier in the spinal meninges and dorsal root ganglia of the rat: immunohistochemical (MHC Class II) and electron microscopic observations. Cell Tissue Res 273: 209–217
Brenan M and Puklavec M (1992) The MRC OX-62 antigen: a useful marker in the purification of rat veiled cells with the biochemical properties of an integrin. J Exp Med 175: 1457–1465
Brown HC, Townsend MJ, Fearn S and Perry VH (1998) An adhesion molecule for macrophages in the brain. J Neurocytol 27: 867–876
Cserr HF and Knopf PM (1992) Cervical lymphatics, the blood-brain barrier and the immunoreactivity of the brain: a new view. Immunology Today 13: 507–512
Diez-Roux G and Lang RA (1997) Macrophages induce apoptosis in normal cells in vivo. Development 124: 3633–3638
Ford AL, Foulcher E, Lemckert FA and Sedgwick,JD (1996) Microglia induce CD4 T lymphocyte final effector function and death. J Exp Med 184: 1737–1745
Giulian D, Li J, Bartel S, Broker J, Li X and Kirkpatrick JB (1995) Cell surface morphology identifies microglia as a distinct class of mononuclear phagocyte. J Neurosci 15: 7712–7726
Gordon LB, Knopf PM and Cserr HF(1992) Ovalbumin is more immunogenic when introduced into brain or cerebrospinal fluid than into extracerebral sites. J Neuroimmunol 40: 81–7
Graeber MB, Streit WJ and Kreutzberg GW (1989) Identity of ED2-positive perivascular cells in rat brain. J Neurosci Res 2: 103–106
Hickey WF and Kimura H (1988) Perivascular microglia are bone marrow derived and present antigen in vivo. Science 239: 290–292
Hickey WF, Vass K and Lassmann H (1992) Bone marrow derived elements in the central nervous system: an immunohistochemical and ultrastructural analysis of rat chimera. J Neuropath Exp Neurol 51: 2466–256
Hoek RM, Ruuls SR, Murphy CA, Wright GJ, Goddard R, Zurawski SM, Blom B, Homola ME, Streit WJ, Brown MH, Barclay AN and Sedgwick JD (2000) Downregulation of the macrophage lineage through interaction with 0X2 (CD200). Science 290: 1768–71
Hughes PM and Perry VH (1999) The role of macrophages in degeneration and regeneration in the peripheral nervous system. In: "Degeneration and Regeneration in the Nervous System". (ed N. R. Saunders and KM Dziegielewska) Harwood, Amsterdam
Innocenti GM, Koppel H and Clarke S (1983) Transitory macrophages in the white matter of the developing visual cortex. I. Light and electron microscopic characteristics and distribution. Dev. Brain Res 11: 39–53
Kaur C, Hao AJ, Wu CH and Ling EA (2001) Origin of microglia. Microsc Res Tech 54: 2–9

Kiefer R, Streit WJ, Toyka KV, Kreutzberg GW and Hartung HP (1995) Transforming growth factor-beta 1: a lesion-associated cytokine of the nervous system. Int J Dev Neurosci 13: 331–339

Konsman JP, Parnet P and Dantzer R (2002) Cytokine-induced sickness behaviour: mechanisms and implications. Trends Neurosci 25: 154–159

Kreutzberg GW (1996) Microglia: a sensor for pathological events in the CNS. Trends Neurosci 19: 312–318

Lang RA and Bishop JM (1993) Macrophages are required for cell death and tissue remodelling in the developing mouse eye. Cell 74: 453–462

Lawson U, Perry VH, Dri P and Gordon S (1990) Heterogeneity in the distribution and morphology of microglia in the normal adult mouse brain. Neuroscience 39: 151–170

Lazarov-Spiegler O, Rapalion O, Agranov G, Schwartz M (1998) Restricted inflammatory reaction in the CNS: a key impediment to axonal regeneration? Nature Medicine 337–342

Ling EA and Wong WC (1993) The origin and nature of ramified and amoeboid microglia: a historical review and current concepts. Glia 7: 9–18

Mato M, Ookawara S, Sakamoto A, Aikawa E, Ogawam T, Mitsuhashi U, Masuzawa T, Suzuki H, Honda M, Yazaki Y, Watanabe E, Luoma J, Yla-Herttuala S, Fraser I, Gordon S and Kodama T (1996) Involvement of specific macrophage-lineage cells surrounding arterioles in barrier and scavenger function in brain cortex. Proc Natl Acad Sci USA 93:3269-3274

Matyszak MK, Lawson LJ, Perry VH and Gordon S (1992) Stromal macrophages of the choroid plexus situated at an interface between the brain and the peripheral immune system constitutively express MHC Class II antigens. J Neuroimmunol 40: 173–182

Matyszak, M.K. & Perry VH (1995) Demyelination in the central nervous system following a delayed-type hypersensitivity response to bacillus Calmette-Guérin. Neuroscience 64: 967–977

Matyszak MK and Perry VH (1996) The potential role of dendritic cells in immune mediated inflammatory responses in the central nervous system. Neuroscience 74: 599–608

McKercher SR, Torbett BE, Anderson KL, Henkel GW, Vestal DJ, Baribault H, Klemsz M, Feeney AJ, Wu GE, Paige CJ, and Maki RA (1996) Targeted disruption of the Pu. 1 gene results in multiple hematopoietic abnormalities. EMBO J 15: 5647–5658

McMenamin PG (1999) Distribution and phenotype of dendritic cells and resident tissue macrophages in the dura mater, leptomeninges, and choroid plexus of the rat brain as demonstrated in wholemount preparations. J Comp Neurol 405: 553–62

Minghetti L, Walsh DT, Levi G and Perry VH (1999) In vivo expression of cyclooxygenase-2 in rat brain following intraparenchymal injection of bacterial endotoxin and inflammatory cytokines J. Neuropath Exp Neurol 58: 1184–1191

Neumann H, Boucraut J, Hahnel C, Misgeld T and Wekerle H (1996) Neuronal control of MHC class II inducibility in rat astrocytes and microglia. Eur J Neurosci 8: 2582–90

Neumann H, Misgeld T, Matsumuro K and Wekerle H (1998) Neurotrophins inhibit major histocompatibility class II inducibility of microglia: involvement of the p75 neurotrophin receptor. Proc Natl Acad Sci U S A 95: 5779–84

Nguyen MD, Julien JP and Rivest S (2002) Innate immunity: the missing link in neuroprotection and neurodegeneration? Nat Rev Neurosci. 3: 216–27

Perry VH (1994) Macrophages and the Nervous System. R.G. Landes Company, Austin pp 1–123

Perry VH (1998) A revised view of the central nervous system microenvironment and major histocompatibility complex class II antigen presentation. J Neuroimmunol 90: 113–21

Perry VH, Crocker PR and Gordon S (1992) The blood-brain barrier regulates the expression of a macrophage scialic acid-binding receptor on microglia. J Cell Sci 101: 201–207

Perry VH, Andersson P-B and Gordon S (1993) Macrophages and inflammation in the central nervous system. Trends Neurosci 16: 268–273

Perry VH and Gordon S (1991) Macrophages and the nervous system. Int Rev Cytol 125: 203–244

Perry VH, Hume DA and Gordon S (1985) Immunohistochemical localization of macrophages and microglia in the adult and developing mouse brain. Neuroscience 15: 313–326

Perry VH, Cunningham C and Boche D (2002) Atypical inflammation in the central nervous system in prion disease. Curr Opin Neurol 15: 349–354

Pow DV, Perry VH, Morris JF and Gordon S (1989) Microglia in the neurohypophysis associate with and endocytose terminal portions of neurosecretory neurons. Neuroscience 33: 567–578

Ransohoff RM and Tani M (1998) Do chemokines mediate leukocyte recruitment in posttraumatic CNS inflammation? Trends Neurosci 21: 154–159

Schmidtmayer J, Jacobsen C, Miksch G, and Sievers J (1994) Blood monocytes and spleen macrophages differentiate into microglia-like cells on monolayers of astrocytes: membrane currents. Glia 12: 259–267

Sievers J, Parwaresch R, Wottge HU (1994) Blood monocytes and spleen macrophages differentiate into microglia-like cells on monolayers of astrocytes: morphology. Glia 12: 245–258

Stevenson PG, Hawke S, Sloan DJ and Banghain CRM (1997) The immunogenicity of intracerebral virus infection depends on anatomical site. J of Virology 71: 145–151

Streit WJ, Graeber MB and Kreutzberg GW (1989) Expression of Ia antigen on perivascular and microglial cells after sublethal and lethal motor neuron injury. Exp Neurol 105: 115–126

Strle K, Zhou JH, Shen WH, Broussard SR, Johnson RW, Freund G G, Dantzer R and Kelley KW (2001) Interleukin-10 in the brain. Crit Rev Immunol 21: 27–49

Walsh DT, Betmouni S and Perry VH (2001) Absence of detectable IL-1β production in murine disease: a model of chronic neurodegeneration. J Neuropath Exp Neurol 60: 173–182

Innate Recognition of Viruses by Macrophage and Related Receptors: Potential Ligands for Antiviral Agents

J. L. Miller · S. Gordon

Sir William Dunn School of Pathology, Oxford University,
South Parks Road, Oxford, OX1 3RE, UK
e-mail: christine.holt@pathology.oxford.ac.uk

Abstract Resident tissue macrophages are ideally placed for a role in the first line of defence against invading viruses and other microorganisms. Binding of viruses to macrophages can occur via a variety of molecules including the direct use of proteins, lipids and oligosaccharides as receptors, as well as opsonic recognition of viruses by the macrophage. Often multiple surface components are involved in virus–macrophage interactions, and interestingly a number of lectin receptors are expressed on macrophages for which viral ligands have been proposed. This review focuses on the initial viral attachment to macrophages and describes the interactions of certain macrophage lectin, proteoglycan and Toll-like receptors with viruses. The study of the role of pattern recognition receptors in the macrophage response to viruses may reveal new features of viral pathogenesis and some of the interactions could provide targets for antiviral agents.

Keywords Attachment, Collectin, Lectin, Macrophage, Proteoglycan, Receptor, Virus

Abbreviations

APC	Antigen-presenting cell
CMV	Cytomegalovirus
CRD	Carbohydrate recognition domain
DC	Dendritic cell
GAG	Glycosaminoglycan
HSV	Herpes simplex virus
LPS	Lipopolysaccharide
MBL	Mannan-binding lectin
MR	Mannose receptor
PAMP	Pathogen-associated molecular pattern
PRR	Pattern recognition receptor
RSV	Respiratory syncytial virus
TLR	Toll-like receptor

1
Introduction

Innate immune mechanisms are critical in limiting the spread of infection within the host and provide protection while specific immunity develops. These processes are rapid and non-specific in the sense of being active against a broad spectrum of microbial pathogens. Innate immunity to pathogens is thought to be triggered by pattern recognition receptors (PRR) on antigen-presenting cells (APC) that detect and respond to conserved structural motifs on invading microorganisms. Toll-like receptors (TLR) have been proposed to play an essential role in linking innate and acquired immunity through their ability to fulfil the role of the PRR on the APC, as they respond to defined lipid, protein and nucleic acid components of microorganisms and signal cytokine production and co-stimulatory molecule induction (Akira et al. 2001). Other PRR which recognise pathogen-associated motifs include complement, CD14, calcium-dependent (C-type) lectins [mannose receptor (MR), mannan-binding lectin (MBL), DC-SIGN], and scavenger receptors. Much of our knowledge of the interaction and function of innate immune receptors in microbial infections has to date focussed on interactions with bacteria, mycobacteria, fungi and parasitic pathogens, whereas the role of these receptors in response to viruses is largely unexplored.

Macrophages are a major cellular element in the clearance and inactivation of viral pathogens. Resident tissue and blood macrophages are ideally placed to play an important role in mediating the innate interaction between the host and virus. Their ability to phagocytose free virus, or virus opsonised by serum collectins, complement or antibody, and hence clear virions from the circulation, constitutes an early line of defence for reducing virus load. Subsequent to phagocytosis, the associated intracellular killing of viruses contributes to non-productive viral clearance. In addition to immediate clearance of invading viruses, macrophages phagocytose virus-infected cells, along with apoptotic cells or necrotic cells, earning their reputation as the most efficient phagocytes in the body.

Viral interaction with, and infection of, macrophages can stimulate these cells to release a number of proinflammatory and immunoregulatory cytokines and chemokines, along with proteolytic enzymes and oxygen radicals in vitro and in vivo (Julkunen et al. 2000; Guidotti and Chisari 2001). These factors act to limit virus spread and initiate a series of immune reactions including the attraction of lymphocytes, neutrophils and mast cells to the site of virus production, inviting them to play a role in host defence. The production of interferon-α and interferon-β by macrophages can further limit virus spread through its ability to induce an antiviral state in uninfected cells.

The activation of the innate immune response is a prerequisite for the triggering of acquired immunity. It is through modulating the threshold of activation of adaptive antigen-recognition receptors and by inducing key costimulatory molecules and cytokines, that innate immunity may shape the nature of the response and determine to which antigens the acquired immune system responds (Fearon and Locksley 1996). Recognition of antigenic arrays or pathogen-associated molecular patterns (PAMPs) by PRR (Medzhitov and Janeway 1997) on macrophages and on dendritic cells (DC) has been hypothesised to induce activation of defence mechanisms in the cell and the maturation of the cell into an efficient APC that subsequently attracts and activates the antigen-specific T cells essential for an adaptive immune response (Fearon and Locksley 1996; Janeway 1989). Mononuclear phagocytes, especially DC, act as APC, taking up, processing and delivering viral antigens to T cells in regional lymph nodes. While DCs are generally regarded as being the most efficient APC, macrophages can also present viral antigens to primed T lymphocytes. Through their function as APCs and their secretion of inflammatory and immunoregulatory cytokines, macrophages form an important bridge between innate and acquired antiviral immune responses.

In addition to their role in viral defense, macrophages are targets for viral infection and may provide an infectious reservoir for persistent viruses such as lentiviruses and cytomegaloviruses (CMV). For example, the cytopathic effects observed following HIV-1 infection of macrophages appear to be minimal, and infected macrophages remain viable; hence, macrophages may provide a mechanism of viral dissemination. In a model system of rhesus macaque monkeys infected with a simian immunodeficiency virus (SIV)/HIV-1 chimera, tissue macrophages were identified as the principal reservoir of virus after the depletion of $CD4^+$ T cells (Igarashi et al. 2001). This system, which is thought to be analogous to the late stages of HIV-1 infections in humans, showed that tissue macrophages sustained high plasma virus loads, implicating tissue macrophages as an important reservoir of virus in vivo. Measles and mumps viruses also avoid killing within macrophages, enabling these viruses to utilise the migratory properties of these cells to disseminate to the respiratory tract and salivary glands, respectively.

Table 1 Examples of some virus receptors found on macrophages

Receptor	Virus	Reference(s)
Protein receptors		
CD4	HIV	Collman et al. 1990; Weiss 2002
Poliovirus receptor	Polioviruses	Mendelsohn et al. 1989; Freistadt and Eberle 2000
Undefined macrophage membrane proteins	Dengue virus	Moreno et al. 2002
TLR3	ds viral RNA	Alexopoulou et al. 2001
Carbohydrate receptors		
Sialic acid-containing oligosaccharides	Influenza virus	Wilson et al. 1981; Weis et al. 1988
Heparan sulphate	Human CMV	Compton et al. 1993
	HSV	Herold et al. 1994
Antibody-dependent enhancement of viral entry		
Fc receptors (via bound immunoglobulin)	Dengue virus	Daughaday et al. 1981
	West Nile virus	Peiris et al. 1981; Cardosa et al. 1986
	HIV	Takeda et al. 1988
	CMV	Inada et al. 1985
	Influenza virus	Ochiai et al. 1988; Tamura et al. 1991

See text for definitions of abbreviations.

Viruses generally infect cells by endocytosis (rather than phagocytosis) or by fusion with the plasma membrane. Entry of a virus is often a multistep process, including initial attachment of the virus to the target cell surface, followed by fusion between the viral and cellular membranes and culminating in the internalisation of the viral genome into the cytosol of the target cells. The extracellular receptors used by viruses for attachment and entry are, presumably, receptors that serve other functions in the host. In some cases, host molecules are discovered as virus receptors prior to determination of their natural ligands and functions in normal host physiology. Given that host proteins are being used by viruses for attachment, choosing such receptors as antiviral targets will provide the challenge to designers of anti-viral drugs, to block viral binding without interfering with the normal functions of the host cell receptors.

Virus particles have multiple avenues for engaging cells and can interact with many molecules on the macrophage cell surface. The composition of a virus allows it to bind in a more complex manner to macrophages, than, for example, molecules such as lipopolysaccharide (LPS). Binding of viral lipid to CD1 or to annexins can occur, as well as protein–protein and lectin-carbohydrate interactions, be they on the cell surface or in endocytic vesicles. Hence, multiple receptors may come into play when examining recognition of viruses by macrophages and, whether multistep or involving a single molecule, the interactions can have a range of possible outcomes. This review will principally focus on the initial vi-

Table 2 Soluble and cellular C-type lectin receptorsfor viruses

Receptor	Location	Viral ligand	Reference(s)
MR	Most macrophages, immature DC	HIV (gp120)	Larkin et al. 1989; Curtis et al. 1992
		Influenza virus	Reading et al. 2000
DC-SIGN	Immature DC, lung and placental macrophages	HIV-1, HIV-2 SIV	Curtis et al. 1992; Geijtenbeek et al. 2000; Pohlmann et al. 2001
		Ebola virus	Alvarez et al. 2002
DC-SIGNR	Endothelial cells in liver and lymph node	HIV-1, HIV-2	Pohlmann et al. 2001
SP-A	Lung surfactant	Influenza virus	Benne et al. 1995; Benne et al. 1997
		RSV	Ghildyal et al. 1999; Hickling et al. 1999; Barr et al. 2000
		HSV	van Iwaarden et al. 1991, 1992
		CMV	Weyer et al. 2000
SP-D	Lung surfactant	Influenza virus	Hartshorn et al. 1994; Reading et al. 1997; Hartshorn et al. 2000
		RSV	Hickling et al. 1999; LeVine et al. 1999
MBL	Serum	Influenza virus	Anders et al. 1990; Hartshorn et al. 1993; Malhotra et al. 1994
		HIV	Ezekowitz et al. 1996

See text for definitions of abbreviations.

ral attachment to macrophages. A selection of macrophage molecules that function as receptors for a number of viruses will be discussed, including receptors involved in both defence mechanisms against viruses and those commandeered by viruses for entry and infection. Identification of cellular receptors for viruses and inhibition of macrophage infection by viruses is of particular interest, since this is an important target for clinical therapy and could be used to prevent viral infection and progression of disease.

2
Interaction of Macrophage Receptors With Viruses

The interaction of macrophages with viruses can be mediated by a diverse set of receptors. Table 1 gives some examples of the types of viral receptors found on macrophages. In some cases the expression of protein receptors can explain the species, tissue and cell tropism of a virus, whereas carbohydrate receptors tend to be more broadly expressed on different cell types and hence do not always directly account for the tropism of a virus. An important mechanism of entry into macrophages for some viruses is via complexes with subneutralising amounts of virus-specific antibody, binding to Fc receptors. Macrophages may also be able to detect entry and replication of virus, using the TLR3 receptor that has recently been shown to recognise double-stranded (ds)RNA, the molecular intermediate associated with many viral infections (Alexopoulou et al. 2001).

A number of viruses bind to multiple receptors during the infection process, a fact which may be important in clustering of different receptors to generate novel molecular assemblies with distinctive properties. For instance, interactions with one receptor may be required to bring about conformational change or endocytosis required for fusion, or for binding to a secondary receptor. Distinguishing between binding of viruses to "productive receptors" (Dimmock 1982) that lead to infection and non-specific binding to cell surface receptors is a key issue that is further complicated by the hypothesis that such non-specific binding to the cell may be important in concentrating virus particles on the cell surface for subsequent interactions with a specific low-affinity host cell receptor.

2.1 Macrophage Lectin Receptors

Lectin receptors can recognise carbohydrate on the surface of micro-organisms, and so selectively alert the innate immune system to the presence of potential pathogens. Such foreign carbohydrate motifs are found on a variety of micro-organisms, and hence lectin receptors may bind to a number of different microbial pathogens. For example, the MR binds to yeast, HIV, influenza virus and a number of bacteria. While bacterial sugars are encoded for and produced by bacteria, viral sugars are, by contrast, produced by the host, for the most part. This provides a greater challenge for recognition and specific targeting of viruses by the immune system, which has evolved to recognise foreign molecules.

The roles of macrophage lectins in host defence against bacteria and yeasts have been well reviewed elsewhere (Linehan et al. 2000) and this review will focus on their known and possible interaction with viruses. Table 2 summarises known interactions between viruses and soluble and cellular C-type lectins, some of which are described in more detail in the following text. The soluble collectins are included in this discussion, although they are not macrophage receptors, as they have similar carbohydrate specificity to that of the macrophage mannose receptor, and they opsonise microbes for their uptake by phagocytes (Holmskov et al. 1994; Hoppe and Reid 1994; Ezekowitz et al. 1996).

2.1.1 Mannose Receptor

The MR (CD206) is an integral membrane C-type lectin expressed on the surface of alveolar and other tissue macrophages. A multilectin receptor, the MR has specificity for two groups of sugar ligands; the 8 carbohydrate recognition domains (CRDs) define the specificity for mannose, fucose and N-acetylglucosamine (Taylor et al. 1990, 1992), while the N-terminal cysteine-rich domain of the MR binds to sulphated terminal N-acetylgalactosamine residues (Fiete et al. 1998; Martinez et al. 1999; Leteux et al. 2000). Capable of binding to a wide range of ligands, the MR is a prototypical PRR (Stahl and Ezekowitz 1998). It is a characteristic marker of macrophages but in addition it can also be expressed on DC (Dong et al. 1999) and selected endothelial cells.

The MR has been implicated as a major endocytic receptor in the infectious entry of influenza virus. The ability of influenza virus to infect macrophages was shown to correlate both with the levels of MR expression on the macrophage and with the level of mannose-containing oligosaccharide present on the haemagglutinin molecule of the virus (Reading et al. 2000). Furthermore, purified glycoproteins from influenza virus inhibited the binding of mannosylated bovine serum albumin (BSA), a ligand of the MR, to peritoneal macrophages. Periodate treatment of the viral glycoproteins to oxidise their carbohydrate, prior to their inclusion in the assay, reduced their inhibitory capacity, consistent with a direct interaction of the influenza virus glycoproteins with the lectin domains of the MR.

The envelope glycoprotein of HIV, gp120, is a heavily glycosylated molecule implicated in binding to the MR (Daughaday et al. 1981; Curtis et al. 1992). Carbohydrate represents approximately half of the molecular weight of gp120 (Allan et al. 1985; Ratner et al. 1985), and when grown in human T cells, the oligosaccharide moieties of gp120 comprise approximately 50% oligomannosidic species, with fucosylated complex- and hybrid-type oligosaccharides also present (Geyer et al. 1988), making this viral glycoprotein a good potential ligand for the MR as well as other lectin receptors. Evidence for the interaction of gp120 with the MR comes from binding studies which showed that gp120 binding to a MR immobilised on beads was inhibitable by mannosylated-BSA but not by soluble CD4 glycoprotein. Additionally, binding of mannosylated-BSA to human peripheral blood monocyte-derived macrophages was inhibited by recombinant gp120, but not by soluble CD4 (Larkin et al. 1989).

Given the wide range of bacteria, yeasts, parasites and mycobacteria recognised by the MR (Fraser et al. 1998, and references therein), it is surprising that so little work has investigated its interaction with viruses. Further research needs to be performed on examining viral ligands for the MR. It should also be noted that the use of mannosylated-BSA as a blocking agent in the work described above does not conclusively demonstrate that these viruses interact with the MR, due to the existence of other mannose-specific lectins on macrophages (Imamura et al. 1984; Fernandes et al. 1999). Available anti-human MR monoclonal antibodies and the recent development of monoclonal antibodies specific for the murine MR (L. Martinez-Pomares, personal communication) will be useful in clarifying viral interactions with this receptor.

2.1.2 DC-SIGN and Also DC-SIGNR

The type II membrane protein DC-SIGN (dendritic cell-specific ICAM-3 grabbing nonintegrin; CD209) has one mannose-binding C-type lectin extracellular domain. This protein is proposed to function normally in binding to ICAM-2 expressed on endothelial cells, to promote transmigration of DC from the blood to lymphoid tissues, and in binding to ICAM-3 on resting T cells, contributing to T-cell activation (Geijtenbeek et al. 2000a,b). DC-SIGN is better studied on immature DC, where it is highly expressed (Geijtenbeek et al. 2000b) but is also found on macrophages (Soilleux et al. 2002). The related molecule, DC-SIGNR, is found on liver sinusoidal endothelium and endothelium of lymph node sinuses and placental villi, and its expression on macrophages has not yet been documented. Both of these lectins, DC-SIGN and DC-SIGNR bind to HIV-1, HIV-2, SIV and Ebola virus (see references in Table 2).

DC-bearing DC-SIGN have been proposed to play a role in the establishment of HIV infection and may be critical in acquiring virus and transmitting it to T lymphocytes. DC-SIGN does not act as a receptor for entry of HIV into cells, but facilitates viral infection in *trans* of target $CD4^{+}$ T cells (Geijtenbeek et al. 2000b; Pohlmann et al. 2001). While mannan very efficiently blocks transmission of HIV-1 by cells transfected with DC-SIGN, mannan (or DC-SIGN-specific monoclonal antibodies) only partially block HIV transmission by DC, suggesting that other factors may be utilised by the virus, in addition to DC-SIGN, for DC-mediated HIV-1 transmission to target cells (Wu et al. 2002b). Additional support for this hypothesis comes from the observation that DC from rhesus macaques do not express DC-SIGN and are still able to efficiently transmit primate lentiviruses (Wu et al. 2002a).

DC-SIGN-mediated transmission of HIV to T cells does not appear to require the ICAM-3 binding activity of the DC-SIGN molecule, and binding to HIV may not be mediated by a lectin interaction. Binding to ICAM-3 requires calcium and is dependent on glycosylation of ICAM-3, whereas in contrast, binding of DC-SIGN to the HIV-1 envelope glycoprotein is independent of N- and O-linked glycosylation (Geijtenbeek et al. 2002). A mutant of DC-SIGN that no longer bound to ICAM-3, but retained a specificity for gp120, was still able to mediate HIV-1 infection of T cells efficiently in *trans*, suggesting that virus transmission was not dependent on DC-SIGN interactions with ICAM-3. Thus, it appears that there are overlapping but distinct sites for DC-SIGN binding to ICAM-3 and to gp120. This may aid the development of inhibitors of HIV attachment to DC-SIGN that have reduced negative effects on the normal function and roles of DC-SIGN.

While research has focussed on DC-mediated transfer of HIV to T cells, now that macrophages have also been shown to bear DC-SIGN, it will be important to assess the degree to which they have roles in carrying virus from the periphery into lymph nodes.

2.1.3
Collectins

Endogenous lectins, including soluble collagenous C-type lectin members of the collectin family, have an important role in innate defence against respiratory viruses (Sastry and Ezekowitz 1993; Malhotra and Sim 1995; Crouch et al. 2000). These proteins are known to bind to influenza virus, respiratory syncytial virus (RSV), herpes simplex virus (HSV), HIV and CMV (see references in Table 1) and display antiviral activity in vitro and in vivo. Members of this family found in humans include serum MBL, and the lung surfactant proteins (SP)-A and -D. Collectins use multivalent lectin-like domains coupled to collagenous stalks for the recognition and opsonisation of microbes for uptake by macrophages. Much of the investigation of the interaction of collectins with viruses has been performed with influenza virus.

The in vitro antiviral activities of collectins against influenza virus include virus neutralisation and aggregation, opsonisation for contact with neutrophils, and lysis of virus-infected cells in the presence of complement (Reading et al. 1995; Hartshorn et al. 1997; Anders et al. 2001) and are consistent with a role for these molecules in first line host defence. MBL and SP-D bind to glycans on the influenza virus haemagglutinin (HA) and neuraminidase glycoproteins, whereas the interaction of SP-A with viruses involves binding of the viruses to SP-A-associated carbohydrates (Benne et al. 1995; van Iwaarden et al. 1991, 1992). Bovine SP-D has been shown to inhibit infectivity of rotaviruses through Ca^{2+}-dependent, mannose-inhibitable attachment to the major viral envelope glycoprotein (Reading et al. 1998b).

Only more recently have roles for collectins in innate defense against influenza virus been demonstrated in vivo. Studies on the sensitivity of a range of strains of influenza virus to collectin-mediated neutralisation revealed a marked inverse correlation between collectin sensitivity and the ability of a virus to replicate in the mouse lung after intranasal inoculation (Reading et al. 1997). Co-administration of mannose-containing oligosaccharides along with virus resulted in markedly increased replication of influenza A virus in the lung. Although MBL levels were undetectable in lavage from normal or influenza-infected mice, SP-D levels were detected and increased several fold after influenza A virus infection. Strains of influenza virus bearing higher levels of glycosylation on the HA molecule grew very poorly in the mouse lung, whereas the A/PR/8/34 strain, which grows to high titres in mouse lung, carries no glycans on the head of its HA molecule and is essentially resistant to neutralisation by SP-D and MBL. Overall, these results provided strong evidence that lung collectins, in particular SP-D, contribute significantly to containment of influenza infection in vivo.

There are many strong indications for roles of SP-D and SP-A in innate immunity. Reading et al. showed that compromise of SP-D, due to elevated glucose levels in the lungs of diabetic mice appeared to be the major factor contributing to the increased susceptibility of diabetic mice to influenza virus (Reading et al.

1998a). The study of infection in the SP-D knockout mouse is hampered by a phenotype sharing features of alveolar lipoproteinosis disease, with activated macrophages and accumulation of surfactant in the alveolar spaces (Botas et al. 1998; Korfhagen et al. 1998). Nevertheless, SP-D-deficient mice showed decreased clearance (compared to wild-type mice), of a heavily influenza virus (LeVine et al. 2001). In contrast to the proteinosis observed in the lung of the SP-D knockout mouse, the SP-A knockout mouse has only marginal defects in surfactant homeostasis and respiratory function, and studies on the immune response to RSV, adenovirus and other pathogens have confirmed its role as an innate immune protein (LeVine et al. 1999; Harrod et al. 1999; Lawson and Reid 2000). Influenza virus and RSV clearance from the lungs of SP-A knockout mice was significantly decreased compared to wild-type SP-$A^{+/+}$ mice and was associated with increased neutrophil and lymphocyte numbers in the bronchial alveolar lavage along with increased titres of proinflammatory cytokines in lung homogenates (LeVine et al. 1999, 2002).

Together these observations strongly indicate that the ability of these respiratory viruses to replicate in the lungs of mice is limited by lectin-mediated defence mechanisms. Furthermore, these studies identify viral carbohydrate as a ligand for recognition of virus by collectins of the innate immune system and are consistent with the hypothesis that glycosylation can affect virulence through mediating susceptibility to such innate mechanisms as collectins and phagocytic lectin receptors. Collectins can be thought of as soluble receptors that bind to PAMPs, to opsonise particles for uptake by macrophages; however, the receptors for collectins are still poorly defined. The antiviral activity of collectins against a number of enveloped viruses supports the hypothesis that collectins represent a general innate defence mechanism of natural resistance.

3 Other Lectin Receptors

Other lectin receptors have been described on the macrophage, and a number of novel lectin receptors have recently been discovered (Balch et al. 1998; Bakker et al. 1999; Bates et al. 1999; Matsumoto et al. 1999; Brown and Gordon 2001). For many of these molecules neither the sugar specificity nor a function has yet been determined. However, by extrapolation from observations with the MR, DC-SIGN and collectins, these lectin receptors are potential PRR that may bind a wide selection of micro-organisms, including viruses. Viral ligands for these macrophage receptors have not been identified as yet. Some receptors that were described initially on DC have subsequently been shown to have broader cell expression. For example, DC-SIGN was originally isolated from human placental cDNA library (Curtis et al. 1992) and was described as being exclusively expressed on DC (Geijtenbeek et al. 2000c); however, it is now realised that this receptor is also expressed on alveolar and placental macrophages, as well as $BDCA2^{+}$ plasmacytoid peripheral blood DC precursors in situ and in vivo (Soilleux et al. 2002).

Of note, DEC-205 (CD205) is topologically very similar to the MR and expressed on DC and macrophages. This multilectin receptor has 10 CRDs and appears to mediate uptake of glycosylated antigens, although to date no ligand specific for any of the CRDs in DEC-205 has been identified. The similarity of DEC-205 to the MR suggested that it too may be involved in recognition of pathogens and play a role in innate immunity and in antigen processing. Interestingly, none of the CRDs of DEC-205 have conserved the key amino acids involved in carbohydrate and calcium binding, consequently it is unlikely that the CRDs of this receptor have lectin activity (Inaba et al. 1995; Jiang et al. 1995; Swiggard et al. 1995).

Finally, viral glycoproteins themselves can also function as lectins, and mediate binding to cell-surface oligosaccharides. Influenza virus initiates infection by binding of the viral HA glycoprotein to terminal N-acetylneuraminic (sialic) acid-containing receptors on the cell surface (Wilson et al. 1981; Weis et al. 1988). The receptor specificity of the influenza virus HA has been well studied and depends on a number of amino acid residues in the receptor-binding pocket of the HA which are involved in either direct or stabilising interactions with sialic acid. Experiments have also suggested that the HIV gp120 glycoprotein may have lectin-like properties. Based on studies examining transcellular transport of gp120-coated particles, Kage et al. suggested that the gp120 glycoprotein of HIV-1 may contain a lectin-like domain that interacts with mannosyl residues on the mucosal surface (Kage et al. 1998).

3.1 TLR

Janeway proposed that leukocytes must have molecules that are able to recognise antigenic arrays or patterns that are generic to micro-organisms and absent from host cells (Janeway 1992). Such PAMPs include bacterial LPS, peptidoglycan and lipoteichoic acid of gram-positive bacteria, unmethylated CpG deoxynucleotide motifs in prokaryotic DNA, and mannans on fungi and viruses, and are recognised by both soluble and cell-associated PRR. Members of the TLR family function as PRR in mammals and have been intensively investigated recently. TLR discriminate between a variety of microbial products, for example TLR2 confers responsiveness to lipoproteins and several gram-positive bacteria, TLR4 binds to LPS, TLR5 mediates responsiveness to flagellin and TLR9 recognises unmethylated DNA containing CpG motifs (CpG-DNA) (Akira et al. 2001).

To date, only TLR4 and TLR3 have been implicated as being involved in innate immunity to viruses. The proinflammatory cytokine response to the fusion protein of RSV in vitro, and cellular responses and viral clearance following intranasal infection with live virus were reduced in mice mutated in the gene encoding TLR4 (Kurt et al. 2000; Haynes et al. 2001). In addition, TLR4 has recently been identified as one of the components mediating activation of B cells in response to the retroviruses mouse mammary tumour virus and Moloney murine leukaemia virus (Reading et al. 1998a). The recognition of dsRNA by TLR3 induces activation of nuclear factor (NF)-κB (in T cells and macrophages) and the production of type 1 interferon (Alexopoulou et al. 2001), suggesting a role for this TLR in the antiviral response; however, this remains to be established. It is of interest that influenza or Sendai infection of human macrophages, or exogenously added interferon-α, enhanced mRNA expression of TLR1, TLR2, TLR3 and TLR7, and downregulated TLR5 mRNA (Miettinen et al. 2001), suggesting that control of TLR expression by type 1 interferon may be a novel mechanism by which interferon can modulate the innate immune response.

Subversion of a host defence mechanism by viruses gives another indication of the importance of that defence mechanism. Vaccinia virus carries two proteins products, A46R and A52R, with similar amino acid structure to Toll/interleukin (IL)-1 receptor domains (that define the TLR family), through which IL-1 and TLR4 signalling can be inhibited (Bowie et al. 2000). The evolution of these antagonistic proteins by vaccinia highlights the importance of TLR-mediated mechanisms in immune defense against this virus.

TLR recognise a heterogeneous variety of ligands; however, it is not known what viral structure might be recognised by TLR. It may be that other molecules and/or receptors assist TLR recognition of viruses, analogous to the manner that cellular CD14 or secreted MD-2 bind to LPS to facilitate its interaction with TLR4. A functional receptor complex may be involved in TLR interaction with virus, which may contain lectins, chemokines, etc. that coordinate to facilitate recognition, binding and signal transduction in response to virus binding. Some TLR have been shown to be located intracellularly, in which case co-receptors may be essential for delivery of virus, or viral antigens, to the TLR. Alternatively the natural pathway of virus infection may result in the virus meeting an intracellular TLR. Future studies in this area should clarify viral interactions with TLR.

3.2 Proteoglycans

Proteoglycan molecules, present on the surface of mammalian cells, and in the extracellular matrix, are composed of a membrane-linked protein core with attached, variously charged sulphated glycosaminoglycans (GAGs). Proteoglycans exist as both pure molecules, which have either only heparan sulphate or chondroitin sulphate GAG chains, and those with mixtures of different GAGs. The GAG chains consist of alternating residues of an amino sugar and an uronic acid, and as the chains polymerise, varying amounts of sulphation and epimerisation result in a large degree of heterogeneity amongst proteoglycans. Proteoglycans can be separated based on the charge characteristics of the GAG chains, with greater levels of sulphation corresponding to a higher charge. Heparin is a highly sulphated version of heparan sulphate and one of the most highly sulphated GAGs.

Proteoglycans on the cell surface are used by a number of viruses as cellular adhesion receptors, with at least 10 human pathogenic virus infections implicated in binding to heparan sulphate (Rostand and Esko 1997; Wadstrom and Ljungh 1999; Bose and Banerjee 2002). For some viruses the initial binding to heparan sulphate-containing proteoglycans appears to function primarily to concentrate the virus particles at the cell surface, before interaction with additional, higher affinity receptors. This is the case for foot-and-mouth disease, which has a primary interaction with heparan sulphate followed by binding to the RGD motif of the $\alpha_V\beta3$ integrin (Jackson et al. 1996). Other viruses, such as HSV, appear to be able to utilise heparan sulphate directly as an entry receptor (Shukla et al. 1999).

HIV and CMV are examples of viruses for which primary binding to cell-surface GAGs appears to assist viral interaction with secondary receptors required for entry. Heparan sulphate on the cell surface is also able to initiate binding of CMV (Neyts et al. 1992; Compton et al. 1993). A soluble form of the human CMV glycoprotein B (gB) showed 40% reduced binding to CHO cells lacking heparan sulphate proteoglycans and to fibroblast cells treated to remove heparan sulphate. However, an undefined non-heparin component of binding remained, suggesting the presence of another class of human CMV cellular receptors (Boyle and Compton 1998). Heparan sulphate has also been proposed to be involved in the initial attachment of HIV-1 to target cells, prior to the known required interaction between gp120 and CD4 and an appropriate chemokine receptor. Only HIV-1 isolates containing highly positively charged V3-loop sequences showed reduced infectivity in cells lacking GAGs (Zhang et al. 2002). This marked strain-dependent difference in the requirement for target cells to express cell-surface heparan sulphate (Ohshiro et al. 1996; Mondor et al. 1998) supports the hypothesis that attachment is likely to be mediated by electrostatic interactions, predominantly between the charged V3 domains of gp120 and cell-surface heparin (Roderiquez et al. 1995).

Evidence to highlight the importance of a primary interaction between HIV and GAGs for the infectious process has recently been highlighted. The gp120–CD4 interaction alone is not thought to be sufficient for tight attachment of HIV to cells such as macrophages, microglia and DC (Sonza et al. 1995; Dick et al. 1997), and the additional binding of virus to cell-surface GAGs, along with other adhesion receptors, is thought to be required for efficient infection of these cells (Ugolini et al. 1999). This was clearly shown when HIV-1 no longer attached to monocyte-derived macrophages after heparitinase treatment, which removes all the cell surface heparan sulphate chains but leaves CD4 unaltered (Saphire et al. 2001).

In the case of HSV-1 and HSV-2, the interaction of glycoproteins gB and gC with cell surface heparan sulphate has been more thoroughly studied. Both clinical and laboratory isolates of alphaviruses initially attach to cells via interactions of gB and gC viral glycoproteins with heparan sulphate proteoglycans (Herold et al. 1991, 1994; Lee and Fuller 1993), and in particular, the sulphate groups at C-2 of the uronic acids and the carboxyl groups were critical for gB binding to heparin (Herold et al. 1995). In the absence of cell-surface heparan sulphate, virus entry is very inefficient. However, with time virus adherence to cells can become irreversible by heparin, suggesting that additional interactions with non-heparan sulphate receptors can occur. Further to their role in viral binding, involvement of proteoglycans in viral entry has been suggested. Binding of gB to heparan sulphate led to fusion of the viral envelope with the host-cell plasma membrane and syncytium formation (Shieh and Spear 1994). Subsequent to virus attachment to heparan sulphate, the viral gD glycoprotein, in concert with gB, gH and gL, interacts with any one of several co-receptors (Spear et al. 2000), to facilitate entry of HSV by a fusion process. Interestingly, one of the classes of co-receptors that gD can interact with to initiate HSV-1 entry are sites in heparan sulphate generated by the action of specific 3-O-sulphotransferases (Shukla et al. 1999).

Both the choice of specific macrophage subsets, and macrophage activation status, can significantly alter expression levels of many receptors on macrophages. For example, freshly isolated monocytes express very low levels of heparan sulphate GAGs, whereas macrophages derived by culture on plastic for 10 days express high heparan sulphate GAG levels (Clasper et al. 1999; Saphire et al. 2001). This may explain why the susceptibility to infection with HIV of freshly isolated monocytes increases during maturation (Gendelman et al. 1986; Rich et al. 1992). Having varied and changing levels of virus receptors on macrophages both increases the complexities of macrophage-virus interactions, and may help explain the tissue-specific susceptibility of cells to viral infection, particularly when infection is mediated via receptors thought to be ubiquitously expressed. For instance, different tissues have been reported to exhibit different heparan sulphate monosaccharide sequences (Lindahl et al. 1998; Jenniskens et al. 2000). The recent data suggesting that specific heparan sulphate sequences are recognised by viruses give us the best clues yet to understanding the tissue and cell-specific tropism of viruses that utilise GAGs for cell attachment and entry (Liu and Thorp 2002). Further analysis of saccharide sequences and structure of GAGs, and viral interactions with them, should improve the understanding of their role in assisting viral infections and will facilitate the future development of intervention strategies.

4 Conclusions

The success of viral invasion is an important factor in determining disease severity, and tissue macrophages are in a position to pose barriers to the establishment and dissemination of virus infection. Viruses bind, both specifically and non-specifically, to a number of cell-surface receptors on macrophages; understanding the biology of host-cell receptors used by viruses for binding and entry into macrophages may reveal new features of viral pathogenesis and lead to new modalities for preventing disease caused by viruses. This review has focussed on macrophage receptors, including lectins and proteoglycans, which are utilised by a number of viruses (along with other intracellular parasites) for attachment and entry. Even though the attachment process is usually very specific, the range of viruses bound by these receptors suggests a degree of pattern recognition, and the spectrum of viruses bound by these receptors is likely to broaden further still with continued research. The outcome of the interaction between viruses and macrophages depends on the state of macrophage differentiation, as it can affect the receptors expressed by the cell, with some viruses potentially utilising alternative receptors on individual or different cells. In particular this highlights the importance of extending studies performed using macrophage cell lines, to include the interaction of viruses with tissue macrophages (Turville et al. 2001), and also to study how receptor expression varies on the different populations of tissue macrophages. As we improve our understanding of the interaction of viruses with their cognate receptors, we improve the opportunity to identify new targets for rationally designed drugs, that are unique and selective for viruses and that minimise the risks of inopportune adverse effects on the host. Knowledge gained from research into the role of host cell molecules as virus receptors may also assist the future development of viral vaccines.

One approach used when designing antiviral agents is to target specific viral replication processes, as has been done in the development of reverse transcriptase and integrase inhibitors for antiretroviral therapy. An alternative approach is competition for the ligand recognised by the virus. This can either be relatively non-specific, as in the case of carbohydrate ligands, or greater selectivity can be achieved using specific protein reagents, such as antibodies. Polyanionic compounds, such as dextran sulphate, have been tested as anti-HIV therapeutic agents to block the non-specific interaction that has been proposed to play a role in attachment of the HIV-1 virion to the cell surface (Abrams et al. 1989; Stafford et al. 1997). Although relatively cheap, sulphated polysaccharides are not specific inhibitors of HIV viral binding, being able to inhibit other viruses also (Leydet et al. 1998). The rational design of sialic acid analogues has led to the development of Relenza and Tamiflu, which bind with high affinity to the influenza virus neuraminidase to inhibit release of virus particles from infected cells. In general, however, sugars and polyanionic compounds do not have high affinities for their ligands and their bioproperties are considered poor. Proteins tend to have higher affinities for their ligands and bind more specifically. In either case, one must hope that the binding site for the viral ligand differs from that of the physiological ligand of a receptor. The mapping of both virus binding and functional domains of receptors may allow targeting of specific areas of receptors for focussed design of antiviral agents; however, we will most likely have to hope for some functional redundancy if we are to successfully target host receptors to prevent viral infection.

Acknowledgements. The authors were supported by funding from the Arthritis Research Campaign (J.L.M.) and the Medical Research Council (S.G.).

5
References

Abrams, D. I., S. Kuno, R. Wong, K. Jeffords, M. Nash, J. B. Molaghan, R. Gorter, and R. Ueno. 1989. Oral dextran sulfate (UA001) in the treatment of the acquired immunodeficiency syndrome (AIDS) and AIDS-related complex. Ann Intern Med 110:183–188

Akira, S., K. Takeda, and T. Kaisho. 2001. Toll-like receptors: critical proteins linking innate and acquired immunity. Nat Immunol 2:675–680

Alexopoulou, L., A. C. Holt, R. Medzhitov, and R. A. Flavell. 2001. Recognition of double-stranded RNA and activation of NF-κB by Toll- like receptor 3. Nature 413:732–738

Allan, J. S., J. E. Coligan, F. Barin, M. F. McLane, J. G. Sodroski, C. A. Rosen, W. A. Haseltine, T. H. Lee, and M. Essex. 1985. Major glycoprotein antigens that induce antibodies in AIDS patients are encoded by HTLV-III. Science 228:1091–1094

Alvarez, C. P., F. Lasala, J. Carrillo, O. Muniz, A. L. Corbi, and R. Delgado. 2002. C-type lectins DC-SIGN and L-SIGN mediate cellular entry by Ebola virus in cis and in trans. J Virol 76:6841–6844

Anders, E. M., C. A. Hartley, and D. C. Jackson. 1990. Bovine and mouse serum beta inhibitors of influenza A viruses are mannose-binding lectins. Proc Natl Acad Sci U S A 87:4485–4489

Anders, E. M., P. C. Reading, and J. L. Miller. 2001. Lectins in innate defence against influenza virus, p. 527–532. *In* A. D. Osterhaus, N. J. Cox, and A. Hampson (ed.), Options for the control of influenza IV. Excerpta Medica, Amsterdam

Bakker, A. B., E. Baker, G. R. Sutherland, J. H. Phillips, and L. L. Lanier. 1999. Myeloid DAP12-associating lectin (MDL)-1 is a cell surface receptor involved in the activation of myeloid cells. Proc Natl Acad Sci U S A 96:9792–9726

Balch, S. G., A. J. McKnight, M. F. Seldin, and S. Gordon. 1998. Cloning of a novel C-type lectin expressed by murine macrophages. J Biol Chem 273:18656–18664

Barr, F. E., H. Pedigo, T. R. Johnson, and V. L. Shepherd. 2000. Surfactant protein-A enhances uptake of respiratory syncytial virus by monocytes and U937 macrophages. Am J Respir Cell Mol Biol 23:586–592

Bates, E. E., N. Fournier, E. Garcia, J. Valladeau, I. Durand, J. J. Pin, S. M. Zurawski, S. Patel, J. S. Abrams, S. Lebecque, P. Garrone, and S. Saeland. 1999. APCs express DCIR, a novel C-type lectin surface receptor containing an immunoreceptor tyrosine-based inhibitory motif. J Immunol 163:1973–1983

Benne, C. A., B. Benaissa-Trouw, J. A. van Strijp, C. A. Kraaijeveld, and J. F. van Iwaarden. 1997. Surfactant protein A, but not surfactant protein D, is an opsonin for influenza A virus phagocytosis by rat alveolar macrophages. Eur J Immunol 27:886–890

Benne, C. A., C. A. Kraaijeveld, J. A. G. van Strijp, E. Brouwer, M. Harmsen, J. Verhoef, L. M. G. van Golde, and J. F. van Iwaarden. 1995. Interactions of surfactant protein A with influenza viruses: binding and neutralization. J. Inf. Dis. 171:335–341

Bose, S., and A. K. Banerjee. 2002. Role of heparan sulfate in human parainfluenza virus type 3 infection. Virology 298:73–83

Botas, C., F. Poulain, J. Akiyama, C. Brown, L. Allen, J. Goerke, J. Clements, E. Carlson, A. M. Gillespie, C. Epstein, and S. Hawgood. 1998. Altered surfactant homeostasis and alveolar type II cell morphology in mice lacking surfactant protein D. Proc. Natl. Acad. Sci. U S A 95:11869–11874

Bowie, A., E. Kiss-Toth, J. A. Symons, G. L. Smith, S. K. Dower, and L. A. O'Neill. 2000. A46R and A52R from vaccinia virus are antagonists of host IL-1 and toll-like receptor signaling. Proc Natl Acad Sci U S A 97:10162–10167

Boyle, K. A., and T. Compton. 1998. Receptor-binding properties of a soluble form of human cytomegalovirus glycoprotein B. J Virol 72:1826–1833

Brown, G. D., and S. Gordon. 2001. Immune recognition. A new receptor for beta-glucans. Nature 413:36–37

Cardosa, M. J., S. Gordon, S. Hirsch, T. A. Springer, and J. S. Porterfield. 1986. Interaction of West Nile virus with primary murine macrophages: role of cell activation and receptors for antibody and complement. J Virol 57:952–959

Clasper, S., S. Vekemans, M. Fiore, M. Plebanski, P. Wordsworth, G. David, and D. G. Jackson. 1999. Inducible expression of the cell surface heparan sulfate proteoglycan syndecan-2 (fibroglycan) on human activated macrophages can regulate fibroblast growth factor action. J Biol Chem 274:24113–24123

Collman, R., B. Godfrey, J. Cutilli, A. Rhodes, N. F. Hassan, R. Sweet, S. D. Douglas, H. Friedman, N. Nathanson, and F. Gonzalez-Scarano. 1990. Macrophage-tropic strains of human immunodeficiency virus type 1 utilize the CD4 receptor. J Virol 64:4468–4476

Compton, T., D. M. Nowlin, and N. R. Cooper. 1993. Initiation of human cytomegalovirus infection requires initial interaction with cell surface heparan sulfate. Virology 193:834–841

Crouch, E., K. Hartshorn, and I. Ofek. 2000. Collectins and pulmonary innate immunity. Immunol Rev 173:52–65

Curtis, B. M., S. Scharnowske, and A. J. Watson. 1992. Sequence and expression of a membrane-associated C-type lectin that exhibits CD4-independent binding of human immunodeficiency virus envelope glycoprotein gp120. Proc Natl Acad Sci USA 89:8356–8360

Daughaday, C. C., W. E. Brandt, J. M. McCown, and P. K. Russell. 1981. Evidence for two mechanisms of dengue virus infection of adherent human monocytes: trypsin-sensitive virus receptors and trypsin-resistant immune complex receptors. Infect Immun 32:469–473

Dick, A. D., M. Pell, B. J. Brew, E. Foulcher, and J. D. Sedgwick. 1997. Direct ex vivo flow cytometric analysis of human microglial cell CD4 expression: examination of central nervous system biopsy specimens from HIV-seropositive patients and patients with other neurological disease. Aids 11:1699–1708

Dimmock, N. J. 1982. Review article initial stages in infection with animal viruses. J Gen Virol 59:1–22

Dong, X., W. J. Storkus, and R. D. Salter. 1999. Binding and uptake of agalactosyl IgG by mannose receptor on macrophages and dendritic cells. J Immunol 163:5427–5434

Ezekowitz, R. A., M. Kuhlman, J. E. Groopman, and R. A. Byrn. 1989. A human serum mannose-binding protein inhibits in vitro infection by the human immunodeficiency virus. J Exp Med 169:185–196

Ezekowitz, R. A., K. Sastry, and K. Reid (ed.). 1996. Collectins and Innate Immunity. RG Landes, Austin, Texas

Fearon, D. T., and R. M. Locksley. 1996. The instructive role of innate immunity in the acquired immune response. Science 272:50–53

Fernandes, M. J., A. A. Finnegan, L. D. Siracusa, C. Brenner, N. N. Iscove, and B. Calabretta. 1999. Characterization of a novel receptor that maps near the natural killer gene complex: demonstration of carbohydrate binding and expression in hematopoietic cells. Cancer Res 59:2709–2717

Fiete, D. J., M. C. Beranek, and J. U. Baenziger. 1998. A cysteine-rich domain of the "mannose" receptor mediates GalNAc-4-SO4 binding. Proc Natl Acad Sci U S A 95:2089–2093

Fraser, I. P., H. Koziel, and R. A. Ezekowitz. 1998. The serum mannose-binding protein and the macrophage mannose receptor are pattern recognition molecules that link innate and adaptive immunity. Semin Immunol 10:363–372

Freistadt, M. S., and K. E. Eberle. 2000. Hematopoietic cells from CD155-transgenic mice express CD155 and support poliovirus replication ex vivo. Microb Pathog 29:203–212

Geijtenbeek, T. B., D. J. Krooshoop, D. A. Bleijs, S. J. van Vliet, G. C. van Duijnhoven, V. Grabovsky, R. Alon, C. G. Figdor, and Y. van Kooyk. 2000a. DC-SIGN-ICAM-2 interaction mediates dendritic cell trafficking. Nat Immunol 1:353–357
Geijtenbeek, T. B., D. S. Kwon, R. Torensma, S. J. van Vliet, G. C. van Duijnhoven, J. Middel, I. L. Cornelissen, H. S. Nottet, V. N. KewalRamani, D. R. Littman, C. G. Figdor, and Y. van Kooyk. 2000b. DC-SIGN, a dendritic cell-specific HIV-1-binding protein that enhances trans-infection of T cells. Cell 100:587–597
Geijtenbeek, T. B., R. Torensma, S. J. van Vliet, G. C. van Duijnhoven, G. J. Adema, Y. van Kooyk, and C. G. Figdor. 2000c. Identification of DC-SIGN, a novel dendritic cell-specific ICAM-3 receptor that supports primary immune responses. Cell 100:575–585
Geijtenbeek, T. B., G. C. van Duijnhoven, S. J. van Vliet, E. Krieger, G. Vriend, C. G. Figdor, and Y. van Kooyk. 2002. Identification of different binding sites in the dendritic cell-specific receptor DC-SIGN for intercellular adhesion molecule 3 and HIV-1. J Biol Chem 277:11314–11320
Gendelman, H. E., O. Narayan, S. Kennedy-Stoskopf, P. G. Kennedy, Z. Ghotbi, J. E. Clements, J. Stanley, and G. Pezeshkpour. 1986. Tropism of sheep lentiviruses for monocytes: susceptibility to infection and virus gene expression increase during maturation of monocytes to macrophages. J Virol 58:67–74
Geyer, H., C. Holschbach, G. Hunsmann, and J. Schneider. 1988. Carbohydrates of human immunodeficiency virus. Structures of oligosaccharides linked to the envelope glycoprotein 120. J Biol Chem 263:11760–11767
Ghildyal, R., C. Hartley, A. Varrasso, J. Meanger, D. R. Voelker, E. M. Anders, and J. Mills. 1999. Surfactant protein A binds to the fusion glycoprotein of respiratory syncytial virus and neutralizes virion infectivity. J Infect Dis 180:2009–2013
Guidotti, L. G., and F. V. Chisari. 2001. Noncytolytic control of viral infections by the innate and adaptive immune response. Annu Rev Immunol 19:65–91
Harrod, K. S., B. C. Trapnell, K. Otake, T. R. Korfhagen, and J. A. Whitsett. 1999. SP-A enhances viral clearance and inhibits inflammation after pulmonary adenoviral infection. Am. J. Physiol. 277:L580-L588
Hartshorn, K. L., E. C. Crouch, M. R. White, P. Eggleton, A. I. Tauber, D. Chang, and K. Sastry. 1994. Evidence for a protective role of pulmonary surfactant protein D (SP-D) against influenza A viruses. J Clin Invest 94:311–319
Hartshorn, K. L., K. Sastry, M. R. White, E. M. Anders, M. Super, R. A. Ezekowitz, and A. I. Tauber. 1993. Human mannose-binding protein functions as an opsonin for influenza A viruses. J Clin Invest 91:1414–1420
Hartshorn, K. L., M. R. White, V. Shepherd, K. Reid, J. C. Jensenius, and E. C. Crouch. 1997. Mechanisms of anti-influenza activity of surfactant proteins A and D: comparison with serum collectins. Am. J. Physiol. 273:L1156-L1166
Hartshorn, K. L., M. R. White, D. R. Voelker, J. Coburn, K. Zaner, and E. C. Crouch. 2000. Mechanism of binding of surfactant protein D to influenza A viruses: importance of binding to haemagglutinin to antiviral activity. Biochem J 351 Pt 2:449–458
Haynes, L. M., D. D. Moore, E. A. Kurt-Jones, R. W. Finberg, L. J. Anderson, and R. A. Tripp. 2001. Involvement of toll-like receptor 4 in innate immunity to respiratory syncytial virus. J. Virol. 75:10730–10737
Herold, B. C., S. I. Gerber, T. Polonsky, B. J. Belval, P. N. Shaklee, and K. Holme. 1995. Identification of structural features of heparin required for inhibition of herpes simplex virus type 1 binding. Virology 206:1108–1116
Herold, B. C., R. J. Visalli, N. Susmarski, C. R. Brandt, and P. G. Spear. 1994. Glycoprotein C-independent binding of herpes simplex virus to cells requires cell surface heparan sulphate and glycoprotein B. J Gen Virol 75 (Pt 6):1211–1222
Herold, B. C., D. WuDunn, N. Soltys, and P. G. Spear. 1991. Glycoprotein C of herpes simplex virus type 1 plays a principal role in the adsorption of virus to cells and in infectivity. J Virol 65:1090–1098

Hickling, T. P., H. Bright, K. Wing, D. Gower, S. L. Martin, R. B. Sim, and R. Malhotra. 1999. A recombinant trimeric surfactant protein D carbohydrate recognition domain inhibits respiratory syncytial virus infection in vitro and in vivo. Eur J Immunol 29:3478–3484

Holmskov, U., R. Malhotra, R. B. Sim, and J. C. Jensenius. 1994. Collectins: collagenous C-type lectins of the innate immune defense system. Immunol Today 15:67–74

Hoppe, H. J., and K. B. Reid. 1994. Collectins–soluble proteins containing collagenous regions and lectin domains–and their roles in innate immunity. Protein Sci 3:1143–1158

Igarashi, T., C. R. Brown, Y. Endo, A. Buckler-White, R. Plishka, N. Bischofberger, V. Hirsch, and M. A. Martin. 2001. Macrophage are the principal reservoir and sustain high virus loads in rhesus macaques after the depletion of CD4+ T cells by a highly pathogenic simian immunodeficiency virus/HIV type 1 chimera (SHIV): Implications for HIV-1 infections of humans. Proc Natl Acad Sci U S A 98:658–663

Imamura, T., S. Toyoshima, and T. Osawa. 1984. Lectin-like molecules on the murine macrophage cell surface. Biochim Biophys Acta 805:235–244

Inaba, K., W. J. Swiggard, M. Inaba, J. Meltzer, A. Mirza, T. Sasagawa, M. C. Nussenzweig, and R. M. Steinman. 1995. Tissue distribution of the DEC-205 protein that is detected by the monoclonal antibody NLDC-145. I. Expression on dendritic cells and other subsets of mouse leukocytes. Cell Immunol 163:148–156

Inada, T., K. T. Chong, and C. A. Mims. 1985. Enhancing antibodies, macrophages and virulence in mouse cytomegalovirus infection. J Gen Virol 66 (Pt 4):871–878

Jackson, T., F. M. Ellard, R. A. Ghazaleh, S. M. Brookes, W. E. Blakemore, A. H. Corteyn, D. I. Stuart, J. W. Newman, and A. M. King. 1996. Efficient infection of cells in culture by type O foot-and-mouth disease virus requires binding to cell surface heparan sulfate. J Virol 70:5282–5287

Janeway, C. A., Jr. 1989. Approaching the asymptote? Evolution and revolution in immunology. Cold Spring Harb Symp Quant Biol 54 Pt 1:1–13

Janeway, C. A. J. 1992. The immune system evolved to discriminate infectious nonself from noninfectious self. Immunol. Today 13:11–16

Jenniskens, G. J., A. Oosterhof, R. Brandwijk, J. H. Veerkamp, and T. H. van Kuppevelt. 2000. Heparan sulfate heterogeneity in skeletal muscle basal lamina: demonstration by phage display-derived antibodies. J Neurosci 20:4099–4111

Jiang, W., W. J. Swiggard, C. Heufler, M. Peng, A. Mirza, R. M. Steinman, and M. C. Nussenzweig. 1995. The receptor DEC-205 expressed by dendritic cells and thymic epithelial cells is involved in antigen processing. Nature 375:151–155

Julkunen, I., K. Melen, M. Nyqvist, J. Pirhonen, T. Sareneva, and S. Matikainen. 2000. Inflammatory responses in influenza A virus infection. Vaccine 19 Suppl 1:S32-S37

Kage, A., E. Shoolian, K. Rokos, M. Ozel, R. Nuck, W. Reutter, E. Kottgen, and G. Pauli. 1998. Epithelial uptake and transport of cell-free human immunodeficiency virus type 1 and gp120-coated microparticles. J Virol 72:4231–4236

Korfhagen, T. R., V. Sheftelyevich, M. S. Burhans, M. D. Bruno, G. F. Ross, S. E. Wert, M. T. Stahlman, A. H. Jobe, M. Ikegami, J. A. Whitsett, and J. H. Fisher. 1998. Surfactant protein-D regulates surfactant phospholipid homeostasis *in vivo*. J. Biol. Chem. 273:28438–28443

Kurt-Jones, E. A., L. Popova, L. Kwinn, L. M. Haynes, L. P. Jones, R. A. Tripp, E. E. Walsh, M. W. Freeman, D. T. Golenbock, L. J. Anderson, and R. W. Finberg. 2000. Pattern recognition receptors TLR4 and CD14 mediate response to respiratory syncytial virus. Nat. Immunol. 1:398–401

Larkin, M., R. A. Childs, T. J. Matthews, S. Thiel, T. Mizuochi, A. M. Lawson, J. S. Savill, C. Haslett, R. Diaz, and T. Feizi. 1989. Oligosaccharide-mediated interactions of the envelope glycoprotein gp120 of HIV-1 that are independent of CD4 recognition. Aids 3:793–798

Lawson, P. R., and K. B. Reid. 2000. The roles of surfactant proteins A and D in innate immunity. Immunol Rev 173:66–78

Lee, W. C., and A. O. Fuller. 1993. Herpes simplex virus type 1 and pseudorabies virus bind to a common saturable receptor on Vero cells that is not heparan sulfate. J Virol 67:5088–5097

Leteux, C., W. Chai, R. W. Loveless, C. T. Yuen, L. Uhlin-Hansen, Y. Combarnous, M. Jankovic, S. C. Maric, Z. Misulovin, M. C. Nussenzweig, and F. Ten. 2000. The cysteine-rich domain of the macrophage mannose receptor is a multispecific lectin that recognizes chondroitin sulfates A and B and sulfated oligosaccharides of blood group Lewis(a) and Lewis(x) types in addition to the sulfated N-glycans of lutropin. J Exp Med 191:1117–1126

LeVine, A. M., J. Gwozdz, J. Stark, M. Bruno, J. Whitsett, and T. Korfhagen. 1999. Surfactant protein-A enhances respiratory syncytial virus clearance *in vivo*. J. Clin. Invest. 103:1015–1021

LeVine, A. M., K. Hartshorn, J. Elliott, J. Whitsett, and T. Korfhagen. 2002. Absence of SP-A modulates innate and adaptive defense responses to pulmonary influenza infection. Am J Physiol Lung Cell Mol Physiol 282:L563-L572

LeVine, A. M., J. A. Whitsett, K. L. Hartshorn, E. C. Crouch, and T. R. Korfhagen. 2001. Surfactant protein D enhances clearance of influenza A virus from the lung *in vivo*. J. Immunol. 167:5868–5873

Leydet, A., C. Moullet, J. P. Roque, M. Witvrouw, C. Pannecouque, G. Andrei, R. Snoeck, J. Neyts, D. Schols, and E. De Clercq. 1998. Polyanion inhibitors of HIV and other viruses. 7. Polyanionic compounds and polyzwitterionic compounds derived from cyclodextrins as inhibitors of HIV transmission. J Med Chem 41:4927–4932

Lindahl, U., M. Kusche-Gullberg, and L. Kjellen. 1998. Regulated diversity of heparan sulfate. J Biol Chem 273:24979–24982

Linehan, S. A., L. Martinez-Pomares, and S. Gordon. 2000. Macrophage lectins in host defence. Microbes Infect 2:279–288

Liu, J., and S. C. Thorp. 2002. Cell surface heparan sulfate and its roles in assisting viral infections. Med Res Rev 22:1–25

Malhotra, R., J. S. Haurum, S. Thiel, and R. B. Sim. 1994. Binding of human collectins (SP-A and MBP) to influenza virus. Biochem J 304 (Pt 2):455–461

Malhotra, R., and R. B. Sim. 1995. Collectins and viral infection. Trends Microbiol 3:240–244

Martinez-Pomares, L., P. R. Crocker, R. Da Silva, N. Holmes, C. Colominas, P. Rudd, R. Dwek, and S. Gordon. 1999. Cell-specific glycoforms of sialoadhesin and CD45 are counter-receptors for the cysteine-rich domain of the mannose receptor. J Biol Chem 274:35211–35218

Matsumoto, M., T. Tanaka, T. Kaisho, H. Sanjo, N. G. Copeland, D. J. Gilbert, N. A. Jenkins, and S. Akira. 1999. A novel LPS-inducible C-type lectin is a transcriptional target of NF-IL6 in macrophages. J Immunol 163:5039–5048

Medzhitov, R., and C. A. Janeway, Jr. 1997. Innate immunity: impact on the adaptive immune response. Curr Opin Immunol 9:4–9

Mendelsohn, C. L., E. Wimmer, and V. R. Racaniello. 1989. Cellular receptor for poliovirus: molecular cloning, nucleotide sequence, and expression of a new member of the immunoglobulin superfamily. Cell 56:855–865

Miettinen, M., T. Sareneva, I. Julkunen, and S. Matikainen. 2001. IFNs activate toll-like receptor gene expression in viral infections. Genes Immun. 2:349–355

Mondor, I., S. Ugolini, and Q. J. Sattentau. 1998. Human immunodeficiency virus type 1 attachment to HeLa CD4 cells is CD4 independent and gp120 dependent and requires cell surface heparans. J Virol 72:3623–3634

Moreno-Altamirano, M. M., F. J. Sanchez-Garcia, and M. L. Munoz. 2002. Non Fc receptor-mediated infection of human macrophages by dengue virus serotype 2. J Gen Virol 83:1123–1130

Neyts, J., R. Snoeck, D. Schols, J. Balzarini, J. D. Esko, A. Van Schepdael, and E. De Clercq. 1992. Sulfated polymers inhibit the interaction of human cytomegalovirus with cell surface heparan sulfate. Virology 189:48–58

Ochiai, H., M. Kurokawa, K. Hayashi, and S. Niwayama. 1988. Antibody-mediated growth of influenza A NWS virus in macrophagelike cell line P388D1. J Virol 62:20–26

Ohshiro, Y., T. Murakami, K. Matsuda, K. Nishioka, K. Yoshida, and N. Yamamoto. 1996. Role of cell surface glycosaminoglycans of human T cells in human immunodeficiency virus type-1 (HIV-1) infection. Microbiol Immunol 40:827–835

Peiris, J. S., S. Gordon, J. C. Unkeless, and J. S. Porterfield. 1981. Monoclonal anti-Fc receptor IgG blocks antibody enhancement of viral replication in macrophages. Nature 289:189–191

Pohlmann, S., F. Baribaud, B. Lee, G. J. Leslie, M. D. Sanchez, K. Hiebenthal-Millow, J. Munch, F. Kirchhoff, and R. W. Doms. 2001. DC-SIGN interactions with human immunodeficiency virus type 1 and 2 and simian immunodeficiency virus. J Virol 75:4664–4672

Pohlmann, S., E. J. Soilleux, F. Baribaud, G. J. Leslie, L. S. Morris, J. Trowsdale, B. Lee, N. Coleman, and R. W. Doms. 2001. DC-SIGNR, a DC-SIGN homologue expressed in endothelial cells, binds to human and simian immunodeficiency viruses and activates infection in trans. Proc Natl Acad Sci U S A 98:2670–2675

Rassa, J. C., J. L. Meyers, Y. Zhang, R. Kudaravalli, and S. R. Ross. 2002. Murine retroviruses activate B cells via interaction with toll-like receptor 4. Proc. Natl. Acad. Sci. U S A 99:2281–2286

Ratner, L., W. Haseltine, R. Patarca, K. J. Livak, B. Starcich, S. F. Josephs, E. R. Doran, J. A. Rafalski, E. A. Whitehorn, K. Baumeister, and et al. 1985. Complete nucleotide sequence of the AIDS virus, HTLV-III. Nature 313:277–284

Reading, P. C., J. Allison, E. C. Crouch, and E. M. Anders. 1998a. Increased susceptibility of diabetic mice to influenza virus infection: compromise of collectin-mediated host defense of the lung by glucose? J. Virol. 72:6884–6887

Reading, P. C., C. A. Hartley, R. A. B. Ezekowitz, and E. M. Anders. 1995. A serum mannose-binding lectin mediates complement-dependent lysis of influenza virus-infected cells. Biochem. Biophys. Res. Commun. 217:1128–1136

Reading, P. C., U. Holmskov, and E. M. Anders. 1998b. Antiviral activity of bovine collectins against rotaviruses. J. Gen. Virol. 79:2255–2263

Reading, P. C., J. L. Miller, and E. M. Anders. 2000. Involvement of the mannose receptor in infection of macrophages by influenza virus. J Virol 74:5190–5197

Reading, P. C., L. S. Morey, E. C. Crouch, and E. M. Anders. 1997. Collectin-mediated antiviral host defense of the lung: evidence from influenza virus infection of mice. J. Virol. 71:8204–8212

Rich, E. A., I. S. Chen, J. A. Zack, M. L. Leonard, and W. A. O'Brien. 1992. Increased susceptibility of differentiated mononuclear phagocytes to productive infection with human immunodeficiency virus-1 (HIV-1). J Clin Invest 89:176–183

Roderiquez, G., T. Oravecz, M. Yanagishita, D. C. Bou-Habib, H. Mostowski, and M. A. Norcross. 1995. Mediation of human immunodeficiency virus type 1 binding by interaction of cell surface heparan sulfate proteoglycans with the V3 region of envelope gp120-gp41. J Virol 69:2233–2239

Rostand, K. S., and J. D. Esko. 1997. Microbial adherence to and invasion through proteoglycans. Infect Immun 65:1–8

Saphire, A. C., M. D. Bobardt, Z. Zhang, G. David, and P. A. Gallay. 2001. Syndecans serve as attachment receptors for human immunodeficiency virus type 1 on macrophages. J Virol 75:9187–9200

Sastry, K., and R. A. B. Ezekowitz. 1993. Collectins: pattern recognition molecules involved in first line host defense. Curr. Opin. Immunol. 5:59–66

Shieh, M. T., and P. G. Spear. 1994. Herpesvirus-induced cell fusion that is dependent on cell surface heparan sulfate or soluble heparin. J Virol 68:1224–1228

Shukla, D., J. Liu, P. Blaiklock, N. W. Shworak, X. Bai, J. D. Esko, G. H. Cohen, R. J. Eisenberg, R. D. Rosenberg, and P. G. Spear. 1999. A novel role for 3-O-sulfated heparan sulfate in herpes simplex virus 1 entry. Cell 99:13–22

Soilleux, E. J., L. S. Morris, G. Leslie, J. Chehimi, Q. Luo, E. Levroney, J. Trowsdale, L. J. Montaner, R. W. Doms, D. Weissman, N. Coleman, and B. Lee. 2002. Constitutive and induced expression of DC-SIGN on dendritic cell and macrophage subpopulations in situ and in vitro. J Leukoc Biol 71:445–457

Sonza, S., A. Maerz, S. Uren, A. Violo, S. Hunter, W. Boyle, and S. Crowe. 1995. Susceptibility of human monocytes to HIV type 1 infection in vitro is not dependent on their level of CD4 expression. AIDS Res Hum Retroviruses 11:769–776

Spear, P. G., R. J. Eisenberg, and G. H. Cohen. 2000. Three classes of cell surface receptors for alphaherpesvirus entry. Virology 275:1–8

Stafford, M. K., D. Cain, I. Rosenstein, E. A. Fontaine, M. McClure, A. M. Flanagan, J. R. Smith, D. Taylor-Robinson, J. Weber, and V. S. Kitchen. 1997. A placebo-controlled, double-blind prospective study in healthy female volunteers of dextrin sulphate gel: a novel potential intravaginal virucide. J Acquir Immune Defic Syndr Hum Retrovirol 14:213–218

Stahl, P. D., and R. A. Ezekowitz. 1998. The mannose receptor is a pattern recognition receptor involved in host defense. Curr Opin Immunol 10:50–55

Swiggard, W. J., A. Mirza, M. C. Nussenzweig, and R. M. Steinman. 1995. DEC-205, a 205-kDa protein abundant on mouse dendritic cells and thymic epithelium that is detected by the monoclonal antibody NLDC-145: purification, characterization, and N-terminal amino acid sequence. Cell Immunol 165:302–311

Takeda, A., C. U. Tuazon, and F. A. Ennis. 1988. Antibody-enhanced infection by HIV-1 via Fc receptor-mediated entry. Science 242:580–583

Tamura, M., R. G. Webster, and F. A. Ennis. 1991. Antibodies to HA and NA augment uptake of influenza A viruses into cells via Fc receptor entry. Virology 182:211–219

Taylor, M. E., K. Bezouska, and K. Drickamer. 1992. Contribution to ligand binding by multiple carbohydrate-recognition domains in the macrophage mannose receptor. J Biol Chem 267:1719–1726

Taylor, M. E., J. T. Conary, M. R. Lennartz, P. D. Stahl, and K. Drickamer. 1990. Primary structure of the mannose receptor contains multiple motifs resembling carbohydrate-recognition domains. J Biol Chem 265:12156–12162

Turville, S. G., J. Arthos, K. M. Donald, G. Lynch, H. Naif, G. Clark, D. Hart, and A. L. Cunningham. 2001. HIV gp120 receptors on human dendritic cells. Blood 98:2482–2488

Ugolini, S., I. Mondor, and Q. J. Sattentau. 1999. HIV-1 attachment: another look. Trends Microbiol 7:144–149

van Iwaarden, J. F., J. A. G. van Strijp, M. J. M. Ebskamp, A. C. Welmers, J. Verhoef, and L. M. G. van Golde. 1991. Surfactant protein A is opsonin in phagocytosis of herpes simplex virus type 1 by rat alveolar macrophages. Am. J. Physiol. 261:L204–209

van Iwaarden, J. F., J. A. G. van Strijp, H. Visser, H. P. Haagsman, J. Verhoef, and L. M. G. van Golde. 1992. Binding of surfactant protein A (SP-A) to herpes simplex virus type 1-infected cells is mediated by the carbohydrate moiety of SP-A. J. Biol. Chem. 267:25019–25043

Wadstrom, T., and A. Ljungh. 1999. Glycosaminoglycan-binding microbial proteins in tissue adhesion and invasion: key events in microbial pathogenicity. J Med Microbiol 48:223–233

Weis, W., J. H. Brown, S. Cusack, J. C. Paulson, J. J. Skehel, and D. C. Wiley. 1988. Structure of the influenza virus haemagglutinin complexed with its receptor, sialic acid. Nature 333:426–431

Weiss, R. A. 2002. HIV receptors and cellular tropism. IUBMB Life 53:201–205

Weyer, C., R. Sabat, H. Wissel, D. H. Kruger, P. A. Stevens, and S. Prosch. 2000. Surfactant protein A binding to cytomegalovirus proteins enhances virus entry into rat lung cells. Am J Respir Cell Mol Biol 23:71–78

Wilson, I. A., J. J. Skehel, and D. C. Wiley. 1981. Structure of the haemagglutinin membrane glycoprotein of influenza virus at 3A resolution. Nature 289:366–373

Wu, L., A. A. Bashirova, T. D. Martin, L. Villamide, E. Mehlhop, A. O. Chertov, D. Unutmaz, M. Pope, M. Carrington, and V. N. KewalRamani. 2002a. Rhesus macaque dendritic cells efficiently transmit primate lentiviruses independently of DC-SIGN. Proc Natl Acad Sci U S A 99:1568–1573

Wu, L., T. D. Martin, R. Vazeux, D. Unutmaz, and V. N. KewalRamani. 2002b. Functional evaluation of DC-SIGN monoclonal antibodies reveals DC-SIGN interactions with ICAM-3 do not promote human immunodeficiency virus type 1 transmission. J Virol 76:5905–5914

Zhang, Y. J., T. Hatziioannou, T. Zang, D. Braaten, J. Luban, S. P. Goff, and P. D. Bieniasz. 2002. Envelope-dependent, cyclophilin-independent effects of glycosaminoglycans on human immunodeficiency virus type 1 attachment and infection. J Virol 76:6332–6343

Macrophage Immunity and *Mycobacterium tuberculosis*

J. D. MacMicking · J. D. McKinney

The Rockefeller University,
1230 York Avenue, Box 21, New York, NY 10021, USA
e-mail: macmicj@rockefeller.edu

Abstract Tuberculosis (TB) is rivalled only by the acquired immunodeficiency syndrome (AIDS) as a communicable cause of death. Yet of an estimated 2 billion individuals who have been infected with the pathogen *Mycobacterium tuberculosis* (*Mtb*), less than 10% will develop disease. For the remainder, natural immunity appears sufficient to limit bacterial growth. An integral component of host protection to TB is the activated macrophage. *Mtb* recognition, phagocytosis, vacuolar trafficking and redox-based killing are all enlisted as part of this cell's anti-tubercular arsenal. When assembled together with lymphocytes and stromal elements as part of the tuberculoid granuloma, macrophages also provide a physical constraint to further dissemination. The liaison between macrophages and T cells in particular forms much of the current basis of vaccination in immunologically naive subjects. Recent experimentation with post-exposure vaccines, however, suggests that cellular immunity may not be fully elicited by the existing single-dose regimen. New approaches that embrace small molecule chemistry to enhance or mimic macrophage effector mechanisms, or which sensitise *Mtb* to further immunologic insult, could help address this issue. Harnessing the macrophage as a therapeutic target could thus prove a useful adjunct to TB vaccination and chemotherapy in the future.

Keywords Cytokine, Macrophage activation, Microbial recognition, Mycobacterium tuberculosis, Tuberculosis, Vaccination

Abbreviations

AFB	Acid-fast bacilli
AM	Alveolar macrophage
BAL	Bronchoalveolar lavage
CGD	Chronic granulomatous disease
CR	Complement receptor
FcR	Fc receptor
GPI	Glycosyl phosphatidylinositol
GTPase	Guanosine 5′-triphosphatase
IFN-γ	Interferon gamma
IL	Interleukin
RF	Interferon regulatory factor
Jak	Janus kinase
LAM	Lipoarabinomannan
LPS	Lipopolysaccharide
LTBI	Latent TB infection
MDR	Multi-drug resistant
MHC	Major histocompatibility complex
MR	Mannose receptor
Mtb	Mycobacterium tuberculosis
MGC	Multinucleated giant cell
NADPH	Nicotinamide adenine dinucleotide phosphate (reduced)
NF-κB	Nuclear factor (NF)-κB
NRAMP	Natural resistance-associated macrophage protein
NO	Nitric oxide

NOS2	Inducible nitric oxide synthase
PAMP	Pathogen-associated molecular pattern
PBMC	Peripheral blood mononuclear cell
PG	Phagosome
PI_3P	Phosphatidyl inositol 3-phosphate
PL	Phagolysosome
PRR	Pattern-recognition receptor
RNI	Reactive nitrogen intermediates
ROI	Reactive oxygen intermediates
SNO	S-nitrosothiol
SP	Surfactant protein
STAT	Signal transducer and activator of transcription
TLR	Toll-like receptor
TNF	Tumour necrosis factor alpha

1 Introduction

1.1 The Relationship Between *Mtb* and Its Human Host: Ancient Origins, Modern Concerns

Exhumation of the 3,203 year-old tomb of Nebwenenef, high priest of Egyptian pharaoh Ramses II, uncovered the mummified remains of a young boy harbouring acid fast-bacilli (AFB) together with blood in his trachea (haemoptysis)—hallmarks of pulmonary tuberculosis (TB) (Zimmermann 1979). A further 31 mummies dated between 3700–1000 B.C. displayed angular kyphosis typical of Pott's disease, wherein TB causes long bone and spinal deformities (Morse et al. 1964). More recently, DNA specific for the pathogenic tubercle bacilli has been detected in the tracheobronchial lymph node of a 1,300-year old Peruvian corpse (Salo et al. 1994). So begins the paleopathological record of mankind's ongoing struggle with one of nature's most durable and successful pathogens, *Mycobacterium tuberculosis* (*Mtb*). This relationship, steeped in antiquity and human suffering, still accounts for nearly 2 million deaths and 8 million active new cases per year (WHO 2000). The introduction of effective chemotherapy in 1952 offered respite but patient non-compliance, governmental neglect and the advent of the AIDS epidemic have again helped raise the spectre of TB in the form of multi-drug resistant (MDR) strains. Add to this a global reservoir of clinically latent TB in nearly 2 billion infected people, an estimate which represents one-third of the earth's population (Dye et al. 1999), and the magnitude of the problem becomes palpable. Little wonder the search for novel anti-microbial drugs and protective vaccines has taken on renewed urgency (McKinney 2000). At the vanguard of this effort will be the quest for understanding how *Mtb* adapts to its human host, an interaction which focuses attention on the bacterium's favoured dwelling and, paradoxically, its chief antagonist: the macrophage.[1]

1.2 Adversarial Profiles

Macrophages and *Mtb* share not only the legacy of co-evolution but also that of contemporaneous discovery. In 1882, the German physician Robert Koch delivered his landmark address to the Berlin Physiological Society in which he provided clear evidence for the tubercle bacillus being the aetiologic agent responsible for "consumption" (Koch 1882). In the same year, Russian-born zoologist Ilya (Elie) Metchnikoff had been watching through his microscope at home in Messina the wandering amoeboid cells in transparent starfish larvae, mobile cells which he knew could ingest solid particles from earlier experiments conducted in coelenterates (Metchnikoff 1880). Out of these observations grew the realisation that such cells may in fact help defend the organism against "noxious intruders" (Metchnikoff 1921). By placing a thorn under the larva's skin, Metchnikoff witnessed the amoeboid cells accumulate at the site of injury. This experience served as the basis for his seminal theory of phagocytosis, a postulate soon formally demonstrated by the ability of gut mesoderm in the water flea, *Daphnia*, to engulf the fungal ascospores of *Monospora bicuspidata* (Metchnikoff 1884). In turning his attention thereafter towards vertebrate immunity, Metchnikoff established the chief importance of macrophages in providing innate defence against invading micro-organisms (Metchnikoff 1905). Among his most significant and enduring findings was one of giant cells from resistant tubercular animals being capable of ingesting and killing Koch's causative agent (Metchnikoff 1888).

Today we appreciate that macrophages represent one of the most highly specialised lineages in all of metazoan immunity. Besides their marked phagocytic profile (shared to a lesser extent by certain other cell types, e.g. retinal epithelia), a number of additional characteristics have helped define their métier as host protectant and are especially relevant to TB. These include: (1) an abundant fixed tissue distribution, with sessile populations in nearly all organ systems, notably within alveoli (alveolar macrophages; AMs) as well as interstitial and intravascular macrophages originally sequestered as monocytes from the pulmonary microcirculation; (2) rapid serosal motility in response to chemotactic gradients generated within cavities such as the pleura; (3) multiple pathways for microbial killing and antigen presentation; and (4) an ability to co-opt other immunocytes to the site of infection, a task accomplished via the elaboration of some ten or more classes of secretory products (e.g. cytokines/chemokines, growth factors, coagulation factors, matrix proteins, bioactive oligopeptides, lipids, sterols, purines, pyridines and oxygen/nitrogen intermediates) (Nathan

[1] The term, "macrophage" as used here refers to all cells of the mononuclear phagocyte system, including monocytes.

1987; Gordon 1999). This degree of plasticity implies much in the way of co-ordinate regulation of a large number of genes. Indeed, macrophages are endowed with enormous biosynthetic capacity. Upon encountering *Mtb*, for example, murine macrophages regulate as many as 600–700 genes (of ~11,000 microarrayed) within the first 24 h (Erht et al. 2001). Some 10% of mRNA transcripts analysed in human macrophages are significantly induced (10- to 100-fold) as early as 6 h and up to 400-fold by 12 h post-infection (Ragno et al. 2001). Given that the vast majority of housekeeping genes remain unchanged and represent nearly 60% of the genomic total (Erht et al. 2001), a 20%–25% response of the remainder illustrates just what a potent signal *Mtb* is for the macrophage transcriptome. When accompanied by activating cytokines such as interferon gamma (IFN-γ), this genetic commitment may reach as high as 40% (Erht et al. 2001).

Why is *Mtb* such a powerful natural stimulus for macrophage activation? The answer, in part, lies with the unique physiology of the organism itself. *Mtb* is a non-motile, rod-shaped actinomycete closely related to saprophytic bacteria such as *M. smegmatis*. Despite staining poorly by the usual gram-stain procedure (owing to the impermeability of its thick, waxy cell wall), the tubercle bacillus is nonetheless grouped together with gram-positive bacteria, since they all possess a single, cytoplasmic membrane (McKinney et al. 1998). Like other gram positives, *Mtb* also shares a peptidoglycan wall yet augments it with an array of complex lipidoglycans, the latter of which may reflect the need to have resisted desiccation in an ancestral soil environment (Russell 2001). This lipid-rich structure avidly retains Carbol fuchsin dye even in the presence of acidic alcohol (hence the term, "acid fast") as well as providing the decidedly foreign (non-self) determinants against which cells of the immune system have evolved.

Where *Mtb* differs markedly from other gram-positive bacteria is in the slow rate at which it replicates: ~20–24 h in synthetic medium or infected mammals (McKinney et al. 1998). This indolent growth contributes to the chronic nature of the disease and undoubtedly provides a continuous source of immune activation, some of which is potentially injurious. Moreover, it imposes lengthy (6–9 month) treatment regimens and substantial obstacles for experimentation. It may also enable the bacilli to reside within the same human host for possibly decades before reactivating later in life (Lillebaek et al. 2002). Such tissue-adapted dormancy could involve metabolic shutdown as a result of the cell-mediated immune response which contains but does not immediately eradicate the infection. Whether *Mtb* acquires significant heritable changes as part of this adaptation remains unclear. Spoligotype and microarray analyses so far depict limited polymorphic diversity among clinical isolates, a genetic invariance which could have several explanations (Kato-Maeda et al. 2001). Multiple mutations and/or deletions, for example, may be poorly tolerated, or their low level representative of recent evolutionary dissemination among human hosts (Musser et al. 2000; Brosch et al. 2002). Alternatively, extant strain similarities may truly reflect a lengthy intracellular quiescence (Musser et al. 2000; Brosch et al. 2002), one which could account for the persistence of certain isogenic strains within communities of low endemicity or reinfection rates (Kato-Maeda et al. 2001).

Reactivation of latent TB occurs in ~5% of infected individuals, while in another 5%, "primary progressive" TB usually ensues within a year or two of transmission (McKinney et al. 1998; Parrish et al. 1998; Flynn and Chan 2001). This still leaves a staggering 90% of people infected with the tubercle bacillus who never develop disease. The inescapable conclusion, often overlooked, is that in the vast majority of cases host immunity appears more than adequate to hold the organism in check—indeed, so successful is it that many human tuberculous lesions are completely sterilised with time (reviewed in McKinney et al. 2001). Part of this success can be ascribed to the activated macrophage, a concept first enunciated and aptly shown in Lurie's classic studies over 60 years ago (Lurie 1939, 1942). Yet despite the insights provided by this work and that of earlier pioneers such as Metchnikoff, neither could have foreseen the complexity of the macrophage anti-tubercular arsenal, an elaborate system of defence that defies its phylogenetically primitive origins.

2 Beyond Metchnikoff: Emerging Complexity of the Macrophage Anti-Tubercular Arsenal

2.1 Sensory Logic of the Macrophage Against *Mtb*

2.1.1 Receptors for *Mtb* Recognition

Located at the interface of gaseous exchange between the outside world and respiring host, AMs appear well situated to sample the incoming repertoire of pathogen-specific motifs belonging to airborne infectious agents like *Mtb*. The necessity of AMs to discriminate not only between self and the external environment, but also between different micro-organisms, poses a serious challenge to innate pulmonary immunity, and one compounded by the high mutation rates of many inhaled pathogens. This challenge, however, has in part been met by an evolving set of receptors capable of recognising invariant microbial structures not found in higher eukaryotes. Janeway first proposed the term "pathogen-associated molecular patterns" (PAMPs) to embrace these conserved motifs and "pattern-recognition receptors" (PRRs) for the host apparatus which detects them (Janeway 1989). PAMPs share common features which allow efficient host recognition: prokaryotic specificity, invariance within a given microbial class (and hence detection by a limited number of germline-encoded PRRs), and lastly, obligate roles in microbial survival, such that alteration or loss would either be lethal or lead to a greatly reduced adaptive fitness. Microbial genes containing PAMPs would therefore not be subject to a high incidence of mutation and "escape mutants" less likely to be selected (Medzhitov 2001).

PAMPs include cell wall components such as yeast mannans, formylated bacterial peptides, trypanosome glycosyl phosphatidylinositol (GPI) linkages, and

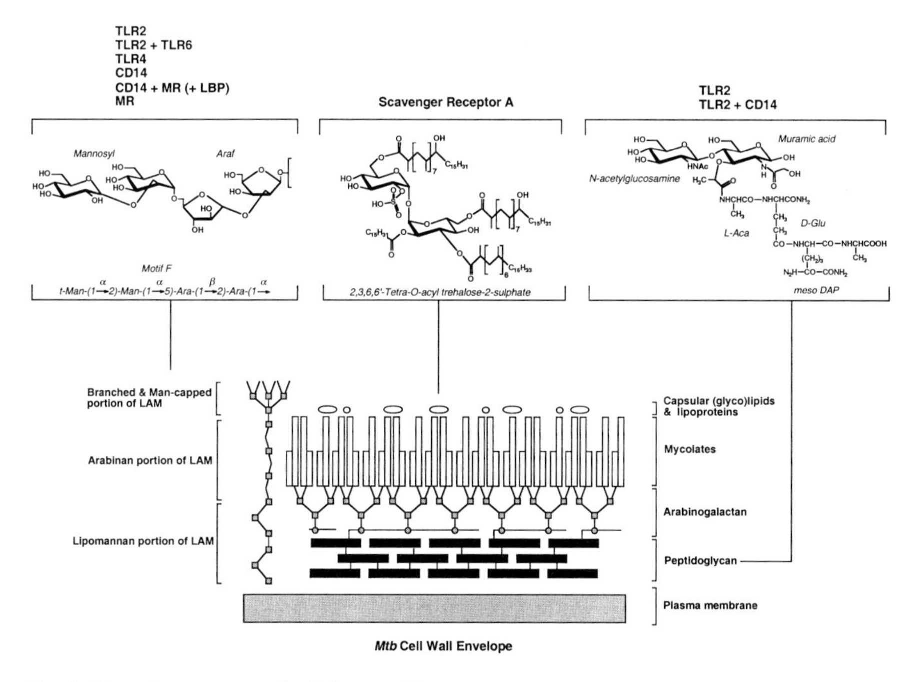

Fig. 1 Macrophage receptors for *Mtb* recognition

lipoteichoic acids and lipopolysaccharides (LPS) on gram-positive and gram-negative organisms, respectively (Aderem and Ulevitch 2000; Medzhitov 2001). Likewise, recognition of *Mtb* by macrophages takes advantage of the unique lipid-rich outer envelope found in all *Mycobacterium* spp., thus satisfying the first two PAMP criteria. That this unusually complex cell wall contributes to the evolutionary fitness of *Mtb*—thereby complying with the third PAMP criterion—is based on a remarkable ability to resist chemical, physical and chemotherapeutic stresses, to withstand macrophage killing mechanisms in order to replicate intracellularly, and to act as a potent adjuvant, the latter of which may elicit immunopathology as an to aid further bronchial spread (Dannenberg and Rook 1994; McKinney et al. 1998).

Largely due to the efforts of Brennan and co-workers, the surface chemical composition of virulent *Mtb* has been well characterised (reviewed in Baulard et al. 1999), enabling some of the molecular signatures buried within to be defined. Prominent among the glycolipids non-covalently attached to the peptidoglycan–arabinogalactan–mycolate scaffold is the mannose-capped lipoarabinomannan (ManLAM) (Fig. 1). ManLAM is recognised in non-opsonised form by several human and mouse macrophage PRRs: Toll-like receptors (TLRs), mannose (C-type lectin) receptors (MRs), and CD14 (Pugin et al. 1994; Schlessinger et al. 1994; Schlessinger et al. 1996; Prigozy et al. 1997; Means et al. 1999; Underhill et al. 1999b). Within this overlap, as well as for other *Mtb* cell wall components, a finer specificity of recognition has begun to emerge. For example, TLRs 2 and 4 both respond to ManLAM, whereas only TLR2 recognises AraLAM from rapidly growing mycobacteria (Means et al. 1999; Underhill et al. 1999b; Means et al. 2001). Class A scavenger receptors ignore both mannosylated and arabinosyl LAM moieties yet scrutinise trehalose 2-sulphate derivatives (Ernst 1998). TLRs 2 and 6 scan peptidoglycan, but only TLR2 detects lipopeptides, phosphatidylinositol mannosides or the *Mtb* 19-kDa lipoprotein (Brightbill et al. 1999; Underhill et al. 1999b; Ozinsky et al. 2000; Jones et al. 2001).

Receptor co-operativity confers another level of sophistication in which discrete signalling modules are assembled to activate certain macrophage responses and direct effector localisation. TLRs 2 and 6 physically associate in order to recognise peptidoglycan and signal tumour necrosis factor alpha (TNF) production, while TLR4 can elicit TNF as a homodimer (Ozinsky et al. 2000). Induction of this cytokine, as well as interleukins (IL)-6 and -12, appears reliant on MyD88 binding to the cytoplasmic tail of each of these TLRs, whereas nitric oxide (NO) secretion does not (Brightbill et al. 1999; Kawai et al. 1999; Underhill et al. 1999a,b; Means et al. 2001). Combinatorial effects also are evident for CD14, a soluble or GPI-linked membrane-bound receptor which lacks a cytoplasmic tail and therefore requires binding partners to help transduce its intracellular signal. Here, CD14 can complex with TLR2 to bind peptidoglycan (Yang et al. 1999b) or with MRs on human macrophages to recognise LAM (Bernardo et al. 1998), a detection couplet which may acquire further sensitivity via the interaction of LAM with the LPS-binding protein (Savedra et al. 1996). Once associated with a particular PRR, the particle-receptor complex could have several destinational fates. Binding of LAM to surface MRs leads to lysosomal delivery (Prigozy et al. 1997) and TLR2, singly or in combination with TLR6, is recruited to phagosomes where it may survey the contents as part of an ongoing homeostatic mechanism (Underhill et al. 1999a; Ozinsky et al. 2000).

2.1.2
Receptors for *Mtb* Uptake

Metchnikoff invoked the term, *fresszellen* (or devouring cells), in his manuscript on *Daphnia* to describe the gustatory activities of primitive mesoderm (Metchnikoff 1884). Having tasted the lipid-laden surface of *Mtb*, macrophages subsequently employ a variety of phagocytic receptors including PRRs to ingest the organism. Unlike pinocytosis, which involves passive solute uptake, or receptor-mediated endocytosis, which enlists clathrin-coated pits for receipt of smaller molecules, phagocytosis uses both receptor diversity to increase its particle range (>0.5 μm) and actin-driven cytoskeletal remodelling to increase the rate of internalisation (Aderem and Underhill 1999). For *Mtb*, a particle 1–4 μm in length and 0.3–0.6 μm in diameter, this task extends primarily to the complement (CR) and Fc (FcR) receptor families, although some assistance may be rendered by CD43 and fibronectin as well (Fratazzi et al. 2000; Pasula et al. 2002). These receptors enable macrophages to capture mycobacteria via complement or antibody (Ab) fixation, respectively. They also indirectly bind different

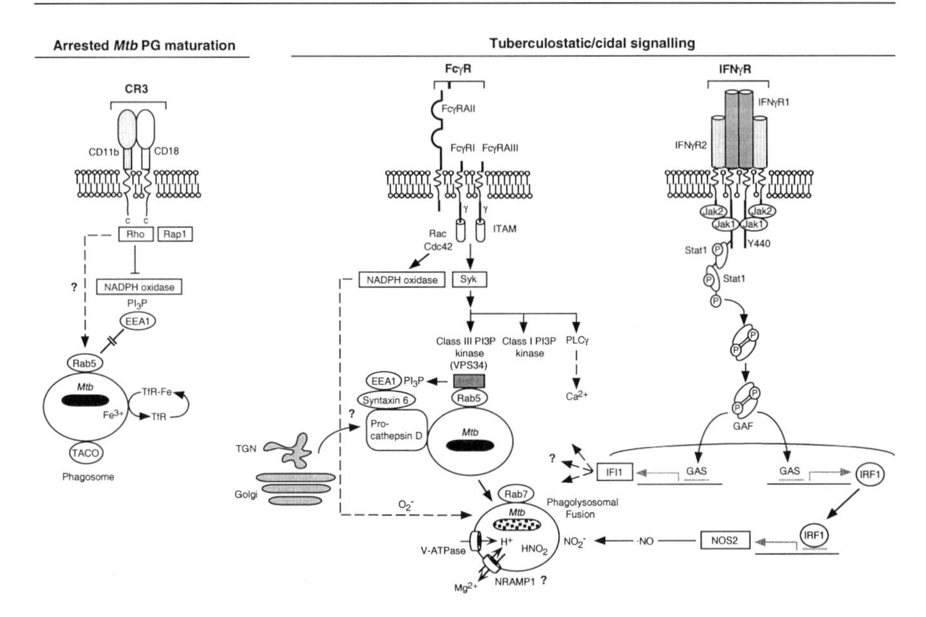

Fig. 2 Receptor-dependent fate of *Mtb* in macrophages

solvent structures on *Mtb*: C3b, C3bi, and C4b cleavage products (recognised by CRs 1, 3 and 4) attach lipid/carbohydrate moieties while Ig domains ensure a (glyco)protein template contact (Ernst 1998).

A role for CRs was first suggested by experiments in which small amounts of fresh (complement-replete) autologous non-immune serum led to enhanced phagocytosis by human monocytes (Schlesinger et al. 1990). Opsonic entry of virulent *Mtb* indicated that as little as 1% serum was sufficient for maximal enhancement (Schlesinger et al. 1990). This percentage is found in the lungs of healthy non-smokers for C3bi and C4b, primarily secreted by AMs, type II epithelia and fibroblasts (Reynolds and Newball 1974; Hill 1993). Human AMs appear particularly reliant on CR4 to mediate mycobacterial uptake (Hirsch et al. 1994; Zaffran et al. 1998) while in human monocytes, *Mtb* internalisation may be blocked by as much as 80% using monoclonal antibodies against CR3 (Schlesinger et al. 1990). Further evidence for CR3 involvement comes from recent studies of CD11b-deficient mice, since this integrin serves as the α-chain in the CD11b/CD18 heterodimer (Fig. 2). Resident macrophages from CD11b$^{-/-}$/- mice exhibit 40%–50% lower levels of serum-mediated *Mtb* uptake and a 50%–60% reduction in non-opsonic binding (Melo et al. 2000), the latter normally mediated via capsular polysaccharides interacting with the CD11b-glucan binding site (Cywes et al. 1997). Despite the reduced internalisation, CD11b$^{-/-}$ mice were no more or less susceptible to TB, like their CR3$^{-/-}$ counterparts (Hu et al. 2000; Melo et al. 2000). Clearly other receptors help control growth of the pathogen.

Engagement of FcRs may lead to better restriction of *Mtb* growth as it usually triggers an oxidative burst, while ligation of CRs does not (Douvas et al. 1986; Kobzik et al. 1990). FcR participation also depends on the maturational state of the cell; it is all but absent in human monocytes (Schlesinger et al. 1990) yet operative in terminally differentiated macrophages (Malik et al. 2000, 2001). Abundant expression of FcγRsI, IIA and IIIA on adult human AMs together with elevated pulmonary IgG levels during TB also raises the likelihood that FcR-mediated uptake of *Mtb* occurs in vivo (Naegel et al. 1984; Fukushima et al. 1991; Sharma et al. 1992). Such uptake leads to a rise in cytosolic Ca^{2+} which promotes phagosome maturation (Armstrong and Hart 1975; Malik et al. 2000), nicotinamide adenine dinucleotide phosphate (reduced) (NADPH) oxidase assembly (Zhou et al. 1997; Melendez et al. 2001), and, at least in the case of human macrophages, killing (Malik et al. 2000) (Fig. 2).

The above studies indicate that the route of *Mtb* entry can influence the organism's fate, and raise the issue of whether *Mtb* internalisation is solely an active host process or assisted by the pathogen. The *Mtb mce1* gene, for example, heterologously conferred invasive properties on *Escherichia coli* for HeLa cells (Arrunda et al. 1993) although its function in *Mtb* awaits confirmation. Preferential uptake by CRs and avoidance of an oxidative burst may be mediated by mycobacteria through salvage of C2a to assemble a C3 convertase on its surface, resulting in cleavage and deposition of C3bi (Schorey et al. 1997). Others have suggested that *Mtb* may selectively use cholesterol-rich caveolae or lipid rafts as portals of entry (Pieters 2001). Mycobacterial products assigned such a function, however, have yet to be identified.

2.2
Cognitive Logic of the Macrophage Against *Mtb*

2.2.1
Intracellular Sorting and the Conduits for Antigen Presentation and Elimination

Almost immediately after uptake, macrophages attempt to dispose of *Mtb* via a series of vesicular transportation pathways leading to lysosomal degradation. These vesicles were among the most striking ultrastructural features of TB granulomata noted by early investigators (Dumont and Sheldon 1965). We now realise they also help establish memory of the encounter through their intersection with major histocompatibility complex (MHC) and CD1 loading compartments.

Podal closure around bacilli and F actin depolymerisation give rise to newly formed phagosomes (PGs) displaying a composition similar to the plasma membrane from which they originate. However, many of the plasma membrane proteins (including FcγRII and MR; Muller et al. 1983) are lost within 3–5 min after PG formation (Pitt et al. 1992; Oh and Swanson 1996). Others, like MHC class II, transferrin receptor, cellubrevin/VAMP3, soluble *N*-ethylmaleimide-sensitive factor attachment protein (SNAP)-23 and syntaxins 3 and 4, remain from

20 min to 72 h on the mycobacterial PG surface (Clemens and Horwitz 1995; Fratti et al. 2000). So does ARF6 (Fratti et al. 2000), which in addition to its involvement in actin rearrangement, recycles endosomal vesicles to the plasma membrane (Al-awar et al. 2000). This protein profile, together with the fact that nascent PGs become accessible to plasma membrane-derived glycosphingolipids and exogenously added tracers like transferrin as early as 5 min after *Mtb* uptake, all suggest an interaction with the recycling endosome network (Clemens and Horwitz 1996; Russell et al. 1996). Fusion with early endosomes similarly results in delivery of the small monomeric guanosine 5′ triphosphatase (GTPase) rab5 (Clemens et al. 2000b; Fratti et al. 2000), considered the critical timer for endosomal docking and fusion (reviewed in Zerial and Mcbride 2001).

It is at this juncture that the mycobacterial PG either proceeds toward a hydrolytically competent phagolysosome (PL) or undergoes maturational arrest (Via et al. 1997; Fratti et al. 2000). The outcome appears largely dictated by the route of *Mtb* entry and activation status of the host cell (Fig. 2). Prevention of PL fusion was first observed by Hart and colleagues using unstimulated macrophages in which viable *Mtb* failed to co-localise with acid phosphatase-rich lysosomes (Armstrong and Hart 1971; Armstrong and Hart 1975). In contrast, *Mtb* that had been coated with Ab or rendered non-viable by irradiation were delivered to lysosomes. Work conducted 25–30 years later suggests the block resides at the level of rab5 and one of its effectors, the early endosome autoantigen 1 (EEA1) (Fratti et al. 2001). EEA1, a tethering molecule that normally couples vesicle docking with SNAP receptor (SNARE) priming, is excluded from unopsonised mycobacterial PGs while present on those encircling latex beads. Moreover, EEA1 is recruited to the latter PGs within 10 min of uptake via its FYVE-domain association with phosphatidyl inositol 3-phosphate (PI_3P), a product generated within the PG membrane (Ellson et al. 2001) by the class III PI_3P kinase VPS34, which directly binds rab5 (Fratti et al. 2001; Vieira et al. 2001). Antibody neutralisation or use of cells deficient for VPS34 have both established the necessity of this kinase for PL development.

The specific involvement of PI_3P could also link Ab-opsonised *Mtb* with the PL maturation initially seen by Armstrong and Hart (1975), an uptake presumably mediated via FcγRs since F'ab-coated *Mtb* does not lead to vacuole acidification (Malik et al. 2000). Clustered FcRs provide docking sites for Syk kinase which has been implicated in lysosomal targeting (Bonnerot et al. 1998). Syk forms the nexus for at least three signalling pathways: (1) class I PI3 kinase, which generates $PI(3,4,5)P_3$ for amphiphysin IIm, dynamin, ARF6 and Rab 11 recruitment, leading to early events such as pseudopod extension and PG closure; (2) PLCγ activation and $PI(1,4,5)P_3$, diacylglycerol and Ca^{2+} mobilisation; and (3) class III PI3 kinase, which generates PI(3)P for EEA1-mediated Rab5/SNARE association and the next stage of PG-lysosome fusion (Crowley et al. 1997; Aderem and Underhill 1999; Greenberg and Grinstein 2002). This stage, characteristically seen in macrophages activated with IFN-γ plus LPS, coincides with a loss of rab5 from the PG (Via et al. 1997) and inaccessibility to transferrin (Schaible et al. 1998). The latter effect may help limit the supply of iron chelated by *Mtb* siderophores for microbial growth (Gobin and Horwitz 1996; De Voss et al. 2000). By 4 h post-uptake, the *Mtb* PL within activated macrophages has acquired the late endosomal marker rab7, lysosomal integral membrane proteins (LGPS or LAMPS), and vacuolar H^+ ATPase (V-ATPase) pump subunits (Xu et al. 1994; Fratti et al. 2000), the last responsible for a drop in lumenal pH from ~6.5 to 5.2 (Schaible et al. 1998). This environment allows for the processing of immature cathepsins and activation of acid phosphatases (Armstrong and Hart 1975; Ullrich et al. 1999) following their likely arrival from the trans-golgi network through the aid of EEA1 binding to syntaxin 6-containing vesicles (Fig. 2; Simonsen et al. 1999; Fratti et al. 2002). Such changes are thought to precede a decrease in mycobacterial viability (Schaible et al. 1998).

How the transition from PG to PL occurs is still unclear. It could enlist repeated, transient vesicular transfer between PGs and late endosomes/lysosomes (a "kiss-and-run" model; Desjardins 1995) or undergo direct fusion with these organelles to form a hybrid compartment (Mullock et al. 1998). The recent discovery of the Rab7 effector RILP (Rab7-interacting lysosomal protein; Cantalupo et al. 2001) has shed some light on the directional nature of PL trafficking. RILP recruits dynein–dynactin motor complexes to Rab7-containing late endosomes/lysosomes, resulting in vesicular transport to the minus end of microtubules rather than towards the macrophage periphery (Jordens et al. 2001). Whether the Rab7-RILP complex is important for polarised maturation of *Mtb* PLs awaits enquiry, although in infected fibroblasts Rab7 appears dispensable (Clemens et al. 2000b). SNARE proteins such as syntaxin 7 and VAMP7, both required for late endosome–lysosome fusion in human AMs (Ward et al. 2000), or the recently cloned Vam6p (Caplan et al. 2001), are among other candidates for this role.

Not all endosomal trafficking proceeds terminally towards lysosomes, which can themselves recycle some of their cargo to the plasma membrane via newly described exosomes (Denzer et al. 2000). For *Mtb*-derived antigens to be presented at the cell surface, loading and sorting must entail retrograde transport. The compartment in which antigens are processed and exported for presentation is very much dependent on their composition. *Mtb* *N*-formyl methionine peptides, for example, associate with H2-M3 MHC class 1b in the endoplasmic reticulum (Chun et al. 2001), while Ag85 is bound to I-A^b in specialised MHC class II compartments (MIICs) (Ramachandra et al. 2001). In humans the latter are enriched for LAMPs 1–3, acid hydrolases, and HLA-DM and HLA-DO which regulate peptide-MHC class II assembly (Geuze 1998). Recycling of antigen-MHC complexes to the plasma membrane is also rapid, with Ag85-bound I-A^b being detected as soon 20 min after uptake (Ramachandra et al. 2001). *Mtb* (glyco)lipids encounter a different set of antigen-presenting molecules: the MHC class I-related CD1 family (Moody and Porcelli 2001). CD1 isoforms are distributed along the endocytic highway; on the plasma membrane/early endosomes (CD1a), early to late endosomes (CD1c), late endosomes/lysosomes (CD1d/CD1e) and MIICs (CD1b). There is thus ample opportunity to sample *Mtb* lipids like LAM, mycolic acids, polyisoprenols, phosphatidyl-inositol mannosides, and lysocardiolipin detected within the macrophage endosomal network (Beatty et al. 2000; Ficsher et al. 2001). Once presented at the surface, human T cells recognising these *Mtb*-derived glycolipids produce cytokines such as IFN-γ (Stenger et al. 1997) to further incite macrophage antimicrobial responses.

2.2.2
Receptors for Activation

Discovery of IFN-γ as the principal macrophage-activating cytokine (Nathan et al. 1983; Schreiber et al. 1983; Nathan et al. 1984) soon led to reports of its effectiveness in curtailing *Mtb* replication (Rook et al. 1986a; Flesch and Kaufmann 1987). It also opened the door to a molecular description of how phagocytes receive and integrate such signals from cell surface to nucleus (Bach et al. 1997).

When both chains of the IFN-γreceptor (IFNGR) bind cytokine, they dimerise, allowing Janus kinases (Jaks) 1 and 2 attached to their cytoplasmic tails to phosphorylate each other. Jaks can further phosphorylate IFNGR1 via the classical pathway (Ramana et al. 2002) to then create docking sites for STAT1 (signal transducer and activator of transcription-1). These latent transcription factors pair to form gamma-activated factors (GAFs) and translocate to the nucleus via the importin-α5 complex (McBride et al. 2002), where they bind IFN-γ activation site (GAS) elements in the promoters of IFN-γ-inducible genes. All this transpires within 5–20 min of receptor engagement (Bach et al. 1997; Darnell 1997). It also initiates waves of secondary transcription via target genes like interferon regulatory factor-1 (IRF-1) and IFN consensus sequence binding protein (ICSBP/IRF-8), which themselves enhance the later expression of macrophage genes such as inducible nitric oxide synthase (NOS2) and FcγR1 (Contursi et al. 2000). Autocrine stimulation of this pathway during TB is also likely, since human AMs are a significant source of IFN-γ (Fenton et al. 1997).

That IFN-γ and IFNGR signalling components are critical for protection against TB is based on several lines of evidence. First, latently infected individuals or exposed asymptomatic household contacts exhibit higher levels of IFN-γ-secreting bronchoalveolar lavage (BAL) or peripheral blood mononuclear cells (PBMC) than do patients with culture-positive pulmonary TB (Schwander et al. 2000; Pathan et al. 2001). Thus IFN-γ expression may correlate with resistance to developing disease. Second, loss-of-function mutations lead to TB susceptibility, as most dramatically demonstrated in IFN-$\gamma^{-/-}$ (Cooper et al. 1993; Flynn et al. 1993), IFN-γR1$^{-/-}$ or STAT1$^{-/-}$ mice (J. MacMicking, unpublished results). The same could apply to humans; hypofunctional and nonfunctional (null) mutations in IFNGRs and STAT1 all lead to greatly impaired anti-mycobacterial immunity, as do mutations in the p40 subunit of the IFN-γ-inducing cytokine, IL-12, or the IL-12 receptor chain (Casanova and Abel 2002). Human macrophages may also be made refractory to IFN-γ signalling by *Mtb* itself, which can interfere with STAT1 binding its transcriptional coactivators CBP and p300 (Ting et al. 1999). Lastly, transgenic IFN-γ reconstitution in the lungs of nullizygous mice had significant ameliorative effects (Collins and Kaufmann 2001), while provision of recombinant IFN-γ alleviated bacterial burdens in patients who were non-responsive to antimicrobials and conferred bactericidal activity on human macrophages in culture (Condos et al. 1997; Bonecini-Almeida et al. 1998).

The effectiveness of IFN-γ in programming macrophages for *Mtb* clearance is augmented by the action of another cytokine, TNF (Chan et al. 1992). Such synergy is a recurring theme in cytokine biology and provides several advantages: (1) increased diversity of the responding host gene repertoire due to alternate transcription factor usage; (2) higher rates of synthesis for those genes with promoters possessing binding sites for both sets of transcription factors; and (3) a lower signalling threshold brought about by shared pre-existing components. The latter is especially germane given the relative paucity of IFNGRs (4,000–12,000/cell) on the macrophage surface (Pace et al. 1983; Finbloom et al. 1985). Examples of the other two benefits are readily appreciated if one examines NOS2 induction. This 260-kDa flavoenzyme requires each of its monomers to be transcribed via STAT1/IRF1-dependent signalling, yet induction is profoundly enhanced by a second signal, such as nuclear factor (NF)-κB originating from TNF or LPS (or LAM) stimulation (MacMicking et al. 1997b; Chan et al. 2001). Moreover, TNF elicits GTP cyclohydrolase 1, which furnishes an essential cofactor (BH_4) for NOS2 catalysis, while NF-κB signals the argininosuccinate synthetase and cationic amino acid transporter 2 genes needed to regenerate and import L-arginine, a NOS2 substrate (Bogdan 2001).

Not only are there co-operative effects between TNF and IFN-γ but also between individual TNF receptors 1 (TNFR1) and 2 (TNFR2); their co-ligation in the presence of IFN-γ leads to a more sustained NO production than with either alone (Riches et al. 1998). This is consistent with a "ligand passing" model in which at low TNF concentrations, the higher affinity and more rapid K_a/K_d of TNFR2 ensures ligand capture for TNFR1 as part of a binary complex (Pickard et al. 1997). TNFR1 then preferentially recruits TNF receptor-associated factor 2 (TRAF2) by virtue of its higher affinity for the TNFR1-associated death domain protein (TRADD) (Baud and Karin 2001). NOS2 induction thereby benefits from the coupling of a stronger intracellular signal issued by TNFR1 (Riches et al. 1998) with the extracellular ligand sensitivity provided by TNFR2.

In mice lacking TNFR1 or TNF, or receiving inhibitory TNFR fusion proteins, *Mtb* grows unabated until death of the host within weeks after infection compared with months for untreated controls (Adams et al. 1995; Flynn et al. 1995; Bean et al. 1999). Some of this marked susceptibility was attributed to diminished macrophage NO production within the first 10 days (Flynn et al. 1995). In humans, administration of infliximab (an anti-TNF mAb) as part of the treatment for rheumatoid and Crohn's disease led to prompt recrudescence of TB in 70 patients who were latently infected (Keane et al. 2001). Natural inhibitory forms of soluble TNFRs (sTNFRI and II) comprising just the extracellular domain could also diminish TNF levels, especially since these are greatly elevated in the serum of active TB patients versus healthy non-TB cohorts, TB contacts or those on antituberculous therapy (Juffermans et al. 1998). Heightened release of sTNFRs has similarly been seen in vitro for *Mtb*-infected AMs (Balcewitz-Sablibska et al. 1998). Each report suggests the TNF/TNFR axis otherwise protects against TB in the human population.

Secreted lymphotoxin (LT-α_3, or TNF-β) serves as another ligand for both TNFRs, and exhibits affinity constants commensurate with those of soluble TNF (Loetscher et al. 1991). Yet unlike TNFR1$^{-/-}$ or TNF$^{-/-}$ mice, LT$\alpha^{-/-}$ chimeras showed uncompromised NOS2 induction despite being equally susceptible to TB (Roach et al. 2001). This highlights two points. First, there exist mechanisms besides NO for *Mtb* containment, notably structural integrity of the granuloma (Roach et al. 2001). Second, lung macrophages probably rely more heavily on autocrine TNFR signalling to induce NO synthesis, since TNF (primarily a monokine) substitutes in the absence of LT (a lymphokine) but not vice versa.

TLR2 and members of the P2 purinergic receptor family have recently been added to the list of receptors involved in activating the macrophage anti-tubercular arsenal. Specific TLR2 engagement using the *Mtb* 19-kDa lipoprotein elicits mouse macrophage NOS2 for restriction of *Mtb* growth, while in human AMs a novel NOS2-, TNF-independent pathway apparently exists (Thoma-Uszynski et al. 2001). P2 purinergic receptor subtypes are likewise engaged in differential fashion: P2X_7 activates cytosolic Ca^{2+} release and phospholipase D for maturation of *Mtb* phagosomes in human macrophages (Kusner and Adams 2000; Kusner and Barton 2001), whereas subtypes other than P2X_7 stimulate *Mtb* killing via NOS2 and possibly reactive oxygen intermediates (ROI) in the mouse (Sikora et al. 1999).

Though not considered a macrophage-activating cytokine in the classical sense, vitamin D_3 [25$(OH)_2$D3] and its ligand-binding receptor (VDR) deserve special mention in the context of TB. Calcitriol [1,25-$(OH)_2$D3] was shown in early studies to be crucial in aiding the tuberculostatic action of IFN-γ in human macrophages (Rook et al. 1986b; Crowle et al. 1987). This bioactive D3 metabolite is synthesised from 25$(OH)_2$D3 by 25$(OH)_2$D3-1 α-hydroxylase and degraded by 24-hydroxylase, enzymes respectively stimulated and inhibited by IFN-γ (Adams and Gacad 1985; Koeffler et al. 1985; Reichel et al. 1987). Calcitriol in turn promotes STAT1-VDR association and the transactivation of IFN-γ-inducible genes (Vidal et al. 2002), suggesting an involvement beyond its well-known effects on macrophage differentiation (Hmama et al. 1999). It is therefore not surprising that some correlation between VDR affinity and the anti-tubercular efficacy of individual metabolites was noted in the initial studies. More recent experiments suggest non-orthodox VDR signalling could account for these observations, because D_3 activates PI3 kinase plus an oxidative burst in PBMCs and NOS2 in human and mouse myelomonocytic cell lines (Rockett et al. 1998; Sly et al. 2001). Moreover, D_3 may contribute to protection in vivo: VDR polymorphisms and TB susceptibility appear to be linked in certain African and Asian populations (Bellamy et al. 1999; Wilkinson et al. 2000).

2.2.3 Deployment of Macrophage Tuberculostatic and Cidal Mechanisms

Nathan and colleagues developed stringent criteria to assess whether a given macrophage effector mechanism operates against a particular pathogen (MacMicking et al. 1997b). They include: (1) correlation of host gene expression with resistance; (2) conferral of direct antimicrobial activity by the host gene or its product (see e.g. Karupiah et al. 1993); (3) loss of host gene function leading to enhanced microbial growth in vitro (via pharmacologic inhibition, RNA interference) or disease exacerbation in vivo (using natural and engineered host mutants, or drug ablation); and (4) evolution of microbial genes resistant to the mechanism in question. The latter would infer its existence through selective pressure. Based on all four criteria, reactive nitrogen intermediates (RNI) arising from the generation of NO by NOS2 emerge as the primary, albeit imperfect, candidates against *Mtb*. Two others, ROI and PL fusion, await further clarification, while more recently a fourth, the β-defensins, has yet to be demonstrated as a natural product of human macrophages (Kisich et al. 2001).

The case for RNI in vitro is compelling. Direct exposure of *Mtb* to as little as 90 ppm of NO gas —approaching the exhaled concentrations found for some pulmonary TB patients (Wang et al. 1998) —kills more than 99% of plated organisms (Long et al. 1999). Unstable derivatives such as nitrogen dioxide (NO_2) appear even more potent (Yu et al. 1999), while the acidified milieu of the macrophage PL may help retrieve additional NO equivalents from stably oxidised forms (e.g. NO_2^-) by protonation to HNO_2 and subsequent dismutation (MacMicking et al. 1997b). Indeed, acidified $NaNO_2$ was as one of the earliest compounds employed in demonstrating a role for NO in direct killing of *Mtb* (Chan et al. 1992). Susceptibility of various *Mtb* strains to acidified $NaNO_2$ correlated inversely with their virulence in guinea-pigs (O'Brien et al. 1994) as well as suggesting the presence of detoxifying pathways on behalf of the pathogen (Rhoades and Orme 1997).

Causal relationships between macrophage NOS2 expression and loss of *Mtb* viability confirm results found in cell-free systems. *Mtb* inhibition in Bcg^r and Bcg^s macrophages correlates with NO production (Arias et al. 1997), while activated murine macrophages treated with NOS inhibitors (Denis 1991a; Chan et al. 1992) or in which the *NOS2* locus has been genetically disrupted (Bodnar et al. 2001; Erht et al. 2001) exhibit little tuberculocidal activity. In human macrophages, eliciting NOS2 via cytokines and/or microbial products appears more complex, akin to human lung epithelia where at least three independent stimuli—IFN-γ, TNF, and IL-1—are required to produce a relatively weak transcriptional response due to STAT1 and NF-κB acting at some distance (~5–8 kbp) from the start site (Ganster et al. 2001). Restricted NOS2 cofactor availability, especially BH_4, could also contribute to the generally lower RNI production of human versus rodent cells (Berholet et al. 1999). Nonetheless, recent evidence shows that macrophages taken from the inflamed lungs or peripheral blood of TB patients, or AMs from healthy donors subsequently infected in vitro, can express NOS2 and produce mycobactericidal amounts of NO (refer to Dlugovitzky et al. 2000; Nathan and Shiloh 2000; Means et al. 2001; Wang et al. 2001). Moreover, where examined, anti-tubercular effects were blocked by NOS inhibitors.

A case for RNI-dependent protection extends to the intact host. Macrophage NOS2 was present in TB granulomas of humans (Fachetti et al. 1999) and resistant wild-type mice (MacMicking et al. 1997a; Mogues et al. 2001); the latter were rendered susceptible with specific NOS2 inhibitors like N^6-(1-iminoethyl)-L-lysine (MacMicking et al. 1997a) and drugs of lower isoform specificity such as aminoguanidine and *N*-methyl-L-arginine (Chan et al. 1995; Flynn et al. 1998). More convincing evidence is provided by NOS2$^{-/-}$ mice, which are extremely sensitive to *Mtb* infection (MacMicking et al. 1997a; Mogues et al. 2001; Scanga et al. 2001). The same is true of other gene-deficient mice (IFN-$\gamma^{-/-}$, TNFR1$^{-/-}$ or TNF$^{-/-}$) as well as glucocorticoid-immunosuppressed and malnourished animals exhibiting secondary defects in macrophage NOS2 expression (reviewed in Nathan and Shiloh 2000). Lastly, the presence of RNI resistance genes may indirectly suggest a role for this pathway in the host response to TB. *NoxR1*, *noxR3*, alkyl hydroperoxide reductase (*ahpC*), peptidyl methionine sulphoxide reductase (*msrA*), dihydrolipoamide dehydrogenase (*Lpd*) and dihydrolipoamide succinyl-transferase (*SucB*) are each posited as *Mtb* RNI resistance genes (Nathan and Shiloh 2000; St. John et al. 2001; Bryk et al. 2002), although chromosomal inactivation has yet to corroborate these claims (Stewart et al. 2000; Springer et al. 2001). Disruption of the oligopeptide permease operon (*oppBCA*), which can transport NO-thiol adducts across the mycobacterial cytoplasmic membrane, confers resistance to *S*-nitrosoglutathione and suggests that NO could be more effective as a congener (Green et al. 2000).

Virulent *Mtb* grown in culture or in murine macrophages are relatively resistant to the effects of ROI, and host cells either treated with ROI scavengers or deficient in the production of superoxide (O_2^-) still managed to restrict *Mtb* replication (Flesch and Kaufmann 1987; O'brien and Andrew 1991; Chan et al. 1992; Adams et al. 1997). Likewise, an oxidative burst triggered by mycobacterial agonists in IFN-γ-primed NOS2$^{-/-}$ macrophages still failed to inhibit microbial growth (MacMicking et al. 1995; Erht et al. 2001). Mice with targeted disruptions in either gp47^{phox} or gp91^{phox} subunits of the periplasmic NADPH oxidase responsible for O_2^- synthesis allow very modest increases in lung bacterial burdens versus wild-type controls (Adams et al. 1997; Cooper et al. 2000). Moreover, humans with crippling mutations in any one of four NADPH oxidase subunits (collectively referred to as chronic granulomatous disease; CGD) do not appear especially vulnerable to infections by mycobacteria (Segal et al. 2000), although anti-tubercular immunity may be impaired in some patients (Lau et al. 1998).

A major reason for the limited potency of ROI against *Mtb* is that the bacterium robustly expresses ROI-detoxifying enzymes (e.g. catalase, superoxide dismutase, peroxiredoxins) both within and outside activated human macrophages (Andersen et al. 1991; Mariani et al. 2000). Such enzymes could also serve to counteract oxidative species arising as by-products of bacterial respiration. Additionally they may work in concert with RNI neutralising pathways to limit the production of compound intermediates such as peroxynitrite ($ONOO^-$), a powerful oxidant that can decompose under acidic conditions to NO_2 and the hydroxyl radical ($OH^·$). Virulent strains of *Mtb*, for example, appear to be more resistant to $ONOO^-$ than non-pathogenic strains (Yu et al. 1999), reminiscent of studies examining sensitivities to hydrogen peroxide (H_2O_2) (Jackett et al. 1978; Laochumroonvorapong et al. 1997). Another reason could include the fact that catalase and peroxiredoxins are also directly effective at scavenging RNI (reviewed in Nathan and Shiloh 2000). Both scenarios may be examined using mice that are doubly deficient for NOS2 and phox (Shiloh et al. 1999) together with the respective *Mtb* mutants, a genetic approach similar to that already employed for the *Mtb sodC* gene (Piddington et al. 2001).

The antimicrobial actions of both RNI and ROI are potentiated by low pH and iron (usually Fe[III]), the latter incorporated into dinitrosyl-iron complexes as well as helping drive the Haber-Weiss reaction for $OH^·$ formation. The usefulness of acid in host defence was first suggested by Metchnikoff (1893) and gained credence from two early observations: that phagocytes acidify their PLs (Rous 1925) and that achlorhydric hosts permit bacterial colonisation of the stomach (Gianella et al. 1973). In the case of mycobacteria, the macrophage PL acidifies to pH ~4.5–5.0 following cytokine stimulation, irrespective of whether the agonist is given before or after bacterial uptake (Schaible et al. 1998; J. MacMicking, unpublished). Acidification coincides with the delivery of V-ATPases, enables processing of lysosomal hydrolases, and correlates with diminished *Mtb* growth (Schaible et al. 1998; Gomes et al. 1999; Ullrich et al. 1999). Conversely, agents that specifically inhibit the ATPase (e.g. omeprezole) allow increased *Mtb* replication in human AMs (Suzuki et al. 2000). As yet there is no reliable information on whether such mechanisms operate in vivo, a void difficult to fill, since natural and engineered ATPase mutants, or pharmacologic manipulation within the intact host, would surely have widespread homeostatic consequences for the cell. Perhaps gain-of-function studies involving forced PL maturation on an RNI/ROI-deficient background may yield information about the requirement for acid per se.

The last point is especially relevant for *Mtb* since it was originally shown to be unique among other *Mycobacteria* spp. in its sensitivity to pH less than 6.5, with a marked attenuation for growth at pH 5.0 (Chapman and Bernard 1962). Later studies, however, found little difference in viability between pH 4.5–7.0 (Jackett et al. 1978; Chan et al. 1992). An explanation for this discordance may rest with the use of complex media varying in divalent cation concentration. When grown at lower pH under restricted Mg^{2+} conditions, for example, *Mtb* fares much less well (Piddington et al. 2000). These conditions could also mimic that found in the host: an *Mtb* mutant lacking the *mgtC* (magnesium transporter) gene replicates poorly at low pH and Mg^{2+} concentration, and its growth is highly attenuated in human macrophages and the lungs of mice (Buchmeier et al. 2000).

Mtb similarly imports other divalent cations—Fe^{2+}, Mn^{2+}, Zn^{2+}, Cu^{2+}—via expression of *Mramp*, a pH-dependent transporter which functions optimally at pH 5.5–6.5 and is upregulated under axenic culture at low pH as well as in human macrophages (Agaroff et al. 1999). In this respect it may compete with its mammalian homologue, natural resistance-associated macrophage protein 1 (NRAMP1, or SLC11A1), a H^{+}/divalent cation antiporter located at the phagosomal membrane (Jabado et al. 2000). A caveat to this relationship is that while NRAMP1's role in protecting the host against mycobacteria other than *Mtb* is well established, its deletion has no effect on TB susceptibility in mice (North et al. 1999). The case in humans is less clear: a lack of association between TB susceptibility and NRAMP1 status was noted for Brazilian populations (Blackwell et al. 1997) while *NRAMP1* polymorphisms in West Africans, Japanese and Koreans may carry an increased risk (Bellamy et al. 1998; Gao et al. 2000; Ryu et al. 2000).

Realisation that *Mtb* could tolerate an acidified, nutrient-poor environment moves understanding beyond the current preoccupation with models of how the bacterium blocks PL fusion. This latter effect, perhaps involving (1) bacterial ammonia generation via urease or sulphatide (Gordon et al. 1980; Goren et al. 1976) or (2) a host tryptophan aspartate-containing coat protein (TACO, or coronin-1; Ferrari et al. 1999), appears relevant only in macrophages that have not been activated (Via et al. 1997; Schaible et al. 1998). Of course, strategies for preventing activation are likely to exist: eliciting deactivating cytokines (e.g. IL-10, transforming growth factor-beta; Murray 1999) or evading antigen presentation altogether (Pancholi et al. 1993) are two such examples. Nonetheless, more attention should perhaps focus on examining bacterial adaptations within a hostile cell and to the combined rather than isolated actions of macrophage tuberculostatic/cidal pathways. Equally, consideration could be given to higher-order structure, where substratum, stromal- and T-cell contact within granulomas provide additional stimuli and where anatomical specialisations (e.g. to low oxygen tension) may be paramount.

2.3
Systems Logic of the Macrophage Against *Mtb*

2.3.1
Macrophage–Lymphocyte Networks: The Role of Granulomas

In humans, tuberculoid granulomas occupy not only the lungs, but anatomical sites as diverse as the larynx, palate, nasal septum, submaxillary and tracheobronchial lymph nodes, spine, bone marrow, genitourinary tract and peritoneal lining (Iseman 2000). Events leading to such widespread dissemination were chronicled in the preantibiotic era by Canetti (1955). From histobacteriologic examination of 30 cadavers, he suggested the following sequelae: a (pre)exudative stage dominated by mononuclear infiltrates and small numbers of AFB; walling off and caseous necrosis at the centre of the granuloma with progressively fewer AFB; lesion resolution by fibrin deposition, sclerosis and calcification accompanied by disappearance of AFB. Alternatively, there could be caseum liquefaction and focal reappearance of AFB in the liquefied areas, the latter of which may rupture into a bronchus with discharge rendering the patient infectious.

The cellular composition and structure of the granuloma appears critical to whether TB lesions regress or become "open". Poorly organised granulomas, as seen in IFN-$\gamma^{-/-}$, TNF$^{-/-}$ and intercellular adhesion molecule (ICAM)-1$^{-/-}$ mice as well as LT$\alpha^{-/-}$ bone-marrow chimeras (Garcia et al. 1997; Sugawara et al. 1998; Bean et al. 1999; Kanedo et al. 1999; Saunders et al. 1999; Roach et al. 2001), correlate with rapid bacterial dissemination and reduced host survival. A characteristic feature in all of these cases is the paucity of epithelioid macrophages and failure to recruit T lymphocytes. The latter's participation in successful *Mtb* containment is underscored by a study of HIV-infected patients suffering from culture-proven tuberculous lymphadenitis (Muller and Kruger 1994). Severely lymphopaenic patients completely lacked epithelioid granulomas, while those HIV patients retaining modest peripheral CD4^{+} counts had granulomas replete with lysozyme-expressing macrophages in apposition with CD25^{+} (IL-2R) lymphocytes, in addition to fewer AFB. The presence of T cells not only ensures a vicinal supply of macrophage-activating T helper (Th)1 cytokines (Robinson et al. 1994; Bergeron et al. 1997; Fachetti et al. 1999; Fenhalls et al. 2000; Wangoo et al. 2001) but also raises the possibility of reciprocal costimulation. AMs in TB granulomas express high levels of the costimulatory molecules B7–1 and B7–2 with nearby T cells being CD28^{+} (Soler et al. 1999). T cell–macrophage liaison may thus engender immunologic memory in addition to microbial killing.

2.3.2
Macrophage-Monocyte Networks: A Case for Multinucleated Giant Cells

Multinucleated giant cells (MGCs) were discovered as large, polykaryonic structures within TB granulomata over 130 years ago (Langhans 1868). Subsequently found to be a common feature in many infectious and foreign body granulomas, MGCs are thought to represent fusions between descendants of the monocyte-macrophage lineage (Anderson 2000). Indeed, MGCs can now routinely be derived in vitro using macrophages alone following the initial demonstration of rabbit AMs to fuse after treatment with supernatants from mycobacteria-sensitised lymph node cells stimulated with antigen (Galindo et al. 1974). The soluble bioactive component in this early study was most likely IFN-γ; use of recombinant IFN-γ when it became available also led human AMs to fuse which could in turn be blocked by anti-IFN-γ mAbs (Nagasawa et al. 1987).

Circulating monocytes are thought to be another source of MGCs (Gillman and Wright 1966) and they, too, appear heavily reliant on soluble factors for their coalescence (Postlethwaite et al. 1982). Again, IFN-γ seems to be the major fusogenic cytokine (Weinberg et al. 1984; Most et al. 1990; Fais et al. 1994; Gasser and Most 1999; Mizuno et al. 2001) although some assistance is rendered by IL-3 (Enelow et al. 1992; Byrd 1998) and IL-4 (Takashima et al. 1993). Multinucleation also requires cell-cell contact in addition to soluble mediators, with lymphocyte function-associated antigen (LFA)-1, ICAM-1, β_2-integrins, CD44, CD47 and the macrophage fusion receptor (MFR) all posited to play a role (Most et al. 1990; Fais et al. 1994; Saginario et al. 1998; Sterling et al. 1998; Gasser and Most 1999).

That multinucleation serves some benefit against *Mtb* is suggested by the ability of IFN-γ/IL-3-treated human monocyte-derived MGCs to severely limit microbial spread (Byrd 1998). How MGCs bring about this restriction remains obscure although it could involve the purinergic $P2X_7$ receptor which is upregulated by IFN-γ (Humphreys and Dubyak 1996). Both an mAb directed against the $P2X_7$ extracellular domain and irreversible ATP analogues (^{o}ATP) block human monocyte fusion (Falzoni et al. 2000). Since stimulation of $P2X_7$ is known to promote PL maturation (see "Receptors for Activation" above), the generation of MGCs via extracellular nucleotides may be tightly coupled with an ability to dispose of the pathogen. This idea is consonant with earlier studies showing elevated lysozymal activity of MCGs within human tuberculoid granulomas (Yamashita et al. 1978; Williams and Williams 1983).

2.4 Macrophage Specialisations Within the Pulmonary Context

2.4.1 Surfactant Receptors

With a surface area approaching 50–100 m^2, the human lung mandates a very broad system of innate defence consisting of ciliary (mechanical) clearance, cough reflex (ironically serving as a vehicle for *Mtb* transmission), and cellular responsiveness. Surfactant proteins (SPs) present within the alveolar lining fluid as a means to reduce surface tension and lung collapse appear increasingly important in this non-adaptive immune repertoire (Holmskov 1999). Their relevance to TB was first noted by the ability of surfactant protein A (SP-A) to promote *Mtb* attachment and phagocytosis by AMs from both normal subjects and HIV patients (Downing et al. 1995; Gaynor et al. 1995). A second surfactant protein, SP-D, has more recently been shown to bind *Mtb* and LAM via its carbohydrate recognition domain, a feature characteristic of the collectin family to which SPs belong (Ferguson et al. 1999). SPs-A and -D have each been localised to endocytic vesicles and lysosomal granules of AMs (Walker et al. 1986; Voorhout et al. 1992), suggesting their uptake was receptor-mediated. This has proved to be the case; the human collectin receptor C1qRp and SP-R210 both bind SP-A while SP-D enters AMs via a scavenger receptor superfamily member, gp-340 (Holmskov 1999). Antibodies directed against SP-R210, in particular, inhibited SP-A-associated uptake of mycobacteria (Weikert et al. 1997) and the subsequent induction of NO and TNF-α synthesis (Weikert et al. 2000). In the presence of IFN-γ, however, SP-A treatment together with *Mtb* actually lowered RNI release (Pasula et al. 1999), while the same was seen in human AMs for peptidoglycan/TLR2-induced TNF-α secretion (Murakami et al. 2001). Thus, SPs could also have macrophage-suppressive functions which help resolve inflammation once the major bacterial burden is cleared.

2.4.2 Hypoxia-Induced Responses

Mtb is an obligate aerobe, despite being equipped with pathways for anaerobic energy metabolism (Cole et al. 1998; McKinney et al. 1998; Wayne and Sohaskey 2001). This may explain why disseminated tuberculous lesions tend to develop more rapidly in the apical lung fields of an upright host where steady-state oxygen tension is highest (Riley 1960; West 2000). Human AMs, on the other hand, remain effective phagocytes even under mildly hypoxic conditions (pO_2 >25 mmHg; ~3% O_2 at sea level) as measured in granulomas or abscess cavities remote from oxygenated blood (Cohen and Cline 1971; Hocking and Golde 1979). The environment within human macrophages may further limit O_2 availability since bacteria appear more sensitive to reduced ambient O_2 tensions when grown intracellularly (Meylan et al. 1992).

Restricted growth of *Mtb* inside macrophages could equally be interpreted from the viewpoint of the host; hypoxia elicits powerful mycobacteriostatic/cidal responses. Approximately 1% O_2 alone or in combination with IFN-γ increased NOS2 expression ~25-fold and TNF secretion 500-fold in murine and human macrophages, respectively (Scannell et al. 1993; Melillo et al. 1996). The molecular basis of these increases is in part due to hypoxia-inducible protein (HIF)-1, a transcriptional activator that functions as master regulator of mammalian O_2 homeostasis (Semenza 2001). HIF-1 binds to hypoxia response elements (HREs) in either 5′ or 3′ flanking regions of target genes following exposure to 0.5%–6% O_2 (Semenza 2001). The murine and rat NOS2 promoters, for example, both contain HREs which are bound in vitro and in vivo (Melillo 1995; Jung et al. 2000). Moreover, the earlier synergy reported for IFN-γ and hypoxia (Mellilo 1995) in promoting macrophage NOS2 expression probably stems from the recent observation that HIF-1 physically associates with IRF-1 as part of a co-activating promoter complex (Tendler et al. 2001). A cycle of stimulation is then propagated since NO enhances HIF-1α expression and DNA binding (Sandau et al. 2001), perhaps through nitrosyl–cysteine modification.

3
Harnessing the Macrophage Anti-tubercular Arsenal

3.1
Intersection of Immunity and Conventional Chemotherapy

When modern chemotherapy against *Mtb* fails, it usually does so for one of two reasons. Firstly, unsuccessful treatment outcomes may be due to acquisition of multi-drug resistance mutations on the part of the pathogen. MDR-TB accounts for only a minor fraction, however, of newly diagnosed cases: ~3% worldwide in the year 2000 (Dye et al. 2002). A second, more common problem is patient non-compliance. TB therapy calls for concomitant administration of 2–3 drugs for a minimum of 6 months. Many patients are unable or unwilling to adhere to such a lengthy drug regimen, and their neglect leads to high relapse rates since short-course therapy is insufficient to eradicate a small subpopulation of mycobacterial "persisters" (Mitchinson 1985). For this reason, the World Health Organisation has recommended that every patient receive directly observed therapy (DOT) to ensure compliance (Who 2000). Extension of the DOT strategy to individuals with latent TB infection (LTBI), however, is not practicable given the enormous number of cases involved. The problem is further aggravated by the fact that LTBI also requires a lengthy drug regimen, usually 6–12 months of isoniazid prophylaxis. To persuade a patient with active TB to complete a protracted course of antibiotics is often challenging: exhorting the same practice in an asymptomatic individual with LTBI may prove even more difficult. Clearly, development of shorter treatment regimens, especially for LTBI, is an urgent priority (Institute of Medicine 2000).

Recent studies in experimental animals suggest that post-exposure vaccination may serve as a useful adjunct to standard drug therapy (Lowrie et al. 1999; Lowrie and Silva et al. 2000). Mice infected with *Mtb* were treated with anti-tuberculosis drugs and then vaccinated with DNA encoding the mycobacterial Hsp65 heat shock protein. Such mice were completely protected from subsequent reactivation of disease, even after being treated with powerful immunosuppressive agents (Lowrie et al. 1999). Thus, in principle, immune modulation in conjunction with conventional chemotherapy could lead to a more rapid and complete cure in humans. Enhancing the efficacy of the immune response in this manner would be particularly apposite in persistent infections like TB, where natural immunity, left unmodified, does not always eliminate the pathogen (McKinney et al. 1998). The observations of Lowrie and colleagues suggest that this failure is not because the host is incapable of sterilising immunity but rather that immunity remains poorly elicited, a situation corrected with vaccination. If one could provide the requisite forms of immunostimulation, not only through vaccination but potentially via macrophage-activating agents based on cytokine signalling or downstream effectors, then new avenues of treatment for LTBI and MDR-TB could be possible in the future.

3.2 Appropriately Activating the Macrophage In Situ

3.2.1 Cytokine and Ab Immunotherapy

Parenteral administration of IFN-γ as adjunctive to antimicrobials or even as replacement therapy in immunodeficiencies such as AIDS or CGD substantiates the virtues extolled for this cytokine in what are largely experimental systems (Murray 1996). In humans, intravascular (i.v.), intramuscular (i.m.) or intradermal (i.d.) IFN-γ treatment appears well tolerated and stimulates the antimicrobial repertoire of PBMCs, circulating neutrophils and tissue macrophages. When given as an aerosol, IFN-γ activates AMs in a lung-restricted fashion (Jaffe et al. 1991). Ameliorative effects have been seen for non-viral infections caused by atypical mycobacteria (e.g. *M. avium, M. chelonei*) (Holland et al. 1994), and cutaneous (e.g. *Leishmania tropica, L. mexicana, L. braziliensis*) or visceral (kala-azar) leishmaniasis, in some cases yielding complete cure, especially if given with other modalities (e.g. antimonial compounds) (Murray 1996). For lepromatous leprosy, i.d. IFN-γ administration shows a clear enhancement of cell-mediated immunity, including oxidative burst, delayed-type hypersensitivity reactions and granuloma formation (Nathan et al. 1986).

In the case of TB, provision of exogenous IFN-γ could be of significant help to those patients refractory to chemotherapy. Benefits have been found for individuals with severe, advanced MDR-TB; 500 μg of aerosolised IFN-γ given thrice weekly for a month led to clear salutary effects in the form of demonstrable weight gain, reversion of positive sputum smears to negative status, and reduction in the size of cavitary lesions (Condos et al. 1997). Similar doses in other studies led to increased respiratory burst capacities and chemokine release from AMs (Jaffe et al. 1991; Halme et al. 1995), suggesting that some of the benefits in people with TB probably operate at the level of the macrophage. An improved outcome was also noted for an MDR patient receiving subcutaneous IFN-γ together with granulocyte-macrophage colony-stimulating factor (GM-CSF) (Raad et al. 1996), as well as in TB and MDR-TB patients given inhaled IFN-α, a benefit largely thought to be due to local induction of IFN-γ (Giosue et al. 1998; Giosue et al. 2000). Whether IFNs could be administered systemically to combat extrapulmonary TB or as a reconstitutive measure in HIV-positive TB patients has yet to be examined.

Inveighing against the use of TNF as potential immunostimulant for TB is its established history as both proinflammatory and pyrogenic; indeed, these are mainly held responsible for certain clinical characteristics of the disease, namely, “phthisis” (wasting/cachexia), night sweats and tissue destruction (Keane et al. 2001). Yet its value as adjunctive agent has been suggested in some (but not all, e.g. Moreira et al. 2002) animal models: mice receiving recombinant human TNF or a non-toxic mimetic peptide (TNF_{70-80}) exhibit increased resistance to mycobacterial infection accompanied by reduced CFU, heightened NO release and better organised granulomas (Denis 1991b; Roach et al. 1999; Briscoe et al. 2000). Small peptide mimics may also have the advantage of bypassing any sTNFR blockage (Juffermans et al. 1998) if composed of epitopes not bound by these truncated receptors. Other TNF-modulating agents (e.g. thalidomide analogues) facilitate in vivo TNF production during TB as shown by recent studies of HIV-infected patients (Bekker et al. 2000; Gotri et al. 2000).

Concern that TNF unduly promotes pulmonary tissue damage is also subject to debate; both acute and persistent TB models suggest that TNF's protective role far outweighs any pathologic involvement (Adams et al. 1995; Flynn et al. 1995; Bean et al. 1999; Mohan et al. 2001; Smith et al. 2002). In humans, too, TNF antagonists (e.g. infliximab) appear to reactivate rather than limit disease (Keane et al. 2001; Martinez et al. 2001; Wagner et al. 2002; Mayordomo et al. 2002). It could even be argued that some degree of TNFR1/2-mediated apoptosis (Mohan et al. 2001) might be useful, either through directly killing *Mtb*-infected macrophages or by liberating drug-sensitive bacteria which are more readily accessed by antimicrobials than inside the cell.

Ab therapy directed against specific *Mtb* cell wall components, e.g. LAM epitopes, is another method which could encourage macrophage activation, this time through uptake by FcγRs. An IgG3 mAb recognising arabinomannan delivered intratracheally has been shown to partially protect wild-type, IFN-$\gamma^{-/-}$ and MHC class II$^{-/-}$ mice against *Mtb* infection (Teitelbaum et al. 1998). All mAb-treated mice exhibited enhanced granuloma formation with a mantle of NOS2-positive macrophages thought to represent a barrier to bacterial dissemination. Whether mAb treatment is effective if begun after clinical signs are evident, however, remains to be determined. So, too, does the issue of whether Ab therapy might hasten immune complex deposition leading to alveolitis and other FcR-mediated lung damage (Clynes et al. 1999). In a recent study of 68 patients with active progressive TB, for example, over 50% had immune complexes in their lungs, although whether this was a primary cause or a secondary consequence of disease cannot be distinguished (Surkova et al. 1999).

As with all recombinant proteins, the utility of cytokines and mAbs is limited by the concerns of cost, bioavailability and generation of neutralising antibodies. However, at least in the case of IFN-γ, long-term administration appears feasible; CGD patients have tolerated repeated prophylactic injections for up to 7 years without noticeable decreases in plasma half-life or side-effects more debilitating than mild fever (Segal 2000). Even so, as a serious avenue for treatment of TB and activator of human macrophages in situ, cytokines and mAbs are impractical in those developing countries most affected by the disease. The real importance of the aforementioned studies is the didactic lesson they provide: enhancing macrophage anti-tubercular mechanisms can have a genuine impact on TB progression, especially in cases where chemotherapy is no longer tenable.

3.2.2
Small Molecule Chemistry: Adaptor Protein Mimics

If macrophage activation is primarily achieved through cytokine receptor signalling, why not obviate the need for a bulky ligand altogether? Small, non-peptidyl mimics could offer advantages both in terms of specificity and pharmacokinetics. Events at or downstream of IFNGRs and novel PI3 K effectors are two examples where such an approach may be applicable to the macrophage-*Mtb* problem.

Elicitation of Jaks or STATs by small molecule modulators has already been demonstrated in the case of Jak3; large-scale chemical screens were based on a known tyrosine kinase inhibitor pharmacophore and homology model derived from kinase crystal structures (Goodman et al. 1998). However, this represents but one tier of a multistage signalling cascade. In searching for modifiers of IFNGR signal transduction, for example, one could extend the screen to several levels: (1) IFNGR1/2 ligand binding and dimerisation, (2) Jak1/2 binding, (3) STAT1 homodimerisation, and (4) STAT1-DNA binding and transcriptional activation (Seidel et al. 2000). At the cell surface, small molecules capable of causing receptor dimerisation have already been found for erythropoietin and the GM-CSF receptor, in the latter case binding to alternate sites which do not compete or interfere with the natural ligand (Qureshi et al. 1999; Seidel et al. 2000).

Beyond receptor clustering lies the problem of how to confer specificity when so many components of the Jak/STAT entourage are shared among different cytokines. Here, a combinatorial approach may prove useful: drugs exhibiting higher specificity for STAT1 could be given with another of known preference for Jaks1/2. Signalling by other cytokines/growth factors would thus be minimised, since additional adaptor proteins (STATs2–5) or kinases (Tyk2, Jak3) obligate for their transduction are either poorly induced or missing. Imiquimod [1-(2-methylpropyl)-1*H*-imidazo[4,5-c]quinolin-4-amine] and its analogue S28463, antiviral drugs used against human papillomavirus and experimentally effective against vacuolar pathogens like *Leishmania* and *M. bovis* BCG, mediate their effects on macrophages preferentially through STAT1 (Bottrel et al. 1999) and MyD88 (Hemmi et al. 2002); they could serve as parent or lead compounds for more soluble derivatives. Selectivity may also be imposed at the level of transcription by targeting the STAT1 linker domain, since this region is critical for promoter complex formation following IFN-γ but not IFN-α/β stimulation (Yang et al. 1999a).

A second signalling intermediate thought to be involved in countering *Mtb* is PI_3P. This lipid directs EEA1 to the mycobacterial PG for subsequent maturation into PLs (see section entitled "Intracellular Sorting and the Conduits for Antigen Presentation and Elimination"). It also appears to bind the NADPH subunit $p40^{phox}$ for oxidase assembly on nascent PGs (Bravo et al. 2001) and could participate in IFN-γ-dependent STAT1 S727 phosphorylation (Nguyan et al. 2001). What makes PI_3P such an attractive biochemical target is its phosphoinositide headgroup interactions with protein FVYE and PX domains; they distinguish

PI_3P from all other lipids, with recognition reliant on its being embedded in a phospholipid bilayer (Misra et al. 2001). Indeed, FYVE domains exclusively bind PI_3P and no other phosphoinositide. The recent co-crystallisation of PI_3P-EEA1 (Dumas et al. 2001) and PI_3P-p40^{phox} (Bravo et al. 2001) has provided structural information on which to design small molecule mimics. Novel PI_3P or p150/VPS34 (class III PI3 K) binding partners could also be isolated by selective capture on phosphoinositide affinity matrices as recently shown for PI3 kinase effectors regulating both Arf and Rho GTPases (Krugmann et al. 2002). High-throughput screens of this type may uncover compounds capable of specifically forcing macrophage PL fusion in the absence of activation.

Exploring the world of small GTPases in PL biogenesis is an avenue that promises much in the way of understanding macrophage immunity during infection with implications for drug development. GTPases are judiciously positioned to direct intracellular traffic and assist the spatial convergence of several effector pathways. For example, IFN-γ-induced Rab5a has recently been shown to remodel the PG environment of engulfed listeriae where it facilitates the translocation of Rac2 (another GTPase) to the PG surface (Prada-Delgardo et al. 2001). With Rac2 comes the NADPH oxidase and potentially NOS2 (Kuncewicz et al. 2001). Another group of IFN-elicited GTPases (the 47-kDa family) help control *Toxoplasma gondii* and *L. monocytogenes* infections in mice (Taylor et al. 2000; Collazo et al. 2001). One of these, interferon-inducible protein 1 (IFI1; LRG47), operates against *Mtb* as well (J. MacMicking, unpublished results). How IFI1 brings about its anti-tubercular effects is unknown, but mechanisms could include assisting PL fusion, Ag presentation or trafficking of lysosomal hydrolases between vesicular compartments (Fig. 2). Once determined, human IFI1 homologues would seem a logical target for pharmacologic intervention, given that its expression is limited to disease or inflammatory states making disruptions of homeostatic processes less likely. Reaching or mimicking vesicle-associated adaptor proteins like IFI1 is probably easier than attempting to deliver drugs into the *Mtb* PG lumen directly. Translocation of xenobiotics across the PG/PL membrane needs take into account substrate-specific porters and efflux pumps (Lloyd 2000); perhaps drug-conjugates utilising the transferrin receptor would enable larger molecules to gain entrance to the interior before PL maturation is complete (Clemens and Horwitz 1996).

3.2.3
Local Provision of Effector Molecules

At present no studies exist describing macrophage effector molecules being directly delivered to the pulmonary tree of TB patients. Nonetheless, one could make a case for NO or its congeners based on indirect findings. Doses as low as 70–90 ppm for 48 h kill both drug-sensitive and drug-resistant strains of *Mtb* (Long et al. 1999); this level is tolerated for up to 2 weeks in neonates with persistent hypertension, where it provides dramatic benefits (Clark et al. 2000). NO gas is preferentially delivered to the well-ventilated areas of the lung, coupling ventilation (V) to perfusion (Q) and acting as a selective pulmonary vasodilator (Moya et al. 2001). Whether it could reach the relatively hypoxic centre of a granuloma, however, remains conjectural. Nor are the amounts of NO needed to sterilise smear-positive patients, some of whom may harbour as many as 10^5-10^6 CFU/ml of saliva or sputum (Yeager et al. 1967), known. Higher doses of NO run the risk of toxicity, especially within an O_2-rich environment where derivatives like $ONOO^-$ arise more frequently. For this reason most NO bioactivity comes in the guise of *S*-nitrosothiols (SNOs) which are relatively resistant to toxic reactions with O_2/O_2^- and are transported across the cell membrane by γ-glutamyl transpeptidases (Moya et al. 2001). It may also explain why free NO in human airway lining fluid and expired air is often below the levels required to dilate blood vessels or relax airways—the majority is complexed with glutathione and SNO-proteins which are themselves sensors of ventilatory hypoxia (Lipton et al. 2001).

An ideal drug would therefore convert NO equivalents into SNOs. It should also avoid potentially harmful peroxidation and allow the desired V/Q matching and pulmonary/systemic activity quotients (i.e. be a gas). Recently, Stamler and colleagues (Moya et al. 2001) used the following criterion to isolate such a compound: (1) high volatility, (2) resistance to oxidative decomposition, (3) low oxidising potential towards haemoglobin, and (4) biocompatibility. *O*-nitrosoethanol (ENO) was found to fulfill these criteria. It is stable at high ambient O_2 concentrations since heterolytic transfer reactions predominate over homolytic decomposition, enabling it to react preferentially with nucleophiles such as sulphurs of glutathione and proteins within the lung. It was also active under hypoxic conditions and did not affect systemic haemodynamics (Moya et al. 2001). ENO could thus represent the prototype of future compounds which allow transnitrosation reactions to occur within pulmonary granulomas in situ. The problem of delivery, however, remains uppermost on the list of limitations needing to be overcome before use against respiratory infections like TB would be considered.

3.3
Beyond Koch: *Mtb* targets as Adjunct to Macrophage Immunity

Besides attempting to enhance the macrophage armamentarium or directly furnish its products, another approach to aiding anti-tubercular immunity could include interference with the putative counter-strategies used by *Mtb* (Table 1).Traditional drugs target bacterial processes that are required for growth and division such as DNA replication or cell-wall biogenesis (McKinney 2000). This may be a critical factor limiting the efficacy of conventional chemotherapy against latent TB, where mycobacteria are thought to be in a metabolically altered and/or essentially non-replicative state (McKinney et al. 1998; Wayne and Sohaskey 2001). Agents that damage *Mtb* indirectly—by ablating RNI/ROI detoxifying enzymes, contesting Fe acquisition and storage, blocking acid tolerance or pathways involved in macrophage deactivation—may be less dependent

Table 1 Some potential *Mtb* counter-immune mechanisms as drug targets

Process	*Mtb* target[a]	Ascribed Function	Reference
1. Iron acquisition and storage	*ideR*	Fe-responsive DNA binding protein—transcriptional regulator	Gold et al. 2001
	mbtA	Siderophore—mycobactin	Gold et al. 2001
	mbtB	Siderophore—mycobactin	de Voss et al. 2000; Gold et al. 2001
	mbtI	Siderophore—mycobactin	Gold et al. 2001
	bfrA	Bacterioferritin	Gold et al. 2001
	bfrB	Ferritin	Gold et al. 2001
2. Carbon utilisation	*icl1*	Isocitrate lysase—key enzyme in the glyoxylate shunt	McKinney et al. 2000
3. Hypoxic shiftdown	*acg*	Putative nitroreductase	Purkayastha et al. 2002
	narK2	Putative nitrite extrusion protein	Sherman et al. 2001
4. Arresting PL maturation	Urease	Converts urea to NH_4 and CO_2	Gordon et al. 1980
	Others?		Reviewed in Russell 2001
5. Acid tolerance/adaptation	Urease	Converts urea to NH_4 and CO_2	Beyrat et al. 1995
	mgtC	Facilitates Mg^{2+} transport	Piddington et al. 2000
	mntH	Divalent cation transporter	Agaroff et al. 1999
	Glutamine synthetase?	NH_4 utilisation	Harth and Horwitz 1999
6. RNI detoxification	*noxR1*	Detoxifies NOx[b], GSNO	Reviewed in Nathan and Shiloh 2000
	noxR3	Detoxifies NOx[b] GSNO	Nathan and Shiloh 2000
	ahpC	Alkyl hydroperoxide subunit—detoxifies organic peroxides, NOx,[b], $ONOO^-$, GSNO	Nathan and Shiloh 2000; Springer et al. 2001
	msrA	Peptidyl methionine sulphoxide reductase	St. John et al. 2001
	lpd	Dihydrolipoamide dehydrogenase	Bryk et al. 2002
	sucB	Dihydrolipoamide succinyl-transferase	Bryk et al. 2002
7. ROI detoxification	*sodC*	Cn,Zn superoxide dismutase	Piddington et al. 2001;
	katG	Catalase-peroxidase	Edwards et al. 2001
	sigH	Alternative sigma factor, required for thioredoxin/thioredoxin reductase induction	Raman et al. 2001

Table 1 (continued)

Process	*Mtb* target[a]	Ascribed Function	Reference
	sulfatide	Cell wall component—ROI scavenger	Pabst et al. 1988
	LAM	Cell wall component—ROI scavenger	Chan et al. 1991
8. DNA repair	*nei*	Endonuclease VIII—repairs oxidative damage	Reviewed in Mizrahi and Andersen 1998
	nth	Endonuclease III—repairs oxidative damage	
	fpg	MutM formamidopyrimidine DNA glcosylase—repairs oxidative damage	
	mutY	AG adenine glycosylase—repairs oxidative damage	
	muT1	8-Oxo-dGTPase—repairs oxidative damage	
	tag1	3-methyladenine DNA glycosylase I—repairs alkylating DNA damage	
	Adl (adaA/alkA fusion)	O^6 methylguanine DNA methyltransferase (Ada domain)/ 3-methyladenine DNA glycosylase II (AlkA domain)—repairs alkylating DNA damage	
	mpg	3 methylpurine DNA glycosylase—repairs alkylating DNA damage	
	Ogt	O^6 alkylguanine DNA transferase—repairs alkylating DNA damage	
9. Avoiding Ag presentation/recognition	Unknown	Unknown	Pancholi et al. 1993; Gercken et al. 1994
	Unknown	Ag85B processing for MHC class II	Ramachandra et al. 2001
	PE-PGRS proteins	Antigenic variation	Banu et al. 2002
10. Antagonising cytokine signalling	Unknown	Disruption of STAT1 binding CBP/p300 co-activators	Ting et al. 1999
	Unknown	Inhibiting 1L-12 production	Nau et al. 2002
11. Macrophage deactivation	LAM	Induction of TGF-β, IL-10	Reviewed in Murray 1999
	LAM	Activation of SHP-1 phosphatase	Knutson et al. 1998

[a] Targets includeboth *Mtb*loci and products.

[b] NOx, unspecified nitrogen oxide intermediates.

on the growth state of the organism for their activity. A cautionary note, however, is that many bacterial species (e.g. *E. coli*, *Salmonella typhimurium*, *L. monocytogenes*) develop a general stress resistance profile upon entry into stationary phase, including mechanisms designed to deal with oxidative stress, acidic pH and other conditions synonymous with the macrophage PL environment (Nystrom 2001). Thus, it will be important to ascertain whether the same counter-strategies are operative in both dividing and non-replicating *Mtb*, a parity not necessarily observed in all bacteria (Kjelleberg 1993; Spector 1998).

Based on the aforementioned approaches, one could envisage future TB drug regimens using conventional antimicrobials, interventions that enhance or supplement endogenous immunity, and agents which further sensitise *Mtb* to killing by its human host. Such novel forms of "combination therapy" could offer several advantages over current protocols: improved effectiveness against MDR-TB, faster (shorter) treatment of active TB, and feasible strategies to address the long-neglected issue of LTBI. With respect to the last point, the successful introduction in 1997 of one-shot therapy for single-lesion paucibacillary leprosy, caused by the related pathogen, *M. leprae*, provides a relevant and inspiring precedent (WHO 1998).

Acknowledgements. John D. MacMicking acknowledges the support of a Life Science Research Foundation Fellowship from the Howard Hughes Medical Institute.

4 References

Adams JS, Gacad MA (1985) Characterization of 1-alpha hydroxylation of vitamin D3 sterols by cultured alveolar macrophages from patients with sarcoidosis. J Exp Med 161: 755–765

Adams LB, Dinauer MC, Morgenstern DE, Krahenbuhl JL (1997) Comparison of the roles of reactive oxygen and nitrogen intermediates in the host response to *Mycobacterium tuberculosis* using transgenic mice. Tuberc Lung Dis 78: 237–246

Adams LB, Mason CM, Kolls JK, Scollard D, Krahenbuhl JL, Nelson S (1995) Exaccerbation of acute and chronic murine tuberculosis by administration of a tumor necrosis factor receptor-expressing adenovirus. J Infect Dis 171: 400–405

Aderem A, Ulevitch RJ (2000) Toll-like receptors in the induction of the innate immune response. Nature 406: 762–787

Aderem A, Underhill DM (1999) Mechanisms of phagocytosis in macrophages. Ann Rev Immunol 17: 593–623

Agaroff D, Monahan IM, Mangan JA, Butcher PD, Krishna S (1999) *Mycobacterium tuberculosis* expresses a novel pH-dependent divalent cation transporter belonging to the Nramp family. J Exp Med 190: 717–724

Al-awar O, Radhakrishna H, Powell NN, Donaldson JG (2000) Separation of membrane trafficking and actin remodelling functions of ARF6 with an effector domain mutant. Mol Cell Biol 20: 3685-3694

Andersen P, Askgaard L, Ljungqvist L, Bennedsen J, Heron I (1991) Proteins released from *Mycobacterium tuberculosis* during growth. Infect Immun 59: 1905–1910

Anderson JM (2000) Multinucleated giant cells. Curr Opin Hematol 7: 40–47

Arias M, Rojas M, Zabaleta JI, Rodriguiz I, Paris SC, Barrera LF, Garcia LF (1997) Inhibition of *Mycobacterium tuberculosis* by Bcg (r) and Bcg (s) macrophages correlates with nitric oxide production. J Infect Dis 176: 1552–1558

Armstrong JA, D'Arcy Hart PD (1971) Response of cultured macrophages to *Mycobacterium tuberculosis*, with observations on fusion of lysosomes with phagosomes. J Exp Med 134: 713-740

Armstrong JA, Hart PD (1975) Phagosome-lysosome interactions in cultured macrophages infected with virulent tubercle bacilli. J Exp Med 142: 1–16

Arrunda S, Bomfim G, Knights R, Huima-Byron T, Riley LW (1993) Cloning of an *M. tuberculosis* DNA fragment associated with entry and survival inside cells. Science 261: 1454–1457

Bach EA, Aguet M, Schreiber RD (1997) The IFN-γ receptor: A paradigm for cytokine receptor signaling. Ann Rev Immunol 15: 563–591

Balcewitz-Sablibska MK, Keane J, Kornfeld H, Remold HG (1998) Pathogenic *Mycobacterium tuberculosis* evades apoptosis of host macrophages by release of TNF-R2, resulting in inactivation of TNF-α. J Immunol 161: 2636–2641

Banu S, Honore N, aint-Joanis B, Philpott D, Prevost MC, Cole ST (2002) Are the PE-PGRS proteins of *Mycobacterium tuberculosis* variable surface antigens? Mol Microbiol 44: 9–19

Baud V, Karin M (2001) Signal transduction by tumor necrosis factor and its relatives. Trends Cell Biol 11: 372–377

Baulard AR, Besra G, Brennan PJ (1999) The cell-wall core of *Mycobacterium*: structure, biogenesis and genetics. In: Ratledge C, Dale J (eds) Mycobacteria: molecular biology and virulence. Blackwell Science, Oxford, pp240–259

Bean AG, Roach DR, Briscoe H, France MP, Korner H, Sedgewick JD, Britton WJ (1999) Structural deficiencies in granuloma formation in TNF gene-targeted mice underlie the heightened susceptibility to aerosol *Mycobacterium tuberculosis* infection, which is not compensated for by lymphotoxin. J Immunol 162: 3504–3511

Beatty WL, Rhoades ER, Ullrich H-J, Chatterjee D, Heuser JE, Russell DG (2000) Trafficking and release of Mycobacterial lipids from infected macrophages. Traffic 1: 235–247

Bekker LG, Haslett P, Maartens G, Steyn L, Kaplan G (2000) Thalidomide-induced antigen-specific immune stimulation in patients with human immunodeficiency virus type I and tuberculosis. J Infect Dis 181: 954–965

Bellamy R, Ruwende C, Corrah R, McAdam KP, Whittle HC, Hill AV (1998) Variations in the *NRAMP1* gene and susceptibility to tuberculosis in west africans. N Eng J Med 338: 640–644

Bellamy R, Ruwende C, Corrah T, McAdam KP, Thursz M, Whittle HC, Hill AV (1999) Tuberculosis and chronic hepatitis B virus infections in the vitamin D receptor gene. J Infect Dis 179: 721–724

Bergeron A, Bonay M, Kambouchner M, Lecossier D, Riquet M, Soler P, Hance A, Tazi A (1997) Cytokine patterns in tuberculous and sarcoid granulomas: correlations with histopathologic features of the granulomatous response. J Immunol 159: 3034–3043

Bernardo J, Billingslea AM, Blumenthal RL, Seetoo KF, Simons ER, Fenton MJ (1998) Differential responses of human mononuclear phagocytes to mycobacterial lipoarabinomannans: role of CD14 and the mannose receptor. Infect Immun 66: 28–35

Bertholet S, Tzeng E, Felley-Bosco E, Mauel J (1999) Expression of inducible nitric oxide synthase in human monocytic U937 cells allows high output nitric oxide synthase. J Leukoc Biol 65: 50–58

Blackwell JM, Black GF, Peacock CS, Miller EN, Sibthorpe D, Gnananandha D, Shaw JJ, Silveira F, Lins-Lainson Z, Ramos F, Collins A, Shaw MA (1997) Immunogenetics of leishmanial and mycobacterial infections: the Belem family study. Philos Trans R Soc Lond B Biol Sci 352: 1331–1345

Bodnar KA, Serbina NV, Flynn JL (2001) Fate of *Mycobacterium tuberculosis* within murine dendritic cells. Infect Immun 69: 800–809

Bogdan C (2001) Nitric oxide and the immune response. Nat Immunol 2: 907–916

Bonecini-Almeida MG, Chitale S, Boutsikakis I, Geng J, Doo H, He S, Ho JL (1998) Induction of in vitro human macrophage anti-*Mycobacterium tuberculosis* activity: requirement for IFN-γ and primed lymphocytes. J Immunol 160: 4490–4499

Bonnerot C, Briken V, Brachet V (1998) Syk protein tyrosine kinase regulates Fc receptor γ-chain mediated transport to lysosomes. EMBO J 17: 4606–4616

Bottrel RL, Yang Y-L, Levy DE, Tomai M, Reis LF (1999) The immune response modifier imiquod requires STAT-1 for induction of interferon, interferon-stimulated genes, and interleukin-6. Antimicrob Agents Chemother 43: 856–861

Bravo J, Karanthanassis D, Pacoid CM, et al (2001) The crystal structure of the PX domain from p40 (phox) bound to phosphoinositol 3-phosphate. Mol Cell 8: 829–839

Brightbill HD, Libraty DH, Krutzik SR, Yang R-B, Belisle JT, Bleharski JR, Maitland M, Norgard MV, Plevy SE, Smale ST, Brennan PJ, Bloom BR, Godowski PJ, Modlin RL (1999) Host defense mechanisms triggered by microbial lipoproteins through Toll-like receptors. Science 285: 732-736

Briscoe H, Roach D, Medows N, Rathjen D, Britton WJ (2000) A novel tumor necrosis factor (TNF) mimetic peptide prevents recrudescence of *Mycobacterium bovis* bacillus Calmette-Guein (BCG) infection in CD4+ T cell-depleted mice. J Leukoc Biol 68: 538–544

Brosch R, Goron SV, Marmiesse M, Brodin P, Buchrieser C, Eiglmeier K, Garnier T, Gutierrez C, Hewinson G, Kremer K, Parsons LM, Pym AS, Samper S., van Soolingen D, Cole ST (2002) A new evolutionary scenario for the *Mycobacterium tuberculosis* complex. Proc Natl Acad Sci USA 99: 3684–3689

Bryk R, Lima CD, Erdjument-Bromage H, Tempst P, Nathan C (2002) Metabolic enzymes of mycobacteria linked to antioxidant defense by a thioredoxin-like protein. Science 295: 1073–1077

Buchmeier N, Blanc-Polard A, Erht S, Piddington D, Riley L, Groisman EA (2000) A parallel intraphagosomal survival strategy shared by *Mycobacterium tuberculosis* and *Salmonella enterica*. Mol Microbiol 35: 1375–1382

Byrd TF (1998) Multinucleated giant cell formation induced by IFN-γ/IL-3 is associated with restriction of *Mycobacterium tuberculosis* cell to cell invasion in human monocyte monolayers. Cell Immunol 188: 89–96

Canetti G (1955). The tubercle bacillus in the pulmonary lesion of man. Springer Publishing, New York, pp 9–20

Cantalupo G, Alifano P, Roberti V, Bruni CB, Bucci C (2001) Rab-interacting lysosomal protein (RILP): the Rab7 effector required for transport to lysosomes. EMBO J 20: 683–693

Caplan S, Hartnell LM, Aguilar RC, Naslavsky N, Bonifacino JS (2001) Human Vam6p promotes lysosomal clustering and fusion in vivo. J Cell Biol 154: 109–121

Casanova J-L, Abel L (2002) Genetic dissection of immunity to mycobacteria: the human model. Ann Rev Immunol 20: 581–620

CDC (2000) Targeted tuberculin testing and treatment of latent tuberculosis infection. M.M.W.R. 49: 1-52

Chan ED, Morris KR, Belisle JT, Hill P, Remigio LK, Brennan PJ, Riches DW (2001) Induction of inducible nitric oxide synthase-NO by lipoarabinomannan of *Mycobacterium tuberculosis* is mediated by MEK1-ERK, MKK7-JNK, and NFκB signaling pathways. Infect Immun 69: 2001-2010

Chan J, Fan XD, Hunter SW, Brennan PJ, Bloom BR (1991) Lipoarabinomannan, a possible virulence factor involved in the persistence of *Mycobacterium tuberculosis* within macrophages. Infect Immun 59: 1755–1761

Chan J, Tanaka K, Carroll D, Flynn J, Bloom BR (1995) Effects of nitric oxide synthase inhibitors on murine infection with *Mycobacterium tuberculosis*. Infect Immun 63: 736–740

Chan J, Xing Y, Magliozzo RS, Bloom BR (1992) Killing of virulent *Mycobacterium tuberculosis* by reactive nitrogen intermediates produced by activated murine macrophages. J Exp Med 175: 1111–1122

Chapman J, Bernard J (1962) The tolerances of unclassified mycobacteria. Am Rev Respir Dis 86: 582-583

Chun T, Serbina NV, Nolt D, Wang B, Chiu NM, Flynn JM, Wang C-R (2001) Induction of M3-restricted cytotoxic T lymphocyte responses by *N*-formylated peptides derived from *Mycobacterium tuberculosis*. J Exp Med 193: 1213–1220

Clark RH, Kueser TJ, Walker MW, Southgate WM, Huckaby JL, Pezez JA, Roy BJ, Keszler M, Kinsella JP (2000) Low-dose nitric oxide therapy for persistent pulmonary hypertension of the newborn. Clinical Inhaled Nitric Oxide Research Group. N Engl J Med 342: 469–474

Clemens DL, Horwitz MA (1995) Characterization of the *Mycobacterium tuberculosis* phagosome and evidence that phagosome maturation is inhibited. J Exp Med 181: 257–270

Clemens DL, Horwitz MA (1996) The *Mycobacterium tuberculosis* phagosome interacts with early endosomes and is accessible to exogenously administered transferrin. J Exp Med 184: 1349-1355

Clemens DL, Lee BY, Horwitz MA (2000a) Deviant expression of Rab5 on phagosomes containing the intracelleular pathogens *Mycobacterium tuberculosis* and *Legionella pneumophilia* is associated with altered phagosomal fate. Infect Immun 68: 2671–2684

Clemens DL, Lee B-Y, Horwitz MA (2000b) *Mycobacterium tuberculosis* and *Legionella pheumophilia* phagosomes exhibit arrested maturation despite acquisition of Rab7. Infect Immun 68: 5154-5166

Clynes R, Maizes J, Guinamard R, Ono M, Takai T, Ravetch J (1999) Modulation of immune complex-induced inflammation in vivo by the coordinate expression of activation and inhibitory Fc receptors. J Exp Med 189: 179–185

Cohen AB, Cline MJ (1971) The human alveolar macrophage: isolation, cultivation in vitro, and studies of morphologic and functional characteristics. J Clin Invest 50: 1390–1398

Cole ST, and forty-one others (1998) Deciphering the biology of *Mycobacterium tuberculosis* from the complete genome sequence. Nature 393: 537–344

Collazo CM, Yap GS, Sempowski GD, Lusby KC, Tessarollo L, Vande Woude GF, Sher A, Taylor GA (2001) Inactivation of LRG-47 and IRG-47 reveals a family of interferon-γ-inducible genes with essential, pathogen-specific roles in resistance to infection. J Exp Med 194: 181–187

Collins HL, Kaufmann SH (2001) The many faces of host responses to tuberculosis. Immunology 103: 1–9

Condos R, Rom WN, Schluger NW (1997) Treatment of multi-drug-resistant pulmonary tuberculosis with interferon-gamma via aerosol. Lancet 349: 1513–1515

Contursi C, Wang I-M, Gabriele L, Gadina M, O'Shea J, Morse III HC, Ozato K (2000) IFN consensus sequence binding protein potentiates STAT1-dependent activation of IFN-γ-responsive promoters of macrophages. Proc Natl Acad Sci USA 97: 91–96

Cooper AM, Dalton DK, Stewart TA, Griffin JP, Russell DG, Orme IM (1993) Disseminated tuberculosis in interferon γ gene-disrupted mice. J Exp Med 178: 2243–2247

Cooper AM, Segal BH, Frank AA, Holland SM, Orme IM (2000) Transient loss of resistance to pulmonary tuberculosis in p47$^{phox-/-}$ mice. Infect Immun 68: 1231–1234

Crowle AJ, Ross EJ, May MH (1987) Inhibition of 1,25(OH)2-vitamin D3 of the mutiplication of virulent tubercle bacilli in cultured human macrophages. Infect Immun 55: 2945–2950

Crowley MT, Costello PS, Fitzer-Attas CJ, Turner M, Meng F, Lowell C, Tybulewicz VLJ, DeFranco AL (1997) A critical role for Syk in signal transduction and phagocytosis mediated by Fc receptors on macrophages. J Exp Med 186: 1027–1039.

Cywes C, Hoppe HC, Daffe M, Ehlers MR (1997) Nonopsonic binding of *Mycobacterium tuberculosis* to complement receptor type 3 is mediated by capsular polysaccharides and is strain dependent. Infect Immun 65: 4258–4266

Dannenberg AM, Rook GAW (1994) Pathogenesis of pulmonary tuberculosis: an interplay of tissue-damaging and macrophage-activating immune responses—dual mechanisms that control bacillary multiplication. In: Bloom BR (ed) Tuberculosis. ASM Press, Washington, pp 459–484

Darnell JE Jr (1997) STATs and gene regulation. Science 277: 1630–1635

De Voss JJ, Rutter K, Schroeder BG, Su H, Zhu Y, Barry CE (2000) The salicylate derived mycobactin siderophores of *Mycobacterium tuberculosis* are essential for growth in macrophages. Proc Natl Acad Sci USA 97: 1252–1257

Denis M (1991a) Interferon-gamma-treated murine macrophages inhibit growth of the tubercle bacilli by generation of reactive nitrogen intermediates. Cell Immunol 132: 150–157

Denis M (1991b) Involvment of cytokines in determining resistance and acquired immunity in murine tuberculosis. J Leukoc Biol 50: 495–501

Denzer K, Kleijmeer M, Heijnen H, Stoovogel W, Geuze H (2000) Exosome: from internal vesicle of the multivesicular body to intercellular signaling device. J Cell Sci 113: 3365

Desjardins M (1995) Biogenesis of phagolysosomes: the "kiss and run" hypothesis. Trends Cell Biol 5: 183–186

Dlugovitzky D, Bay ML, Rateni L, Fiorenza G, Vietti L, Farroni MA, Bottasso OA (2000) Influence of disease severity on nitrite and cytokine production by peripheral blood monouclear cells (PBMC) from patients with pulmonary tuberculosis (TB). Clin Exp Immunol 122: 343–349

Douvas GS, Berger EM, Repine JE, Crowle AJ (1986) Natural mycobacteriostatic activity in human monocyte-derived adherent cells. Am Rev Respir Dis 134: 44–48

Downing JF, Pasula R, Wright JR, Twigg HL, Martin JW (1995) Surfactant protein A promotes attachment of *Mycobacterium tuberculosis* to alveolar macrophages during infection with human immunodeficiency virus. Proc Natl Acad Sci USA 92: 4848–4852

Dumas JJ, Merithew E, Sudharshan E, Rajamani D, Hayes S, Lawe D, Covera S, Lambright DG (2001) Multivalent endosome targeting by homodimeric EEA1. Mol Cell 8: 947–958

Dumont A, Sheldon H (1965) Changes in the fine structure of macrophages in experimentally produced tuberculous granulomas in hamsters. Lab Invest 14: 2034–2055

Dye C, Scheele S, Dolin P, Pathania V, Raviglione MC (1999) Global burden of tuberculosis—estimated incidence, prevalence, and mortality by country. JAMA 282: 677–686

Dye C, Williams BG, Espinal MA, Raviglione MC (2002) Erasing the world's slow stain: strategies to beat multidrug-resistant tuberculosis. Science 295: 2042–2046

Edwards KM, Cynamon MH, Voladri RK, Hager CC, Destefano MS, Tham KT, Lakey DL, Bochan MR, Kernodle DS (2001) Iron-cofactored superoxide dismutase inhibits host responses to *Mycobacterium tuberculosis*. Am J Respir Crit Care Med 164: 2213–2219

Ellson CD, Anderson KE, Morgan G, Chilvers ER, Lipp P, Stephens LR, Hawkins PT (2001) Phosphatidylinositol 3-phosphate is generated in phagosomal membranes. Curr Biol 11: 1631-1635

Enelow RI, Sullivan GW, Carper HT, Mandell GL (1992) Induction of multinucleated giant cell formation from in vitro culture of human monocytes with interleukin-3 and interferon-γ: comparison with other stimulating factors. Am J Respir Cell Mol Biol 6: 57–62

Erht S, Schnappinger D, Bekiranov S, Drenkow J, Shi S, Gingeras TR, Gaasterland T, Schoolnik G, Nathan C (2001) Reprogramming the macrophage transcriptome in re-

sponse to interferon-γ and *Mycobacterium tuberculosis*: signaling roles of nitric oxide synthase-2 and phagocyte oxidase. J Exp Med 194: 1123–1139
Ernst J (1998) Macrophage receptors for *Mycobacterium tuberculosis*. Infect Immun 66: 1277–1281
Fachetti F, Vermi W, Fiorentini S, Chilosi M, Caruso A, Duse M, Notarangelo LD, Badolato R (1999) Expression of inducible nitric oxide synthase in human granulomas and histiocytic reactions. Am J Path 154: 145–152
Fais S, Burgio VL, Silvestri M, Capobianchi MR, Pacchiarotti A, Pallone F (1994) Multinucleated giant cells generation by interferon-γ. Changes in expression and distribution of intracellular adhesion molecule-1 during macrophages fusion and multinucleated giant cell formation. Lab Invest 71: 737–744
Falzoni S, Chiozzi P, Ferrari D, Buell G, Di Virgilio F (2000) $P2X_7$ receptor and polykarion formation. Mol Biol Cell 11: 3169–3176
Fenhalls G, Wong A, Bezuidenhout J, van Helden P, Bardin P, Lukey PT (2000) In situ production of gamma interferon, interleukin-4, and tumor necrosis factor alpha mRNA in human lung tuberculous granulomas. Infect Immun 68: 2827–2836
Fenton MJ, Vermeulen MW, Kim S, Burdick M, Streiter RM, Kornfeld H (1997) Induction of gamma interferon production in human alveolar macrophages by *Mycobacterium tuberculosis*. Infect Immun 65: 5149–5156
Ferguson JS, Voelker DR, McCormack FX, Schlesinger LS (1999) Surfactant protein D binds to *Mycobacterium tuberculosis* bacilli and lipoarabinomannan via carbohydrate-lectin interactions resulting in reduced phagocytosis of the bacteria by macrophages. J Immunol 163: 312–321
Ferrari G, Langen H, Naito M, Pieters J (1999) A coat protein on phagosomes involved in the intracellular survival of mycobacteria. Cell 97: 435–447
Ficsher K, Chatterjee D, Torrelles J, Brennen PJ, Kaufmann SH, Schaible UE (2001) Mycobacterial lysocardiolipin is exported from phagosomes upon cleavage of cardiolipin by a macrophage-derived lysosomal phospholipase A2. J Immunol 167: 2187–2192
Finbloom DS, Hoover DL, Wahl LM (1985) The characteristics of binding of human recombinant interferon-gamma to its receptor on human monocytes and human monocyte-like lines. J Immunol 135: 300–305
Flesch I, Kaufmann SH (1987) Mycobacterial growth inhibition by interferon-γ activated bone marrow macrophages and differential susceptibility among strains of *Mycobacterium tuberculosis*. J Immunol 138: 4408–4413
Flynn J, Chan J, Triebold K, Dalton D, Stewart T, Bloom B (1993) An essential role for interferon γ in resistance to *Mycobacterium tuberculosis* infection. J Exp Med 178: 2249–2254
Flynn JL, Goldstein MM, Chan J, Triebold KJ, Pfeffer K, Lowenstein CJ, Schreiber R, Mak TW, Bloom BR (1995) Tumor necrosis factor-α is required in the protective immune response against *Mycobacterium tuberculosis* in mice. Immunity 2: 561–572
Flynn JL, Scanga CA, Tanaka KE, Chan J (1998) Effects of aminoguanidine on latent murine tuberculosis. J Immunol 160: 1796–1803
Flynn JL, Chan J (2001) Tuberculosis: latency and reactivation. Infect Immun 69: 4195–4201
Frattazzi C, Manjunath N, Arbeit RD, Carini C, Gerken TA, Ardman B, Remold-O'Donnell E, Remold HG (2000) A macrophage invasion mechanism of mycobacteria implicating the extracellular domain of CD43. J Exp Med 192: 183–192
Fratti RA, Backer JM, Gruenberg J, Corvera S, Deretic V (2001) Role of phosphatidylinositol 3-kinase and Rab5 effectors in phagosomal biogenesis and mycobacterial phagosome maturation arrest. J Cell Biol 154: 631–644
Fratti RA, Chua J, Deretic V (2002) Cellubrevin alterations and *Mycobacterium tuberculosis* phagosome : maturation arrest. J Biol Chem 277: 17320–17326

Fratti RA, Vergne I, Chua I, Skidmore J, Deretic V (2000) Regulators of membrane trafficking and *Mycobacterium tuberculosis* phagosome maturation block. Elecrophoresis 21: 3378–3385
Fukushima K, Hiratani K, Kadota J, Komori K, Hirota M, Hara K (1991) Analysis of cellular and biochemical contents of bronchoalveolar lavage fluid from patients with pulmonary tuberculosis. Kekkaku 66: 589–598.
Galindo B, Lazdins J, Castillo R (1974) Fusion of normal rabbit alveolar macrophages induced by supernatant fluids from BCG-sensitized lymph node cells after elicitation by antigen. Infect Immun 9: 212–216
Ganster RW, Taylor BS, Shao L, Geller DA (2001) Complex regulation of human inducible nitric oxide synthase gene transcription by Stat 1 and NF-κB. Proc Natl Acad Sci USA 98: 8638–8643
Gao P-S, Fujishima S, Mao X-Q, et al. (2000) Genetic variants of *NRAMP1* and active tuberculosis in Japanese populations. Clin Genet 58: 74–76
Garcia I, Miyazaki Y, Marchal G, Lesslauer W, Vassalli P (1997) High sensitivity of transgenic mice expressing soluble TNFR1 fusion protein to mycobacterial infections: synergistic action of TNF and IFN-gamma in the differentiation of protective granulomas. Eur J Immunol 27: 3182–3190
Gasser A, Most J (1999) Generation of multinucleated giant cells in vitro by culture of human monocytes with *Mycobacterium bovis* BCG in combination with cytokine-containing supernatants. Infect Immun 67: 395–402
Gaynor CD, McCormack FX, Voelker DR, McGowan SE, Schlesinger LS (1995) Pulmonary surfactant protein A mediates enhanced phagocytosis of *Mycobacterium tuberculosis* by a direct interaction with human macrophages. J Immunol 155: 5343–5351
Gercken J, Pryjma J, Ernst M, Flad HD (1994) Defective antigen presentation of *Mycobacterium tuberculosis*-infected monocytes. Infect Immun 62: 3472–3478
Geuze HJ (1998) The role of endosomes and lysosomes in MHC class II functioning. Immunol Today 19: 282–287
Giannella RA, Broitman SA, Zamcheck N (1973) Influence of gastric acidity on bacterial and parasitic enteric infections. Ann Int Med 78: 271–276
Gillman T, Wright LJ (1966) Probable in vivo origin of multi-nucleated giant cells from circulating mononuclears. Nature 209: 263–265
Giosue S, Casarini M, Alemanno L, Galluccio G, Mattia P, Pedicelli G, Rebek L, Bisetti A, Ameglio F (1998) Effects of aerosilized interferon-alpha in patients with pulmonary tuberculosis. Am J Respir Crit Care Med 158: 1156–1162
Giosue S, Casarini M, Ameglio F, Zangrilli P, Palla M, Altieri AM, Bisetti A (2000) Aerosilized interferon-alpha treatment in patients with multi-drug-resistant pulmonary tuberculosis. Eur Cytokine Netw 11: 99–104
Gobin J, Horwitz MA (1996) Exochelins of *Mycobacterium tuberculosis* remove iron from human iron-binding proteins and donate iron to mycobactins in the *M. tuberculosis* cell wall. J Exp Med 183: 1527–1532
Gold B, Rodriguez GM, Marras SA, Pentecost M, Smith I (2001) The *Mycobacterium tuberculosis* IdeR is a dual function regulator that controls transcription of genes involved in iron acquisition, iron storage and survival in macrophages. Mol Microbiol 42: 851–865
Gomes MS, Paul S, Moreira AL, Appelberg R, Rabinovitch M, Kaplan G (1999) Survival of *Mycobacterium avium* and *Mycobacterium tuberculosis* in acidified vacuoles of murine macrophages. Infect Immun 67: 3199–3206
Goodman PA, Niehoff LB, Uckun FM (1998) Role of tyrosine kinases in induction of the c-jun proto-oncogene in irradiated B-lineage cells. J Biol Chem 273: 17742–17748
Gordon AH, D'Arcy Hart PD, Young MR (1980) Ammonia inhibits phagosome-lysosome fusion in macrophages. Nature 286: 79–81
Gordon S (1999) Macrophages and the immune response. In: Paul WE (ed) Fundamental Immunology. Lippencott-Raven, Philadelphia, pp 533–545

Goren MB, D'Arcy Hart P, Young MR, Armstrong JA (1976) Prevention of phagosome-lysosome fusion in cultured macrophages by sulfatides of *Mycobacterium tuberculosis*. Proc Natl Acad Sci USA 73: 2510-2514

Gotri A, Rossi M, Trabattoni D, Marchetti G, Fusi M, Molteni C, Clerici M, Franzetti F (2000) Tumor necrosis factor-α increased production during thalidomide treatment in patients with tuberculosis and human immunodeficiency virus coinfection. J Infect Dis 182: 639

Green RM, Seth A, Connell ND (2000) A peptide permease mutant of *Mycobacterium bovis* BCG resistant to the toxic peptides glutathione and *S*-nitrosoglutathione. Infect Immun 68: 429–436

Greenberg S, Grinstein S (2002) Phagocytosis and innate immunity. Curr Opin Immunol 14: 136–145

Halme M, Maasilta P, Repo H, Ristola M, Takinen E, Mattson K, Cantell K (1995) Inhaled recombinant interferon gamma in patients with lung cancer: pharmacokinetics and effects on chemiluminescence responses of alveolar macrophages and peripheral blood neutrophils and monocytes. Int J Radiat Biol Phys 31: 93–101

Harth G, Horwitz MA (1999) An inhibitor of exported *Mycobacterium tuberculosis* glutamine synthetase selectively blocks the growth of pathogenic mycobacteria in axenic culture and human monocytes: extracellular proteins as potential novel drug targets. J Exp Med 189: 1425–1435

Hemmi H, Kaisho T, Takeuchi O, Sato S, Sanjo H, Hoshino K, Horiuchi T, Tomizawa H, Takeda K, Akira S (2002) Small anti-viral compounds activate immune cells via the TLR7 MyD88-dependent signaling pathway. Nat Immunol 3: 196–200

Hill LD, Sun L, Leuschen MP, Zack TL (1993) C3 synthesis by A549 alveolar epithelial cells is increased interferon-gamma and dexamethasone. Immunology 79: 236–240

Hirsch CS, Ellner JJ, Russell DG, Rich EA (1994) Complement receptor-mediated uptake and tumor necrosis factor-mediated growth inhibition of *Mycobacterium tuberculosis* by human alveolar macrophages. J Immunol 152: 743–753

Hmama Z, Nandan D, Sly L, Knutson KL, Herrera-Velit P, Reiner NE (1999) 1α,25-dihydroxyvitamin D_3-induced myeloid cell differentiation is regulated by a vitamin D receptor-phosphatidylinositol 3-kinase signaling complex. J Exp Med 190: 1583–1594

Hocking WG, Golde DW (1979) The pulmonary alveolar macrophage. N Eng J Med 301: 580, 639

Holland SM, Eisenstein EM, Kuhns DB, Turner ML, Fleisher TA, Strober W, Gallin JI (1994) Treatment of refractory disseminated nontuberculous mycobacterial infection with interferon gamma. N Engl J Med 330: 1348–1355

Holmskov U (1999) Lung surfactant proteins (SP-A and SP-D) in non-adaptive host responses to infection. J Leukoc Biol 66: 747–752

Hu C, Mayadas-Norton T, Tanaka K, Chan J, Salgame P (2000) *Mycobacterium tuberculosis* infection in complement receptor 3-deficient mice. J Immunol 165: 2596–2602

Humphreys BD, Dubyak GR (1996) Induction of the P2z/P2X_7 nucleotide receptor and associated phosholipase D activity by lipopolysaccharide and IFN-gamma in the human THP-1 monocytic cell line. J Immunol 157: 5627–5637

Institute of Medicine (2000) Ending neglect: The elimination of tuberculosis in the United States. National Academy Press, Washington, DC.

Iseman MD (2000) Extrapulmonary tuberculosis in adults. In: Iseman, MD (ed) A Clinician's Guide to Tuberculosis. Lippincott Williams and Wilkins, Philadelphia, pp 145–197

Jabado N, Jankowski A, Dougaparsad S, Picard V, Grinstein S, Gros P (2000) Natural resistance to intracellular infections: natural resistance-associated macrophage protein 1 (NRAMP1) functions as a pH-dependent manganese transporter at the phagosomal membrane. J Exp Med 192: 1237-1247

Jackett PS, Aber V, Lowrie DB (1978) Virulence and resistance to superoxide, low pH and hydrogen peroxide among strains of *Mycobacterium tuberculosis.* J Gen Microbiol 107: 37–45

Jaffe HA, Buhl R, Mastrangeli A, Holroyd KJ, Saltini C, Czerski D, Jaffe HS, Kramer S, Sherwin S, Crystal RG (1991) Organ-specific cytokine therapy. Local activation of monouclear phagocytes by delivery of an aerosol of recombinant interferon-γ to the human lung. J Clin Invest 88: 297-302

Janeway CJ (1989) Approaching the asymptote? Evolution and revolution in immunology. Cold Spring Harb Symp Quant Biol 54: 1–13

Jones BW, Heldwin KA, Means TK, Saukkonen JJ, Fenton MJ (2001) Differential roles of Toll-like receptors in the elicitation of proinflammatory responses by macrophages. Ann Rheum Dis 60 (Suppl 3): 6–12

Jordens I, Fernandez-Borja M, Marsman M, Dusseljee S, Janssen L, Calafat J, Janssen H, Wubbolts R, Neefjes J (2001) The Rab7 effector protein RILP controls lysosomal transport by inducing the recruitment of dynein-dynactin motors. Curr Biol 11: 1680–1685

Juffermans NP, Verbon A, van Deventer SJ, van Deutekom H, Speelman P, van der Poll T (1998) Tumor necrosis factor and interleukin-1 inhibitors as markers of disease activity in tuberculosis. Am J Respir Crit Care Med 157: 1328–1331

Jung F, Palmer LA, Zhou N, Johns RA (2000) Hypoxic regulation of inducible nitric oxide synthase via hypoxia inducible factor-1 in cardiac myocytes. Circ Res 86: 319–325

Kanedo H, Yamada H, Mizuno S, Udagawa T, Kazumi Y, Sekikawa K, Sugawara I (1999) Role of tumor necrosis factor-alpha in *Mycobacterium*-induced granuloma formation in tumor necrosis factor-alpha-deficient mice. Lab Invest 79: 379–386

Karupiah G, Xie QW, Buller RM, Duarte C, Nathan C, MacMicking JD (1993) Inhibition of viral replication by interferon-gamma-induced nitric oxide synthase. Science 261: 1445–1448

Kato-Maeda M, Bifani PJ, Kreiswirth BN, Small PM (2001) The nature and consequence of genetic variability within *Mycobacterium tuberculosis.* J Clin Invest 107: 533–537

Kawai T, Adachi O, Ogawa T, Takeda K, Akira S (1999) Unresponsiveness of MyD88-deficient mice to endotoxin. Immunity 9: 143–150

Keane J, Gershon S, Wise R, Mirabile-Evans E, Kasznica J, Scwieterman WD, Siegel JN, Braun MM (2001) Tuberculosis associated with infliximab, a tumor necrosis factor-α-neutralizing agent. N Engl J Med 345: 1098–1104

Kisich KO, Heifets L, Higgins M, Diamond G (2001) Antimycobacterial agent based on mRNA encoding human beta-defensin 2 enables primary macrophages to restrict growth of *Mycobacterum tuberculosis.* Infect Immun 69: 2692–2699

Knutson KL, Hmama Z, Hererra-Velit P, Rochford R, Reiner NE (1998) Lipoarabinomannan of *Mycobacterium tuberculosis* promotes protein tyrosine phosphorylation and inhibition of mitogen-activated protein kinase in human mononuclear phagocytes. Role of the Src homology 2 containing tyrosine phophatase 1. J Biol Chem 273: 645–652

Kobzik L, Godleski J, Brain J (1990) Selective down-regulation of alveolar macrophage oxidative response to opsonin-independent phagocytosis. J Immunol 144: 4312–4319

Koch R (1882) Die atiologie der tuberkulose. Berliner Klin Wochenschr 19: 221–230

Koeffler HP, Reichel H, Bishop JE, Norman AW (1985) Gamma interferon stimulates production of 1,25-dihydroxyvitamin D3 by normal human macrophages. Biochem Biophys Res Comm 127: 596–603

Krugmann S, Anderson KE, Ridley SH, et al (2002) Identification of ARAP3, a novel PI3 K effector regulating both Arf and Rho GTPases, by selective capture on phosphoinositide affinity matrices. Mol Cell 9: 95–108

Kuncewicz T, Balakrishnan P, Snuggs MB, Kone BC (2001) Specific association of nitric oxide synthase-2 with Rac isoforms in activated murine macrophages. Am J Physiol 281: F326-F336

Kusner DJ, Adams J (2000) ATP-induced killing of virulent *Mycobacterium tuberculosis* within human macrophages requires phospholipase D. J Immunol 164: 379–388
Kusner DJ, Barton JA (2001) ATP stimulates human macrophages to kill intracellular virulent *Mycobacterium tuberculosis* via calcium-dependent phagosome-lysosome fusion. J Immunol 167: 3308–3315
Langhans T (1868) Ueber Riesenzellen mit wandestandigen Kernen in Tuberkeln und die fibrose Form des Tuberkels. Virchows Arch Pathol Anat 42: 382–404
Laochumroonvorapong P, Paul S, Manca C, Freedman VH, Kaplan G (1997) Mycobacterial growth and sensitivity to H_2O_2 killing in human monocytes in vitro. Infect Immun 65: 4850–4857
Lau YL, Chan CF, Ha SY, Hui YF, Yuen KY (1998) The role of phagocytic respiratory burst in host defense against *Mycobacterium tuberculosis*. Clin Infect Dis 26: 226–227
Lillebaek T, Dirksen A, Baess I, Strunge B, Thomsen VO, Andersen AB (2002) Molecular evidence of endogenous reactivation of *Mycobacterium tuberculosis* after 33 years of latent infection. J Infect Dis 185: 401–404
Lipton AJ, Johnson MA, MacDonald T, Lieberman MW, Gozal D, Gaston B (2001) *S*-nitrosothiols signal ventilatory response to hypoxia. Nature 413: 171–174
Lloyd J (2000) Lysosome membrane permeability: implications for drug delivery. Adv Drug Delivery Rev 41: 189–200
Loetscher H, Gentz R, Zulauf M, Lustig A, Tabuchi H, Schlaeger EJ, Brockhaus M, Gallati H, Manneberg M, Lesslauer W (1991) Recombinant 55-kDa tumor necrosis factor (TNF) receptor. Stoichiometry of binding to TNF alpha and TNF beta and inhibition of TNF activity. J Biol Chem 266: 18324–18329
Long R, Light B, Talbot JA (1999) Mycobacteriocidal action of exogenous nitric oxide. Antimicrobial Agents Chemo. 43: 403–405
Lowrie DB, Tascon RE, Bonato VL, Lima VM, Faccioli LH, Stavropoulas E, Colston MJ, Hewinson RG, Moelling K, Silva CL (1999) Therapy of tuberculosis in mice by DNA vaccination. Nature 400: 269–271
Lowrie DB, Silva CL (2000) Enhancement of immunocompetence in tuberculosis by DNA vaccination. Vaccine 18: 1712–1716
Lurie M (1939) Studies on the mechanism of immunity in tuberculosis. The mobilisation of mononuclear phagocytes in normal and immunized animals and their relative capacities for division and phagocytosis. J Exp Med 579–605
Lurie M (1942) Studies on the mechanism of immunity in tuberculosis. The fate of tubercle bacilli ingested by mononuclear phagocytes derived from normal and immunized animals. J Exp Med 75: 247–268
MacMicking JD, Nathan C, Hom G, Chartrain N, Fletcher DS, Trumbauer M, Stevens K, Xie Q-w, Sokol K, Hutchinson N, Chen H, Mudgett JS (1995) Altered responses to bacterial infection and endotoxic shock in mice lacking inducible nitric oxide synthase. Cell 81: 641–650
MacMicking JD, North RJ, LaCourse R, Mudgett JS, Shah SK, Nathan CF (1997a) Identification of nitric oxide synthase as a protective locus against tuberculosis. Proc Natl Acad Sci USA 94: 5243–5248
MacMicking J, Xie Q-w, Nathan C (1997b) Nitric oxide and macrophage function. Ann Rev Immunol 15: 323–350
Malik ZA, Denning GM, Kusner DJ (2000) Inhibition of Ca^{2+} signaling by *Mycobacterium tuberculosis* is associated with reduced phagolysosomal fusion and increased survival within human macrophages. J Exp Med 191: 287–302
Malik ZA, Iyer SS, Kusner DJ (2001) *Mycobacterium tuberculosis* phagosomes exhibit altered calmodulin-dependent signal transduction: contribution to inhibition of phagosome-lysosome fusion and intracellular survival in human macrophages. J Immunol 166: 3392–3401

Mariani F, Cappelli G, Riccardi G, Colizzi V (2000) *Mycobacterium tuberculosis* H37Rv comparative gene-expression analysis in synthetic medium and human macrophage. Gene 253: 281–291

Martinez O, Noiseux C, Martin J, Lara V (2001) Reactivation tuberculosis in a patient with anti-TNF-α treatment. Am J Gastroenterol 96: 1665–1666

Mayordomo L, Marenco JL, Gomez-Mateos J, Rejon E (2002) Pulmonary miliary tuberculosis in a patient with anti-TNF-alpha treatment. Scand J Rheumatol 31: 44–45

McBride KM, Banninger G, McDonald C, Reich NC (2002) Regulated nuclear import of the STAT1 transcription factor by direct binding of importin-α. *EMBO J* 21: 1754–1763

McKinney JD (2000) In vivo *veritas*: the search for TB drug targets goes live. Nat Med 6: 1330–133: 1754–1763

McKinney JD, Jacobs WR Jr, Bloom BR (1998) Persisting problems in tuberculosis. In: Krause, R, Galin JI, Fauci AS (eds) Emerging Infections. Academic Press, New York, pp 51–146

McKinney JD, Honer zu Bentrup K, Munoz-Elias EJ, Miczak A, Chen b, Chan WT, Swenson D, Sacchettini JC, Jacobs WR, Russell DG (2000) Persistence of *Mycobacterium tuberculosis* in macrophages and mice requires the glyoxylate shunt enzyme isocitrate lyase. Nature 406: 735-738

McKinney JD, Bloom BR, Modlin RL (2001) Tuberculosis and leprosy. In: Austen KF, Frank MM, Atkinson JP, Cantor H (eds) Samter's Immunologic Diseases, 6th Ed. Lippincott Williams & Wilkins, Philadelphia, pp 985–1002

Means TK, Jones BW, Schromm AB, Shurtleff BA, Smith JA, Keane J, Golenbock DT, Vogel SN, Fenton MJ (2001) Differential effects of a Toll-like receptor antagonist on *Mycobacterium tuberculosis*-induced macrophage responses. J Immunol 166: 4074–4082

Means TK, Wang S, Lien E, Yoshimura A, Golenbock DT, Fenton MJ (1999) Human Toll-like receptors mediate cellular activation by *M. tuberculosis*. J Immunol 163: 3920–3926

Medzhitov R (2001) Toll-like receptors and innate immunity. Nat Rev Immunol 1: 135–144

Melendez AJ, Bruetschy L, Floto RA, Harnett MM, Allen JM (2001) Functional coupling of FcγRI to nicotinamide adenine dinucleotide phosphate (reduced form) oxidative burst and immune complex trafficking requires the activation of phospholipase D1. Blood 98: 3421–3428

Melillo G, Musso T, Sica A, Taylor LS, Cox GW, Varesio L (1995) A hypoxia-responsive element mediates a novel pathway of activation of the inducible nitric oxide synthase promoter. J Exp Med 182: 1683–1693

Melillo G, Taylor LS, Brooks A, Cox GW, Varesio L (1996) Regulation of inducible nitric oxide synthase expression in IFN-gamma-treated murine macrophages cultured under hypoxic conditions. J Immunol 157: 2638–2644

Melo M, Catchpole I, Hagger G, RW S (2000) Utilization of CD11b knockout mice to characterize the role of complement receptor 3 (CR3, CD11b/CD18) in the growth of *Mycobacterium tuberculosis* in macrophages. Cell Immunol 10: 13–23

Metchnikoff E (1880) Uber die intracellulare Verdauung bei coelenteraten. Zool Anzeiger 3: 261–263

Metchnikoff E (1884) Untersuchungen uber die intracellulare Verdauung bei Wirbelosen Thieren. Arb Zool Inst Wein 5: 141–168

Metchnikoff E (1888) Uber die phagocytaere Rolle der Tuberkeliesenzellen. Arch Pathol Anat 113: 63-94

Metchnikoff E (1893) Lectures on the comparative pathology of inflammation. Kegan, Paul, Trübner, Trench, London.

Metchnikoff E (1905) Immunity in the Infectious Diseases. MacMillan, New York.

Metchnikoff O (1921) Life of Elie Mecthnikoff. Houghton Mifflin, Boston. pp116–117

Meylan PP, Richman DD, Korthbluth RS (1992) Reduced intracellular growth of mycobacteria in human macrophages cultivated at physiologic oxygen pressure. Am Rev Respir Dis 145: 947–953

Misra S, Miller GJ, Hurley JH (2001) Recognizing phosphatidylinositol 3-phosphate. Cell 107: 559–562

Mitchison DA (1985) The action of anti-tuberculous drugs in short course-chemotherapy. Tubercle 66: 219–225.

Mizrahi V, Andersen SJ (1998) DNA repair in *Mycobacterium tuberculosis*: what have we learnt from the genome sequence? Mol Microbiol 29: 1331–1339

Mizuno K, Okamoto H, Horio T (2001) Muramyl dipeptide and mononuclear cell supernatant induce Langhans-type cells from human monocytes. J Leukoc Biol 70: 386–394

Mogues T, Goodrich ME, Ryan L, LaCourse R, North RJ (2001) The relative importance of T cell subsets in immunity and immunopathology of airborne *Mycobacterium tuberculosis* infection in mice. J Exp Med 193: 271–280

Mohan VP, Scanga CA, Yu K, Scott HM, Tanaka KE, Tsang E, Tsai MM, Flynn JL, Chan J (2001) Effects of tumor necrosis factor alpha on host immune response in chronic persistent tuberculosis: possible role for limiting pathology. Infect Immun 69: 1847–1855

Moody DB, Porcelli SA (2001) CD1 trafficking: invariant chain gives a new twist to the tale. Cell 15: 861–865

Moreira AL, Tsenova L, Aman MH, Bekker L-G, Freeman S, Mangaliso B, Shroder U, Jagirdar J, Rom WN, Tovey MG, Freedman VH, Kaplan G (2002) Mycobacterial antigens exacerbate disease manifestations in *Mycobacterium tuberculosis*-infected mice. Infect Immun 70: 2100-2002

Morse D, Brothmed D, Ucko P (1964) Tuberculosis in ancient Egypt. Am Rev Respir Dis 90: 524–541

Most J, Neumayer HP, Dierich MP (1990) Cytokine-induced generation of multinucleated giant cells in vitro requires interferon-γ and expression of LFA-1. Eur J Immunol 20: 1661–1667

Moya MP, Gow AJ, McMahon TJ, Toone EJ, Cheifetz IM, Goldberg RN, Stamler JS (2001) *S*-nitrosothiol repletion by an inhaled gas regulates pulmonary function. Proc Natl Acad Sci USA 98: 5792–5797

Muller H, Kruger S (1994) Immunohistochemical analysis of cell composition and *in situ* cytokine expression in HIV- and non-HIV-associated tuberculous lymphadenitis. Immunobiology 191: 354–368

Muller WM, Steinman R,M Cohn ZA (1983) Membrane proteins of the vacuolar system. III. Further studies on the composition and recycling of endocytic vacuole membrane in cultured macrophages. J Cell Biol 96: 29–36

Mullock BM, Bright NA, Fearon CW, Gray SR, Luzio JP (1998) Fusion of lysosomes with late endosomes produces a hybrid organelle with intermediate density and is NSF dependent. J Cell Biol 140: 591–601

Murakami S, Iwaki D, Mitsuzawa H, Sano H, Takahashi H, Voelker DR, Akino T, Kuroni Y (2001) Surfactant protein A inhibits peptidoglycan-induced TNF-alpha secretion in U937 cells and alveolar macrophages by direct inhibition with toll-like receptor 2. J Biol Chem 277: 6830–6837

Murray HW (1996) Current and future clinical applications of interferon-gamma in host antimicrobial defense. Intensive Care Med 22: S456-S461

Murray PJ (1999) Defining the requirements for immunological control of mycobacterial infections. Trends Microbiol 7: 366–371

Musser JM, Amin A, Ramaswamy S (2000) Negligible genetic diversity of *Mycobacterium tuberculosis* host immune system protein targets: evidence for limited selective pressure. Genetics 155: 7–16

Naegel GP, Young KR, Reynolds HY (1984) Receptors for human IgG subclasses on human alveolar macrophages. Am Rev Resp Dis 129: 413–418
Nagasawa H, Miyaura C, Abe E, Suda T, Horiguchi M, Suda T (1987) Fusion and activation of human alveolar macrophages induced by recombinant interferon-γ and their suppression by dexamethasone. Am Rev Respir Dis 136: 916–921
Nathan CF (1987) Secretory products of macrophages. J Clin Invest 79–319–326
Nathan CF, Kaplan G, Levis WR, Nusrat A, Witmer MD, Sherwin SA, Job CK, Horowitz CR, Steinman R,M Cohn ZA (1986) Local and systemic effects of intradermal recombinant interferon-γ in patients with lepromatous leprosy. N Engl J Med 315: 6–15
Nathan CF, Murray HW, Weibe ME, Rubin BY (1983) Identification of interferon-γ as the lymphokine that activates human macrophage oxidative metabolism and antimicrobial activity. J Exp Med 158: 670–685
Nathan CF, Prendergast TJ, Weibe ME, Stanley ER, Platzer E, Remold HG, Welte K, Rubin BY, Murray HW (1984) Activation of human macrophages: comparison of other cytokines with interferon-γ. J Exp Med 1600: 600–605
Nathan C, Shiloh MU (2000) Reactive oxygen and nitrogen intermediates and the relationship between mammalian hosts and microbial pathogens. Proc Natl Acad Sci USA 97: 8841–8848
Nau GJ, Richmond JFL, Sclesinger A, Jennings EG, Lander ES, Young RA (2002) Human macrophage activation programs induced by bacterial pathogens. Proc Natl Acad Sci USA 99: 1503–1508
Nguyan H, Ramana CV, Bayes J, Stark GR (2001) Roles of phosphatidylinosotol 3-kinase in interferon-γ-dependent phosphorylation of STAT1 on serine 727 and activation of gene expression. J Biol Chem 276: 33361–33368
North RJ, LaCourse R, Ryan L, Gros P (1999) Consequence of *Nramp1* deletion to *Mycobacterium tuberculosis* infection in mice. Infect Immun 67: 5811–5814
Nystrom T (2001) Not quite dead enough: on bacterial life, culturability, senescence, and death. Arch Microbiol 176: 159–164
O'Brien L, Carmichael J, Lowrie DB, Andrew PW (1994) Strains of *Mycobacterium tuberculosis* differ in susceptibility to reactive nitrogen intermediates in vitro. 62: 5187–5190
O'Brien S, Andrew PW (1991) Guinea pig alveolar macrophage killing of *Mycobacterium tuberculosis* in vitro does not require hydrogen peroxide or hydroxyl radical. Microbiol Path 11: 229–236
Oh YK, Swanson JA (1996) Different fates of phagocytosed particles after delivery into macrophage lysosomes. J Cell Biol 132: 585–593
Ozinsky A, Underhill DM, Fontenot JD, Hajjar AM, Smith KD, Wilson CB, Schroeder L, Aderem A (2000) The repertoire for pattern recognition of pathogens by the innate immune system is defined by cooperation between Toll-like receptors. Proc Natl Acad Sci USA 97: 13766–13771
Pabst MJ, Gross JM, Brozna JP, Goren MB (1988) Inhibition of macrophage priming by sulfatide from *Mycobacterium tuberculosis.* J Immunol 140: 634–640
Pace JL, Russell SW, Schreiber RD, Altman A, Katz DH (1983) Macrophage activation: priming activity from a T-cell hybridoma is attributable to interferon-gamma. Proc Natl Acad Sci USA 80: 3782-3786
Pancholi P, Mizra A, Bhardwaj N, Steinman RM (1993) Sequestration from immune CD4+ T cells of mycobacteria growing in human macrophages. Science 260: 984–986
Parrish NM, Dick JD, Bishai WR (1998) Mechanism of latency in *Mycobacterium tuberculosis.* Trends Microbiol. 6: 107–112.
Pasula R, Wright JR, Kachel DL, Martin WJ (1999) Surfactant protein A suppresses reactive nitrogen intermediates by alveolar macrophages in response to *Mycobacterium tuberculosis.* J Clin Invest 103: 483–490
Pasula R, Wisniowski P, Martin WJ (2002) Fibronectin facilitates *Mycobacterium tuberculosis* attachment to murine alveolar macrophages. Infect Immun 70: 1287–1292

Pathan AA, Wilkinson KA, Klenerman P, McShane H, Davidson RN, Pasvol G, Hill AV, Lalvani A (2001) Direct ex vivo analysis of antigen-specific IFN-gamma-secreting CD4 T cells in *Mycobacterium tuberculosis*-infected individuals: associations with clinical disease state and effect of treatment. J Immunol 167: 5217–5225

Pinckard JK, Sheehan KC, Schreiber RD (1997) Ligand-induced formation of p55 and p75 tumor necrosis factor heterocomplexes in intact cells J Biol Chem 272: 10784–10789

Piddington DL, Fang FC, Laessig T, Cooper AM, Orme IM, Buchmeier NA (2001) Cu,Zn superoxide dismutase of *Mycobacterium tuberculosis* contributes to survival in activated macrophages that are generating an oxidative burst. Infect Immun 69: 4980–4987

Piddington DL, Kashkouli A, Buchmeier NA (2000) Growth of *Mycobacterium tuberculosis* in a defined medium is very restricted by acid pH and Mg^{2+} levels. Infect Immun 68: 4518–4522

Pieters J (2001) Entry and survival of pathogenic mycobacteria in macrophages. Microbe Infect 3: 249-255

Pitt A, Mayorga LS, Schwartz AL, Stahl PD (1992) Transport of phagosomal components to an endosomal compartment. J Biol Chem 267: 126–132

Postlethwaite AE, Jackson BK, Beachey EH, Kang AH (1982) Formation of multinucleated giant cells from human monocyte precursors. Mediation by a soluble protein from antigen- and mitogen-stimulated lymphocytes. J Exp Med 155: 168–178

Prada-Delgado A, Carrasco-Marin E, Bokoch G, Alvarez-Dominiguez C (2001) Interferon-γ listericidal action is mediated by novel Rab5a functions at the phagosomal environment. J Biol Chem 276: 19059–19065

Prigozy TI, Seiling PA, Clemens D, Stewart PL, Behar SM, Porcelli SA, Brenner MB, Modlin RL, Kronenberg M (1997) The mannose receptor delivers lipoglycan antigens to endosomes for presentation to T cells by CD1b molecules. Immunity 6: 187–197

Pugin J, Heumann ID, Tomasz A, Kravchenko VV, Akamatsu Y, Nishijima M, Glauser MP, Tobias PS, Ulevitch RJ (1994) CD14 is a pattern recognition receptor. Immunity 1: 509–516

Purkayastha A, McCue LA, McDonough KA (2002) Identification of a *Mycobacterium tuberculosis* putative classical nitroreductase gene whose expression is coregulated with that of the *acr* gene within macrophages, in standing versus shaking cultures and under low oxygen conditions. Infect Immun 70: 1518–1529

Qureshi SA, Kim RM, Konteatis Z, Biazzo DE, Motamedi H, Rodriguez R, Boice JA, Calaycay JR, Bednarek MA, Griffin P, Gao YD, Chapman K, Mark DF (1999) Mimicry of erythropoietin by a non-peptide molecule. Proc Natl Acad Sci USA 96: 12156–12161

Raad I, Hachem R, Leeds N, Sawaya R, Salem Z, Atweh S (1996) Use of adjunctive treatment with interferon-γ in an immunocompromised patient who had refractory multi-drug resistant tuberculosis of the brain. Clin Infect Dis 22: 572–574

Ragno S, Romano M, Howell S, Pappin DJC, Jenner PJ, Colston MJ (2001) Changes in gene expression in macrophages infected with *Mycobacterium tuberculosis*: a combined transcriptomic and proteomic approach. Immunology 104: 99–108

Ramachandra L, Noss E, Bomm WH, Harding CV (2001) Processing of *Mycobacterium tuberculosis* antigen 85B involves intraphagosomal formation of peptide-major histocompatibility complex II complexes and is inhibited by live bacilli that decrease phagosome maturation. J Exp Med 194: 1421–1432

Raman S, Song T, Puyang X, Bardarov S, Jacobs WR, Husson RN (2001) The alternative sigma factor SigH regulates major components of oxidative and heat stress responses in *Mycobacterium tuberculosis*. J Bacteriol 183: 6119–6125

Ramana CV, Gil MP, Scheiber RD, Stark GR (2002) Stat1-dependent and independent pathways in IFN-γ-dependent signalling. Trends Immunol 23: 96–101

Reichel H, Koeffler HP, Norman AW (1987) Synthesis in vitro of 1,25-dihydroxyvitamin D3 and 24,25-dihydroxyvitamin D3 by interferon-γ-stimulated normal human bone marrow and alveolar macrophages. J Biol Chem 262: 10931–10987

Reynolds HY, Newball HH (1974) Analysis of proteins and respiratory cells obtained from human lungs by bronchial lavage. J Clin Lab Med 84: 559–573

Reyrat J-M, Berthet F-X, Giquel B (1995) The urease locus of *Mycobacterium tuberculosis* and its utilisation for the demonstration of allelic exchange in *Mycobacterium bovis* bacillus Calmette-Guerin. Proc Natl Acad Sci USA 92: 8768–8772

Rhoades ER, Orme IM (1997) Susceptibility of a panel of virulent strains of *Mycobacterium tuberculosis* to reactive nitrogen intermediates. Infect Immun 65: 1189–1195

Riches DW, Chan ED, Zahradka EA, Winston BW, Remigio LK, Lake FR (1998) Cooperative signaling by tumor necrosis factor receptors CD120a (p55) and CD120b (p75) in the expression of nitric oxide and inducible nitric oxide synthase by mouse macrophages. J Biol Chem 273: 22800-22806

Riley RL (1960) Apical localization of pulmonary tuberculosis. Bull Johns Hopkins Hospital 106: 232-239

Roach DR, Briscoe H, Saunders B, France MP, Riminton S, Britton WJ (2001) Secreted lymphotoxin-α is essential for the control of an intracellular bacterial infection. J Exp Med 193: 239–246

Roach DR, BrIscoe H, Baumgart K, Rathjen DA, Britton WJ (1999) Tumor necrosis factor (TNF) and a TNF-mimetic peptide modulate the granulomatous response to *Mycobacterium bovis* BCG infection in vivo. Infect Immun 67: 5473–5476

Robinson DS, Ying S, Taylor IK, Wangoo A, Mitchell DM, Kay AB, Hamid Q, Shaw RJ (1994) Evidence for a Th1-bronchoalveolar T-cell subset and predominance of interferon-gamma gene activation in pulmonary tuberculosis. Am J Respir Crit Care Med 149: 989–993

Rockett KA, Brookes R, Udalova I, Vidal V, Hill AV, Kwiatkowski D (1998) 1,25-dihydroxyvitamin D3 induces nitric oxide synthase and supresses growth of *Mycobacterium tuberculosis* in a human macrophage-like cell line. Infect Immun 66: 5314–5321

Rook GA, Steele J, Ainsworth M, Champion Br (1986a) Activation of macrophages to inhibit proliferation of *Mycobacterium tuberculosis*: comparison of the effects of recombinant gamma interferon on human monocytes and murine peritoneal macrophages. Immunology 59: 333–338

Rook GA, Steele J, Fraher L, Barker S, Karmali R, O'Riordan J, Stanford J (1986b) Vitamin D3, gamma interferon, and control of the proliferation of *Mycobacterium tuberculosis* by human macrophages. Immunology 57: 159–163

Rous P (1925) The relative reaction within living mammalian tissues. II. On the mobilization of acid material within cells, and the reaction as influenced by the cell state. J Exp Med 41: 399–411

Russell DG (2001) *Mycobacterium tuberculosis*: Here today, and here tomorrow. Nat Rev Mol Cell Biol 2: 569–586

Russell DG, Dant J, Sturgill-Koszycki S (1996) *Mycobacterium avium*- and *Mycobacterium tuberculosis*-containing vacuoles are dynamic, fusion competent vesicles that are accessible to glycosphingolipids from the host cell plasmalemma. J Immunol 156: 4764–4773

Ryu S, Park Y, Bai G, Kim S, Park S, Kang S (2000) 3'UTR polymorphisms in the *NRAMP1* gene are associated with susceptibility to tuberculosis in Koreans. Int J Tuberc Lung Dis 4: 577–580

Saginario C, Sterling H, Beckers C, Kobyashi R, Solimena M, Ullu E, Vigery A (1998) MFR, a putative receptor mediating fusion of macrophages. Mol Cell Biol 18: 6213–6223

Salo WL, Aufderheide AC, Buikstra J, Holcomb TA (1994) Identification of *Mycobacterium tuberculosis* DNA in a pre-Columbian Peruvian mummy. Proc Natl Acad Sci USA 91: 2091–2094

Sandau KB, Fandrey J, Brune B (2001) Accumulation of HIF-alpha under the influence of nitric oxide. Blood 97: 1009–1015
Saunders BM, Frank AA, Orme IM (1999) Granuloma formation is required to contain bacillus growth and delay mortality in mice chronically infected with *Mycobacterium tuberculosis*. Immunology 98: 324–328
Savedra RJ, Delude R, Ingalls R, Fenton M, Golenbock D (1996) Mycobacterial lipoarabinomannan recognition requires a receptor that shares components of the endotoxin signaling system. J Immunol 157: 2549–2554
Scanga CA, Mohan VP, Tanaka K, Alland D, Flynn JL, Chan J (2001) The inducible nitric oxide synthase locus confers protection against aerogenic challenge of both clinical and laboratory strains of *Mycobacterium tuberculosis* in mice. Infect Immun 69: 7711–7717
Scannell G, Waxman K, Kaml GJ, Ioli G, Gatanaga T, Yamamoto R, Granger GA (1993) Hypoxia induces a human macrophage cell line to release tumor necrosis factor-α and its soluble receptors in vitro. J Surg Res 54: 281–285
Schaible UE, Sturgill-Koszycki S, Schlesinger PH, Russell DG (1998) Cytokine activation leads to acidification and increases maturation of *Mycobacterium avium*-containing phagosomes in murine macrophages. J Immunol 160: 1290–1296
Schlesinger LS, Bellinger-Kawahara CG, Payne NR, Horwitz MA (1990) Phagocytosis of *Mycobacterium tuberculosis* is mediated by human monocyte complement receptors and complement component C3. J Immunol 144: 2771–2780
Schlessinger LS, Hull SR, Kaufman TM (1994) Binding of the terminal mannosyl units of lipoarabinomannan from a virulent strain of *Mycobacterium tuberculosis* to human macrophages. J Immunol 152: 4070–4079
Schlessinger LS, Kaufman TM, Iyer S, Hull SR, Marchiando LK (1996) Differences in mannose receptor-mediated uptake of lipoarabinomannan from virulent and attenuated strains of *Mycobacterium tuberculosis* by human macrophages. J Immunol 157: 4568–4575
Schorey JS, Carroll MC, Brown EJ (1997) A macrophage invasion mechanism of pathogenic mycobacteria. Science 277: 1091–1093
Schreiber RD, Pace JL, Russell SW, Altman A, Katz DH (1983) Macrophage-activating factor produced by a T cell hybridoma: physiochemical and biosynthetic resemblance to gamma-interferon. J Immunol 131: 826–832
Schwander SK, Torres M, Carranza CC, Escobedo D, Tary-Lehman M, Anderson P, Toossi Z, Ellner JJ, Rich EA, Sada E (2000) Pulmonary mononuclear cell responses to antigens of *Mycobacterium tuberculosis* in healthy household contacts of patients with active tuberculosis and healthy controls from the community. J Immunol 165: 1479–1485
Segal B, Leto T, Gallin J, Malech H, Holland S (2000) Genetic, biochemical, and clinical features of chronic granulomatous disease. Medicine 79: 170–200
Seidel HM, Lamb P, Rosen J (2000) Pharmaceutical intervention in the Jak/STAT signaling pathway. Oncogene 19: 2645–2656
Semenza GL (2001) HIF-1 and mechanisms of oxygen sensing. Curr Opin Cell Biol 13: 167–171
Sharma SK, Pande JN, Singh YN, Verma K, Kathait SS, Khare SD, Malaviya AN (1992) Pulmonary and immunologic abnormalities in miliary tuberculosis. Am Rev Respir Dis 145: 1167–1171
Sherman DR, Voskuil M, Schnappinger D, Liao R, Harrell MI, Schoolnik GK (2001) Regulation of the *Mycobacterium tuberculosis* gene encoding alpha-crystallin. Proc Natl Acad Sci U S A. 98:7534–9
Shiloh MU, MacMicking JD, Nicholson S, Brause JE, Potter S, Marino M, Fang F, Dinauer M, Nathan C (1999) Phenotype of mice and macrophages deficient in both phagocyte oxidase and inducible nitric oxide synthase. Immunity 10: 29–38

Sikora A, Lui J, Brosnan C, Buell G, Chessel I, Bloom BR (1999) Purinergic signaling regulates radical-mediated bacterial killing mechanisms in macrophages through a $P2X_7$-independent mechanism. J Immunol 163: 558–561

Simonsen A, Gaulier J-M, D'Arrigo A, Stenmark H (1999) The Rab5 effector EEA1 interacts directly with syntaxin-6. J Biol Chem 274: 28857–28860

Sirakoya TD, Thirumala AK, Dubey VS, Sprecher H, Kolattukudy PE (2001). The *Mycobacterium tuberculosis* pks2 gene encodes the synthase for the hepta- and octamethyl-branched fatty acids required for sulfolipid synthesis. J Biol Chem 276: 16833–16839

Sly LM, Lopez M, Nauseef WN, Reiner NE (2001) 1α,25-dihydroxyvitamin D3-induced monocyte antimycobacterial activity is regulated by phosphatidylinositol 3-kinase and mediated by the NADPH-dependent phagocyte oxidase. J Biol Chem 276: 35482–35493

Smith S, Liggitt D, Jeromsky E, Tan X, Skerrett SJ, Wilson CB (2002) Local role of tumor necrosis factor alpha in the pulmonary inflammatory response to *Mycobacterium tuberculosis* infection. Infect Immun 70: 2082–2089

Soler P, Boussaud V, Moreau J, Bergeron A, Bonnette P, Hance AJ, Tazi A (1999) In situ expression of B7 and CD40 costimulatory molecules by normal human lung macrophages and epitheloid cells in tuberculoid granulomas. Clin Exp Immunol 116: 332–339

Spector, M.P. (1998) The starvation-stress response (SSR) of *Salmonella*. Adv Microb Physiol 40: 233-279

Springer B, Master S, Sander P, Zahrt T, McFalone M, Song J, Papavinasasundaram KG, Colston MJ, Boettger E, Deretic V (2001) Silencing of oxidative stress response in *Mycobacterium tuberculosis*: expression patterns of *ahpC* in virulent and avirulent strains and effect of *ahpc* inactivation. Infect Immun 69: 5967–5973

St John G, Brot N, Ruan J, Erdjument-Bromage H, Tempst P, Weissbach H, Nathan C (2001) Peptide methionine sulfoxide reductase from *Escherichia coli* and *Mycobacterium tuberculosis* protects bacteria against oxidative damage from reactive nitrogen intermediates. Proc Natl Acad Sci USA 98: 9901–9906

Stenger S, Mazzaccaro RJ, Uyemura K, Cho S, Barnes PF, Rosat JP, Sette A, Brenner MB, Porcelli SA, Bloom BR, Modlin R (1997) Differential effects of cytolytic T cell subsets on intracellular infection. Science 276: 1684–1687

Sterling H, Saginario C, Vignery A (1998) CD44 occupancy prevents macrophage multinucleation. J Cell Biol 143: 837–847

Stewart GR, Erht S, Riley L,W Dale JW, McFadden J (2000) Deletion of the putative antioxidant noxR1 does not alter the virulence of *Mycobacterium tuberculosis* H37Rv. Tuber Lung Dis 80: 237–242

Sugawara I, Yamada H, Kazumi Y, Doi N, Otomo K, Aoki T, Mizuno S, Udagawa T, Tagawa Y, Iwakura Y (1998) Induction of granulomas in interferon-gamma-deficient mice by avirulent but not by virulent strains of *Mycobacterium tuberculosis.* J Med Micobiol 47: 87–877

Surkova LK, Shpakovskaia NS, Duis'mikeeva ML (1999) Contribution of immune complex pathological reactions to immunogenesis of pulmonary tuberculosis. Probl Tuberk 6: 46–50

Suzuki K, Tsuyuguchi K, Matsumoto H, Niimi A, Tanaka E, Amitani R (2000) Effect of proton pump inhibitor alone or in combination with clathrinomycin on mycobacterial growth in human macrophages. FEMS Microbiol Lett 182: 69–72

Takashima T, Ohnisji K, Tsuyuguchi I, Kishimoto S (1993) Differential regulation of formation of multinucleated giant cells from concanavalin A-stimulated human monocytes by IFN-γ and IL-4. J Immunol 150: 3002–3010

Taylor GA, Collazo CM, Yap GS, Nguyen K, Gregorio TA, Taylor LS, Eagleston B, Secrest L, Southon EA, Reid SW, Tessarollo L, Bray M, McVicar DW, Kommschlies KL, Young HA, Biron CA, Sher A, Vande Woude GF (2000) Pathogen-specific loss of host resis-

tance in mice lacking the IFN-gamma-inducible gene IGTP. Proc Natl Acad Sci USA 97: 751–755
Teitelbaum R, Glatman-Freedman A, Chen B, Robbins JB, Unanue E, Casadevall A, Bloom BR (1998) A mAb recognizing a surface antigen of *Mycobacterium tuberculosis* enhances host survival. Proc Natl Acad Sci USA 95: 15688–15693
Tendler DS, Bao C, Wang T, Huang EL, Ratovitski EA, Pardoll DA, Lowenstein CJ (2001) Intersection of interferon and hypoxia signal transduction pathways in nitric oxide-induced tumor apoptosis. Cancer Res 61: 3682–3688
Thoma-Uszynski S, Stenger S, Takeuchi O, Ochoa M, Engele MT, Sieling PA, Barnes PF, Rollinghoff M, Bolcskei PL, Wagner M, Akira S, Norgard MV, Belisle JT, Godowski PJ, Bloom BR, Modlin RL (2001) Induction of direct antimicrobial activity through mammalian Toll-like receptors. Science 291: 1544–1547
Thomas ED, Ramberg RE, Sale GE, Sparkes RS, Golde DW (1976) Direct evidence for a bone marrow origin of the alveolar macrophage in man. Science 192: 1016–1018
Ting LM, Kim AC, Cattamanchi A, Ernst JD (1999) *Mycobacterium tuberculosis* inhibits IFN-gamma transcriptional responses without inhibiting activation of STAT1. J Immunol 163: 3898–3906
Ullrich H-J, Beatty WL, Russell DG (1999) Direct delivery of procathepsin D to phagosomes: implications for phagosome biogenesis and parasitism by *Mycobacterium*. Eur J Cell Biol 78: 739–748
Underhill DM, Ozinsky A, Hajjar AM, Stevens A, Wilson CB, Bassetti M, Aderem A (1999a) The Toll-like receptor 2 is recruited to macrophage phagosomes and discriminates between pathogens. Nature 401: 811–815
Underhill DM, Ozinsky A, Smith KD, Aderem A (1999b) Toll-like receptor-2 mediates mycobacteria-induced proinflammatory signaling in macrophages. Proc Natl Acad Sci USA 96: 14459–14463
Via LE, Fratti RA, McFalcone M, Pagan-Ramos E, Deretic D, Deretic V (1997) Effects of cytokines on mycobacterial phagosome maturation. J Cell Sci 111: 897–905
Vidal M, Ramana CV, Dusso AS (2002) Stat1-vitamin D receptor interactions antagonize 1,25-dihydroxyvitamin D transcriptional activity and enhance stat1-mediated transcription. Mol C ell Biol 22: 2777–2787
Vieira OV, Botelho R,J Rameh L, Brachmann SM, Matsuo T, Davidson HW, Schreiber A, Backer JM, Cantley LC, Grinstein S (2001) Distinct roles of class I and class III phosphatidyl 3-kinases in phagosome formation and maturation. J Cell Biol 155: 19–25
Voorhout WF, Veenendaal T, Kuroki Y, Ogasawa Y, van Golde LM, Geuze HJ (1992) Immunocytochemical localization of surfactant protein D (SP-D) in type II cells, Clara cells, and alveolar macrophages of rat lung. J Histochem Cytochem 40: 1589–1597
Wagner TE, Huseby ES, Huseby JS (2002) Exaccerbation of *Mycobacterium tuberculosis* enteritis masquerading as crohn's disease after treatment with a tumor necrosis factor-α inhibitor. Am J Med 112: 67–69
Walker SR, Williams MC, Benson B (1986) Immunocytochemical localization of the major surfactant apoproteins in type II cells, Clara cells, and alveolar macrophages of rat lung. J Histochem Cytochem 34: 1137–1148
Wang, C-h, Lui C-Y, Lin H-C, Yu C-T, Chung K, Kuo H (1998) Increased exhaled nitric oxide in active pulmonary tuberculosis due to inducible nitric oxide synthase upregulation in alveolar macrophages. Eur J Respir 11: 809–815
Wang C, Lin H, Lui C, Huang K, Huang T, Yu C, Kuo H (2001) Upregulation of inducible nitric oxide synthase and cytokine secretion in peripheral blood monocytes from pulmonary tuberculosis patients. Int J Tuberc Lung Dis 5: 283–291
Wangoo A, Sparer T, Brown IN, Snewin VA, Janssen R, Thole J, Cook HT, Shaw RJ, Young DB (2001) Contribution of Th1 and Th2 cells to protection and pathology in experimental models of granulomatous lung disease. J Immunol 166: 3432–3439
Ward DM, Pevsner J, Scullion MA, Vaughn M, Kaplan J (2000) Syntaxin 7 and VAMP7 are soluble attachment *N*-ethylmaleimide-sensitive factor attachment protein receptors

required for late endosome-lysosome and homotypic lysosome fusion in alveolar macrophages. Mol Biol Cell 11: 2327–2333

Wayne LG, Sohaskey CD (2001) Nonreplicating persistence of *Mycobacterium tuberculosis*. Ann Rev Microbiol 55: 139–163.

Weikert LF, Edwards K, Chroneos ZC, Hager C, Hoffman L, Shepherd VL (1997) SP-A enhances uptake of bacillus *Calmette-Guerin* by macrophages through a specific SP-A receptor. Am J Physiol 272: L989-L995

Weikert LF, Lopez JP, Abdolrasulnia R, Chroneos ZC, Shepherd VL (2000) Surfactant protein A enhances mycobacterial killing by rat macrophages through a nitric oxide-dependent pathway. Am J Physiol Lung Cell Mol Physiol 279: L216-L223

Weinberg JB, Hobbs MB, Misukonis MA (1984) Recombinant human γ-interferon induced human monocyte prokaryon formation. Proc Natl Acad Sci USA 81: 4554–4557

West JB (2000) Respiratory physiology. Lippencot Williams & Wilkins, Philadelphia, pp 45–61

World Health Organization (1998) WHO Expert Committee on Leprosy. World Health Organ Tech Rep Ser 874: 1–43

World Health Organization (2000) Global tuberculosis report. WHO Press, Geneva

Wilkinson RJ, Llewelyn M, Toossi Z, Patel P, Pasvol G, Lalvani A, Wright D, Latif M, Davidson RN (2000) Influence of vitamin D deficiency and vitamin D receptor polymorphisms on tuberculosis among Gujarati Asians in west London: a case-control study. Lancet 355: 618–621

Williams G, Williams W (1983) Granulomatous inflammation—a review. Clin Path 36: 723–733

Xu S, Cooper A, Sturgill-Koszycki S, van Heyningen T, Chatterjee D, Orme I, Allen P, Russell DG (1994) Intracellular trafficking in *Mycobacterium tuberculosis* and *Mycobacterium avium*-infected macrophages. J Immunol 153: 2568–2578

Yamashita K, Iwamoto T, Iijima S (1978) Immunohistochemical observation of lysozyme in macrophages and giant cells in human granulomas. Acta Pathol Jpn 28: 689–695

Yang E, Wen Z, Haspel R, Zhang J, Darnell JE Jr (1999a) The linker domain of Stat1 is required for gamma interferon-driven transcription. Mol Cell Biol 19: 5106–5112

Yang RB, Mark MR, Gurney AL, Godowski PJ (1999b) Signaling events induced by lipopolysaccharide-activated toll-like receptor 2. J Immunol 163: 639–643

Yeager H, Lacey J, Smith LR, LeMaistre CA (1967) Quantitative studies of mycobacterial populations in sputum and saliva. Am Rev Respir Dis 98: 998–1004

Yu K, Mitchell C, Xing Y, Magliozzo RS, Bloom BR, Chan J (1999) Toxicity of nitrogen oxides and related oxidants on mycobacteria: *M. tuberculosis* is resistant to peroxynitrite anion. Tuberc Lung Dis 79: 191–198

Zaffran Y, Zhang L, Ellner JJ (1998) Role of CR4 in *Mycobacterium tuberculosis*-human macrophages binding and signal transduction in the absence of serum. Infect Immun 66: 4541–4544

Zerial M, McBride H (2001) Rab proteins as membrane organizers. Nat Rev Mol Cell Biol 2: 107–119

Zhou H, Duncan RF, Robinson TW, Gao L, Forman HJ (1997) Ca^{2+}-dependent $p47^{phox}$ translocation in hydroperoxide modulation of the alveolar macrophage respiratory burst. Am J Physiol 273: L1042–1047

Zimmermann M (1979) Pulmonary and osseous tuberculosis in an Egyptian mummy. Bull NY Acad Med 55: 604–608

Detection and Control of Fungi by Macrophages: The Role of Carbohydrates and Antifungal Agents

J. A. Willment · S. Gordon · G. D. Brown

Sir William Dunn School of Pathology,
University of Oxford,
South Parks Road, Oxford, OX1 3RE, UK
e-mail: Gbrown@molbiol.ox.ac.uk

Abstract Macrophages play an important role in the innate immune response to fungal pathogens. They express receptors which recognise a variety of fungal molecular patterns, many of which are conserved cell wall carbohydrates. We present an overview of the macrophage receptors shown to be involved in fungal recognition and binding, the various antifungal mechanisms utilised by these cells, and demonstrate strategies that fungal pathogens have evolved to escape these mechanisms. We also provide an overview of the current clinical anti-fungal agents, as well as strategies which are being developed to enhance the anti-microbial mechanisms of the macrophages themselves. Finally, we discuss fungal-derived carbohydrates and their potential use as immunomodulators.

Keywords Anti-fungal agents, Beta-glucan, Carbohydrate, Fungi, Macrophage receptor

Abbreviations

H. MW	High molecular weight
CLP	Caecal ligation and puncture
$\downarrow$	Decrease
$\uparrow$	Increase
$\leftrightarrow$	No change

1 Introduction

Macrophages (MΦs) play a central role in the innate recognition of a range of pathogens, including fungi, and in the modulation of the subsequent effector mechanisms. Recognition generally leads to phagocytosis and killing of the invading pathogen through a variety of passive and active mechanisms. However, some fungal pathogens have learned to subvert their host's anti-microbial defence mechanisms requiring the clinical administration of anti-fungal compounds. Furthermore, with the emergence of drug-resistant strains, a greater understanding of how the immune system deals with fungal pathogens is crucial if novel anti-fungal strategies are to be developed. In this chapter, we have placed particular emphasis on the MΦ receptors involved in the recognition of fungal pathogens, as well as the killing mechanisms employed. We also discuss both current and potential strategies used to control fungal pathogens, and the role of fungal derived carbohydrates in modulating MΦ and immune responses.

Although this chapter will focus specifically on the contribution of MΦs to the innate recognition and control of fungal pathogens, these cells are not the only line of defence. Neutrophils, dendritic cells (DCs) and some lymphocytes have all been shown to play a role in the immune response to various fungal pathogens (see Vazquez-Torres and Balish 1997 for a review). Furthermore, although not discussed, the reader should bear in mind that MΦ and their specialised relatives, DCs, have a crucial role in instructing the adaptive immune response, which also plays an important role in the control of fungal pathogens (see Romani 2002 for a review).

2 Macrophage Receptors

Contact between the MΦ and fungal pathogen occurs when MΦ receptors recognise structures, such as polysaccharides, which are displayed on the fungal cell wall. The fungal cell wall, apart from giving the cell its rigidity and providing protection against environmental stress, contributes to virulence by providing a platform for adhesion to the host. The structure of the cell wall, which has been derived from the non-pathogenic fungus *Saccharomyces cerevisiae*, consists mostly of polysaccharides, including mannoproteins, β-glucans and chitin (see Fig. 1 and Chaffin et al. 1998; Lipke and Ovalle 1998 for reviews). In total, about

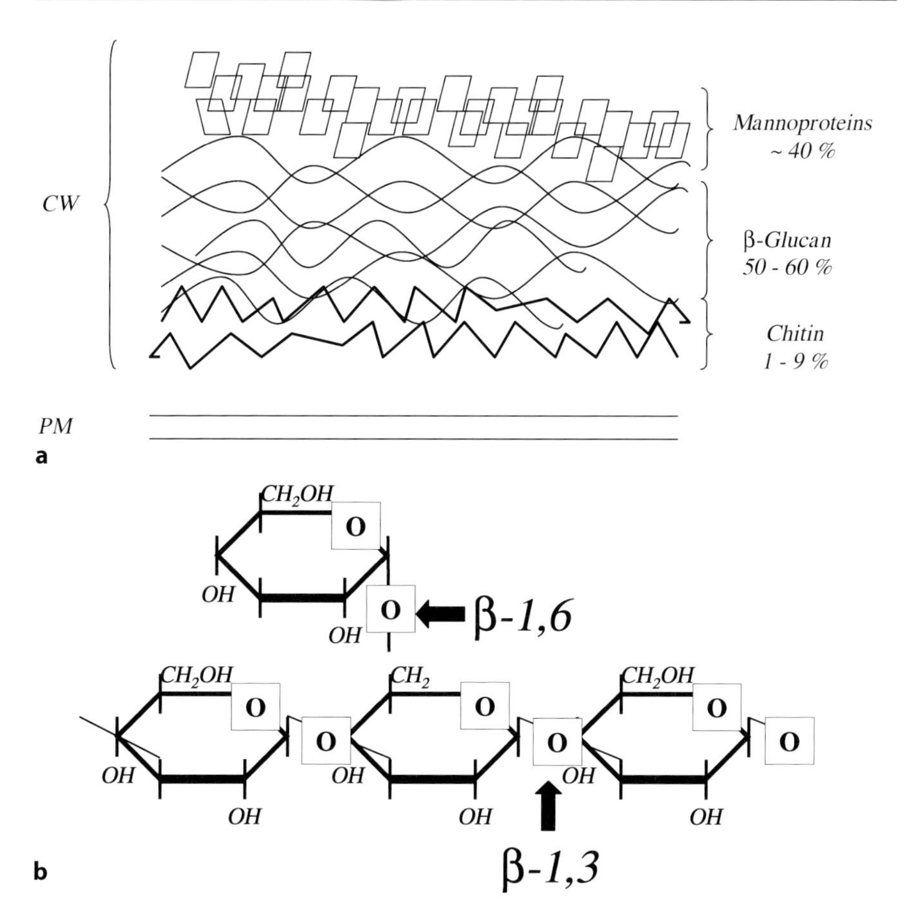

Fig. 1a, b The fungal cell wall (CW) is modular in structure and composed of various polysaccharide layers. **a** The plasma membrane (PM) is surrounded by a chitin layer (unbranched acetyl-glucosamine polymers) linked to β-glucans. Mannan is connected via di-*N*-acetylchitobiose residues to the CW protein fraction. (see Chaffin et al. 1998 and Lipke and Ovalle 1998 for reviews). **b** β-glucans form linear chains via β-1,3-linked glucose residues and branched chains via β-1,6 linkage(s)

80%–90% of the *Candida albicans* fungal cell wall is composed of carbohydrates, some of which have immunomodulatory activities, discussed in more detail below (see Tzianabos 2000 for a review).

The recognition of fungal pathogens occurs either directly via specific MΦ surface receptors or indirectly through the recognition of opsonins, such as C3 or antibodies, which coat the microbial cell surface (see Aderem and Underhill 1999; Underhill and Ozinsky 2002a for reviews). These receptors bind conserved molecular patterns that are often part of immutable structural components, such as conserved polysaccharides found within the cell wall (Medzhitov and Janeway 1997). Although complex and simple carbohydrate structures are recognised by a range of MΦ receptors (see Table 1), we will focus on those specifically shown to have the ability to recognise intact fungi; the mannose receptor (MR), the β-glucan receptor (βGR), and the complement receptor (CR)3. The Toll-like receptors, which mediate intracellular signalling and resultant pro-inflammatory response, will also be discussed briefly (see Chap. 5). These receptors should not be thought of in isolation, but rather that they are simultaneously contributing to the recognition of fungal pathogens. Finally, it should be noted that the function of these receptors and their interactions remains to be fully elucidated. It is also likely that other MΦ receptors, which recognise fungal pathogens, have yet to be identified.

2.1 Mannose Receptor

The MR, often referred to as the mannose-fucose receptor, is a type I transmembrane glycoprotein, containing an extracellular cysteine-rich domain, a fibronectin type II-like domain, and eight tandem carbohydrate recognition domains (CRDs) (see Fig. 2) (see Stahl and Ezekowitz 1998 and Martinez-Pomares et al. 2001 for reviews). The MR is expressed on the majority of differentiated mononuclear cell phagocytes (see Table 1), with weak or no expression on most MΦ cell lines (Pontow et al. 1992).

The MR exhibits preferences for oligosaccharides with the following terminal residues, l-fucose>d-mannose≥d-*N*-acetyl-glucosamine>>>d-galactose (Stahl et al. 1978), and displays high affinity for endogenous ligands containing branched α-linked oligo-mannoses (Kery et al. 1992; Linehan et al. 2001). MR transfectant CHO cells are capable of mediating the phagocytosis of yeast, such as *Pneumocystis carinii* and *C. albicans*, the pinocytosis of mannosylated glycoproteins (Ezekowitz et al. 1990; Ezekowitz et al. 1991) and the binding of *Cryptococcus neoformans* mannoproteins (Mansour et al. 2002). The MR also exists as a soluble cleaved form capable of binding the *S. cerevisiae*-derived particle, zymosan, as well as *C. albicans* (Martinez-Pomares et al. 1998). While both endogenous and exogenous ligands have been identified for the MR (see Table 1), its full role in the binding of yeast remains unresolved (see Linehan et al. 2000 and Martinez-Pomares et al. 2001 for reviews). Experimental conditions under which fungal binding was assayed often did not take into account the presence of other mannose-binding lectins, such as Nkcl/Dectin-2 (Fernandes et al. 1999; Ariizumi et al. 2000a).

The MR is believed to signal via its 45-amino acid cytoplasmic tail, via as yet unknown mechanisms (Ezekowitz et al. 1990). The pro-inflammatory cytokines interleukin (IL)-6, granulocyte-MΦ colony-stimulating factor (GM-CSF), tumour necrosis factor (TNF)-α and IL-12 can all be generated after phagocytosis of fungi (Stein and Gordon 1991; Garner et al. 1994; Shibata et al. 1997; Yamamoto et al. 1997). The MR is not, however, responsible for the production of certain chemokines, such as MΦ inflammatory protein (MIP)-1β, MIP-2 and KC (a platelet factor 4 neutrophil chemoattractant family member) during *C. albicans* infection, indicating that other receptors are involved in fungal-mediated

Table 1 Macrophage lectinswith known and potential fungal derived carbohydrate binding ability (modifiedfrom Linehan et al. 2000)

Receptor	Expression	Ligands	Carbohydrate	Reference
β-glucan receptor (Dectin-1)	Myeloid cells	*S. cerevisiae, C. albicans*, unidentified ligand on T cells,	β-1,6-Glucans and β-1,3-glucans	Ariizumi et al. 2000b; Brown and Gordon 2001; Willment et al. 2001
Mannose receptor (mannosyl-fucosyl receptor)	Differentiated MΦs, lymphatic and sinusoidal endothelium, cultured DCs, perivascular microglia, and mesangial cells	Sialoadhesin, lysosomal proteases, glycosidases, peroxidases, yeasts and fungi	Fucose, mannose, N-GlcAc, galactose	Ezekowitz and Stahl 1988; Linehan et al. 2001; Linehan et al. 1999; Martinez-Pomares et al. 2001; Stahl 1992
MΦ galactose receptor	Peritoneal and tumoricidal MΦs	Tumour Tn antigen microbes	Galactose or N-GlcAc terminal oligosaccharides	Suzuki et al. 1996
CR3 (Mac-1, CD11b/CD18)	Lymphocytes, MΦs, NK cells, neutrophils	Microbes, ICAM-1, fibrinogen	Non-specific sugar binding	Muto et al. 1993; Ross 2000; Ross et al. 1985a; Ross and Vetvicka 1993; Thornton et al. 1996
Nkcl (Dectin-2)	MΦs, monocytes, neutrophils, langerhans cells	Unknown	Mannose-sepharose	Ariizumi et al. 2000a; Fernandes et al. 1999
Mouse MΦ C-type lectin	Primary MΦs and cell lines	Unknown	Unknown	Balch et al. 1998
Galectin-3 (Mac-2)	MΦs	IgE and other host molecules, *C. albicans*	β-1,2-Linked oligomannosides	Cherayil et al. 1989; Fradin et al. 2000

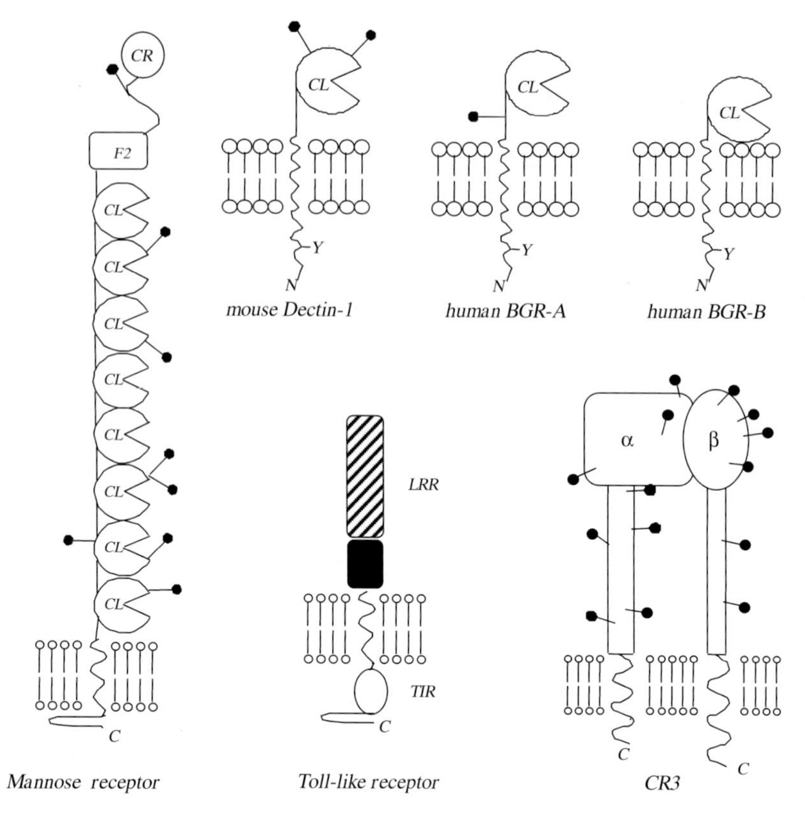

Fig. 2 Schematic representation of the MΦ receptors involved in fungal recognition. The following abbreviations and symbols represent protein domains, their various modifications, and important motifs: *CL*, lectin C-type; α, CD11b integrin subunit; β, CD18 integrin subunit; ●-, N-glycosylation sites; *F2*, fibronectin type II-like repeats; *Y*, ITAM motif; *LLR*, leucine-rich repeat C-terminal domain; *CR*, cysteine-rich domain; *TIR*, Toll-interleukin 1-resistance intracellular signalling domain. The protein orientation relative to the lipid bilayer (*open lollipops*) is indicated by the *N* (N-terminal) and *C* (C-terminal) ends

signalling (Yamamoto et al. 1997). These data support the concept that MΦ receptors act in concert to generate the appropriate signalling response.

2.2
The β-Glucan Receptor

The existence of a non-opsonic zymosan and yeast-binding β-glucan inhibitable monocyte/MΦ receptor was first demonstrated two decades ago (Czop and Austen 1985a,c; Czop and Kay 1991). Recently, a trypsin sensitive C-type lectin receptor, Dectin-1, was demonstrated to be a MΦ βGR (Brown and Gordon 2001; Brown et al. 2002; Willment et al. 2001). Dectin-1 is a type II transmembrane protein with a single CRD, and a cytoplasmic tail containing an immunomodulatory tyrosine-activating motif (ITAM) (see Fig. 2) (Ariizumi et al. 2000b). Murine Dectin-1 is expressed in a number of cell types; including DCs, monocytes, MΦs, neutrophils and even a small sub-population of T cells (Brown and Gordon 2001; Taylor et al. 2002).

Dectin-1 recognises a variety of carbohydrates containing β-1,3 and/or β-1,6 linked glucans, as well as an unidentified endogenous ligand, so far described only on T cells (Ariizumi et al. 2000b; Brown and Gordon 2001; Willment et al. 2001). Murine NIH3T3 fibroblasts transduced with the human and murine form of Dectin-1 were shown to recognise and internalise intact *C. albicans* and *S. cerevisiae* blastospores, in a β-glucan dependant fashion (Brown and Gordon 2001; Willment et al. 2001). As most fungi contain β-glucans within their cell wall (see Douglas 2001 for a review), it is likely that they interact with Dectin-1. Indeed, a number of pathogens have been reported to interact with β-glucan binding receptors, including *P. carinii*, *C. neoformans* and *Aspergillus fumigatus* (Kan and Bennett 1991; Cross and Bancroft 1995; Vassallo et al. 2000).

The interaction of fungi with the βGR results in the activation of MΦs and the release of pro-inflammatory mediators and cytokines (see Czop 1986 and Williams 1997 for reviews). These stimulatory effects appear to be linked to the structure of the β-glucans (see below) and may be mediated by the ITAM motif in the cytoplasmic tail of this receptor (Brown and Gordon 2001). Furthermore, the β-glucan content of fungi varies depending on their morphological state which results in the production of different cytokine and chemokine profiles (for example see Torosantucci et al. 2000). A more comprehensive list of the effects of β-glucans on the immune system is presented in Table 2.

2.3
Complement Receptor 3

CR3 (Mac-1) is a heterodimer composed of two chains, CD11b, which is unique to CR3, and CD18, which is common to all $\beta 2$ integrins (see Fig. 1). CR3 is widely expressed in monocytes, MΦ, DCs, neutrophils, eosinophils, natural killer (NK) cells, some $CD8^+$ T cells and $CD5^+$ B cells. It has diverse functions (see Table 1) ranging from mediating migration of myeloid leukocytes and NK cells, to the phagocytosis and killing of complement opsonised microbes (see Ross and Vetvicka 1993; Ross et al. 1999, 2000 for reviews). Unlike the β-glucan or mannose receptors, CR-mediated recognition does not directly result in the release of pro-inflammatory responses, such as the production of reactive oxygen intermediates and arachidonic acid, implying that other receptors are involved in mediating these responses (Wright and Silverstein 1983; Aderem et al. 1985; Aderem and Underhill 1999).

Most fungi become opsonised with C3b by activating the alternative and mannose-binding protein pathways, or though the classical (antibody-mediated) pathway (see Kozel 1996, 1998 for reviews). As CR3 has a binding site for the C3b component of complement, it plays an important role in the recognition of opsonised fungi. In addition, CR3 possesses a lectin domain on CD11b, which binds a broad range of carbohydrates as well as unopsonised yeast particles

Table 2 Immunomodulatory effectsof β-glucans

Model	β-Glucan treatment	Structure	Post-treatment	Response	Reference
J774a.1	Glucan-phosphate	Soluble	LPS	$\downarrow$ NF-κB	Williams et al. 2000
RAW 264.7	Grifolan	H. MW	None	$\uparrow$ IL-6, $\uparrow$ IL-1 and $\uparrow$ TNF-α	Adachi et al. 1994
RAW 264.7	Grifolan	Insoluble	None	$\uparrow$ TNF-α	Ishibashi et al. 2001
Murine resident peritoneal	β-Glucan	Particulate	None	Superoxide release, $\uparrow$ IL-1α	Gallin et al. 1992
MΦ	β-Glucan	Soluble	Particulate glucan	No effect	
Mice	β-Glucan	Soluble	LPS, staphylococcal enterotoxin B, toxic shock syndrome toxin1	$\downarrow$ TNF-α, $\downarrow$ IL-6	Soltys and Quinn 1999
Mice	Curdlan sulphate	Soluble	LPS	$\downarrow$ TNF-α, $\downarrow$ IL-1β, $\leftrightarrow$ IL-10, $\leftrightarrow$ IL-6	Masihi et al. 1997
Mice	Grifolan	H. MW	LPS	$\uparrow\uparrow$ TNF-α	Ohno et al. 1995
	SSG	H. MW			
	OL-2	H. MW			
Mice	β-Glucan	Soluble	CLP	$\downarrow$ NF-κB, $\downarrow$ IL-6	Williams et al. 1999
Rats	PGG-glucan	Soluble	*S. aureus*	$\leftrightarrow$ TNF-α, $\leftrightarrow$ IL-1β, $\uparrow$ clearance of infection	Liang et al. 1998
U937	Zymosan	Particulate	None	$\uparrow$ NF-κB	Kadish et al. 1986
Human monocytes	Yeast particles	Particulate	None	$\uparrow$ TNF-α, $\uparrow$ IL-1β	Abel and Czop 1992
Human monocytes	Zymosan	Particulate	None	$\uparrow$ leukotrienes	Janusz et al. 1987
	Yeast-derived	Soluble	Particulate glucans	Decrease	
Human mononuclear cells	β-Glucan	Particulate	None	$\uparrow$ IL-1R agonist, $\leftrightarrow$ IL-1β	Poutsiaka et al. 1993
	β-Glucan	Soluble	Particulate	Decrease IL-1R agonist	
Human dermal fibroblasts	Glucan-phosphate	Soluble	None	$\uparrow$ NF-κB, $\uparrow$ IL-6	Kougias et al. 2001
Rabbit alveolar MΦ	Zymosan	Particulate	None	$\uparrow$ arachidonic acid	Daum and Rohrbach 1992
	β-Glucan	Soluble	Particulate glucan	Decrease	
Rabbit alveolar MΦ	β-Glucan	Particulate	None	$\uparrow$ TNF-α	Olson et al. 1996

(Ross et al. 1985a; Thornton et al. 1996). The importance of complement and CR3 in the recognition and control of fungal infections is highlighted by genetic defects which result in decreased resistance to a variety of fungal pathogens, including *C. neoformans*, *C. albicans*, *A. fumigatus* and *Paracoccidioides brasiliensis* (Ross et al. 1985b; Kozel 1996, 1998). However, as uptake via CR3 also bypasses the cellular defence mechanisms mentioned above, this receptor is also targeted by a number of fungal pathogens, including *H. capsulatum*, *C. albicans*, and *Blastomyces dermatitidis*, as a means of invading the cell (Hogan et al. 1996).

2.4 Toll-Like Receptors

Innate immune specificity is thought to be determined by homo- or heterodimers of Toll-like receptors (TLR) (see Fig. 2), which cooperate with phagocytic receptors, such as the MR, to trigger a pathogen-specific response (Ozinsky et al. 2000; Takeuchi and Akira 2001; Underhill and Ozinsky 2002b; Vasselon and Detmers 2002). Toll receptors were originally identified as key components of the antifungal response in Drosophila (Lemaitre et al. 1996). Although a number of TLRs have now been identified in mammals (Medzhitov 2001), only a few have so far been implicated in fungal recognition (Underhill et al. 1999). TLR-2 was shown to participate in the immune response to zymosan as well as gram-positive bacteria, peptidoglycans and lipoarabinomannan. However, heterodimers of TLR-2, and TLR-6 are required for the complete activation of the transcription factor nuclear factor-kappa B (NF-κB), and the subsequent TNF-α production during the detection of zymosan (Underhill et al. 1999; Ozinsky et al. 2000).

TLR-4 was originally thought to be involved in the recognition of bacterial components, such as lipopolysaccharide (LPS) and lipoteichoic acid (Underhill and Ozinsky 2002b), but has recently been shown to be involved in the recognition of fungal pathogens as well (Shoham et al. 2001). The requirements for TLR-4-induced signalling, however, vary depending on the pathogen and cell type examined (Wang et al. 2001; Netea et al. 2002). In *Candida*-treated peripheral blood mononuclear cells and murine MΦs, TLR-2, but not TLR-4, is involved in the production of the pro-inflammatory cytokines, TNF-α and IL-1β (Netea et al. 2002). However, TLR-4-defective mice are more susceptible to *C. albicans* infection, perhaps as a result of the reduction in the production of various chemokines required for neutrophil recruitment in these mice (Netea et al. 2002). Similarly *C. neoformans* glucuronoxylomannan activates NF-κB via TLR-4 and CD14, as examined in transfected fibroblast cells, but the activation does not result in TNF-α production (Shoham et al. 2001). In contrast, *A. fumigatus* hyphae induce TNF-α, IL-1β and IL-6 production in human leukocytes via TLR-4 and CD14, but not via TLR-2 mediated pathways (Wang et al. 2001). Overall, these data suggest that a wide range of fungi, and consequently a diverse range of potential ligands, can be recognised by a variety of Toll-like receptors, a process which leads to the appropriate immune response.

3 MΦ Killing and Fungal Avoidance Mechanisms

The recognition and binding of fungal pathogens, via the various MΦ surface receptors, leads to phagocytosis and the implementation of a variety of anti-microbial mechanisms. Killing is achieved by a combination of oxygen-independent mechanisms, such as the low pH, degradative environment and limitation of nutrients within the phagosome, as well as oxygen-dependent mechanisms, including the production of reactive oxygen and nitrogen intermediates.

The production of reactive oxygen intermediates is an effective antimicrobial mechanism against fungi, such as *C. albicans* (Hampton et al. 1998; Clark 1999). Superoxide (O_2^-) is generated by the activated membrane-associated enzyme complex phagocyte NADPH oxidase (phox), which transfers electrons from NADPH to O_2. Toxic oxidants, such as hydroxyl radicals and hydrogen peroxidase (H_2O_2), are then generated within the phagosome killing the internalised organism. Mutations of the phox proteins, resulting in chronic granulomatous disease, are defined by the lack of in vitro phagocyte killing and the recurrence of bacterial and fungal infections, highlighting the importance of this antimicrobial mechanism (Leijh et al. 1977; Borgato et al. 2001). The role of superoxide acting as a signal for the activation of neutrophil granule proteases, and not as a toxic intermediate, was recently shown to be essential for bacterial killing and may also prove to be important in fungal killing (Reeves et al. 2002).

Myeloperoxidase (MPO) catalyses the production of hypochlorous acid, a potent microbicidal agent, from H_2O_2 (see Hoy et al. 2002 for a review). MPO is located within the granules of monocytes and neutrophils and is involved in their fungicidal and anti-microbial functions (Marodi et al. 1991; Hampton et al. 1998). It may function as a regulator of the oxidative degradation process, by the sequestration of H_2O_2 in the phagosome, thereby facilitating granule protease activity (Reeves et al. 2002). Although MΦs do not synthesise MPO, recombinant human MPO has been shown to enhance killing of *Candida* by GM-CSF-activated MΦs (Marodi et al. 1998). There may also be a link between the ability of the MΦ MR to bind MPO and its role in antifungal mechanisms (Shepherd and Hoidal 1990).

The inducible nitric oxide (NO) synthase (iNOS) plays a key role in the killing of pathogens (see MacMicking et al. 1997 for a review). iNOS enables the production of NO via the oxidative deamination of l-arginine. The NO then reacts with superoxide, or with thiol groups to produce peroxynitrite and nitrosothiols. A combination of NO, superoxide and peroxynitrite seems to be required for the full MΦ fungicidal activity (Vazquez-Torres and Balish 1997). Induction of iNOS occurs in the presence of microbial products, such as LPS and zymosan, or pro-inflammatory cytokines, such as TNF-α, IFN-γ and IL-1β. The importance of iNOS in controlling infections is evident by the use of specific

NOS inhibitors, which generally exacerbate infections (MacMicking et al. 1997). Furthermore, NOS knock-out mouse models display an increased susceptibility to fungal infections, such as *C. neoformans* (Rivera et al. 2002).

In addition to these oxygen-dependent killing mechanisms, the non-oxygen-dependent mechanisms also play an important role in the control of fungal infections. The confinement of microbes within the phagosome acts as a physical barrier limiting the availability of nutrients, whose supply is further restricted by active mechanisms. For example, the transferrin receptor, which transports iron via the endosomal pathway, is down-regulated during infection, while the recruitment of the natural resistance-associated MΦ protein (Nramp)-1 to the phagosome ensures the removal of iron and other essential divalent cations (Gruenheid et al. 1997; Blackwell and Searle 1999; Sunder-Plassmann et al. 1999). The degradative enzymes and acidic pH of the phagosome contribute to fungal killing, although these mechanisms have yet to be fully characterised (Vazquez-Torres and Balish 1997). The iron content and pH of the phagosome have also been targeted for anti-fungal therapy, discussed below.

Although MΦs are very efficient at killing microbial invaders, fungal pathogens have subverted many of these killing mechanisms to ensure their survival. Some fungi actively seek intracellular residence in a modified phagosome, such as *H. capsulatum*, which survives and replicates intracellularly by regulating the phagosomal pH (Newman 1999). Other fungi ensure that they avoid recognition, such as *C. neoformans* which has a viscous polysaccharide capsule predominantly composed of glucuronoxylomannan that protects against recognition by phagocytes (Perfect et al. 1998). Other avoidance mechanisms include the ability to suppress iNOS, as occurs in *C. albicans* infections (Vazquez-Torres and Balish 1997; Schroppel et al. 2001), and the production of compounds that interfere with phagocytosis, such as the aflatoxins and gliotoxins of *A. fumigatus* (Tomee and Kauffman 2000).

4 Antifungal Agents

While only few fungi are considered "professional pathogens", many are opportunistic pathogens capable of infecting immunocompromised individuals, such as AIDS patients (van Burik and Magee 2001). To combat these infections, two general antifungal strategies have been pursued, those that target the fungus itself and those that enhance the microbicidal activity of phagocytes. Often the simultaneous use of these strategies results in improved efficacy (for example see Tanida et al. 2001). Most drugs in clinical use directly debilitate the fungus and include those that target fungal cell membrane sterols, such as the polyenes (including amphotericin B), azoles (including fluconazole), allylamines and morpholines. Drugs which target fungal cell-wall synthesis include compounds such as the echinocandins and pneumocandins, which target β-glucan synthesis, and nikkomycin Z, which inhibits chitin synthesis. Other antifungal drugs include 5-fluorouracil, which interferes with RNA, DNA and protein synthesis.

For a more complete review on these and other antifungal agents see Ghannoum and Rice (1999).

Although not in routine clinical use, a number of studies have indicated that modulation of phagocyte function by immunotherapy can be beneficial for the management of systemic fungal infections. A number of cytokines, including GM-CSF, G-CSF and IFN-γ, have been shown to enhance phagocyte antifungal activity (Marodi et al. 1993; Farmaki and Roilides 2001; Roilides and Farmaki 2001). Furthermore, pretreatment with cytokines, such as GM-CSF, could reduce the risk of disseminated fungal infections in patients therapeutically immunosuppressed with drugs such as corticosteroids (Brummer et al. 2002). Other potential antifungal drugs include compounds which modify the phagolysosome, such as chloroquine, which by altering the phagolysosomal pH and iron availability, is effective against fungal pathogens such as *Histoplasma capsulatum* (Weber et al. 2000). Finally, as mentioned above, carbohydrates derived from the cell wall of the fungi themselves also show promise as immunomodulators for the control of fungal infection (Williams et al. 1978; Williams et al. 1991; Garner and Hudson 1996).

5 Immunomodulation by Fungal-Derived Carbohydrates

There is a considerable body of research documenting the effects of fungal-derived carbohydrates on the immune system, and an interest in generating novel therapeutics based on these compounds. These effects were initially noted over 40 years ago with zymosan, which is rich in α-mannan and β-glucans, and now is widely used as a particulate activator of MΦs (Benacerraf et al. 1959; Czop et al. 1989). More recent studies have shown that mannoproteins have immunomodulatory properties (see below), whilst soluble β-glucans can have mitogenic, anti-infective, anti-sepsis, and anti-tumourigenic effects (Riggi and Di Luzio 1961; Williams and Di Luzio 1980; Browder et al. 1987; Sherwood et al. 1987; Sandula et al. 1995; Williams et al. 1996; Kogan et al. 1997; Ross et al. 1999).

Fungal mannans have both suppressive and stimulatory activities on the immune system (Garner et al. 1990; Delfino et al. 1996; Delfino et al. 1997). Mannan is often associated with protein, which may be contributing to the immunomodulatory effects (Palma et al. 1992; Gomez et al. 1996; Tzianabos 2000; Mansour et al. 2002). As a stimulator, mannans can interact with phagocyte receptors, such as the mannose receptor, inducing the production of pro-inflammatory cytokines, such as TNF-α and IL-6 (Garner and Hudson 1996; Tzianabos 2000). The suppressive activities of mannans have been documented in patients suffering from *Candida* or *Cryptococcus* infections (Tzianabos 2000). Although the suppressive mechanism is not clearly understood, it is thought to involve interactions with T lymphocytes (Garner et al. 1990; Tzianabos 2000).

β-Glucans have potent stimulatory effects on the immune system, although the degree of immunomodulation is linked to structure of these carbohydrates

(Bohn and BeMiller 1995; Williams et al. 1996; Kogan et al. 1997; Williams 1997). The tertiary structure, degree of branching, polymer length and carbohydrate content have all been implicated in the ability of a particular β-glucan to stimulate the immune system. These stimulatory effects are probably related to the ability of the glucan to cross-link the βGRs on the phagocyte surface (Okazaki et al. 1995; Mueller et al. 2000). Unfortunately, the detailed structure of most immunomodulatory β-glucans and the molecular mechanisms by which they exert their effects are unknown, hampering efforts to develop these compounds for therapeutic use. In general, the administration of β-glucans results in phagocyte activation and the production of pro-inflammatory mediators, such as TNF-α (Czop and Austen 1985b,c; Browder et al. 1990). A list of some of these glucans and their immunomodulatory effects is presented in Table 2.

6 Conclusions

MΦs have an important role in the innate immune response to fungal pathogens, and consequently they express a number of receptors, which are able to recognise fungal molecular patterns. The identification of these receptors has greatly aided the understanding of the mechanisms by which the immune system recognises these pathogens and how the ensuing cellular and immune response is generated. These studies have also provided insight into the mechanisms by which fungal pathogens subvert the immune system and have led to the development of therapeutics, including the use of fungal-derived carbohydrates as immunomodulators.

Acknowledgements. We thank the Wellcome Trust, Arthritis Research Campaign, and the Medical Research Council, UK for funding.

7 References

Abel, G and Czop, JK. (1992) Stimulation of human monocyte beta-glucan receptors by glucan particles induces production of TNF-alpha and IL-1 beta. Int J Immunopharmacol, 14:1363–73

Adachi, Y, Okazaki, M, Ohno, N and Yadomae, T. (1994) Enhancement of cytokine production by macrophages stimulated with (1–3)-beta-D-glucan, grifolan (GRN), isolated from Grifola frondosa. Biol Pharm Bull, 17:1554–60

Aderem, A and Underhill, DM. (1999) Mechanisms of phagocytosis in macrophages. Annu Rev Immunol, 17:593–623

Aderem, AA, Wright, SD, Silverstein, SC and Cohn, ZA. (1985) Ligated complement receptors do not activate the arachidonic acid cascade in resident peritoneal macrophages. J Exp Med, 161:617–22

Ariizumi, K, Shen, GL, Shikano, S, Ritter, R, 3rd, Zukas, P, Edelbaum, D, Morita, A and Takashima, A. (2000a) Cloning of a second dendritic cell-associated C-type lectin (dectin-2) and its alternatively spliced isoforms. J Biol Chem, 275:11957–63

Ariizumi, K, Shen, GL, Shikano, S, Xu, S, Ritter, R, 3rd, Kumamoto, T, Edelbaum, D, Morita, A, Bergstresser, PR and Takashima, A. (2000b) Identification of a novel, dendritic cell-associated molecule, dectin-1, by subtractive cDNA cloning. J Biol Chem, 275:20157–67

Balch, SG, McKnight, AJ, Seldin, MF and Gordon, S. (1998) Cloning of a novel C-type lectin expressed by murine macrophages. J Biol Chem, 273:18656–64

Barclay, AN, Brown, MH, Law, SKA, McKnight, AJ, Tomlinson, MG and van der Merwe, PA. (1997) The Leucocyte Antigen FactsBook. Academic Press Ltd, London

Benacerraf, B, Thorbecke, GJ and Jacoby, D. (1959) Effect of zymosan on endotoxin toxicity in mice. Proc. Soc. Exp. Biol. Med., 100:796–799

Blackwell, JM and Searle, S. (1999) Genetic regulation of macrophage activation: understanding the function of Nramp1 (=Ity/Lsh/Bcg). Immunol Lett, 65:73–80

Bohn, JA and BeMiller, JN. (1995) (1–3)- beta-D-Glucans as biological response modifiers: a review of structure-functional activity relationships. Carbohydrate Polymers, 28:3–14

Borgato, L, Bonizzato, A, Lunardi, C, Dusi, S, Andrioli, G, Scarperi, A and Corrocher, R. (2001) A 1.1-kb duplication in the p67-phox gene causes chronic granulomatous disease. Hum Genet, 108:504–10

Browder, W, Williams, D, Pretus, H, Olivero, G, Enrichens, E, Mao, P and Franchello, A. (1990) Beneficial effect of enhanced macrophage function in the trauma patient. Ann. Surg., 211:605–613

Browder, W, Williams, D, Sherwood, E, McNamee, R, Jones, E and DiLuzio, N. (1987) Synergistic effect of nonspecific immunostimulation and antibiotics in experimental peritonitis. Surgery, 102:206–14

Brown, GD and Gordon, S. (2001) Immune recognition. A new receptor for beta-glucans. Nature, 413:36–7

Brown, GD, Taylor, PR, Reid, DM, Willment, JA, Williams, DL, Martinez-Pomares, L, Wong, SYC and Gordon, S. (2002) Dectin-1 is a major beta-glucan receptor on macrophages. J. Exp. Med., In press

Brummer, E, Maqbool, A and Stevens, DA. (2002) Protection of peritoneal macrophages by granulocyte/macrophage colony-stimulating factor (GM-CSF) against dexamethasone suppression of killing of Apergillus, and the effect of human GM-CSF. Microbes and Infection, 4:133–138

Chaffin, WL, Lopez-Ribot, JL, Casanova, M, Gozalbo, D and Martinez, JP. (1998) Cell wall and secreted proteins of Candida albicans: identification, function, and expression. Microbiol Mol Biol Rev, 62:130–80

Cherayil, BJ, Weiner, SJ and Pillai, S. (1989) The Mac-2 antigen is a galactose-specific lectin that binds IgE. J Exp Med, 170:1959–72

Clark, RA. (1999) Activation of the neutrophil respiratory burst oxidase. J Infect Dis, 179 Suppl 2:S309–17

Cross, CE and Bancroft, GJ. (1995) Ingestion of acapsular Cryptococcus neoformans occurs via mannose and beta-glucan receptors, resulting in cytokine production and increased phagocytosis of the encapsulated form. Infect Immun, 63:2604–11

Czop, JK. (1986) The role of beta-glucan receptors on blood and tissue leukocytes in phagocytosis and metabolic activation. Pathol Immunopathol Res, 5:286–96

Czop, JK and Austen, KF. (1985a) A beta-glucan inhibitable receptor on human monocytes: its identity with the phagocytic receptor for particulate activators of the alternative complement pathway. J Immunol, 134:2588–93

Czop, JK and Austen, KF. (1985b) Generation of leukotrienes by human monocytes upon stimulation of their beta-glucan receptor during phagocytosis. Proc Natl Acad Sci U S A, 82:2751–5

Czop, JK and Austen, KF. (1985c) Properties of glycans that activate the human alternative complement pathway and interact with the human monocyte beta-glucan receptor. J Immunol, 135:3388–93

Czop, JK and Kay, J. (1991) Isolation and characterization of beta-glucan receptors on human mononuclear phagocytes. J Exp Med, 173:1511–20

Czop, JK, Valiante, NM and Janusz, MJ. (1989) Phagocytosis of particulate activators of the human alternative complement pathway through monocyte beta-glucan receptors. Prog Clin Biol Res, 297:287–96

Daum, T and Rohrbach, MS. (1992) Zymosan induces selective release of arachidonic acid from rabbit alveolar macrophages via stimulation of a beta-glucan receptor. FEBS Lett, 309:119–22

Delfino, D, Cianci, L, Lupis, E, Celeste, A, Petrelli, ML, Curro, F, Cusumano, V and Teti, G. (1997) Interleukin-6 production by human monocytes stimulated with Cryptococcus neoformans components. Infect Immun, 65:2454–6

Delfino, D, Cianci, L, Migliardo, M, Mancuso, G, Cusumano, V, Corradini, C and Teti, G. (1996) Tumor necrosis factor-inducing activities of Cryptococcus neoformans components. Infect Immun, 64:5199–204

Douglas, CM. (2001) Fungal beta(1,3)-D-glucan synthesis. Med Mycol, 39:55–66

Ezekowitz, RA, Sastry, K, Bailly, P and Warner, A. (1990) Molecular characterization of the human macrophage mannose receptor: demonstration of multiple carbohydrate recognition-like domains and phagocytosis of yeasts in Cos-1 cells. J Exp Med, 172:1785–94

Ezekowitz, RA and Stahl, PD. (1988) The structure and function of vertebrate mannose lectin-like proteins. J Cell Sci Suppl, 9:121–33

Ezekowitz, RA, Williams, DJ, Koziel, H, Armstrong, MY, Warner, A, Richards, FF and Rose, RM. (1991) Uptake of Pneumocystis carinii mediated by the macrophage mannose receptor. Nature, 351:155–8

Farmaki, E and Roilides, E. (2001) Immunotherapy in patients with systemic mycoses: a promising adjunct. BioDrugs, 15:207–14

Fernandes, MJ, Finnegan, AA, Siracusa, LD, Brenner, C, Iscove, NN and Calabretta, B. (1999) Characterization of a novel receptor that maps near the natural killer gene complex: demonstration of carbohydrate binding and expression in hematopoietic cells. Cancer Res, 59:2709–17

Fradin, C, Poulain, D and Jouault, T. (2000) beta-1,2-linked oligomannosides from Candida albicans bind to a 32-kilodalton macrophage membrane protein homologous to the mammalian lectin galectin-3. Infect Immun, 68:4391–8

Gallin, EK, Green, SW and Patchen, ML. (1992) Comparative effects of particulate and soluble glucan on macrophages of C3H/HeN and C3H/HeJ mice. Int J Immunopharmacol, 14:173–83

Garner, RE, Childress, AM, Human, LG and Domer, JE. (1990) Characterization of Candida albicans mannan-induced, mannan-specific delayed hypersensitivity suppressor cells. Infect Immun, 58:2613–20

Garner, RE and Hudson, JA. (1996) Intravenous injection of Candida-derived mannan results in elevated tumor necrosis factor alpha levels in serum. Infect Immun, 64:4561–6

Garner, RE, Rubanowice, K, Sawyer, RT and Hudson, JA. (1994) Secretion of TNF-alpha by alveolar macrophages in response to Candida albicans mannan. J Leukoc Biol, 55:161–8

Ghannoum, MA and Rice, LB. (1999) Antifungal agents: mode of action, mechanisms of resistance, and correlation of these mechanisms with bacterial resistance. Clinical Microbiology Reviews, 12:501–517

Gomez, MJ, Torosantucci, A, Arancia, S, Maras, B, Parisi, L and Cassone, A. (1996) Purification and biochemical characterization of a 65-kilodalton mannoprotein (MP65), a main target of anti-Candida cell-mediated immune responses in humans. Infect Immun, 64:2577–84

Gruenheid, S, Pinner, E, Desjardins, M and Gros, P. (1997) Natural resistance to infection with intracellular pathogens: the Nramp1 protein is recruited to the membrane of the phagosome. J Exp Med, 185:717–30

Hampton, MB, Kettle, AJ and Winterbourn, CC. (1998) Inside the neutrophil phagosome: oxidants, myeloperoxidase, and bacterial killing. Blood, 92:3007–17

Hogan, LH, Klein, B and Levitz, SM. (1996) Virulence factors of medically important fungi. Clinical Microbiological Reviews, 9:469–488

Hoy, A, Leininger-Muller, B, Kutter, D, Siest, G and Visvikis, S. (2002) Growing significance of myeloperoxidase in non-infectious diseases. Clin Chem Lab Med, 40:2–8

Ishibashi, K, Miura, NN, Adachi, Y, Ohno, N and Yadomae, T. (2001) Relationship between solubility of grifolan, a fungal 1,3-beta-D-glucan, and production of tumor necrosis factor by macrophages in vitro. Biosci Biotechnol Biochem, 65:1993–2000

Janusz, MJ, Austen, KF and Czop, JK. (1987) Lysosomal enzyme release from human monocytes by particulate activators is mediated by beta-glucan inhibitable receptors. J Immunol, 138:3897–901

Kadish, JL, Choi, CC and Czop, JK. (1986) Phagocytosis of unopsonized zymosan particles by trypsin-sensitive and beta-glucan-inhibitable receptors on bone marrow-derived murine macrophages. Immunol Res, 5:129–38

Kan, VL and Bennett, JE. (1991) Beta 1,4-oligoglucosides inhibit the binding of Aspergillus fumigatus conidia to human monocytes. J Infect Dis, 163:1154–6

Kery, V, Krepinsky, JJ, Warren, CD, Capek, P and Stahl, PD. (1992) Ligand recognition by purified human mannose receptor. Arch Biochem Biophys, 298:49–55

Kogan, G, Machova, E and Sandula, J. (1997) Immunomodulating activity of the beta-glucan from *Saccharomyces cerevisiae* and its soluble derivatives. In Suzuki, S. and Suzuki, M. (eds.), *Fungal Cells in Biodefense Mechanisms*. Saikon Publishing Co. Ltd., Tokoyo, pp. 301–6

Kougias, P, Wei, D, Rice, PJ, Ensley, HE, Kalbfleisch, J, Williams, DL and Browder, IW. (2001) Normal human fibroblasts express pattern recognition receptors for fungal (1–3)-beta-D-glucans. Infect Immun, 69:3933–8

Kozel, TR. (1996) Activation of the complement system by pathogenic fungi. Clin Microbiol Rev, 9:34–46

Kozel, TR. (1998) Complement activation by pathogenic fungi. Res Immunol, 149:309–20; discussion 514–5

Leijh, PC, van den Barselaar, MT and van Furth, R. (1977) Kinetics of phagocytosis and intracellular killing of Candida albicans by human granulocytes and monocytes. Infect Immun, 17:313–8

Lemaitre, B, Nicolas, E, Michaut, L, Reichhart, JM and Hoffmann, JA. (1996) The dorsoventral regulatory gene cassette spatzle/Toll/cactus controls the potent antifungal response in Drosophila adults. Cell, 86:973–83

Liang, J, Melican, D, Cafro, L, Palace, G, Fisette, L, Armstrong, R and Patchen, ML. (1998) Enhanced clearance of a multiple antibiotic resistant Staphylococcus aureus in rats treated with PGG-glucan is associated with increased leukocyte counts and increased neutrophil oxidative burst activity. Int J Immunopharmacol 20:595–614

Linehan, SA, Martinez-Pomares, L, da Silva, RP and Gordon, S. (2001) Endogenous ligands of carbohydrate recognition domains of the mannose receptor in murine macrophages, endothelial cells and secretory cells; potential relevance to inflammation and immunity. Eur J Immunol, 31:1857–66

Linehan, SA, Martinez-Pomares, L and Gordon, S. (2000) Macrophage lectins in host defence. Microbes Infect, 2:279–88

Linehan, SA, Martinez-Pomares, L, Stahl, PD and Gordon, S. (1999) Mannose receptor and its putative ligands in normal murine lymphoid and nonlymphoid organs: In situ expression of mannose receptor by selected macrophages, endothelial cells, perivascular microglia, and mesangial cells, but not dendritic cells. J Exp Med, 189:1961–72

Lipke, PN and Ovalle, R. (1998) Cell wall architecture in yeast: new structure and new challenges. J Bacteriol, 180:3735–40

MacMicking, J, Xie, QW and Nathan, C. (1997) Nitric oxide and macrophage function. Annu Rev Immunol, 15:323–50
Mansour, MK, Schlesinger, LS and Levitz, SM. (2002) Optimal T cell responses to Cryptococcus neoformans mannoprotein are dependent on recognition of conjugated carbohydrates by mannose receptors. J Immunol, 168:2872–9
Marodi, L, Forehand, JR and Johnston, RB, Jr. (1991) Mechanisms of host defense against Candida species. II. Biochemical basis for the killing of Candida by mononuclear phagocytes. J Immunol, 146:2790–4
Marodi, L, Schreiber, S, Anderson, DC, MacDermott, RP, Korchak, HM and Johnston, RB, Jr. (1993) Enhancement of macrophage candidacidal activity by interferon-gamma. Increased phagocytosis, killing, and calcium signal mediated by a decreased number of mannose receptors. J Clin Invest, 91:2596–601
Marodi, L, Tournay, C, Kaposzta, R, Johnston, RB, Jr. and Moguilevsky, N. (1998) Augmentation of human macrophage candidacidal capacity by recombinant human myeloperoxidase and granulocyte-macrophage colony-stimulating factor. Infect Immun, 66:2750–4
Martinez-Pomares, L, Linehan, SA, Taylor, PR and Gordon, S. (2001) Binding properties of the mannose receptor. Immunobiology 204:527–35
Martinez-Pomares, L, Mahoney, JA, Kaposzta, R, Linehan, SA, Stahl, PD and Gordon, S. (1998) A functional soluble form of the murine mannose receptor is produced by macrophages in vitro and is present in mouse serum. J Biol Chem, 273:23376–80
Masihi, KN, Madaj, K, Hintelmann, H, Gast, G and Kaneko, Y. (1997) Down-regulation of tumor necrosis factor-alpha, moderate reduction of interleukin-1beta, but not interleukin-6 or interleukin-10, by glucan immunomodulators curdlan sulfate and lentinan. Int J Immunopharmacol 19:463–8
Medzhitov, R. (2001) Toll-like receptors and innate immunity. Nature Rev Immunol, 1:135–45
Medzhitov, R and Janeway, CA, Jr. (1997) Innate immunity: impact on the adaptive immune response. Curr Opin Immunol, 9:4–9
Medzhitov, R, Preston-Hurlburt, P and Janeway, CA, Jr. (1997) A human homologue of the Drosophila Toll protein signals activation of adaptive immunity. Nature, 388:394–7
Mueller, A, Raptis, J, Rice, PJ, Kalbfleisch, JH, Stout, RD, Ensley, HE, Browder, W and Williams, DL. (2000) The influence of glucan polymer structure and solution conformation on binding to (1–3)-beta-D-glucan receptors in a human monocyte-like cell line. Glycobiology, 10:339–46
Muto, S, Vetvicka, V and Ross, GD. (1993) CR3 (CD11b/CD18) expressed by cytotoxic T cells and natural killer cells is upregulated in a manner similar to neutrophil CR3 following stimulation with various activating agents. J Clin Immunol, 13:175–84
Netea, MG, Van Der Graaf, CA, Vonk, AG, Verschueren, I, Van Der Meer, JW and Kullberg, BJ. (2002) The Role of Toll-like Receptor (TLR) 2 and TLR4 in the Host Defense against Disseminated Candidiasis. J Infect Dis, 185:1483–9
Newman, SL. (1999) Macrophages in host defense against *Histoplasma capsulatum*. Trends in Microbiology, 7:67–71
Ohno, N, Asada, N, Adachi, Y and Yadomae, T. (1995) Enhancement of LPS triggered TNF-alpha (tumor necrosis factor-alpha) production by (1–3)-beta-D-glucans in mice. Biol Pharm Bull, 18:126–33
Okazaki, M, Adachi, Y, Ohno, N and Yadomae, T. (1995) Structure-activity relationship of (1–3)-beta-D-glucans in the induction of cytokine production from macrophages, in vitro. Biol Pharm Bull, 18:1320–7
Olson, EJ, Standing, JE, Griego-Harper, N, Hoffman, OA and Limper, AH. (1996) Fungal beta-glucan interacts with vitronectin and stimulates tumor necrosis factor alpha release from macrophages. Infect Immun, 64:3548–54

Ozinsky, A, Underhill, DM, Fontenot, JD, Hajjar, AM, Smith, KD, Wilson, CB, Schroeder, L and Aderem, A. (2000) The repertoire for pattern recognition of pathogens by the innate immune system is defined by cooperation between toll-like receptors. Proc Natl Acad Sci U S A, 97:13766–71

Palma, C, Serbousek, D, Torosantucci, A, Cassone, A and Djeu, JY. (1992) Identification of a mannoprotein fraction from Candida albicans that enhances human polymorphonuclear leukocyte (PMNL) functions and stimulates lactoferrin in PMNL inhibition of candidal growth. J Infect Dis, 166:1103–12

Perfect, JR, Wong, B, Chang, YC, Kwon-Chung, KJ and Williamson, PR. (1998) Cryptococcus neoformans: virulence and host defences. Med Mycol, 36:79–86

Pontow, SE, Kery, V and Stahl, PD. (1992) Mannose receptor. Int Rev Cytol:221–44

Poutsiaka, DD, Mengozzi, M, Vannier, E, Sinha, B and Dinarello, CA. (1993) Cross-linking of the beta-glucan receptor on human monocytes results in interleukin-1 receptor antagonist but not interleukin-1 production. Blood, 82:3695–700

Reeves, EP, Lu, H, Jacobs, HL, Messina, CG, Bolsover, S, Gabella, G, Potma, EO, Warley, A, Roes, J and Segal, AW. (2002) Killing activity of neutrophils is mediated through activation of proteases by K+ flux. Nature, 416:291–7

Riggi, SJ and Di Luzio, NR. (1961) Identification of a reticuloendothelial stimulating agent in zymosan. American Journal of Physiology 200:297–300

Rivera, J, Mukherjee, J, Weiss, LM and Casadevall, A. (2002) Antibody efficacy in murine pulmonary Cryptococcus neoformans infection: a role for nitric oxide. J Immunol, 168:3419–27

Roilides, E and Farmaki, E. (2001) Granulocyte colony-stimulating factor and other cytokines in antifungal therapy. Clin Microbiol Infect, 7:62–7

Romani, L. (2002) Immunobiology of invasive Candidiasis. In Calderone, R.A. (ed.) *Candida and Candidiasis.* ASM Press, Washington, pp. 223–241

Ross, GD. (2000) Regulation of the adhesion versus cytotoxic functions of the Mac-1/CR3/alphaMbeta2-integrin glycoprotein. Crit Rev Immunol 20:197–222

Ross, GD, Cain, JA and Lachmann, PJ. (1985a) Membrane complement receptor type three (CR3) has lectin-like properties analogous to bovine conglutinin as functions as a receptor for zymosan and rabbit erythrocytes as well as a receptor for iC3b. J Immunol, 134:3307–15

Ross, GD, Thompson, RA, Walport, MJ, Springer, TA, Watson, JV, Ward, RH, Lida, J, Newman, SL, Harrison, RA and Lachmann, PJ. (1985b) Characterization of patients with an increased susceptibility to bacterial infections and a genetic deficiency of leukocyte membrane complement receptor type 3 and the related membrane antigen LFA-1. Blood, 66:882–90

Ross, GD and Vetvicka, V. (1993) CR3 (CD11b, CD18): a phagocyte and NK cell membrane receptor with multiple ligand specificities and functions. Clin Exp Immunol, 92:181–4

Ross, GD, Vetvicka, V, Yan, J, Xia, Y and Vetvickova, J. (1999) Therapeutic intervention with complement and beta-glucan in cancer. Immunopharmacology, 42:61–74

Sandula, J, Machova, E and Hribalova, V. (1995) Mitogenic activity of particulate yeast beta-(1–3)-D-glucan and its water-soluble derivatives. Int J Biol Macromol, 17:323–6

Schroppel, K, Kryk, M, Herrmann, M, Leberer, E, Rollinghoff, M and Bogdan, C. (2001) Suppression of type 2 NO-synthase activity in macrophages by Candida albicans. Int J Med Microbiol, 290:659–68

Shepherd, VL and Hoidal, JR. (1990) Clearance of neutrophil-derived myeloperoxidase by the macrophage mannose receptor. Am. J. Respir. Cell Mol. Biol., 2:335–340

Sherwood, ER, Williams, DL, McNamee, RB, Jones, EL, Browder, IW and Di Luzio, NR. (1987) In vitro tumoricidal activity of resting and glucan-activated Kupffer cells. J Leukoc Biol, 42:69–75

Shibata, Y, Metzger, WJ and Myrvik, QN. (1997) Chitin particle-induced cell-mediated immunity is inhibited by soluble mannan: mannose receptor-mediated phagocytosis initiates IL-12 production. J Immunol, 159:2462–7

Shoham, S, Huang, C, Chen, JM, Golenbock, DT and Levitz, SM. (2001) Toll-like receptor 4 mediates intracellular signaling without TNF-alpha release in response to Cryptococcus neoformans polysaccharide capsule. J Immunol, 166:4620–6

Soltys, J and Quinn, MT. (1999) Modulation of endotoxin- and enterotoxin-induced cytokine release by in vivo treatment with beta-(1,6)-branched beta-(1,3)-glucan. Infect Immun, 67:244–52

Stahl, PD. (1992) The mannose receptor and other macrophage lectins. Curr Opin Immunol, 4:49–52

Stahl, PD and Ezekowitz, RA. (1998) The mannose receptor is a pattern recognition receptor involved in host defense. Curr Opin Immunol, 10:50–5

Stahl, PD, Rodman, JS, Miller, MJ and Schlesinger, PH. (1978) Evidence for receptor-mediated binding of glycoproteins, glycoconjugates, and lysosomal glycosidases by alveolar macrophages. Proc Natl Acad Sci U S A, 75:1399–403

Stein, M and Gordon, S. (1991) Regulation of tumor necrosis factor (TNF) release by murine peritoneal macrophages: role of cell stimulation and specific phagocytic plasma membrane receptors. Eur J Immunol, 21:431–7

Sunder-Plassmann, G, Patruta, SI and Horl, WH. (1999) Pathobiology of the role of iron in infection. Am J Kidney Dis, 34:S25–9

Suzuki, N, Yamamoto, K, Toyoshima, S, Osawa, T and Irimura, T. (1996) Molecular cloning and expression of cDNA encoding human macrophage C-type lectin. Its unique carbohydrate binding specificity for Tn antigen. J Immunol, 156:128–35

Takeuchi, O and Akira, S. (2001) Toll-like receptors; their physiological role and signal transduction system. Int Immunopharmacol, 1:625–35

Tanida, T, Rao, F, Hamada, T, Ueta, E and Osaki, T. (2001) Lactoferrin peptide increases the survival of Candida albicans-inoculated mice by upregulating neutrophil and macrophage functions, especially in combination with amphotericin B and granulocyte-macrophage colony-stimulating factor. Infect Immun, 69:3883–90

Thornton, BP, Vetvicka, V, Pitman, M, Goldman, RC and Ross, GD. (1996) Analysis of the sugar specificity and molecular location of the beta-glucan-binding lectin site of complement receptor type 3 (CD11b/CD18). J Immunol, 156:1235–46

Tomee, JF and Kauffman, HF. (2000) Putative virulence factors of *Aspergillus fumigatus*. Clinical and Experimental Allergy, 30:476–484

Torosantucci, A, Chiani, P and Cassone, A. (2000) Differential chemokine response of human monocytes to yeast and hyphal forms of Candida albicans and its relation to the beta-1,6 glucan of the fungal cell wall. J Leukoc Biol, 68:923–32

Taylor, PR., Brown, GD, Reid, DM, Willment, JA, Martinez-Pomares, L, Gordon, S. and Wong, SYC. (2002) The beta-glucan receptor, Dectin-1, is predominantly expressed on the surface of cells of the monocyte/macrophage and neutrophil lineages. J Immunol 269, 3876-82

Tzianabos, AO. (2000) Polysaccharide immunomodulators as therapeutic agents: structural aspects and biologic function. Clin Microbiol Rev, 13:523–33

Underhill, DM and Ozinsky, A. (2002a) PHAGOCYTOSIS OF MICROBES: Complexity in Action. Annu Rev Immunol 20:825–52

Underhill, DM and Ozinsky, A. (2002b) Toll-like receptors: key mediators of microbe detection. Curr Opin Immunol, 14:103–10

Underhill, DM, Ozinsky, A, Hajjar, AM, Stevens, A, Wilson, CB, Bassetti, M and Aderem, A. (1999) The Toll-like receptor 2 is recruited to macrophage phagosomes and discriminates between pathogens. Nature, 401:811–5

van Burik, JA and Magee, PT. (2001) Aspects of fungal pathogenesis in humans. Annu Rev Microbiol, 55:743–72

Vassallo, R, Standing, JE and Limper, AH. (2000) Isolated Pneumocystis carinii cell wall glucan provokes lower respiratory tract inflammatory responses. J Immunol, 164:3755–63

Vasselon, T and Detmers, PA. (2002) Toll receptors: a central element in innate immune responses. Infect Immun, 70:1033–41

Vazquez-Torres, A and Balish, E. (1997) Macrophages in resistance to candidiasis. Microbiol Mol Biol Rev, 61:170–92

Wang, JE, Warris, A, Ellingsen, EA, Jorgensen, PF, Flo, TH, Espevik, T, Solberg, R, Verweij, PE and Aasen, AO. (2001) Involvement of CD14 and toll-like receptors in activation of human monocytes by Aspergillus fumigatus hyphae. Infect Immun, 69:2402–6

Weber, SM, Levitz, SM and Harrison, TS. (2000) Chloroquine and the fungal phagosome. Curr Opin Microbiol, 3:349–53

Williams, DL. (1997) Overview of (1–3)-beta-D-glucan immunobiology. Mediators of Inflammation, 6:247–250

Williams, DL, Cook, JA, Hoffmann, EO and Di Luzio, NR. (1978) Protective effect of glucan in experimentally induced candidiasis. J Reticuloendothel Soc, 23:479–490

Williams, DL and Di Luzio, NR. (1980) Glucan-induced modification of murine viral hepatitis. Science 208:67–9

Williams, DL, Ha, T, Li, C, Kalbfleisch, JH, Laffan, JJ and Ferguson, DA. (1999) Inhibiting early activation of tissue nuclear factor-kappa B and nuclear factor interleukin 6 with (1->3)-beta-D-glucan increases long-term survival in polymicrobial sepsis. Surgery, 126:54–65

Williams, DL, Ha, T, Li, C, Laffan, J, Kalbfleisch, J and Browder, W. (2000) Inhibition of LPS-induced NFkappaB activation by a glucan ligand involves down-regulation of IKKbeta kinase activity and altered phosphorylation and degradation of IkappaBalpha. Shock, 13:446–52

Williams, DL, Mueller, A and Browder, W. (1996) Glucan-based macrophage stimulators. Clin. Immunother., 5:392–399

Williams, DL, Pretus, HA, McNamee, RB, Jones, EL, Ensley, HE, Browder, IW and Di Luzio, NR. (1991) Development, physicochemical characterization and preclinical efficacy evaluation of a water soluble glucan sulfate derived from Saccharomyces cerevisiae. Immunopharmacology, 22:139–55

Willment, JA, Gordon, S and Brown, GD. (2001) Characterisation of the human {beta}-glucan receptor and its alternatively spliced isoforms. J Biol Chem 20:20

Wright, SD and Silverstein, SC. (1983) Receptors for C3b and C3bi promote phagocytosis but not the release of toxic oxygen from human phagocytes. J Exp Med, 158:2016–23

Yamamoto, Y, Klein, TW and Friedman, H. (1997) Involvement of mannose receptor in cytokine interleukin-1beta (IL-1beta), IL-6, and granulocyte-macrophage colony-stimulating factor responses, but not in chemokine macrophage inflammatory protein 1beta (MIP-1beta), MIP-2, and KC responses, caused by attachment of Candida albicans to macrophages. Infect Immun, 65:1077–82

Subject Index

Printing: Saladruck Berlin
Binding: Stürtz AG, Würzburg